DYNAMICS
in ENGINEERING
STRUCTURES

DYNAMICS
in ENGINEERING
STRUCTURES

Vladimír Koloušek, DSc.

Professor, Technical University, Prague

Edited by:

R. F. McLean, MSc, PhD, ARCST, C Eng, M I Mech E
Senior Lecturer, Department of Dynamics and Control,
University of Strathclyde

and

J. F. Fleming, BSc
Lecturer, Department of Dynamics and Control,
University of Strathclyde

HALSTED PRESS
A Division of JOHN WILEY & SONS, Inc.,
605 Third Avenue, New York, N.Y. 10016

The Butterworth Group

ENGLAND
Butterworth & Co (Publishers) Ltd
London: 88 Kingsway, WC2B 6AB

AUSTRALIA
Butterworths Pty Ltd
Sydney: 586 Pacific Highway, NSW 2067
Melbourne: 343 Little Collins Street, 3000
Brisbane: 240 Queen Street, 4000

CANADA
Butterworth & Co (Canada) Ltd
Toronto: 14 Curity Avenue, 374

NEW ZEALAND
Butterworths of New Zealand Ltd
Wellington: 26—28 Waring Taylor Street, 1

SOUTH AFRICA
Butterworth & Co (South Africa) (Pty) Ltd
Durban: 152—154 Gale Street

ENGLISH EDITION FIRST PUBLISHED IN 1973
BY BUTTERWORTH & CO LTD, 88 KINGSWAY,
LONDON IN CO-EDITION
WITH ACADEMIA — PUBLISHING HOUSE OF THE
CZECHOSLOVAK ACADEMY OF SCIENCES, PRAGUE

Translated by D. Hajšmanová

ISBN 0 408 70160 9

Printed in Czechoslovakia

CONTENTS

PREFACE

In the field of structural design the engineering profession is called upon to deal with problems of an increasingly complex nature. Modern civil engineering structures are required to span larger distances than in the past yet stay within the economic limits set down for them. The latter stipulation leads to slender structures which are very sensitive to dynamic effects. In a number of these structures the situation is further aggravated by the fact that the dynamic loads they are expected to support grow continually during their working lives as a consequence of the ever increasing speeds of vehicles moving over them, intensified soil vibration induced by incessantly heavier traffic, and by the dynamic effects of the machinery they carry, etc. Static analysis in itself can no longer ensure the safety of such structures and an all embracing dynamic analysis becomes the order of the day.

When functioning as a design engineer, the author first came in contact with serious dynamic problems when designing antenna masts, structures in which wind-induced dynamic loads play a vital part. A thorough examination of the dynamic behaviour of various types of railway bridges of different materials, notably steel, reinforced concrete and prestressed concrete, which he undertook subsequently showed him yet more facets of the same problems. His later activities were concerned with the design of foundations for various machines, design and critical appraisal of buildings with installed machinery, and studies of miscellaneous problems of structural dynamics. This book is an outgrowth of his theoretical analyses and teaching experience in the field.

Many of the practical problems the author met during his career were amenable to the application of the conventional methods of structural dynamics. But just as many of the problems required the development of a new method of approach or a modification made to known techniques. It is the problems of the latter kind that form the contents of Chapters 2–10, Chapter 1 representing merely a general introduction to the subject matter.

The author, having had a wide range of industrial experience has evolved his theories with practical solutions in mind. Practical examples are used extensively throughout the book to allow practising engineers, as well as students to understand the application of the theory. By following these practical examples they can then apply suitable techniques to their own particular problems.

8

The analysis is presented in such a manner as to allow tabular methods to be used for the simple cases, and can easily be extended in a form suitable for digital computers for the more lengthy calculations. The inclusion of the comprehensive sets of tables and numerical values of the frequency functions further accentuates the practical nature of this book.

In preparing the book the author has been aided by several of his associates whose assistance is hereby gratefully acknowledged. Mr. Jiří Vaněk helped the author in computing Example 1.1. Section 4.1.1 is the work of Messrs. Zdeněk Němeček and Petr Řeřicha who at the time of its writing attended the author's seminar on dynamics. Examples 4.3 and 5.9 were solved by Mr. Jiří Šejnoha. Section 6.2 was prepared in collaboration with Ivo Babuška, DSc.

FOREWORD

This book does not aspire to cover the whole broad field of engineering structural dynamics. It deals with only those topics of the discipline that come within the range of the author's practice as a project engineer and research scientist. The main subjects of the analysis are structures containing continuous beams of both constant and varying cross-section, two-dimensional as well as three-dimensional frame structures, and bridges of various types subjected to moving loads. Particular attention is accorded to damping, static axial forces and other effects.

The first chapter of the book is of an introductory character and reviews the methods of analysis which have proved to be the most effective for solving the problems within systems that have a finite number of degrees of freedom. The problems of symmetry, antisymmetry and cyclosymmetry receive special treatment.

Chapter 2 deals with the vibration of straight bars of uniform cross-section. The calculation of end forces and moments, normalisation of natural modes and the determination of other pertinent quantities are greatly facilitated by the introduction of frequency functions, written symbolically as $F_i(\lambda)$ and $\Phi_i(\lambda)$.

The third chapter turns to the analysis of frame structures, both two- and three-dimensional, composed of prismatic members. The treatment is based on the use of the slope-deflection method in the form developed by the author. There follows a discussion of the possibilities and advantages of the frequency functions in their application to other methods, e.g. the energy method. The approach outlined in this chapter has been found to be very useful for the vibration analyses of industrial buildings, masts and towers, etc.

The subject matter of the fourth chapter consists of non-prismatic and curved bars, and systems embodying them. After the direct solution of a straight non-prismatic bar comes the derivation of the frequency functions $F(\lambda_g, \lambda_h)$ — in two parameters this time — and the application of these functions to an analysis of end forces. Having found the direct approach fairly complicated, the author devises two approximate methods: the slope-deflection method and the simplified slope-deflection method which he proves to be equally suitable for systems with straight bars as for systems with curved bars of plain or skew curvature. Another method of importance discussed in this chapter is the method of discrete mass points. Large-span bridges and structures are practical examples of such systems.

10

Chapter 5 is devoted to three-dimensional axially symmetric systems and to two-dimensional systems with repeated elements. The theoretical solutions of the two systems are formally similar. Due to the benefits of standardisation and precasting, systems with repeated elements are becoming increasingly popular. Dynamic analysis benefits in a similar manner to manufacturing processes in that the repetition of elements makes for considerable simplification.

The sixth chapter deals with some of the problems of damped vibration. Its first section analyses the damped vibration of frame systems on the usual assumption that damping is proportional to the velocity of motion. Both the slope-deflection method and the frequency functions — generalised to take care of the damping — are made use of in the solution. The second section presents a penetrating analysis of the nature of damped vibration, and formulates a more general theory of this phenomenon.

The seventh chapter studies the dynamic effects of moving loads, with attention focused on railway bridges. Detailed treatment is accorded to the vibration of steel bridges with continuous main girders, and to reinforced concrete arch bridges. The theoretical solution is complemented by numerical examples in which the results of dynamic analysis are compared with experimental data obtained via measurements on actual structures. In the case of continuous steel bridges, the comparison shows excellent agreement between the two sets of results. In the case of reinforced concrete arch bridges, the experimental results brought to light new facts which the author incorporated in his theoretical approach.

Chapter 8 turns to some of the secondary effects, for example, those of axial forces, shear deflection and rotatory inertia. Using the vibration of a guyed mast as an example, the author shows that axial forces mainly affect the frequencies and modes of free vibrations of slender-type structures. On the other hand, shear deflections and moments of inertia forces come into play in the vibration of systems with massive members, such as the frame foundations of large-capacity machinery. Two-parameter frequency functions $F(a, d)$ are again applied to simplify the solution.

Chapter 9 points out some further uses of the frequency functions as evolved by the author. The two illustrative dynamic problems to which they are applied, deal respectively with simple and continuous plates, and with the analysis of the effects of impact on beams.

Chapter 10 is an appendix containing comprehensive tables as an aid to practical computations, and also tables of the numerical values of the frequency functions.

NOTATION

(A) Symbols

a	distance; length; abscissa of point $x = a$ at which load is applied
a	factor defined by equations 8.9, 8.49 or 9.48
$\bar{a}$	factor defined by equations 8.14 or 9.54
a, b, c, d	coefficients; constants
b	width of rectangular cross-section
b	as subscript denotes damping
c	velocity
d	factor defined by equations 8.10, 8.50 or 9.49
$\bar{d}$	factor defined by equations 8.15 or 9.55
e	base of the system of natural logarithms
f	frequency
$f_0, f_{(j)}$	natural frequency, natural frequency of the j-th natural mode
g	acceleration of gravity
g, h	letters denoting points of system
h	height; depth of beam or plate section
h	number of points of a cyclosymmetric system
i	$\sqrt{(-1)}$
i, j, k	as subscripts; integers denoting a sequence of points, terms of series, natural frequencies, etc.
k	coefficient; constant
l	length of bar; length of span
m	concentrated mass
n	integer; constant; rev/s
$p(x, t)$	continuous load varying with x and t
$p(x)$	load amplitude defined as $p(x, t) = p(x) \sin \omega t$ for $\sin \omega t = 1$
p	uniform load
$p_{(j)}$	load characteristic of vibrations in the j-th mode (modal analysis)
p	point of force application; letter denoting a point
q_1, q_2	generalised coordinates
$q_{(j)}$	principal generalised coordinates of the j-th natural mode
r	integer; radius; gyration radius; displacement
$r(t, \tau)$	relaxation curve

r	displacement vector
r, s	respectively antisymmetric, symmetric component of displacement
r_j, s_j	coefficients of series representing displacement of cyclosymmetric systems
s	letter denoting a point; abscissa of point; centre; centre of gravity
Δs	length of curved bar element
t	time
u, v, w	displacement respectively in the X, Y, Z direction
$v_{(j)}(x)$	displacement amplitude at point x for system vibrating in the j-th natural mode
x, y, z	coordinates
$^1y_k, {}^2y_k$	successive approximations of deflection y at point k
A	algebraic complement
$A(t)$	envelope curve
A, B, C, D	constants; coefficients
$\mathbf{A}, \mathbf{B}$	matrices; operators
C	elasticity coefficient
E	modulus of elasticity in tension, compression
$\mathbf{E}$	unit matrix
F	function
$F(\lambda)$	frequency function
G	weight; shear modulus
H	cylindrical stiffness of plate
J	cross-sectional moment of inertia
I	mass moment of inertia
$\mathbf{I}$	identity operator
$K, \overline{K}$	coefficients; constants
K	moment about axis X
$\mathscr{K}$	component of external moment about axis X
L	dissipated energy; area of hysteresis loop
L	moment about axis Y
$\mathscr{L}$	component of external moment about axis Y
M	moment; moment about axis Z
$\mathscr{M}$	component of external moment about axis Z
N	normal force; axial force; rev/s
P	force; centrifugal force
P	amplitude of harmonic force $P(t) = P \sin \omega t$ for $\sin \omega t = 1$
Q	total weight of bridge
R	amplitude of time dependent force $R(t) = R \sin \omega t$
R_j, S_j	coefficients of series representing load in case of cyclosymmetric systems
S	cross-sectional area
T	period
$T(x, t)$	shear force in section x at time t

$U_{(j)}$	normalised displacement of mass point after impact on beam
V	potential energy
$\left.\begin{array}{l} V_{i(j)} \\ V_{(j)}(x) \end{array}\right\}$	normalised displacement components in free vibration
X, Y, Z	coordinate axes; forces acting in their directions
$X_i(x)$	function
$\mathscr{X}, \mathscr{Y}, \mathscr{Z}$	components of external load applied at a frame joint in X, Y, Z direction
α	factor defined by equations 1.109 or 8.51
$\bar{\alpha}$	factor defined by equation 1.168
α, β, γ	angles
β	damping constant defined by equation 6.39
β, γ	variables used in equations 4.12, 4.13
γ	factor defined by equation 8.11 and 10.1
δ	factor defined by equation 10.6
δ	logarithmic decrement of damping
δ_{ik}	flexibility coefficient defined as displacement of point k due to unit static force applied to point i
ε	unit normal strain
ε	ratio of coefficients
$\varepsilon, \bar{\varepsilon}$	factors used in equation 7.11
ε	angular displacement of storey or panel, equation 8.70
ζ	rotation about axis Z
η	rotation about axis Y
ϑ	factor defined by equation 2.75
$\varkappa$	coefficient; amount of energy dissipated per cycle of damped vibration
$\varkappa$	buckling coefficient; factor of safety against buckling
λ	factor defined by equation 2.23
μ	mass per unit length or unit area
ν	Poisson's ratio, $\nu < 1$
ξ	variable, as in equation 4.9
ξ	rotation about axis X
ϱ	density; distance
σ	force per unit area
τ	time
φ	phase shift
χ	mass moment of inertia per unit length (torsional vibration)
ψ	parameter defined by equation 2.64
ψ	damping factor as in equation 6.67
ω	angular velocity; angular or circular frequency (frequency for short)
$\omega_{(j)}, \omega_0$	circular frequency of free vibration
ω_b	circular frequency of damping
ω_0	circular frequency of damped free vibration

$\Gamma(\alpha)$, $\Gamma(\delta)$	buckling functions
Δ	difference; determinant
Θ	$= 2\pi k/T$ for $k \to \infty$, $T \to \infty$ (equation 6.93)
Λ, $\bar{\Lambda}$	parameters defined by equation 6.8
Φ	angle
$\Phi(\lambda)$	frequency function
Ψ, $\bar{\Psi}$	factors defined by equation 6.22
Ω	circular frequency of moving harmonic force

(B)   Coordinates and displacements of points

The right-handed system of rectangular Cartesian coordinates with the positive Y-axis (or the positive Z-axis in the case of plates) pointing downward, is used throughout. A point is located by its coordinates x, y, z. A change of this position in a given time interval is denoted by $u(t)$, $v(t)$, $w(t)$. For the direction of displacements, forces, rotations and moments refer to Fig. 3.19.

(C)   Variables

Displacements and forces vary, not only with time but also with the position of the point considered. In concentrated-mass systems having a finite number of degrees of freedom, the displacements and forces are denoted by $v_i(t)$, $P_i(t)$, respectively, where i is the serial number of the relevant point. In continuous-mass systems where the displacement and load are both continuous with respect to position, the quantities are denoted by $v(x, t)$, $p(x, t)$, etc. The functional dependence upon location and time is always written out in full. When it is not specified, it means that the respective quantities are constant.

Harmonic motion and harmonic forces are written as

$$v(x, t) = v(x) \sin \omega t$$

$$P(t) \quad = P \sin \omega t$$

where $v(x)$ and P denote amplitude, that is to say, quantities which do not vary with time. It would be more correct — but also more complicated — to write

$$v(x, t) = v(x, t = \text{const.}) \sin \omega t$$

or

$$v(x, t) = v(x)_{\text{max}} \sin \omega t$$

It is hoped that the brevity of notation will not result in an incorrect interpretation.

Derivatives with respect to time are denoted by a dot placed over the respective symbol. Thus $dv(x, t)/dt = \dot{v}(x, t)$ is the velocity in the Y-direction at section x and at time t. Similarly $\ddot{u}(t) = [d^2u_i(t)]/dt^2$ is the acceleration of point i in the X-direction. Derivatives with respect to x are marked with primes, e.g. $v''(x) = [d^2v(x)]/dx^2$, etc.

(D)   Matrices and vectors

Matrices and vectors are set out in bold type or denoted by brackets containing the general term, $[\delta_{ik}]$. Bold type is also used for designating the operators in Chapter 6.2.

(E)   Subscripts

The meaning of the use of subscripts is explained in the text. In order to distinguish the subscripts of frequency and modes of free vibration from other serial numbers, e.g. those denoting the points of the system, the former are written in parentheses. Thus, for example, $f_{(2)}$ indicates the second natural frequency and $v_{3(1)}$ the displacement of point 3 in free vibration in the first natural mode. Generally, however, the natural frequency is written as f_0.

(F)   Signs

The displacements and force components are positive when they act in the direction of the respective positive coordinate axes. The rotation and moment components are positive if they rotate clockwise when viewed in the positive direction of the respective axis. In the solution of frame systems by the slope-deflection method this rule also applies to the determination of the signs of end moments and forces.

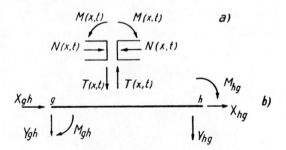

Fig. 0.1 Positive sense of forces and moments. (a) Internal force components, bending moments, normal and shear forces (b) end forces.

It does not, however, apply to the signs of the internal stress components, i.e. bending moments and shear forces, which always act in couples; these are established in the same way as in the theory of elasticity. The positive sense is indicated in Fig. 0.1.

(G) Systems of units used

In numerical computations A Force-Length-Time system is used, i.e. the unit of force is the megapond — a force equal on magnitude to the weight (in standard gravitational field) of a body of mass 1000 kg i.e. Mp = 9810 N. The unit of length is the metre (m) and of time, the second (s).

1 SYSTEMS WITH SEVERAL DEGREES OF FREEDOM

1.1 Introductory notes

Betti's repricocal theorem states: if an elastic body is subjected to the two loading systems A and B which maintain equilibrium and produce the respective displacements a and b, the work that would be done by the first set of forces A in acting through the second set of displacements b is equal to the work that would be done by the second set of forces B in acting through the first set of displacements a.

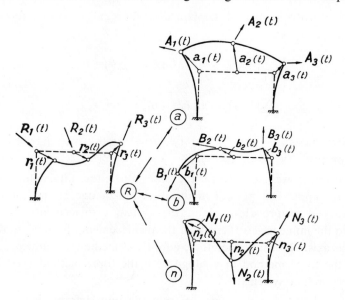

Fig. 1.1 Derivation of the equations of motion. R — actual state of motion, a, b, n — auxiliary states

The theorem can be extended to cover bodies in a state of motion or vibration by the addition of the inertia forces to the externally applied loads. In a form devised by Maxwell the reciprocal theorem is used in the formation of the influence coefficients which are used extensively in the theory of structures.

The equations describing the free vibration of systems with n degrees of freedom may be obtained by applying the reciprocal theorem n times in succession to two states of the system. One of the states is invariably the actual state of motion as characterised by displacements $r_i(t)$, external forces $R_i(t)$ and inertia forces $-m_i\ddot{r}_i(t)$; the other, an arbitrary fictitious 'auxiliary' state with known displacements and associated forces. The latter may be either the state of static deformation or the instantaneous state of motion. The idea is illustrated in Fig. 1.1. On applying the reciprocal theorem to states r and a, we get the equations

$$\sum_i [R_i(t) - m_i\ddot{r}_i(t)]\, a_i(t) = \sum_i [A_i(t) - m_i\ddot{a}_i(t)]\, r_i(t) \tag{1.1}$$

which state that the work done by the forces of the first state on the displacements of the second state equals the work done by the forces of the second state on the displacements of the first state. The products in the equations are scalar. If it is the state of static deformation that we choose for state a, eq. 1.1 simplifies to

$$\sum_i [R_i(t) - m_i\ddot{r}_i(t)]\, a_i = \sum_i A_i\, r_i(t) \tag{1.2}$$

The application of the reciprocal theorem to states r and b up to r and n will result in similar equations, viz.

$$\sum_i [R_i(t) - m_i\ddot{r}_i(t)]\, b_i = \sum_i B_i\, r_i(t)$$

$$\vdots \qquad\qquad \vdots \qquad\qquad \vdots$$

$$\sum_i [R_i(t) - m_i\ddot{r}_i(t)]\, n_i = \sum_i N_i\, r_i(t) \tag{1.3}$$

The auxiliary states a to n are arbitrary virtual states which may be chosen to suit the line of approach we intend to use.

Thus, for example, we may take for the states a to n the static deformation caused by a unit force applied at point i in the direction of one of the coordinate axes. According to the reciprocal theorem, the system displacements thus produced correspond to the influence coefficients of the displacement of point i in the direction of the chosen axis. We shall, therefore, adopt the method of influence coefficients (also called the force method in analogy with the force method used in the static analysis of redundant systems).

Or, we may choose for the states a to n the unit deformation (displacement) of a point of the system in the direction of one of the coordinate axes, whilst keeping the other displacements zero. In this case we shall make use of a method which we shall call the method of elasticity coefficients, known also as the deformation method — again analogous to the static case. It is clear that in such a simple form this method cannot be applied unless the virtual displacements of the individual points are mutually independent.

Both the method of influence coefficients and the method of elasticity coefficients are well suited to the analysis of free vibration as well as to the forced vibration of concentrated-mass systems.

However, once we know the modes of free vibration, we can solve for the case of forced vibration more expediently by expanding it in a series of natural modes. For auxiliary states, represented by natural modes, this method — the so-called modal analysis method — yields directly a set of n equations that are mutually independent.

1.2   Method of influence coefficients

1.2.1   Free vibration

Fig. 1.2 shows a simple system with four degrees of freedom — an elastic massless beam with four concentrated masses. For free vibration, the system displacements at time t are to the left, and the auxiliary states with unit loads are to the right of Fig. 1.3. The unit loads of these auxiliary states are independent of time and are

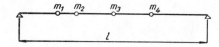

Fig. 1.2 System with a finite number of degrees of freedom

applied in the vertical plane at points 1, 2, 3, 4, respectively. The work done by the forces of state R on the displacements of state 1 will give us the left-hand side, while the work done by the forces of state 1 on the displacements of state R will give us the right-hand side of the first of eqs 1.4. The remaining three eqs 1.4 are obtained by applying the reciprocal theorem to states $R - 2$, $R - 3$ and $R - 4$:

$$v_1(t) = -m_1 \frac{d^2 v_1(t)}{dt^2} \delta_{11} - m_2 \frac{d^2 v_2(t)}{dt^2} \delta_{12} - m_3 \frac{d^2 v_3(t)}{dt^2} \delta_{13} - m_4 \frac{d^2 v_4(t)}{dt^2} \delta_{14}$$

$$v_2(t) = -m_1 \frac{d^2 v_1(t)}{dt^2} \delta_{21} - m_2 \frac{d^2 v_2(t)}{dt^2} \delta_{22} - m_3 \frac{d^2 v_3(t)}{dt^2} \delta_{23} - m_4 \frac{d^2 v_4(t)}{dt^2} \delta_{24}$$

$$v_3(t) = -m_1 \frac{d^2 v_1(t)}{dt^2} \delta_{31} - m_2 \frac{d^2 v_2(t)}{dt^2} \delta_{32} - m_3 \frac{d^2 v_3(t)}{dt^2} \delta_{33} - m_4 \frac{d^2 v_4(t)}{dt^2} \delta_{34}$$

$$v_4(t) = -m_1 \frac{d^2 v_1(t)}{dt^2} \delta_{41} - m_2 \frac{d^2 v_2(t)}{dt^2} \delta_{42} - m_3 \frac{d^2 v_3(t)}{dt^2} \delta_{43} - m_4 \frac{d^2 v_4(t)}{dt^2} \delta_{44}$$

$$(1.4)$$

The general form of the above equations is

$$v_i(t) = -\sum_k m_k \frac{d^2 v_k(t)}{dt^2} \delta_{ik} \qquad (1.4a)$$

and the matrix form (in the notation of Fig. 1.3) is

$$\mathbf{v}(t) = -\left[m_k \delta_{ik}\right] \ddot{\mathbf{v}} \tag{1.5}$$

where

$$\left[m_k \delta_{ik}\right] = \left[\delta_{ik}\right] \mathbf{m}$$

Eqs 1.4a and 1.5 apply, as does the complete procedure to be described presently, to systems with any number n of degrees of freedom.

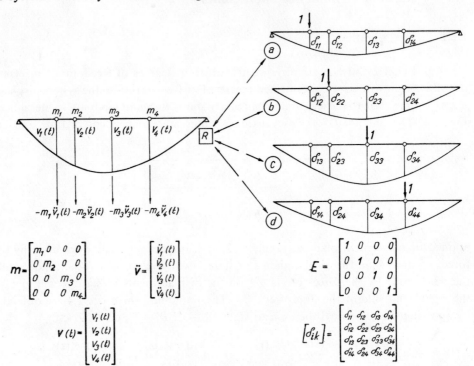

Fig. 1.3 *Method of influence coefficients. R — actual motion, a, b, c, d — unit states*

Eq. 1.4a has a particular solution in the form of harmonic motion

$$v_k(t) = v_k^0 \sin \omega_0 t \quad \text{for} \quad k = 1, 2, \ldots \tag{1.6}$$

which, on substitution in eq. 1.4 and cancellation of $\sin \omega_0 t$, gives the following equations for the amplitudes:

$$v_1^0 - \sum_k m_k \omega_0^2 v_k^0 \delta_{1k} = 0$$

$$v_2^0 - \sum_k m_k \omega_0^2 v_k^0 \delta_{2k} = 0$$

The general form of the amplitude equations is

$$v_i^0 - \sum_k m_k \omega_0^2 v_k^0 \delta_{ik} = 0 \tag{1.7}$$

or in matrix form

$$\mathbf{v}^0 - \omega_0^2 [m_k \delta_{ik}] \mathbf{v}^0 = \mathbf{0} \tag{1.8}$$

where

$$\mathbf{v}^0 = \begin{bmatrix} v_1^0 \\ v_2^0 \\ \vdots \\ v_n^0 \end{bmatrix} \tag{1.9}$$

is the column matrix (vector) of displacement.

Eq. 1.8 may also be written as

$$\left[\mathbf{A} - \frac{1}{\omega_0^2} \mathbf{E} \right] \mathbf{v}^0 = \mathbf{0} \tag{1.10}$$

where

$$\mathbf{A} = [m_k \delta_{ik}] = \begin{bmatrix} m_1 \delta_{11} & m_2 \delta_{12} & \cdots & m_n \delta_{1n} \\ m_1 \delta_{12} & m_2 \delta_{22} & \cdots & m_n \delta_{2n} \\ \vdots & \vdots & & \vdots \\ m_1 \delta_{1n} & m_2 \delta_{2n} & \cdots & m_n \delta_{nn} \end{bmatrix} \tag{1.11}$$

and $\mathbf{E}$ is the unit matrix.

The homogeneous set of eqs 1.7 possesses both a zero trivial solution corresponding to the at-rest state of the system, and a non-trivial solution with non-zero amplitudes which, however, is possible only on the condition that the determinant of the coefficients is zero. This condition leads to the so-called secular equation of the n-th degree in ω_0^2, n denoting the number of degrees of freedom of our vibrating system. After some manipulation the secular equation may be written in the form

$$\begin{vmatrix} m_1 \delta_{11} - \dfrac{1}{\omega_0^2} & m_2 \delta_{12} & \cdots & m_n \delta_{1n} \\ m_1 \delta_{12} & m_2 \delta_{22} - \dfrac{1}{\omega_0^2} & \cdots & m_n \delta_{2n} \\ \vdots & \vdots & & \vdots \\ m_1 \delta_{1n} & m_2 \delta_{2n} & \cdots & m_n \delta_{nn} - \dfrac{1}{\omega_0^2} \end{vmatrix} = 0 \tag{1.12}$$

or simply

$$\det \left[\mathbf{A} - \frac{1}{\omega_0^2} \mathbf{E} \right] = 0 \tag{1.13}$$

22

It can be solved by any of the usual methods, for example by Newton's method of approximation, 'regula falsi', etc.

By solving the secular equation we get n positive roots for the circular frequency of free vibrations, ω_0, viz. $\omega_{(1)}, \omega_{(2)} \dots \omega_{(j)} \dots \omega_{(n)}$.

Since eqs 1.7 are homogeneous they yield − rather than the absolute values of v_i^0 − the ratios $v_2^0/v_1^0 \dots v_n^0/v_1^0$, which are different, of course, for different $\omega_{(j)}$. In the case

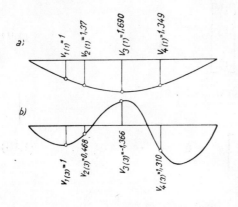

Fig. 1.4 Modes of free vibration of system shown in Fig. 1.2. (a) First natural mode, (b) third natural mode

considered we write these ratios as $v_{2(1)}/v_{1(1)} \dots v_{2(2)}/v_{1(2)}$, etc. and refer the reader to Fig. 1.4 for the meaning of the notation used.

The natural modes satisfy the conditions of orthogonality, viz.

$$\sum_i m_i v_{i(j)} v_{i(h)} = 0 \quad \text{for} \quad j \neq h \tag{1.14}$$

$$\sum_i m_i v_{i(j)}^2 \quad \neq 0 \tag{1.15}$$

The vibration intensity is not specified by eqs 1.7 but it can be assumed to be so large as to make deflections $v_{i(j)}$ − denoted by $V_{i(j)}$ in our case − satisfy the condition

$$\sum_i m_i V_{i(j)}^2 = 1 \tag{1.16}$$

Deflections $V_{i(j)}$, spoken of as 'normalised' deflections, are computed from the non-normalised ones by means of the formula

$$V_{i(j)} = \frac{v_{i(j)}}{\left[\sum (m_i v_{i(j)}^2)\right]^{1/2}} \tag{1.17}$$

In matrix form, the orthogonality conditions may be written as

$$\mathbf{V}'_{(j)} \mathbf{m} \mathbf{V}_{(j)} = \mathbf{E} \tag{1.18}$$

where

$$\mathbf{V}_{(j)} = \begin{bmatrix} V_{1(1)} & V_{1(2)} & \cdots & V_{1(n)} \\ V_{2(1)} & V_{2(2)} & \cdots & V_{2(n)} \\ \vdots & \vdots & & \vdots \\ V_{n(1)} & V_{n(2)} & \cdots & V_{n(n)} \end{bmatrix} \tag{1.19}$$

$\mathbf{V}'_{(j)}$ is the transpose of matrix $\mathbf{V}_{(j)}$

$$\mathbf{V}'_{(j)} = \begin{bmatrix} V_{1(1)} & V_{2(1)} & \cdots & V_{n(1)} \\ V_{1(2)} & V_{2(2)} & \cdots & V_{n(2)} \\ \vdots & \vdots & & \vdots \\ V_{1(n)} & V_{2(n)} & \cdots & V_{n(n)} \end{bmatrix} \tag{1.20}$$

m is the diagonal matrix of mass, and $\mathbf{E}$ the unit matrix.

A straightforward and expeditious method of solving the set of equations

$$\frac{v_{i(1)}}{\omega_{(1)}^2} = \sum_{k=1}^{n} m_k \delta_{ik} v_{k(1)} \tag{1.21}$$

is the so-called Stodola's method of successive approximations[72]. When applying it to the problem of finding the frequencies and shapes of natural modes we proceed as follows: choose a suitable curve ^1v_i as the probable form of the first natural mode. Apply to the system loads $m_k \, {}^1v_k$, and compute the static deflection ^1y_i at point i from the equation

$$^1y_i = \sum_{k=1}^{n} m_k \delta_{ik} {}^1v_k \quad \text{for} \quad i = 1 \text{ to } n \tag{1.22}$$

or in its matrix form

$$^1\mathbf{y} = [m_k \delta_{ik}] \, {}^1\mathbf{v} \tag{1.23}$$

where

$$^1\mathbf{y} = \begin{bmatrix} {}^1y_1 \\ \vdots \\ {}^1y_n \end{bmatrix} \quad \text{and} \quad {}^1\mathbf{v} = \begin{bmatrix} {}^1v_1 \\ \vdots \\ {}^1v_n \end{bmatrix} \tag{1.24}$$

Since

$$\sum_k m_k \delta_{ik} {}^1v_k \approx \sum_k m_k \delta_{ik} v_{k(1)}$$

it is by eq. 1.7 that

$$^1y_i \approx \frac{v_{i(1)}}{\omega_{(1)}^2} \approx \frac{{}^1v_i}{\omega_{(1)}^2} \tag{1.25}$$

and from this follows the first approximation to the first natural circular frequency

$$^1\omega_{(1)}^2 = \frac{{}^1v_i}{{}^1y_i} \tag{1.26}$$

In the next step the system is subjected to loads $m\,^1y_k$, or better still, to loads $m\,^2v_k$ where $^2v_k = k_1\,^1y_k$, k_1 being an appropriately chosen constant (see below). ^2v_k represents the second approximation of the first natural mode. Similarly as in the first step

$$^2y_i = \sum_k m_k \delta_{ik}\,^2v_k$$

$$^2\omega^2_{(1)} = \frac{^2v_i}{^2y_i}.$$

$$(1.27)$$

It is readily proved (cf. Ref. 14) that the procedure converges and after several steps locates the first natural frequency to the desired degree of accuracy. Since it is advantageous to choose the first approximation of the natural mode such that it will give $^1v_1 = 1$, set $k_1 = 1/^1y_1$ in the second approximation to get $^2v_1 = 1$, and continue along these lines in all succeeding approximations.

For the computation of higher natural frequencies the procedure just outlined must be complemented by the orthogonality conditions since it would otherwise always lead to the same result, i.e. the value of the first natural mode.

Let us now write eqs 1.7 for the second natural mode leaving out the first of the equations and replacing it by the orthogonality condition 1.14:

$$0 = \sum_{k=1}^{n} m_k v_{k(1)} v_{k(2)}$$

$$(1.28)$$

$$\frac{v_{i(2)}}{\omega^2_{(2)}} = \sum_{k=1}^{n} m_k \delta_{ik} v_{k(2)} \quad \text{for} \quad i = 2 \text{ to } n$$

$$(1.29)$$

Elimination of $v_{1(2)}$ reduces the original set of n equations to one of $n-1$ equations in the form

$$\frac{v_{i(2)}}{\omega^2_{(2)}} = \sum_{k=2}^{n} m_k \delta_{ik(2)} v_{k(2)} \quad \text{for} \quad i = 2 \text{ to } n$$

$$(1.30)$$

where

$$\delta_{ik(2)} = \delta_{ik} - \delta_{i1} a_{k(1)}$$

$$(1.31)$$

and

$$a_{k(1)} = \frac{v_{k(1)}}{v_{1(1)}}$$

$$(1.32)$$

In the matrix form eqs 1.30 are

$$\frac{\mathbf{V}_{(2)}}{\omega^2_{(2)}} = \mathbf{A}_{(2)} \mathbf{V}_{(2)}$$

$$(1.33)$$

where

$$\mathbf{A}_{(2)} = [m_i \delta_{ik(2)}]$$

Eqs 1.30 are solved by the same method of successive approximations as that used to solve eqs 1.21. This gives us the deflections $v_{2(2)}$ to $v_{n(2)}$ and $\omega_{(2)}^2$. The remaining deflection $v_{1(2)}$ is found from eq. 1.28:

$$m_1 v_{1(2)} = -m_2 \frac{v_{2(1)}}{v_{1(1)}} v_{2(2)} \cdots - m_n \frac{v_{n(1)}}{v_{1(1)}} v_{n(2)} = \sum_{k=2}^{n} m_k a_{k(1)} v_{k(2)} \qquad (1.34)$$

For the computation of the third natural mode, eqs 1.7 are written for only $n-2$ points, the first two equations of the set being replaced by the orthogonality conditions 1.14, viz.

$$\frac{v_{i(3)}}{\omega_{(3)}^2} = \sum_{k=1}^{n} m_k \delta_{ik} v_{k(3)} \quad \text{for} \quad i = 3 - n \qquad (1.35)$$

$$0 = \sum_{k=1}^{n} m_k v_{k(1)} v_{k(3)} \qquad (1.36)$$

$$0 = \sum_{k=1}^{n} m_k v_{k(2)} v_{k(3)} \qquad (1.37)$$

Elimination of $v_{1(3)}$ and $v_{2(3)}$ reduces the original set of n equations to one of $(n-2)$ equations

$$\frac{v_{i(3)}}{\omega_{(3)}^2} = \sum_{k=3}^{n} m_k \delta_{ik(3)} v_{k(3)} \quad \text{for} \quad i = 3 - n \qquad (1.38)$$

where

$$\delta_{ik(3)} = \delta_{ik(2)} - \delta_{i2(2)} a_{k(2)} \qquad (1.39)$$

and

$$a_{k(2)} = \frac{\begin{vmatrix} v_{1(1)} & v_{1(2)} \\ v_{k(1)} & v_{k(2)} \end{vmatrix}}{\begin{vmatrix} v_{1(1)} & v_{1(2)} \\ v_{2(1)} & v_{2(2)} \end{vmatrix}} \qquad (1.40)$$

Eqs 1.38 give $v_{i(3)}$ for $i = 3$ to n, while $v_{2(3)}$ is computed from the equation

$$m_2 v_{2(3)} = -m_3 a_{3(2)} v_{3(3)} - \cdots - m_n a_{n(2)} v_{n(3)} = -\sum_{k=3}^{n} m_k a_{k(2)} v_{k(3)} \qquad (1.41)$$

and $v_{1(3)}$ from the equation

$$m_1 v_{1(3)} = -m_2 \frac{v_{2(1)}}{v_{1(1)}} v_{2(3)} - \cdots - m_n \frac{v_{n(1)}}{v_{1(1)}} v_{n(3)} = -\sum_{k=2}^{n} m_k a_{k(1)} v_{k(3)} \qquad (1.42)$$

The fourth mode is computed analogously from the equations

$$\frac{v_{i(4)}}{\omega_{(4)}^2} = \sum_{k=4}^{n} m_k \delta_{ik(4)} v_{k(4)} \quad \text{for} \quad i = 4 \text{ to } n \qquad (1.43)$$

where

$$\delta_{ik(4)} = \delta_{ik(3)} - \delta_{i3(3)}a_{k(3)} \tag{1.44}$$

and

$$a_{k(3)} = \frac{\begin{vmatrix} v_{1(1)} & v_{1(2)} & v_{1(3)} \\ v_{2(1)} & v_{2(2)} & v_{2(3)} \\ v_{k(1)} & v_{k(2)} & v_{k(3)} \end{vmatrix}}{\begin{vmatrix} v_{1(1)} & v_{1(2)} & v_{1(3)} \\ v_{2(1)} & v_{2(2)} & v_{2(3)} \\ v_{3(1)} & v_{3(2)} & v_{3(3)} \end{vmatrix}} \tag{1.45}$$

and from the equations

$$m_3 v_{3(4)} = -\sum_{k=4}^{n} m_k a_{k(3)} v_{k(4)}$$

$$m_2 v_{2(4)} = -\sum_{k=3}^{n} m_k a_{k(2)} v_{k(4)}$$

$$m_1 v_{1(4)} = -\sum_{k=2}^{n} m_k a_{k(1)} v_{k(4)}$$

$$\tag{1.46}$$

It therefore holds generally for the j-th natural mode that

$$\frac{v_{i(j)}}{\omega_{(j)}^2} = \sum_{k=j}^{n} m_k \delta_{ik(j)} v_{k(j)} \tag{1.47}$$

$$\delta_{ik(j)} = \delta_{ik(j-1)} - \delta_{i,j-1(j-1)} a_{k(j-1)} \tag{1.48}$$

$\delta_{ik(1)} = \delta_{ik}$ being the influence coordinates of displacement (influence coefficients).

$$a_{k(j)} = \frac{\begin{vmatrix} v_{1(1)} & \cdots & v_{1(j)} \\ \vdots & & \vdots \\ v_{j-1(1)} & \cdots & v_{j-1(j)} \\ v_{k(1)} & \cdots & v_{k(j)} \end{vmatrix}}{\begin{vmatrix} v_{1(1)} & \cdots & v_{1(j)} \\ \vdots & & \vdots \\ v_{j(1)} & \cdots & v_{j(j)} \end{vmatrix}} \tag{1.49}$$

$$m_i v_{i(j)} = -\sum_{k=i+1}^{n} m_k a_{k(i)} v_{k(j)} \quad \text{for} \quad i = 1 \text{ to } j-1 \tag{1.50}$$

N o t e : In the foregoing exposition the individual points were eliminated wholly mechanically, i.e. in the order they were numbered — 1, 2, 3, 4. For enhanced accuracy it is recommended that in numerical computations the elimination procedure should start with the point whose diagonal ordinate is either the largest, or at least not the smallest of all. This suggestion is followed in Example 1.1 where the elimination, though starting with point 1 continues through points 4, 2, 3. More advantageous still would be the order 3, 4, 2, 1.

The general solution to eqs 1.4 then consists of the particular solutions 1.6 with arbitrarily large phase shift $\varphi^0_{(j)}$ i.e.

$$v_i(t) = \sum_{j=1}^{n} v_{i(j)} \sin\left(\omega_{(j)}t + \varphi^0_{(j)}\right) \tag{1.51}$$

or

$$v_i(t) = \sum_{j}^{n} \left[A_{i(j)} \cos \omega_{(j)}t + B_{i(j)} \sin \omega_{(j)}t\right] \tag{1.52}$$

The integration constants $v_{i(j)}$, $\varphi^0_{(j)}$ of eqs 1.51 and $A_{i(j)}$, $B_{i(j)}$ of eqs 1.52 are bound by the relations

$$v_{i(j)} = \left(A^2_{i(j)} + B^2_{i(j)}\right)^{1/2}$$

$$\varphi^0_{(j)} = \tan^{-1} \frac{A_{i(j)}}{B_{i(j)}}$$

$$\tag{1.53}$$

and

$$A_{i(j)} = v_{i(j)} \sin \varphi^0_{(j)}$$

$$B_{i(j)} = v_{i(j)} \cos \varphi^0_{(j)}$$

$$\tag{1.54}$$

They are computed from the initial conditions of motion by means of eqs 1.71, 1.73 — see below — in which we set $P_{(k)} = 0$.

1.2.2 Forced vibration

Whenever a system is acted upon by a time-varying external load $P_i(t)$ (Fig. 1.5), forced vibration is set up. In such a case the reciprocal theorem is applied

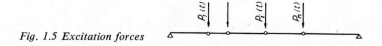

Fig. 1.5 Excitation forces

to state R (Fig. 1.6a) and to states $i = 1$ to n (Fig. 1.6b). The equations thus obtained are

$$v_i(t) = \sum_{k} \delta_{ik} \left[P_k(t) - m_k \frac{\mathrm{d}^2 v_k(t)}{\mathrm{d}t^2}\right] \tag{1.55}$$

or in matrix form

$$\mathbf{v} = [\delta_{ik}] (\mathbf{P} - \mathbf{m\ddot{v}}) \qquad (1.56)$$

where $\mathbf{P}$ is the column matrix (force vector)

$$\mathbf{P} = \begin{bmatrix} P_1(t) \\ \vdots \\ P_n(t) \end{bmatrix} \qquad (1.57)$$

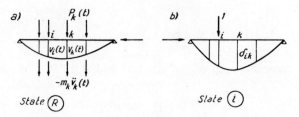

Fig. 1.6 Forced vibration (actual motion — left, unit state — right)

Eqs 1.55 may be solved by the so-called modal analysis, a method in which both the deflections and the forces are expanded in a series of the form:

$$v_k(t) = \sum_{j=1}^{n} q_{(j)}(t)\, V_{k(j)} \qquad (1.58)$$

$$\frac{P_k(t)}{m_k} = \sum_{j=1}^{n} p_{(j)}(t)\, V_{k(j)} \qquad (1.59)$$

Knowing the variations with time of forces $P_i(t)$ we may compute the coefficient $p_{(j)}(t)$ of the second series by multiplying both sides of eq. 1.59 by $m_k V_{k(1)}$ and summing for all k s:

$$\sum_k P_k(t)\, V_{k(1)} = \sum p_{(1)}(t)\, m_k V_{k(1)}^2 + \sum p_{(2)}(t)\, m_k V_{k(1)} V_{k(2)}$$
$$+ \dots \sum p_{(n)}(t)\, m_k V_{k(1)} V_{k(n)} \qquad (1.60)$$

After substitution of the orthogonality conditions 1.14 and 1.16 and some manipulation, eq. 1.60 gives

$$p_{(1)}(t) = \sum_k P_k(t)\, V_{k(1)}$$

Similar relations apply to the remaining coefficients so that generally

$$p_{(j)}(t) = \sum_k P_k(t)\, V_{k(j)} \quad \text{for} \quad j = 1 \text{ to } n \qquad (1.61)$$

The coefficients of the first series, eqs 1.58 are computed from eqs 1.55 by the substitution of eqs 1.58 and 1.59, i.e.

$$\sum_j q_{(j)}(t)\, V_{i(j)} = \sum_j \sum_k m_k \left(p_{(j)}(t) - \frac{d^2 q_{(j)}(t)}{dt^2} \right) V_{k(j)} \delta_{ik} \qquad (1.62)$$

At the j-th natural mode it holds, however, by eqs 1.7 that

$$\sum_k m_k V_{k(j)} \delta_{ik} = \frac{V_{i(j)}}{\omega_{(j)}^2}$$

and this substituted in eq. 1.62 gives after some manipulation

$$\sum_j \left(\frac{d^2 q_{(j)}(t)}{dt^2} + \omega_{(j)}^2 q_{(j)}(t) - p_{(j)}(t) \right) V_{i(j)} = 0 \qquad (1.63)$$

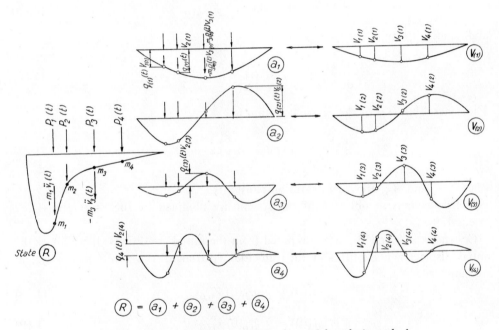

Fig. 1.7 Solution of forced vibration by modal analysis method

Eq. 1.63 can be satisfied for any arbitrary i only on the condition that the expression in the parentheses is equal to zero. Consequently

$$\frac{d^2 q_{(j)}(t)}{dt^2} + \omega_{(j)}^2 q_{(j)}(t) = p_{(j)}(t) \quad \text{for} \quad j = 1 \text{ to } n \qquad (1.64)$$

Given the variations in quantity $p_{(j)}(t)$ we can compute $q_{(j)}(t)$ from eqs 1.64.

Eqs 1.64 follow directly from the equality of the work done by the forces of states $R - V_{(1)}$ to $R - V_{(4)}$ (Fig. 1.7). State R is obtained by the superposition of loads and displacements shown in a_1 to a_4. Since the natural modes are orthogonal, the work done by the forces of state a_1 on displacements $V_{(2)}$, $V_{(3)}$, $V_{(4)}$ is zero, so that the resultant equations are

$$\sum_i [p_{(j)}(t) - \ddot{q}_{(j)}(t)] \, m_i V_{i(j)} V_{i(j)} = \sum_i \omega_{(j)}^2 m_i V_{i(j)} V_{i(j)} q_{(j)}(t)$$

and from there eqs 1.64 can be formulated.

Harmonic load

If the external forces are harmonic variables, $P_k \sin \omega t$, it holds by eqs 1.61 that

$$p_{(j)}(t) = \sin \omega t \sum_k P_k V_{k(j)} \tag{1.65}$$

and hence also by eqs 1.64

$$\frac{\mathrm{d}^2 q_{(j)}(t)}{\mathrm{d}t^2} + \omega_{(j)}^2 \, q_{(j)}(t) = \sin \omega t \sum_k P_k V_{k(j)} \tag{1.66}$$

The general integral of the above equation is

$$q_{(j)}(t) = \sum_k \frac{P_k V_{k(j)}}{\omega_{(j)}^2 - \omega^2} \sin \omega t + A_{(j)} \cos \omega_{(j)} t + B_{(j)} \sin \omega_{(j)} t \tag{1.67}$$

where $A_{(j)}$ and $B_{(j)}$ are the integration constants.

The deflections at points k of the system are obtained by substituting eqs 1.67 in eqs 1.58.

Constants $A_{(j)}$ and $B_{(j)}$ are computed from the initial conditions. If we know the initial displacements $v_k(0)$ and the velocities $\dot{v}_k(0)$ at all the points in the system we may write

$$v_k(0) = \sum_j q_{(j)}(0) \, V_{k(j)} \tag{1.68}$$

$$\dot{v}_k(0) = \sum_j \dot{q}_{(j)}(0) \, V_{k(j)} \tag{1.69}$$

where $q_{(j)}(0)$, $\dot{q}_{(j)}(0)$ denote $q_{(j)}(t)$, $\mathrm{d}q_{(j)}(t)/\mathrm{d}t$ at time $t = 0$. Expressions 1.68 and 1.69 are analogous to eqs 1.59 and can, therefore, be used for computing $q_{(j)}(0)$ and $\dot{q}_{(j)}(0)$ by the procedure applied in obtaining eqs 1.61:

$$q_{(j)}(0) = \sum_k m_k \, v_k(0) \, V_{k(j)}$$

$$\dot{q}_{(j)}(0) = \sum_k m_k \, \dot{v}_k(0) \, V_{k(j)}$$

$$\tag{1.70}$$

Eqs 1.67 then give

$$A_{(j)} = q_{(j)}(0) \tag{1.71}$$

and the derivative of eqs 1.67 is

$$\dot{q}_{(j)}(t) = \sum_k \frac{P_k V_{k(j)}\omega}{\omega_{(j)}^2 - \omega^2} \cos \omega t - A_{(j)}\omega_{(j)} \sin \omega_{(j)}t + B_{(j)}\omega_{(j)} \cos \omega_{(j)}t$$

$$\dot{q}_{(j)}(0) = \sum_k \frac{P_k V_{k(j)}\omega}{\omega_{(j)}^2 - \omega^2} + B_{(j)}\omega_{(j)} \tag{1.72}$$

and

$$B_{(j)} = \frac{1}{\omega_{(j)}} \left(\dot{q}(0) - \sum_k \frac{P_k V_{k(j)}\omega}{\omega_{(j)}^2 - \omega^2} \right) \tag{1.73}$$

External load varying generally with time

The general integral of eq. 1.64 is

$$q_{(j)}(t) = \frac{1}{\omega_{(j)}} \int_0^t P_{(j)}(\tau) \sin \omega_{(j)}(t - \tau) \, d\tau + q_{(j)}(0) \cos \omega_{(j)}t + \frac{\dot{q}_{(j)}(0)}{\omega_{(j)}} \sin \omega_{(j)}t \tag{1.74}$$

where τ is any time instant whithin the interval 0 and t. To check the correctness of solution 1.74 we substitute it in eqs 1.64. Since in eqs 1.74 the variable t is contained both in the integrand and in the upper limit of the integral, the first derivative with respect to t is

$$\frac{dq_{(j)}(t)}{dt} = \lim_{\Delta t \to 0} \frac{1}{\omega_{(j)}} \frac{\int_0^{t+\Delta t} P_{(j)}(\tau) \sin \omega_{(j)}(t + \Delta t - \tau) \, d\tau - \int_0^t P_{(j)}(\tau) \sin \omega_{(j)}(t - \tau) \, d\tau}{\Delta t}$$

$$- q_{(j)}(0) \omega_{(j)} \sin \omega_{(j)}t + \dot{q}(0) \cos \omega_{(j)}t$$

Since

$$\lim_{\Delta t \to 0} \frac{\int_t^{t+\Delta t} P_{(j)}(\tau) \sin \omega_{(j)}(t + \Delta t - \tau) \, d\tau}{\Delta t} = 0$$

then

$$\frac{dq_{(j)}(t)}{dt} = \int_0^t P_{(j)}(\tau) \cos \omega_{(j)}(t - \tau) \, d\tau - q_{(j)}(0) \omega_{(j)} \sin \omega_{(j)}t + \dot{q}_{(j)}(0) \cos \omega_{(j)}t$$

The second derivative of eqs 1.74 is

$$\frac{\mathrm{d}^2 q_{(j)}(t)}{\mathrm{d}t^2} = \lim_{\Delta t \to 0} \frac{\displaystyle\int_0^t p_{(j)}(\tau)\cos\omega_{(j)}(t + \Delta t - \tau)\,\mathrm{d}\tau - \int_0^t p_{(j)}(\tau)\cos\omega_{(j)}(t - \tau)\,\mathrm{d}\tau}{\Delta t}$$

$$+ \lim_{\Delta t \to 0} \frac{\displaystyle\int_t^{t+\Delta t} p_{(j)}(\tau)\cos\omega_{(j)}(t + \Delta t - \tau)\,\mathrm{d}\tau}{\Delta t} - q_{(j)}\omega_{(j)}^2 \cos\omega_{(j)}t$$

$$- \dot{q}_{(j)}(0)\,\omega_{(j)}\sin\omega_{(j)}t = -\omega_{(j)}\int_0^t p_{(j)}(\tau)\sin\omega_{(j)}(t - \tau)\,\mathrm{d}\tau + p_{(j)}(t)$$

$$- q_{(j)}\omega_{(j)}^2 \cos\omega_{(j)}t - \dot{q}_{(j)}(0)\,\omega_{(j)}\sin\omega_{(j)}t$$

Eq. 1.64 is satisfied by the incorporation of the second derivative in eq. 1.74.

1.3   Method of elasticity coefficients

1.3.1   Free vibration

Let us apply the reciprocal theorem to the actual state R of a system executing free vibration, and to the auxiliary states 1, 2, 3, etc. shown in Fig. 1.8 in which

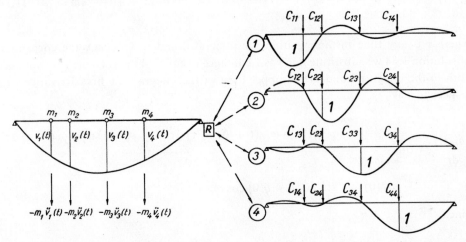

Fig. 1.8 Method of elasticity coefficients. R — actual state, 1, 2, 3, 4 — unit states

the system is statically deformed, with unit displacement always applied at one point only, the displacements of the remaining points being zero. The resulting equations of motion are

$$m_i \frac{\mathrm{d}^2 v_i(t)}{\mathrm{d}t^2} + \sum_k C_{ik}\, v_k(t) = 0 \quad \text{for} \quad i = 1, 2, \dots \tag{1.75}$$

where the elasticity constants C_{ik} are forces causing the respective auxiliary deformations as denoted in Fig. 1.8.

Using the notation of that figure we may write eqs 1.75 in matrix form as follows

$$m\ddot{v} + Cv = 0 \qquad (1.76)$$

Assuming that the motion is harmonic, that is

$$v_i(t) = v_i^0 \sin \omega_0 t$$

eq. 1.75 gives after rearrangement

$$m_i \omega_0^2 v_i^0 - \sum_k C_{ik} v_k^0 = 0 \qquad (1.77)$$

In matrix form eq. 1.77 is

$$\left(C - m\omega_0^2\right) v^0 = 0 \qquad (1.78)$$

or

$$\left(A^{-1} - E\omega_0^2\right) v^0 = 0 \qquad (1.79)$$

where

$$A^{-1} = \begin{bmatrix} \dfrac{C_{11}}{m_1} & \dfrac{C_{12}}{m_1} & \cdots & \dfrac{C_{1n}}{m_1} \\ \dfrac{C_{12}}{m_2} & \dfrac{C_{22}}{m_2} & \cdots & \dfrac{C_{2n}}{m_2} \\ \vdots & \vdots & & \vdots \\ \dfrac{C_{1n}}{m_n} & \dfrac{C_{2n}}{m_n} & \cdots & \dfrac{C_{nn}}{m_n} \end{bmatrix} \qquad (1.80)$$

A non-trivial solution to eq. 1.79 is possible only if

$$\det\left[A^{-1} - E\omega_0^2\right] = 0 \qquad (1.81)$$

Matrix A according to eq. 1.11 is the inverse of matrix A^{-1} corresponding to eq. 1.80; therefore eq. 1.13 is identical to eq. 1.81. By solving eq. 1.81 we again get n positive roots for ω_0, viz $\omega_{(1)} \ldots \omega_{(n)}$.

Eq. 1.77 can also be solved by the method of successive approximations. It holds for the highest natural mode that

$$v_{i(n)}\omega_{(n)}^2 = \sum_k \frac{C_{ik}}{m_i} v_{k(n)} \qquad (1.82)$$

As with the method of influence coefficients, we choose $^1v_{k(n)}$, and compute

$$^1\bar{y}_i = \sum_k \frac{C_{ik}}{m_i} \, ^1v_{k(n)} \qquad (1.83)$$

34

and

$$^1\omega^2_{(n)} = \frac{^1\bar{y}_k}{^1v_{k(n)}} \tag{1.84}$$

In the second step

$$^2v_{k(n)} = \frac{^1\bar{y}_k}{^1\bar{y}_n}$$

$$^2\bar{y}_k = \sum_k \frac{C_{ik}}{m_i} {}^2v_{k(n)}$$

$$^2\omega^2_{(n)} = \frac{^2\bar{y}_k}{^2v_{k(n)}} \quad \text{etc.}$$

$$\tag{1.85}$$

We shall now prove that this procedure will always lead to the highest natural mode: Let us expand ^1v_k in a series in terms of the natural modes

$$^1v_{k(n)} = \sum_j q_{(j)}v_{k(j)}$$

After substituting the above in eq. 1.83 it is found that because of eq. 1.77

$$^1\bar{y}_k = \sum_j q_{(j)}v_{k(j)}\omega^2_{(j)}$$

and from eq. 1.85

$$^3v_{k(n)} = \frac{1}{^1\bar{y}_n\,^2\bar{y}_n} \sum_j q_{(j)}v_{k(j)}\omega^4_{(j)}$$

It becomes clear after repeated application of the procedure that the last term grows far faster than the remaining ones because $\omega_{(n)} > \omega_{(n-1)} \ldots > \omega_{(1)}$. At the r-th step we may therefore write

$$^r\bar{y}_k = \frac{1}{^1\bar{y}_n\,^2\bar{y}_n \ldots\,^{r-1}\bar{y}_n} \sum_j q_{(j)}v_{k(j)}\omega^{2r}_{(j)} \approx {}^rv_{k(n)}\omega^2_{(n)} \tag{1.86}$$

if r is sufficiently large.

The system of equations for computing the $(n-1)$-th natural mode consists of one orthogonality equation

$$0 = \sum_{k=1}^n m_k v_{k(n)}v_{k(n-1)} \tag{1.87}$$

and $(n-1)$ equations

$$v_{i(n-1)}\omega^2_{(n-1)} = \sum_{k=1}^n \frac{C_{ik}}{m_i} v_{k(n-1)} \quad \text{for} \quad i = 1 \text{ to } n-1 \tag{1.88}$$

Elimination of $v_{n(n-1)}$ leads to the system of equations

$$v_{i(n-1)}\omega_{(n-1)}^2 = \sum_{k=1}^{n-1} \frac{C_{ik(n-1)}}{m_i} v_{k(n-1)} \quad \text{for} \quad i = 1 \text{ to } n-1 \tag{1.89}$$

where

$$C_{ik(n-1)} = C_{ik} - C_{in}b_{k(n)} \tag{1.90}$$

$$b_{k(n)} = \frac{m_k}{m_n} \frac{v_{k(n)}}{v_{n(n)}} \tag{1.91}$$

System 1.89, again solved by iteration, gives $v_{1(n-1)}$ to $v_{n-1(n-1)}$ and $\omega_{(n-1)}^2$.
   Deflection $v_{n(n-1)}$ is found from the equation

$$v_{n(n-1)} = -\sum_{k=1}^{n-1} b_{k(n)}v_{k(n-1)} \tag{1.92}$$

The equation for the $(n-2)$-th natural mode is

$$v_{i(n-2)}\omega_{(n-2)}^2 = \sum_{k=1}^{n-2} \frac{C_{ik(n-2)}}{m_i} v_{k(n-2)} \tag{1.93}$$

where

$$C_{ik(n-2)} = C_{ik(n-1)} - C_{i,n-1(n-1)}b_{k(n-1)} \tag{1.94}$$

and

$$b_{k(n-1)} = \frac{m_k}{m_{n-1}} \frac{\begin{vmatrix} v_{n(n)} & v_{n(n-1)} \\ v_{k(n)} & v_{k(n-1)} \end{vmatrix}}{\begin{vmatrix} v_{n(n)} & v_{n(n-1)} \\ v_{n-1(n)} & v_{n-1(n-1)} \end{vmatrix}} \tag{1.95}$$

1.3.2   Forced vibration

When the system according to Fig. 1.8 is subjected to external forces, the application of the reciprocal theorem to state R (Fig. 1.7) and states 1 to 4 (Fig. 1.8) results in the equations

$$m_i \frac{d^2v_i(t)}{dt^2} + \sum_{k=1}^{n} C_{ik} v_k(t) = P_i(t) \tag{1.96}$$

Expanding deflections $v_i(t)$ as well as forces $P_i(t)$ in a series of the kind used in eqs 1.58 and 1.59, we get

$$\sum_{j=1}^{n} \left(\frac{d^2q_{(j)}(t)}{dt^2} V_{i(j)} + q_{(j)}(t)\sum_{k=1}^{n} \frac{C_{ik}}{m_i} V_{k(j)} - p_{(j)}(t) V_{i(j)} \right) = 0 \tag{1.97}$$

Since

$$\sum_{k=1}^{n} \frac{C_{ik}}{m_i} V_{k(j)} = \omega_{(j)}^2 V_{i(j)}$$

and since eq. 1.97 holds good for all i, it must be the case that

$$\frac{d^2 q_{(j)}(t)}{dt^2} + \omega_{(j)}^2 \, q_{(j)}(t) = p_{(j)}(t)$$

which is the same as eq. 1.64.

Note: In the simple form outlined above, the method of elasticity coefficients is applicable only to systems whose mass points can be considered to move without causing displacements of the remaining points, as is the case shown in Fig. 1.8.

Example 1.1

We are going to solve the system considered in our foregoing exposition (Fig. 1.2) and evaluate its natural modes and circular frequencies by means of successive approximations, applying first the method of influence coefficients and then the method of elasticity coefficients.

(a)   Method of influence coefficients

The matrix of influence coefficients appertaining to the system is set out in Table A_0. The first four rows of the table give the reduced influence coefficients:

$$\delta_{ik} = \frac{EJ \cdot 10^3}{l^3} \, \delta_{ik}^* \tag{a}$$

the last row gives the reduced masses:

$$m_k = \frac{m_k^*}{m_1^*} \tag{b}$$

The asterisk marks the actual values of the quantities at points $k = 1, 2, 3, 4$.

Table A_0

i \\ k	1	2	3	4
$[\delta_{ik}] =$ 1	8·534	10·966	11·834	8·600
2	10·966	14·700	16·500	12·300
3	11·834	16·500	20·834	16·500
4	8·600	12·300	16·500	14·700
m_k	1·0	1·5	2·0	1·5

Because of eqs (a) and (b), the evaluated circular frequencies $\omega_{(k)}$ will also be reduced. The actual value $\omega_{(k)}^*$ is given by the relation

$$\omega_{(k)}^{*2} \leqq \frac{EJ}{m_1^*} \frac{10^3}{l^3} \omega_{(k)}^2 \tag{c}$$

Multiplying matrix $[\delta_{ik}]$ by mass matrix $\boldsymbol{m}$ gives matrix $\boldsymbol{A}_{(1)} = [m_k \delta_{ik}]$ (Table A_1)

Table A_1

$\underset{i}{\overset{k}{\diagdown}}$	1	2	3	4
$\boldsymbol{A}_{(1)} =$ ⟶ 1	8·534	16·449	23·667	12·900
$[m_k \delta_{ik}]$ ⟶ 2	10·966	22·050	33·000	18·450
3	11·834	24·750	41·667	24·750
4	8·600	18·450	33·000	22·050

Next, matrix $\boldsymbol{A}_{(1)}$ is iterated successively by vectors $^1\boldsymbol{v}_{(1)}$, $^2\boldsymbol{v}_{(1)}$, etc. and the results are set out in a table.

$k =$	1	2	3	4
$^1v_{k(1)} =$	1·0	1·37	1·69	1·37
$^1y_{k(1)} =$	88·739	122·221	150·066	119·855
$1 : {}^1\omega_{(1)}^2 =$	88·739	89·212	88·770	87·485
$^2v_{k(1)} =$	1·0	1·3773	1·6911	1·3506
$^2y_{k(1)} =$	88·6352	122·0603	149·8126	119·5982
$1 : {}^2\omega_{(1)}^2 =$	88·6352	88·6229	88·5888	88·5519
$^3v_{k(1)} =$	1·0	1·3771	1·6902	1·3492
$^3y_{k(1)} =$	88·5938	122·0022	149·7380	119·5362
$1 : {}^3\omega_{(1)}^2 =$	88·5938	88·5936	88·5920	88·5914

Continuing the computation beyond this point would have no effect on the results. Hence the first natural mode

$$\boldsymbol{v}_{(1)} = \begin{bmatrix} v_{1(1)} \\ v_{2(1)} \\ v_{3(1)} \\ v_{4(1)} \end{bmatrix} = \begin{bmatrix} 1·0 \\ 1·3771 \\ 1·6902 \\ 1·3492 \end{bmatrix}$$

$\dfrac{1}{\omega_{(1)}^2} = 88·594$ (rounded off to three decimal places)

The orthogonality condition $v'_{(1)}[m]\,v_{(j)} = 0$, i.e. $\displaystyle\sum_{k=1}^{k=4} v_{k(1)}m_k v_{k(j)} = 0$ yields the component $v_{1(j)}$ in terms of the other components $(j \neq 1)$

$$v_{1(j)} = -\left(2{\cdot}0656 v_{2(j)} + 3{\cdot}3804 v_{3(j)} + 2{\cdot}0238 v_{4(j)}\right)$$

This expression together with rows $i = 2$, 3 and 4 of matrix $A_{(1)}$ gives the reduced matrix $A_{(2)}$ set out in Table A_2.

Table A_2

	$\begin{array}{c} k \\ \hline i \end{array}$	2	3	4
	2	−0·602	−4·069	−3·743
$A_{(2)}$	3	0·305	1·663	0·800
	4	0·685	3·929	4·646

The method of successive iterations used on matrix $A_{(1)}$ is now applied to matrix $A_{(2)}$ and the results are as follows

$k =$	2	3	4
$^1v_{k(2)} =$	1·0	−0·21	−1·20
$^1y_{k(2)} =$	4·7441	−1·0042	−5·7153
$1 : {}^1\omega^2_{(2)} =$	4·7441	4·7819	4·7628
$^2v_{k(2)} =$	1·0	−0·2117	−1·2047
$^2y_{k(2)} =$	4·7686	−1·0108	−5·7438
$1 : {}^2\omega^2_{(2)} =$	4·7686	4·7748	4·7678
$^3v_{k(2)} =$	1·0	−0·2120	−1·2045
$^3y_{k(2)} =$	4·7691	−1·0112	−5·7441
$1 : {}^3\omega^2_{(2)} =$	4·7691	4·7698	4·7689
$^4v_{k(2)} =$	1·0	−0·2120	−1·2044
$^4y_{k(2)} =$	4·7688	−1·0111	−5·7436
$1 : {}^4\omega^2_{(2)} =$	4·7688	4·7690	4·7690

The first component of vector $v_{(2)}$ is found from the orthogonality condition used in the reduction of matrix $A_{(1)}$ to matrix $A_{(2)}$ i.e.

$$v_{1(2)} = -2{\cdot}0656 \times 1{\cdot}0 + 3{\cdot}3804 \times 0{\cdot}2120 + 2{\cdot}0238 \times 1{\cdot}2044 = 1{\cdot}0885$$

Normalising the second vector to its first component gives the second natural mode

$$\mathbf{v}_{(2)} = \begin{bmatrix} v_{1(2)} \\ v_{2(2)} \\ v_{3(2)} \\ v_{4(2)} \end{bmatrix} = \begin{bmatrix} 1\cdot0 \\ 0\cdot9187 \\ -0\cdot1948 \\ -1\cdot1065 \end{bmatrix}$$

$$\frac{1}{\omega_{(2)}^2} \approx 4\cdot769$$

Matrix $\mathbf{A}_{(2)}$ can be further reduced, either by means of formulae 1.39, 1.40, or directly by means of the orthogonality conditions

$$\mathbf{v}_{(1)}'[m]\,\mathbf{v}_{(j)} = 0 \quad \text{and} \quad \mathbf{v}_{(2)}'[m]\,\mathbf{v}_{(j)} = 0 \quad \text{where} \quad j = 3, 4$$

Using the latter we eliminate the first component $v_{1(j)}$. With the substitution of the numerical values we find for the remaining three components

$$v_{2(j)}, v_{3(j)} \quad \text{and} \quad v_{4(j)}$$

that

$$0\cdot6876v_{2(j)} + 3\cdot7700v_{3(j)} + 3\cdot6836v_{4(j)} = 0$$

Expressing component $v_{4(j)}$ in terms of the other two, i.e.

$$v_{4(j)} = -0\cdot1867v_{2(j)} - 1\cdot0235v_{3(j)}$$

and using this expression with the last two rows $(i = 3, 4)$ of matrix $\mathbf{A}_{(2)}$ we get the reduced matrix $\mathbf{A}_{(3)}$ (Table A_3)

Table A_3

	$\dfrac{k}{i}$	2	3
$\mathbf{A}_{(3)}$	2	+0·096	−0·238
	3	+0·156	+0·842

Matrix $\mathbf{A}_{(3)}$ belonging to the reduced system with two degrees of freedom is best solved directly. Setting the determinant 1.13 equal to zero we get the frequency equation

$$\left(\frac{1}{\omega^2}\right)^2 + 0\cdot938\,\frac{1}{\omega^2} - 0\cdot118 = 0$$

and the frequencies are given by

$$\frac{1}{\omega_{(3)}^2} \approx 0\cdot789 ; \quad \frac{1}{\omega_{(4)}^2} \approx 0\cdot149$$

(correct to three decimal places). With these values we find the corresponding components of vectors $\mathbf{v}_{(3)}$ and $\mathbf{v}_{(4)}$ by direct substitution in either of the two equations in matrix $\mathbf{A}_{(3)}$. Having components $v_{2(j)}$, $v_{3(j)}$ for $j = 3, 4$, we can compute components $v_{1(j)}$, $v_{4(j)}$ from the orthogonality conditions applied in the reduction of matrix $\mathbf{A}_{(2)}$ to $\mathbf{A}_{(3)}$. Normalising them to their first components gives the shapes of the remaining two modes, viz. for the third natural mode

$$\mathbf{v}_{(3)} = \begin{bmatrix} v_{1(3)} \\ v_{2(3)} \\ v_{3(3)} \\ v_{4(3)} \end{bmatrix} = \begin{bmatrix} 1{\cdot}0 \\ 0{\cdot}4707 \\ -1{\cdot}3706 \\ 1{\cdot}3149 \end{bmatrix}$$

$$\frac{1}{\omega_{(3)}^2} \approx 0{\cdot}789$$

and for the fourth natural mode

$$\mathbf{v}_{(4)} = \begin{bmatrix} v_{1(4)} \\ v_{2(4)} \\ v_{3(4)} \\ v_{4(4)} \end{bmatrix} = \begin{bmatrix} 1{\cdot}0 \\ -0{\cdot}7162 \\ 0{\cdot}1595 \\ -0{\cdot}0295 \end{bmatrix}$$

$$\frac{1}{\omega_{(4)}^2} \approx 0{\cdot}149$$

Checks

(1) The sum of the frequencies $\sum_{j=1}^{4} 1/\omega_{(j)}^2$ ought to be equal to the sum $\sum_{i=1}^{4} a_{ii}$ of the diagonal elements of matrix $\mathbf{A}_{(1)}$ of the system. Since

$$\sum_{i=1}^{4} a_{ii} = 94{\cdot}301 = \sum_{j=1}^{4} \frac{1}{\omega_{(j)}^2}$$

we find that this condition is satisfied.

(2) Any two of the natural mode shapes should be orthogonal to one another, i.e. the condition

$$\mathbf{v}_{(i)}'[m]\,\mathbf{v}_{(j)} = 0 \quad (i = 1, 2, 3, 4; \; j = 1, 2, 3, 4)$$

must be satisfied for any two of the above values of i and j. Using the numerical values arrived at in the foregoing example we find that

$$\mathbf{v}_{(1)}'[m]\,\mathbf{v}_{(2)} = -0{\cdot}0001 \qquad \mathbf{v}_{(2)}'[m]\,\mathbf{v}_{(3)} = +0{\cdot}0002$$

$$\mathbf{v}_{(1)}'[m]\,\mathbf{v}_{(3)} = +0{\cdot}0002 \qquad \mathbf{v}_{(2)}'[m]\,\mathbf{v}_{(4)} = -0{\cdot}0001$$

$$\mathbf{v}_{(1)}'[m]\,\mathbf{v}_{(4)} = +0{\cdot}0001 \qquad \mathbf{v}_{(3)}'[m]\,\mathbf{v}_{(4)} = -0{\cdot}0009$$

Considering that the computation has been carried out to four decimal places, we may regard the second condition to be satisfied with adequate accuracy.

(b) *Method of elasticity coefficients*

The matrix of elasticity coefficients is defined by the relation

$$[C_{ik}] = [\delta_{ik}]^{-1}$$

The numerical values of the elasticity coefficients and the inverse of the mass matrix are set out in Table C.

Table C

		k	1	2	3	4	$1/m_k$
	i						
$[C_{ik}]$	1		3·9120	−3·7948	0·7278	0·0667	$1/1 = 1·0$
	2		−3·7948	4·4550	−1·6208	0·3052	$1/1·5 = 0·667$
	3		0·7278	−1·6208	1·6330	−0·9033	$1/2·0 = 0·500$
	4		0·0667	0·3052	−0·9033	0·7800	$1/1·5 = 0·667$

Multiplying by the inverse mass matrix gives matrix $C_{(4)} = m^{-1}[C_{ik}]$. (Note that this time multiplication is carried out in the rows).

Table D

		k	1	2	3	4
	i					
$C_{(4)} = A^{-1}$	1		3·9120	−3·7948	0·7278	0·0667
	2		−2·5298	2·9700	−1·0805	0·2035
	3		0·3639	−0·8104	0·8165	−0·4516
	4		0·0445	0·2035	−0·6022	0·5200

As in the preceding solution, the natural frequencies will be reduced according to the relationship

$$\omega_{(j)}^{*2} = \omega_{(j)}^{2} \frac{EJ \times 10^3}{m_1^* l^3}$$

(the actual values are again marked with an asterisk).

The procedure that follows is analogous to that used with the method of influence coefficients except that the iteration first yields the highest natural frequency and its corresponding vector, i.e. $\omega_{(4)}$ and $\mathbf{v}_{(4)}$:

k	1	2	3	4
$^1 v_{k(4)} =$	1·0	−0·716	0·160	−0·030
$^1 \bar{y}_{k(4)} =$	6·7435	−4·8353	1·0883	−0·2132
$^1 \omega^2_{(4)} =$	6·7435	6·7532	6·8021	7·1053
$^2 v_{k(4)} =$	1·0	−0·7170	0·1614	−0·0316
$^2 \bar{y}_{k(4)} =$	6·7482	−4·8401	1·0910	−0·2150
$^2 \omega^2_{(4)} =$	6·7482	6·7505	6·7596	6·8037
$^3 v_{k(4)} =$	1·0	−0·7173	0·1617	−0·0319
$^3 \bar{y}_{k(4)} =$	6·7500	−4·8414	1·0916	−0·2153
$^3 \omega^2_{(4)} =$	6·750	6·750	6·751	6·753

(As before, continuing the computation beyond this point would not affect the results.)

In this way we obtain the (reduced) highest natural frequency

$$\omega^2_{(4)} = 6·750$$

and its reciprocal

$$1/\omega^2_{(4)} = 1/6·570 = 0·1481$$

and find that it agrees well with that arrived at by the method of influence coefficients, viz. $1/\omega^2_{(4)} = 0·149$.

To reduce matrix $\mathbf{C}_{(4)}$ we use the orthogonality condition

$$\mathbf{v}'_{(4)}[m]\,\mathbf{v}_{(j)} = 0 \quad (j = 1, 2, 3)$$

which leads to the expression

$$v_{1(j)} = 1·076v_{2(j)} - 0·3234v_{3(j)} + 0·0479v_{4(j)}$$

defining component $v_{1(j)}$ in terms of the other three. Using this expression with the rows $i = 2, 3, 4$ of matrix $\mathbf{C}_{(4)}$ gives the reduced matrix $\mathbf{C}_{(3)}$ (Table D).

Table E

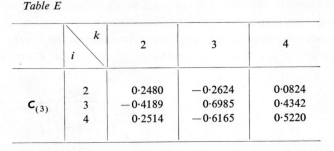

	k	2	3	4
$\mathbf{C}_{(3)}$	2	0·2480	−0·2624	0·0824
	3	−0·4189	0·6985	0·4342
	4	0·2514	−0·6165	0·5220

The iterative procedure applied to matrix $\mathbf{C}_{(3)}$ now converges to the third natural mode. Its steps are shown in the table below:

k	2	3	4	Step
$^{1}v_{k(3)} =$	0·340	−1·0	0·961	
$^{1}\bar{y}_{k(3)} =$	0·4259	−1·2582	1·2036	1
$^{1}\omega_{(3)}^{2} =$	1·2527	1·2582	1·2525	
·	·	·	·	2
$^{3}v_{k(3)} =$	0·3386	−1·0	0·9564	
$^{3}\bar{y}_{k(3)} =$	0·4252	−1·2556	1·2009	3
$^{3}\omega_{(3)}^{2} =$	1·2557	1·2556	1·2556	

Thus we get for the third circular frequency $\omega_{(3)}^{2} \approx 1·256$, a value that again compares well with that obtained by the method of influence coefficients $(\omega_{(3)}^{2} \approx 1·268)$. Normalised to the first component, the third mode shape (in transposed form) is

$$\mathbf{v}'_{(3)} = (1·0;\ 0·4616;\ -1·3634;\ 1·3039)$$

Note: The methods discussed in the foregoing sections are only some of those available for the solution of the equations of motion. For other methods of interest the reader is referred to References 10, 14, 16, 25, 29, 54, 70.

1.4 Systems vibrating in several directions

In all the cases considered so far, the displacements were vertical. Evidently, however, either of the two methods presented may be generalised to arbitrary displacements, i.e. displacements in any direction and rotations about any axis.

44

By way of an example, consider a foundation block of mass m for which the plane of the drawing is a plane of symmetry. The motion in the plane of the drawing has three components: displacement $u(t)$ in the direction of the X axis, displacement $v(t)$ in the direction of the Y axis, and rotation $\zeta(t$ about the Z axis at right angle

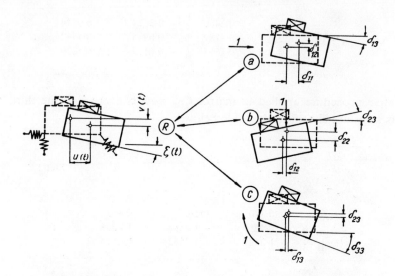

Fig. 1.9 Vibration of a foundation block — solution by the method of influence coefficients

to the plane of the drawing. I is the mass moment of inertia, the remaining symbols being as indicated in Fig. 1.9. Applying the reciprocal theorem to the states marked out in the figure, we get the following system of equations:

$$u(t) = -m\frac{d^2u(t)}{dt^2}\delta_{11} - m\frac{d^2v(t)}{dt^2}\delta_{12} - I\frac{d^2\zeta(t)}{dt^2}\delta_{13}$$

$$v(t) = -m\frac{d^2u(t)}{dt^2}\delta_{12} - m\frac{d^2v(t)}{dt^2}\delta_{22} - I\frac{d^2\zeta(t)}{dt^2}\delta_{23}$$

$$\zeta(t) = -m\frac{d^2u(t)}{dt^2}\delta_{13} - m\frac{d^2v(t)}{dt^2}\delta_{23} - I\frac{d^2\zeta(t)}{dt^2}\delta_{33}$$

$$(1.98)$$

With notation $u(t) = r_1(t)$, $v(t) = r_2(t)$, $\zeta(t) = r_3(t)$, $m_1 = m_2 = m$, $m_3 = I$, eqs 1.98 will take on the form

$$r_i(t) = -\sum_k m_k\delta_{ik}\ddot{r}_k \qquad (1.99)$$

where

$$\ddot{r}_k = \frac{d^2r_k(t)}{dt^2}$$

In matrix form, eq. 1.99 is written as

$$\boldsymbol{r} = -\left[\delta_{ik}\right] \boldsymbol{m}\ddot{\boldsymbol{r}}$$

(1.100)

where $\boldsymbol{m}$ is the diagonal mass matrix

$$\begin{bmatrix} m & 0 & 0 \\ 0 & m & 0 \\ 0 & 0 & I \end{bmatrix}$$

(1.101)

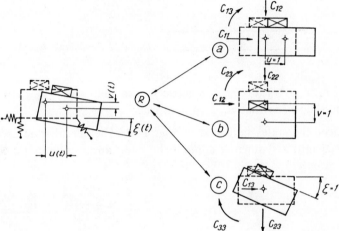

Fig. 1.10 Vibration of a foundation block — solution by the method of elasticity coefficients

The problem may also be solved by the method of elasticity coefficients (Fig. 1.10). Then

$$\boldsymbol{m}\ddot{\boldsymbol{r}} + \left[C_{ik}\right] \boldsymbol{r} = \boldsymbol{0}$$

(1.102)

which is analogous to eq. 1.76.

Eq. 1.102 is the matrix expression of the following three equations

$$m \frac{\mathrm{d}^2 u(t)}{\mathrm{d}t^2} + C_{11}\, u(t) + C_{12}\, v(t) + C_{13}\, \zeta(t) = 0$$

$$m \frac{\mathrm{d}^2 v(t)}{\mathrm{d}t^2} + C_{12}\, u(t) + C_{22}\, v(t) + C_{23}\, \zeta(t) = 0$$

$$I \frac{\mathrm{d}^2 \zeta(t)}{\mathrm{d}t^2} + C_{13}\, u(t) + C_{23}\, v(t) + C_{33}\, \zeta(t) = 0$$

(1.103)

1.5   Symmetry and cyclosymmetry of systems

1.5.1   Simple symmetry

Consider a system symmetric to the vertical axis which passes through the centre of the beam (Fig. 1.11). It is clear that any shape of symmetric deformation of such a system will be orthogonal to all shapes of antisymmetric deformation

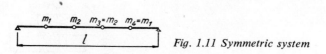

Fig. 1.11 Symmetric system

because $v_1 = v_4$, $v_2 = v_3$ for the symmetric shapes and $v_1 = -v_4$, $v_2 = -v_3$ for the antisymmetric shapes, so that eq. 1.16 is satisfied.

Consider now the states indicated in Fig. 1.12. The actual state of deformation can be resolved into a symmetric component s and an antisymmetric component r.

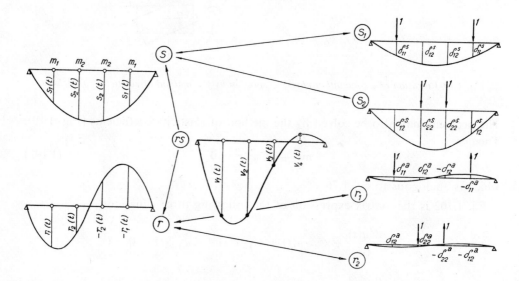

Fig. 1.12 Solution of a symmetric system. Left — resolution of motion in symmetric and antisymmetric components. Right — symmetric and antisymmetric unit states

The work performed mutually by state s and auxiliary states r_1 and r_2 is evidently zero, and the same holds true for state r and auxiliary states s_1 and s_2. Applying the reciprocal theorem we get the following equations

$$s_i(t) = -\sum_{k=1}^{2} m_k \delta_{ik}^s \ddot{s}_k(t) \quad \text{for} \quad i = 1, 2 \tag{1.104}$$

$$r_i(t) = -\sum_{k=1}^{2} m_k \delta_{ik}^r \ddot{r}_k(t) \quad \text{for} \quad i = 1, 2 \tag{1.105}$$

Eqs 1.104 and 1.105 are mutually independent and can be solved separately.

It is, therefore, reasonable to classify the modes of vibration of symmetrical systems in two groups, one containing the symmetric, the other the antisymmetric mode shapes.

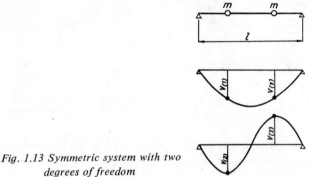

Fig. 1.13 Symmetric system with two
degrees of freedom

If a symmetrical system has only two degrees of freedom (such as the one shown in Fig. 1.13), its natural modes can be drawn directly, without any numerical solution.

Systems with an even number of degrees of freedom possess $n/2$ symmetric and $n/2$ antisymmetric mode shapes. Systems with an odd number of degrees of freedom can have either $(n + 1)/2$ or $(n - 1)/2$ symmetric mode shapes, depending on the configuration of the system.

Note: The above considerations apply only on the condition that the system is symmetric, not only as regard the distribution of vibrating masses but also as regard its elastic properties. Otherwise the conditions for dynamic equilibrium according to d'Alembert's principle would not be satisfied in symmetric motion.

1.5.2   Cyclosymmetry of three-dimensional systems

1.5.2.1   Free vibration

The solution simplifies still further in the case of three-dimensional systems with several planes of symmetry or two-dimensional systems with identical repeated elements.

Consider a system as shown in Fig. 1.14. Apply the reciprocal theorem to the actual state of motion, R (Fig. 1.15a) and to the unit state k (Fig. 1.15b). After some manipulation the equation of motion becomes

$$w_{k-1}(t) + 2w_k(t) + w_{k+1}(t) + \frac{m}{C \sin^2 \pi/h} \ddot{w}_k(t) = 0 \qquad (1.106)$$

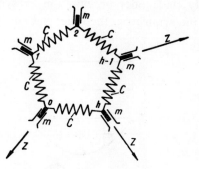

Fig. 1.14 Cyclosymmetric system

where C is the spring constant of one spring and h is the number of points. Eq. 1.106 is a difference equation with constant coefficients. Assume first that the motion is harmonic, viz.

$$w_k(t) = w_k^0 \sin \omega_0 t \qquad (1.107)$$

Substitution of eq. 1.107 in eq. 1.106 and cancellation of $\sin \omega_0 t$ give

$$w_{k-1}^0 + \left(2 - \frac{m\omega_0^2}{C \sin^2 \pi/h}\right) w_k^0 + w_{k+1}^0 = 0 \qquad (1.108)$$

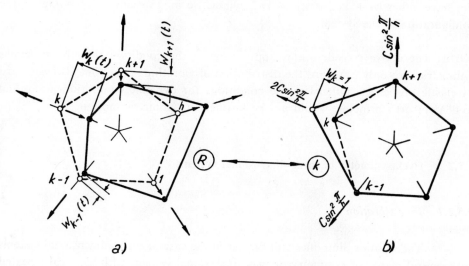

Fig. 1.15 Vibration of a cyclosymmetric system. R — actual motion, k — unit state

For $h = 5$, eq. 1.108 is written out in the form of the system of equations given in Table 1.1, with

$$\alpha = \frac{m\omega_0^2}{2C \sin^2 \pi/h} \tag{1.109}$$

Table 1.1

w_1^0	w_2^0	w_3^0	w_4^0	w_5^0	
$2 - 2\alpha$	1			1	$= 0$
1	$2 - 2\alpha$	1			$= 0$
	1	$2 - 2\alpha$	1		$= 0$
		1	$2 - 2\alpha$	1	$= 0$
1			1	$2 - 2\alpha$	$= 0$

The equation is satisfied if

$$w_k^0 = s_j \sin\left(\frac{2\pi kj}{h} + \Phi_j\right) \tag{1.110}$$

where $j = 0, 1, 2 \ldots h/2$ for even hs and $j = 0, 1, 2 \ldots (h - 1)/2$ for odd hs. Here s_j and Φ_j are arbitrary constants. Substitution in eq. 1.108 gives

$$\sin\left[\frac{2\pi j}{h}(k - 1) + \Phi_j\right] + 2(1 - \alpha)\sin\left[\frac{2\pi j}{h}k + \Phi_j\right]$$
$$+ \sin\left[\frac{2\pi j}{h}(k + 1) + \Phi_j\right] = 0 \tag{1.111}$$

Further,

$$\sin\left[\frac{2\pi j}{h}(k - 1) + \Phi_j\right] + \sin\left[\frac{2\pi j}{h}(k + 1) + \Phi_j\right] = 2\cos\frac{2\pi j}{h}\sin\left(\frac{2\pi j}{h}k + \Phi_j\right) \tag{1.112}$$

and substitution of eq. 1.112 in eq. 1.111 and subsequent rearrangement give

$$\alpha = 1 + \cos\frac{2\pi j}{h} \tag{1.113}$$

Replacing α in the above by eq. 1.109 we get, for the circular frequency, the expression

$$\omega_j = \sin\frac{\pi}{h}\left(\frac{2C}{m}\alpha\right)^{1/2} = \sin\frac{\pi}{h}\left[\frac{2C}{m}\left(1 + \cos\frac{2\pi j}{h}\right)\right]^{1/2} = 2\sin\frac{\pi}{h}\cos\frac{\pi j}{h}\left(\frac{C}{m}\right)^{1/2} \tag{1.114}$$

If h is even, eq. 1.114 gives $(1 + h/2)$ values of ω_j. Since expression 1.110 contains an arbitrary constant Φ_j, there are clearly an infinite number of j-th natural shapes, each resulting from the superposition of two mutually orthogonal components.

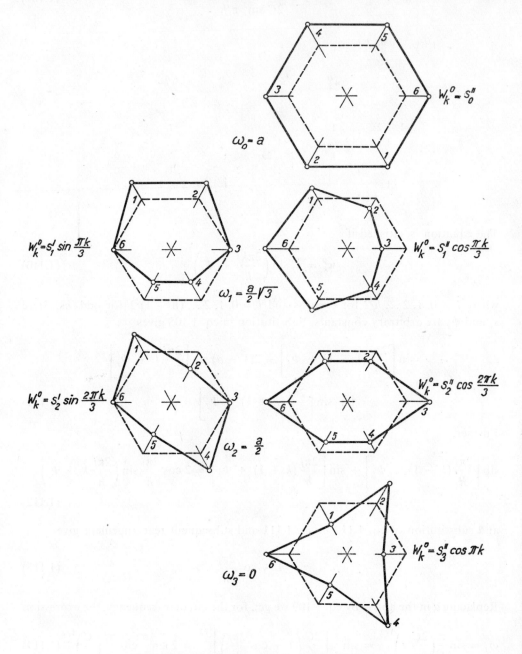

Fig. 1.16 Fundamental modes of a hexagonal system

One of the components is obtained by setting $\Phi_j = 0$ in eq. 1.110,

$$w_k^0 = s_j' \sin \frac{2\pi k j}{h} \tag{1.115}$$

the other by setting $\Phi_j = \pi/2$

$$w_k^0 = s_j'' \cos \frac{2\pi k j}{h} \tag{1.116}$$

s_j' and s_j'' are again arbitrary constants. However, there is only one natural mode for $j = 0$ and $j = h/2$.

Thus, for example, for the system shown in Fig. 1.16 with $h = 6$, we get

$$2 \sin \frac{\pi}{h} = 2 \sin \frac{\pi}{6} = 1$$

Writing

$$2 \sin \frac{\pi}{h} \left(\frac{C}{m}\right)^{1/2} = a$$

we find that

$$\omega_{00} = a$$

$$\omega_1 = a \cos \frac{\pi}{6} = \frac{a}{2} 3^{1/2}$$

$$\omega_2 = a \cos \frac{2\pi}{6} = \frac{a}{2}$$

$$\omega_3 = a \cos \frac{\pi}{2} = 0$$

The modes of free vibration computed from formulae 1.115 and 1.116 are indicated in Fig. 1.16.

Denoting the sequence from the lowest to the highest natural frequency by $\omega_{(1)}$, $\omega_{(2)} \ldots \omega_{(n)}$, we see that

$$\omega_{(1)} = \omega_3 ; \quad \omega_{(2)} = \omega_{(3)} = \omega_2 ; \quad \omega_{(4)} = \omega_{(5)} = \omega_1 ; \quad \omega_{(6)} = \omega_{00}$$

Note that for a given system some of the frequencies can turn out zero. In our case of $h = 6$, it is frequency ω_3 that is zero. We have, of course, assumed throughout

52

that the vibration amplitudes are infinitely small. Were they of finite values, the vibration would be non-linear.

If a system has an odd number of mass points, eq. 1.114 gives $(h + 1)/2$ values of natural frequencies. Frequency ω_{00} corresponds to a single mode of free vibration, the remaining frequencies always corresponding to the two modes defined by eqs 1.115 and 1.116. Accordingly we get for a system with $h = 5$

$$a = 2 \sin \frac{\pi}{h} \left(\frac{C}{m}\right)^{1/2} = 1 \cdot 176 \left(\frac{C}{m}\right)^{1/2}$$

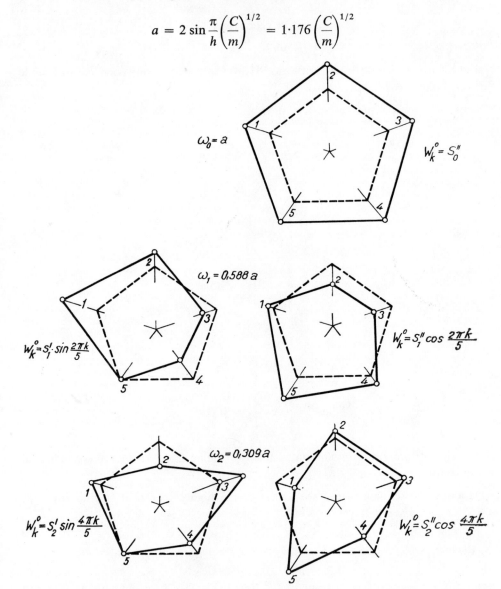

Fig. 1.17 Fundamental modes of a pentagonal system

and

$$\omega_{00} = a \qquad\qquad\qquad = \omega_{(5)}$$

$$\omega_1 = a \cos \frac{\pi}{5} \; = 0{\cdot}809a = \omega_{(3)} = \omega_{(4)}$$

$$\omega_2 = a \cos \frac{2\pi}{5} = 0{\cdot}309a = \omega_{(1)} = \omega_{(2)}$$

The corresponding modes of free vibration are illustrated in Fig. 1.17; they were computed from formulae 1.115 and 1.116.

For $j = 0$, eq. 1.115 gives $w_1^0 = w_2^0 = w_3^0 = w_4^0 = w_5^0 = s_0$.

For $j = 1$, eqs 1.115 and 1.116 with $s_1' = s_2'' = 1$ give

$$w_1^0 = \sin \frac{2\pi}{5} = \quad 0{\cdot}951 \quad \text{or} \quad w_1^0 = \cos \frac{2\pi}{5} = \quad 0{\cdot}309$$

$$w_2^0 = \sin \frac{4\pi}{5} = \quad 0{\cdot}578 \quad \text{or} \quad w_2^0 = \cos \frac{4\pi}{5} = -0{\cdot}809$$

$$w_3^0 = \sin \frac{6\pi}{5} = -0{\cdot}578 \quad \text{or} \quad w_3^0 = \cos \frac{6\pi}{5} = -0{\cdot}809$$

$$w_4^0 = \sin \frac{8\pi}{5} = -0{\cdot}951 \quad \text{or} \quad w_4^0 = \cos \frac{8\pi}{5} = \quad 0{\cdot}309$$

$$w_5^0 = 0 \qquad\qquad\qquad \text{or} \quad w_5^0 = 1$$

and for $j = 2$

$$w_1^0 = \sin \frac{4\pi}{5} = \quad 0{\cdot}578 \quad \text{or} \quad w_1^0 = \cos \frac{4\pi}{5} = -0{\cdot}809$$

$$w_2^0 = \sin \frac{8\pi}{5} = -0{\cdot}951 \quad \text{or} \quad w_2^0 = \cos \frac{8\pi}{5} = \quad 0{\cdot}309$$

$$w_3^0 = \sin \frac{12\pi}{5} = \quad 0{\cdot}951 \quad \text{or} \quad w_3^0 = \cos \frac{12\pi}{5} = \quad 0{\cdot}309$$

$$w_4^0 = \sin \frac{16\pi}{5} = -0{\cdot}578 \quad \text{or} \quad w_4^0 = \cos \frac{16\pi}{5} = -0{\cdot}809$$

$$w_5^0 = 0 \qquad\qquad\qquad \text{or} \quad w_5^0 = 1$$

A somewhat more complicated case of a cyclosymmetric system is shown in Fig. 1.18. The ring of the system consists of massless springs with spring constant C (as in the preceding case). It rests, however, on elastic columns of stiffness EJ_x/l^3 in the radial, and stiffness EJ_z/l^3 in the tangential (X) direction. J_x, J_z are the cross-sectional moments of inertia of the column section with respect to axes X and Z, and l is the column length.

The unit state (state of unit displacement) depicted in Fig. 1.19a is brought about by forces

$$Z_k = \frac{3EJ_x}{l^3} + 2C \sin^2 \frac{\pi}{h} = a$$

$$Z_{k-1,k} = Z_{k+1,k} = C \sin^2 \frac{\pi}{h} = c$$

$$X_{k-1,k} = -X_{k+1,k} = -C \sin \frac{\pi}{h} \cos \frac{\pi}{h} = b$$

$$(1.117)$$

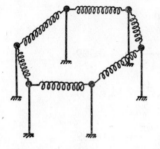

Fig. 1.18 Cyclosymmetric system on elastic columns

Applying the reciprocal theorem to the actual state of motion, R, and to the unit state (Fig. 1.19a) we get the difference equation

$$cw_{k-1}(t) + aw_k(t) + cw_{k+1}(t) + bu_{k-1}(t) - bu_{k+1}(t) = -m\ddot{w}_k(t) \quad (1.118)$$

The unit state shown in Fig. 1.19b is induced by forces

$$X_k = \frac{3EJ_z}{l^3} + 2C \cos^2 \frac{\pi}{h} = d$$

$$X_{k-1,k} = X_{k+1,k} = -C \cos^2 \frac{\pi}{h} = e$$

$$Z_{k-1,k} = -Z_{k+1,k} = -b$$

$$(1.119)$$

The reciprocal theorem now results in

$$-bw_{k-1}(t) + bw_{k+1}(t) + eu_{k-1}(t) + du_k(t) + eu_{k+1}(t) = -m\ddot{u}_k(t) \quad (1.120)$$

Assuming harmonic motion

$$u_k(t) = u_k^0 \sin \omega t$$
$$w_k(t) = w_k^0 \sin \omega t$$

$$(1.121)$$

we get from eqs 1.118 and 1.120

$$cw_{k-1}^0 + (a - m\omega^2)\,w_k^0 + cw_{k+1}^0 + bu_{k-1}^0 - bu_{k+1}^0 = 0 \qquad (1.122)$$

$$-bw_{k-1}^0 + bw_{k+1}^0 + eu_{k-1}^0 + (d - m\omega^2)\,u_k^0 + eu_{k+1}^0 = 0 \qquad (1.123)$$

The equations are written out in Table 1.2.

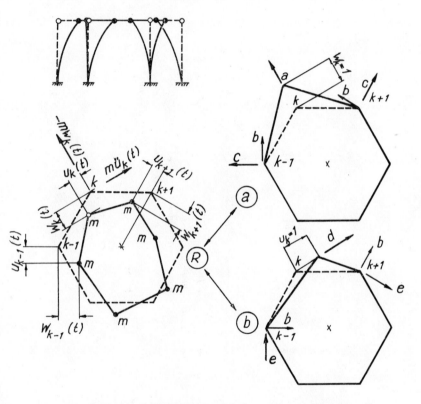

Fig. 1.19 (a), (b) Unit states of system shown in Fig. 1.18

Their solution is in the form

$$w_k^0 = s_j \sin\left(\frac{2\pi jk}{h} + \Phi_j\right)$$

$$u_k^0 = r_j \cos\left(\frac{2\pi jk}{h} + \Phi_j\right)$$

$$(1.124)$$

Table 1.2.

w_1^0	w_2^0	w_3^0	w_4^0	w_5^0	u_1^0	u_2^0	u_3^0	u_4^0	u_5^0	
$a-m\omega^2$	c			c	b				$-b$	$=0$
c	$a-m\omega^2$	c			$-b$	b				$=0$
	c	$a-m\omega^2$	c			$-b$	b			$=0$
		c	$a-m\omega^2$	c			$-b$	b		$=0$
c			c	$a-m\omega^2$				$-b$	b	$=0$
b	$-b$				$d-m\omega^2$	e			e	$=0$
	b	$-b$			e	$d-m\omega^2$	e			$=0$
		b	$-b$			e	$d-m\omega^2$	e		$=0$
			b	$-b$			e	$d-m\omega^2$	e	$=0$
$-b$				b	e			e	$d-m\omega^2$	$=0$

As in eqs 1.112 the relations

$$\cos\left[\frac{2\pi j}{h}(k-1) + \Phi_j\right] - \cos\left[\frac{2\pi j}{h}(k+1) + \Phi_j\right] = 2\sin\frac{2\pi j}{h}\sin\left(\frac{2\pi jk}{h} + \Phi_j\right)$$

$$\cos\left[\frac{2\pi j}{h}(k-1) + \Phi_j\right] + \cos\left[\frac{2\pi j}{h}(k+1) + \Phi_j\right] = 2\cos\frac{2\pi j}{h}\cos\left(\frac{2\pi jk}{h} + \Phi_j\right)$$

$$-\sin\left[\frac{2\pi j}{h}(k-1) + \Phi_j\right] + \sin\left[\frac{2\pi j}{h}(k+1) + \Phi_j\right] = 2\sin\frac{2\pi j}{h}\cos\left(\frac{2\pi jk}{h} + \Phi_j\right)$$

$$(1.125)$$

can be used and eqs 1.122 and 1.123 then give

$$s_j\left[a + 2c\cos\frac{2\pi j}{h} - m\omega^2\right] + 2r_jb\sin\frac{2\pi j}{h} = 0$$

$$2s_jb\sin\frac{2\pi j}{h} + r_j\left[d + 2e\cos\frac{2\pi j}{h} - m\omega^2\right] = 0$$

$$(1.126)$$

The natural frequencies are computed from the condition that the determinant of the coefficients of eq. 1.126 should be equal to zero. For each j there are two

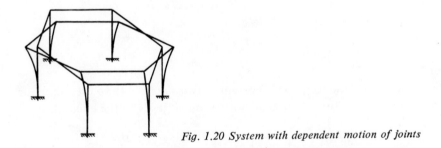

Fig. 1.20 System with dependent motion of joints

positive roots for circular frequency ω_j; the two natural mode shapes, belonging to each of the frequencies are computed as in the preceding example (see eqs 1.115, 1.116). This gives for the first mode shape

$$w_k^0 = s_j'\sin\frac{2\pi jk}{h}; \quad u_k^0 = r_j'\cos\frac{2\pi jk}{h} \qquad (1.127)$$

and for the second

$$w_k^0 = s_j''\cos\frac{2\pi jk}{h}; \quad u_k^0 = -r_j''\sin\frac{2\pi jk}{h} \qquad (1.128)$$

The ratios of the coefficients

$$\frac{r_j}{s_j} = \frac{r'_j}{s'_j} = \frac{r''_j}{s''_j} = \varepsilon_j \tag{1.129}$$

are determined by eqs 1.126.

The system illustrated in Fig. 1.20 resembles that of Fig. 1.18 except that the elements of the top ring are rigid. All the ring elements must satisfy the condition of zero elongation. This condition may be expressed by the equation

$$(w_k + w_{k-1}) \sin \frac{\pi}{h} + (u_k - u_{k-1}) \cos \frac{\pi}{h} = 0 \tag{1.130}$$

from which

$$\frac{u_k - u_{k-1}}{w_k + w_{k-1}} = -\tan \frac{\pi}{h} \tag{1.131}$$

The modes of free vibration are described by expressions 1.124 and eq. 1.131. Substituting the former in the latter we get

$$\frac{r_j[\cos (2\pi jk/h + \Phi_j) - \cos (2\pi j(k-1)/h + \Phi_j)]}{s_j[\sin (2\pi jk/h + \Phi_j) + \sin (2\pi j(k-1)/h + \Phi_j)]} = -\tan \frac{\pi}{h} \tag{1.132}$$

and by formulae 1.125

$$-\tan \frac{\pi}{h} = -\frac{r_j \sin \pi j/h}{s_j \cos \pi j/h} \tag{1.133}$$

so that

$$\varepsilon_j = \frac{r_j}{s_j} = \tan \frac{\pi}{h} \cot \frac{\pi j}{h} \tag{1.134}$$

The application of the reciprocal theorem to two states, both of the j-th mode — one a free vibration, the other a static deformation — leads to the equation

$$\sum_k m\omega^2(w_{k(j)}^2 + u_{k(j)}^2) = \sum_k \left(\frac{3EJ_x}{l^3} w_{k(j)}^2 + \frac{3EJ_z}{l^3} u_{k(j)}^2\right) \tag{1.135}$$

from which

$$\omega_j^2 = \frac{3E}{ml^3} \frac{\sum_k (J_x w_{k(j)}^2 + J_z u_{k(j)}^2)}{\sum_k (w_{k(j)}^2 + u_{k(j)}^2)}$$

$$= \frac{3E}{ml^3} \frac{J_x s_j^2 + J_z r_j^2}{s_j^2 + r_j^2}$$

$$= \frac{3E}{ml^3} \frac{J_x + J_z \tan^2 \pi/h \cot^2 \pi j/h}{1 + \tan^2 \pi/h \cot^2 \pi j/h} \tag{1.136}$$

1.5.2.2 Forced vibration

Once the modes of free vibration are known, forced vibration can be solved by the expansion of the natural modes. In this procedure advantage is taken of the possibility that the natural modes may be normalised in accordance with eq. 1.17. Because of the relation

$$\sum_{k=0}^{h-1} \sin^2\left(\frac{2\pi jk}{h} + \Phi\right) = \frac{h}{2}$$ (1.137)

applicable when $0 < j < h/2$, formula 1.18 simplifies in such a manner that, for example, for the system in Fig. 1.16

$$W_{ij} = \frac{\sin\left(2\pi ji/h + \Phi\right)}{\left[\sum m \sin^2\left(2\pi jk/h + \Phi\right)\right]^{1/2}} = \left(\frac{2}{hm}\right)^{1/2} \sin\left(\frac{2\pi ji}{h} + \Phi\right)$$ (1.138)

where W_{ij} is the displacement of point i in the j-th normalised natural modes, and the two shapes, one at $\Phi = 0$, the other at $\Phi = \pi/2$ are represented by eq. 1.138.

For steady-state forced vibration eq. 1.67 gives at $\Phi = 0$ (the j-th mode of the first kind)

$$q_j(t) = \frac{1}{\omega_j^2 - \omega^2} \sum_k P_k \left(\frac{2}{hm}\right)^{1/2} \sin\frac{2\pi jk}{h} \sin \omega t$$ (1.139)

and at $\Phi = \pi/2$ (the j-th mode of the second kind)

$$q_j(t) = \frac{1}{\omega_j^2 - \omega^2} \sum_k P_k \left(\frac{2}{hm}\right)^{1/2} \cos\frac{2\pi jk}{h} \sin \omega t$$ (1.140)

Formulae 1.137 to 1.140 apply only to $0 < j < h/2$.

For $j = 0$, $\Phi = 0$, $W_{ij} = 0$; for $j = 0$, $\Phi = \pi/2$, $W_{ij} = (hm)^{-1/2}$
For $j = h/2$ (even h), $\Phi = 0$, $W_{ij} = 0$;
For $j = h/2$, $\Phi = \pi/2$, $W_{ij} = (-1)^i (hm)^{-1/2}$

The displacement at point i is then given by

$$w_i(t) = \sum q_j(t) W_{ij}$$ (1.141)

where the summation sign includes, for both kinds, all of the modes j. Therefore

$$w_i(t) = \frac{\sin \omega t}{hm} \left\{ 2 \sum_{0<j<h/2} \frac{1}{\omega_j^2 - \omega^2} \left[\sin\frac{2\pi ji}{h} \sum_{k=0}^{h-1} P_k \sin\frac{2\pi jk}{h} + \cos\frac{2\pi ji}{h} \sum_{k=0}^{h-1} P_k \right. \right.$$
$$\left. \left. \cdot \cos\frac{2\pi jk}{h} \right] + \frac{1}{\omega_{00}^2 - \omega^2} \sum_{k=0}^{h-1} P_k + \frac{(-j)^i}{\omega_{h/2}^2 - \omega^2} \sum_{k=0}^{h-1} P_k(-1)^k \right\}$$ (1.142)

The last term holds only when h is even.

1.5.3 Cyclosymmetry of two-dimensional systems

The difference equations can also be used for two-dimensional systems with repeated identical elements. Fig. 1.21 shows a system consisting of rigid, massless

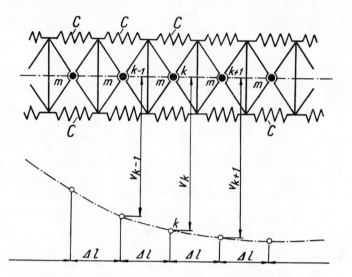

Fig. 1.21 System composed of rigid panels and springs (beam model)

frames interconnected by springs of stiffness constant C. The mass points have identical masses, m. The motion of point k is described by the equation

$$m \frac{d^2 v_k(t)}{dt^2} + \frac{Ca^2}{2 \, \Delta l^2} \left[v_{k-2}(t) - 4v_{k-1}(t) + 6v_k(t) - 4v_{k+1}(t) + v_{k+2}(t) \right] = 0 \quad (1.143)$$

Assuming harmonic vibration

$$v_k(t) = v_k \sin \omega t \quad (1.144)$$

we get after some manipulation

$$-m\omega^2 v_k + \frac{Ca^2}{2 \, \Delta l^2} \Delta^4 v_k = 0 \quad (1.145)$$

where

$$\Delta^4 v_k = \Delta^2 v_{k-1} - 2 \, \Delta^2 v_k + \Delta^2 v_{k+1} \quad (1.146)$$

$$\Delta^2 v_k = v_{k-1} - 2v_k + v_{k+1} \quad (1.147)$$

Eq. 1:145 is satisfied by the particular solutions

$$v_k = \cos \varkappa_1 k \qquad (1.148)$$

$$v_k = \sin \varkappa_1 k \qquad (1.149)$$

$$v_k = \cosh \varkappa_2 k \qquad (1.150)$$

$$v_k = \sinh \varkappa_2 k \qquad (1.151)$$

where $\varkappa_1$ and $\varkappa_2$ are coefficients yet to be determined.

Thus, for example, it holds for the last expression 1.151 that

$$\Delta^2 v_k = \sinh \varkappa_2(k-1) - 2\sinh \varkappa_2 k + \sinh \varkappa_2(k+1)$$
$$= 2(\cosh \varkappa_2 - 1)\sinh \varkappa_2 k \qquad (1.152)$$

$$\Delta^4 v_k = 2(\cosh \varkappa_2 - 1)\left[\sinh \varkappa_2(k-1) - 2\sinh \varkappa_2 k + \sinh \varkappa_2(k+1)\right]$$
$$= 4(\cosh \varkappa_2 - 1)^2 \sinh \varkappa_2 k \qquad (1.153)$$

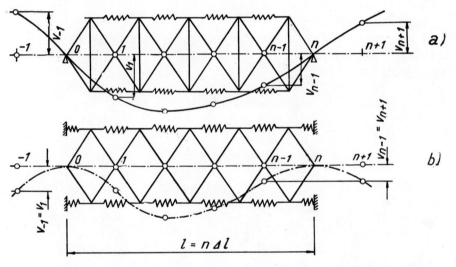

Fig. 1.22 System shown in Fig. 1.21. (a) Simply supported, (b) fixed

On substituting the above in eq. 1.145 we get

$$\omega^2 = \frac{2Ca^2}{m\,\Delta l^2}(\cosh \varkappa_2 + 1)^2 \qquad (1.154)$$

i.e. the expression that relates the circular frequency, ω, and coefficient $\varkappa_2$. A similar relation, viz.

$$\omega^2 = \frac{2Ca^2}{m\,\Delta l^2}(\cos \varkappa_1 - 1)^2 \qquad (1.155)$$

is obtained for the first and second particular solutions. Since the latter defines the relation between ω and $\varkappa_1$, it must hold that

$$\cos \varkappa_1 + \cosh \varkappa_2 = 2$$

The complete solution consists of a linear combination of all partial solutions, i.e. of eq. 1.148 through to eq. 1.151,

$$v_k = A \cos \varkappa_1 k + B \sin \varkappa_1 k + C \cosh \varkappa_2 k + D \sinh \varkappa_2 k \qquad (1.156)$$

Eq. 1.156 makes it possible to solve the free vibration of the system shown in Fig. 1.22 for any boundary condition. Consider, for example, a system like that shown in Fig. 1.22a, i.e. one simply supported at both ends. The boundary conditions are then

$$\begin{aligned} v_0 \quad &= v_n = 0 \\ v_{-1} &= -v_1 \\ v_{n+1} &= -v_{n-1} \end{aligned}$$

$$(1.157)$$

What the last two conditions tell us is that the springs at the supports remain unstretched and may be omitted.

Substituting eq. 1.157 in eq. 1.156 gives

$$A + C = 0$$

$$A \cos \varkappa_1 - B \sin \varkappa_1 + C \cosh \varkappa_2 \neg D \sinh \varkappa_2$$

$$= -A \cos \varkappa_1 - B \sin \varkappa_1 - C \cosh \varkappa_2 - D \sinh \varkappa_2$$

and from there

$$A = C = 0 \qquad (1.158)$$

further

$$B \sin (\varkappa_1 n) + D \sinh (\varkappa_2 n) = 0$$

$$B \sin [\varkappa_1(n + 1)] + D \sinh [\varkappa_2(n + 1)] = -B \sin [\varkappa_1(n - 1)] - D \sinh [\varkappa_2(n - 1)]$$

or

$$B \cos \varkappa_1 \sin (\varkappa_1 n) + D \cosh \varkappa_2 \sinh (\varkappa_2 n) = 0$$

and from there

$$D \sin (\varkappa_2 n) = 0 \quad \text{and} \quad D = 0 \qquad (1.159)$$

and

$$B \sin (\varkappa_1 n) = 0 \qquad (1.160)$$

The last equation can be satisfied either on the condition that $B = 0$ — but then all the constants of eq. 1.156 would be zero and consequently the system would be at rest — or on the condition that

$$\sin (\varkappa_1 n) = 0 \tag{1.161}$$

so that

$$\varkappa_1 n = \pi, \quad 2\pi \ldots, \quad \ldots \text{ up to } n \tag{1.162}$$

If this is so, eq. 1.156 takes on the form

$$v_{kj} = B_j \sin \frac{\pi j k}{n} \tag{1.163}$$

where B_j is an arbitrary constant.

The natural circular frequency of the system is then defined by eq. 1.155 in the form

$$\omega_j^2 = \frac{2Ca^2}{m \, \Delta l^2} \left(\cos \frac{\pi j}{n} - 1 \right)^2 \tag{1.164}$$

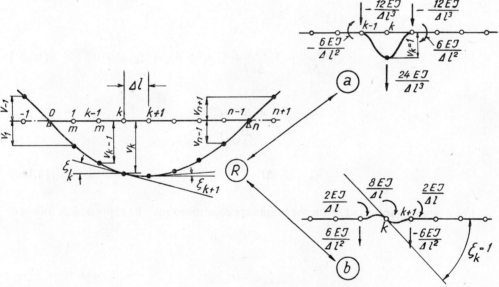

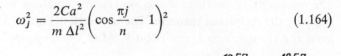

Fig. 1.23 Vibration of a simply supported beam with discrete points. (a) Actual motion, (b) unit states

Were the system supported as shown in Fig. 1.22b, we would write the boundary conditions as

$$v_0 \quad = v_n = 0$$
$$v_{-1} = v_1$$
$$v_{n+1} = v_{n-1}$$

$$\tag{1.165}$$

and on substituting them in eq. 1.156 get

$$A + C = 0$$

$$B \sin \varkappa_1 + D \sinh \varkappa_2 = 0$$

$$A[\cos(\varkappa_1 n) - \cosh(\varkappa_2 n)] + B\left[\sin(\varkappa_1 n) - \frac{\sin \varkappa_1}{\sinh \varkappa_2} \sinh(\varkappa_2 n)\right] = 0$$

$$-A[\sin \varkappa_1 \sin(\varkappa_1 n) + \sinh \varkappa_2 \sinh(\varkappa_2 n)] + B \sin \varkappa_1[\cos(\varkappa_1 n) - \cosh(\varkappa_2 n)] = 0$$
$$(1.166)$$

Together with eq. 1.154 to eq. 1.156 the last two equations completely determine the modes of free vibration of the system being considered.

Fig. 1.23 shows a beam with identical masses placed at regular intervals. The equations of motion of the system are obtained in the familiar way, i.e. by applying the reciprocal theorem. The unit states (Fig. 1.23a, b) are the unit displacement and the unit rotation at point k. After some manipulation we get

$$-6\frac{v_{k-1}}{\Delta l} + (12 - \bar{\alpha})\frac{v_k}{\Delta l} - 6\frac{v_{k+1}}{\Delta l} - 3\zeta_{k-1} + 3\zeta_{k+1} = 0$$

$$3\frac{v_{k-1}}{\Delta l} - 3\frac{v_{k+1}}{\Delta l} + \zeta_{k-1} + 4\zeta_k + \zeta_{k+1} = 0 \qquad (1.167)$$

where

$$\bar{\alpha} = \frac{m \, \Delta l^3 \, \omega^2}{2EJ} \qquad (1.168)$$

and

$$\Delta l = \frac{l}{n} \qquad (1.169)$$

For a simply supported beam the boundary conditions may be expressed as follows:

$$v_0 = v_n = 0$$
$$v_{-1} = -v_1$$
$$v_{n+1} = -v_{n-1}$$
$$\zeta_{-1} = \zeta_1$$
$$\zeta_{n+1} = \zeta_{n-1} \qquad (1.170)$$

Eqs 1.167 and 1.170 are satisfied by the solution

$$\frac{v_k}{\Delta l} = s'_j \sin \frac{\pi j k}{n}$$

$$\zeta_k = r'_j \cos \frac{\pi j k}{n} \qquad (1.171)$$

With reference to the fundamental relationships 1.112 and 1.125 we get after substituting eq. 1.171 in eq. 1.167

$$\left[12\left(1 - \cos\frac{\pi j}{n}\right) - \bar{\alpha}\right]s'_j - 6\sin\frac{\pi j}{n}r'_j = 0$$

$$-6\sin\frac{\pi j}{n}s'_j + \left(4 + 2\cos\frac{\pi j}{n}\right)r'_j = 0$$

$$(1.172)$$

where constants s'_j and r'_j have non-zero values only if the determinant of the coefficients of the last equation is zero. Accordingly

$$12\left(1 - \cos\frac{\pi j}{n}\right) - \bar{\alpha} - \frac{18\sin^2\pi j/n}{2 + \cos\pi j/n} = 0$$

and from there

$$\bar{\alpha} = 6\frac{1 - 2\cos\pi j/n + \cos^2\pi j/n}{2 + \cos\pi j/n} \qquad (1.173)$$

By eq. 1.168

$$\omega_j = \left[\frac{12EJn^3}{ml^3}\frac{(1 - \cos\pi j/n)^2}{2 + \cos\pi j/n}\right]^{1/2} \qquad (1.174)$$

A system with both end supports permitting vertical displacement (Fig. 1.24) will have the same frequencies of free vibration as the system just discussed, but

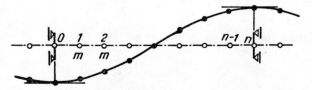

Fig. 1.24 Beam with discrete mass points on supports permitting vertical displacement but no rotation

will have different natural mode shapes. Even though its equations of motion will again be eqs 1.167, the boundary conditions will now change to

$$\zeta_0 = \zeta_n = 0 \qquad (1.175)$$

Eqs 1.167 and 1.175 will be satisfied by the solution

$$\frac{v_k}{\Delta l} = s''_j\cos\frac{\pi jk}{n}$$

$$\zeta_k = -r''_j\sin\frac{\pi jk}{n}$$

$$(1.176)$$

Constants s_j'' and r_j'' will again be defined by eq. 1.172, and the natural frequency by eq. 1.174. But in this case we have to consider yet another motion, i.e. that at $j = 0$ at which the circular frequency $\omega_0 = 0$ and all the points undergo the same vertical displacements. The system as a whole will therefore execute a uniform motion in the vertical direction.

The system of eqs 1.167 can be reduced to difference equations of the fourth order. The latter are set out in Table 1.3 for three points in succession, i.e. $k - 1$, k and $k + 1$.

Table 1.3

$v_{k-2}/\Delta l$	$v_{k-1}/\Delta l$	$v_k/\Delta l$	$v_{k+1}/\Delta l$	$v_{k+2}/\Delta l$	ζ_{k-2}	ζ_{k-1}	ζ_k	ζ_{k+1}	ζ_{k+2}	
-6	$12 - \bar{\alpha}$	-6			-3		3			$= 0$
	-6	$12 - \bar{\alpha}$	-6			-3		3		$= 0$
		-6	$12 - \bar{\alpha}$	-6			-3		3	$= 0$
3	-3				1	4	1			$= 0$
	3	-3				1	4	1		$= 0$
		3	-3				1	4	1	$= 0$

Expressing first the variable ζ in terms of the displacement

$$v_{k-2} - 4v_{k-1} + 6v_k - 4v_{k+1} + v_{k+2} - \frac{\bar{\alpha}}{3}(v_{k-1} + 4v_k + v_{k-1}) = 0 \qquad (1.177)$$

and then using the notation of eq. 1.146, we get

$$\Delta^4 v_k - \frac{\bar{\alpha}}{3}(6v_k + \Delta^2 v_k) = 0 \qquad (1.178)$$

By analogy with eqs 1.152 and 1.153

$$\Delta^2 v_k = 2\left(\cos\frac{\pi j}{n} - 1\right)\sin\frac{\pi jk}{n}$$

$$\Delta^4 v_k = 4\left(\cos\frac{\pi j}{n} - 1\right)^2 \sin\frac{\pi jk}{n}$$

$$(1.179)$$

therefore, after substitution in eq. 1.178, the latter turns out to be

$$4\left(\cos\frac{\pi j}{n} - 1\right)^2 - \frac{\bar{\alpha}}{3}\left[6 + 2\left(\cos\frac{\pi j}{n} - 1\right)\right] = 0 \qquad (1.180)$$

which is identical to eq. 1.173.

1.6   Approximate methods

One of the best known and widely used methods of approximately determining the fundamental frequency of vibrating systems is the energy method evolved by Rayleigh[61]. We shall not dwell upon it except to show how the pertinent formulae

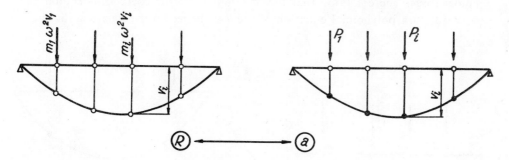

Fig. 1.25 The energy method

may be derived through the application of the reciprocal theorem on which our considerations have been based so far. Let us take for the unit state any deformation that may be assumed to be close to the fundamental mode of free vibration (e.g. that shown in Fig. 1.25a). Next, let us introduce a plausible, though not entirely correct, assumption that this deformation represents the natural mode of the system's vibration (Fig. 1.25 R).

Application of the reciprocal theorem results in the equation

$$\sum_i m_i \omega_{(1)}^2 v_i^2 = \sum_i P_i v_i \qquad (1.181)$$

from which the approximate value of the first natural circular frequency is obtained as

$$\omega_{(1)}^2 \approx \frac{\sum\limits_i P_i v_i}{\sum\limits_i m_i v_i^2} = \frac{2W}{\sum\limits_i m_i v_i^2} \qquad (1.182)$$

where $W = \frac{1}{2}\sum\limits_i P_i v_i$ and is the potential energy of deformation of the system.

Galerkin's and Ritz's methods. More accurate results are arrived at by considering several unit states, for example, those indicated in Fig. 1.26. The mode of free vibration is assumed to be approximately a linear combination of the deformations shown in the figure. The displacement at point k is then

$$v_k = q_1 a_{k1} + q_2 a_{k2} + \dots q_h a_{kh} \qquad (1.183)$$

68

Applying the reciprocal theorem to this state and to the unit states we get

$$\sum_{k=1}^{n} m_k \omega^2 \sum_{j=1}^{h} q_j a_{kj} a_{ki} = \sum_{k=1}^{n} P_{ki} \sum_{j=1}^{h} q_j a_{kj} \quad \text{for} \quad i = 1 - h \tag{1.184}$$

Expression 1.184 represents a system of linear equations in q_j. The equations are homogeneous. The condition that the determinant of the coefficients should equal zero leads to a polynomial equation of the h-th degree for the square of the circular

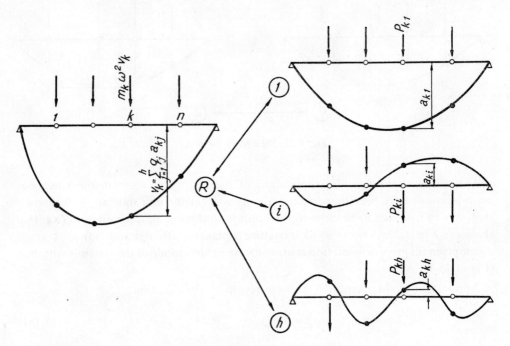

Fig. 1.26 Methods of Ritz and Galerkin

frequency, ω^2, whose solution yields h positive roots, $\omega_{(1)} \ldots \omega_{(h)}$. The ratios $q_1 : q_2 : \ldots q_h$ computed from eq. 1.184 for each of the frequencies and substituted in eq. 1.183 help us to find the displacements corresponding to the individual modes of free vibration. The results as we have just shown are obtained by Galerkin's[22] as well as by Ritz's[64] method although the two methods rest on completely different principles.

2 PRISMATIC BARS WITH UNIFORM MASS DISTRIBUTION

2.1 Method of influence coefficients; integral equations of motion

In Section 1.2.1 the application of the method of influence coefficients to the computation of free vibration resulted in eq. 1.4, valid for a massless beam with discrete mass points, viz.

$$v_i(t) = -\sum_k m_k \frac{d^2 v_k(t)}{dt^2} \delta_{ik} \tag{2.1}$$

Consider now a beam with a uniformly distributed mass. If the number of mass points placed on the beam increases beyond all limits while the total mass of the system remains constant, the present case is clearly the limiting case of the previous one. Denoting the mass per unit length by the symbol μ, we get the mass of an element of length dx as $\mu \, dx$. The sum in eq. 2.1 will change to an integral, and the equation will take on the following form

$$v(x, t) = -\int_0^l \mu \frac{\partial^2 v(s, t)}{\partial t^2} \delta(x, s) \, ds \tag{2.2}$$

In the above

$v(x, t)$ is the vertical displacement at point x at time t,
$v(s, t)$ is the vertical displacement at point s at time t,
$\delta(x, s)$ is the influence coefficient of deflection at point x,

    i.e. the vertical deflection at point x produced by unit vertical force at point s.

Assuming that the free vibration of the beam is an harmonic one for which

$$v(x, t) = v(x) \sin \omega_0 t \tag{2.3}$$

eq. 2.2 turns out after some rearrangement to be

$$v(x) = \mu \omega_0^2 \int_0^l \delta(x, s) \, v(s) \, ds \tag{2.4}$$

It holds according to the Maxwell reciprocal theorem that

$$\delta(x, s) = \delta(s, x) \tag{2.5}$$

For a simply supported beam of constant cross-section the influence coefficients are

$$\delta(x, s) = \frac{(l - s) x}{6EJl} [s(2l - s) - x^2] \quad \text{for} \quad s \geqq x$$

$$\delta(x, s) = \frac{(l - x) s}{6EJl} [x(2l - x) - s^2] \quad \text{for} \quad s \leqq x$$

$$(2.6)$$

In the homogeneous integral equation in eq. 2.4 expression $\delta(x, s)$ represents the kernel, symmetric in this case because of eq. 2.5. The value $\mu\omega_0^2$ is termed the eigenvalue of the integral equation. A solution of eq. 2.4 is obtained by changing the equation into a differential one. Triple differentiation with respect to x gives

$$v'''(x) = \mu\omega_0^2 \int_0^l \delta'''(x, s) \, v(s) \, ds \tag{2.7}$$

where

$$v'''(x) = \frac{d^3 v(x)}{dx^3}$$

$$\delta'''(x, s) = \frac{\partial^3 \delta(x, s)}{\partial x^3}$$

Because of (2.6)

$$\delta'''(x, s) = -\frac{(l - s)}{EJl} \quad \text{for} \quad s \geqq x$$

$$\delta'''(x, s) = \frac{s}{EJl} \quad \text{for} \quad s \leqq x$$

From (2.7)

$$v'''(x) = \frac{\mu\omega_0^2}{EJl} \left[\int_0^x s \, v(s) \, ds - \int_x^l (l - s) \, v(s) \, ds \right] \tag{2.8}$$

Differentiation of the above with respect to x gives

$$v''''(x) = \frac{\mu\omega_0^2}{EJl} \lim_{\Delta x \to 0} \left[\frac{\int_0^{x+\Delta x} s \, v(s) \, ds - \int_0^x s \, v(s) \, ds}{\Delta x} \right.$$

$$\left. - \frac{\int_{x+\Delta x}^l (l - s) \, v(s) \, ds - \int_x^l (l - s) \, v(s) \, ds}{\Delta x} \right] = \frac{\mu\omega_0^2}{EJ} v(x) \tag{2.9}$$

The resultant differential equation is thus

$$v''''(x) - \frac{\mu \omega_0^2}{EJ} v(x) = 0 \tag{2.10}$$

In the case of a simply supported beam the applicable boundary conditions state that the deflection and bending moment at the end points must be zero, viz.

$$\begin{aligned} v(0) &= 0 & v(l) &= 0 \\ M(0) &= 0 & M(l) &= 0 \end{aligned}$$

$$\tag{2.11}$$

Since the bending moment is defined by the familiar relation

$$M(x) = -EJ\, v''(x) \tag{2.12}$$

the last two of the boundary conditions may be expressed as

$$v''(0) = 0 \quad v''(l) = 0 \tag{2.13}$$

The differential equation 2.10 holds good for prismatic bars with all kinds of end supports (e.g. clamped at one or both ends, etc.). The effect of the mode of support is expressed by the boundary conditions which vary from one case to another, as do the influence coefficients $\delta(x, s)$. Since, however, the latter are always a cubic function of x, the expression for $v''''(x)$ always turns out the same.

2.2 Differential equation of free lateral vibration of a uniform beam

The differential equation 2.10 can also be arrived at through the following considerations: Consider an element of mass $\mu\, dx$ of a beam of constant cross-section

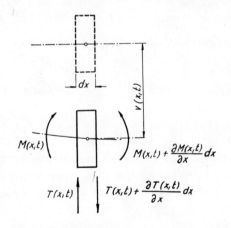

Fig. 2.1 Element of length of a bar

performing transverse vibrations in the absence of external forces. Fig. 2.1 shows the deflections of, and the forces acting on, the element and the bending moments.

Recalling that in elasticity theory the latter two are defined by the relations

$$M(x, t) = -EJ \frac{\partial^2 v(x, t)}{\partial x^2}$$

$$T(x, t) = -EJ \frac{\partial^3 v(x, t)}{\partial x^3} \tag{2.14}$$

and applying d'Alembert's principle to the equilibrium of vertical forces

$$\frac{\partial T(x, t)}{\partial x} \, dx - \mu \, dx \frac{\partial^2 v(x, t)}{\partial t^2} = 0 \tag{2.15}$$

we get on combining the last two equations

$$EJ \frac{\partial^4 v(x, t)}{\partial x^4} + \mu \frac{\partial^2 v(x, t)}{\partial t^2} = 0 \tag{2.16}$$

If the motion is harmonic — no matter whether the vibration is free or forced — with circular frequency ω,

$$v(x, t) = v(x) \sin \omega t \tag{2.17}$$

substitution in eq. 2.16 will give us

$$EJ \frac{d^4 v(x)}{dx^4} - \mu \omega^2 \, v(x) = 0$$

which is identical with eq. 2.10. Since no boundary conditions have been considered so far, the equation holds good for bars with any kind of end supports.

The differential equation 2.10 may also be obtained as the limiting case of the difference equations arrived at through the method of elasticity coefficients in the preceding chapter. Thus from eq. 1.178 we get as the limit of the expression

$$\frac{\Delta^4 v_k}{\Delta l^4} - \frac{m\omega^2}{6EJ \, \Delta l} \left(6 v_k + \frac{\Delta^2 v_k}{\Delta l^2} \Delta l^2 \right) = 0 \tag{2.18}$$

on setting $\Delta l = dx$; $m/\Delta l = \mu$, the equation 2.10, i.e.

$$\frac{d^4 v(x)}{dx^4} - \frac{\mu \omega^2}{EJ} v(x) = 0$$

In yet another way, eq. 2.10 is obtained as the limiting case of the system illustrated in Fig. 1.21. The system then represents the Bernoulli-Euler beam model. Writing $m = \mu \, \Delta l$ and assuming that the stiffness constant of the spring is given by

$$C = \frac{2EJ}{a^2 \, \Delta l} \tag{2.19}$$

we see that in the limiting case of $\Delta l \to dx$, eq. 1.145 becomes eq. 2.10.

2.3 Solution of the differential equation of free lateral vibration of a uniform beam

The complete integral of the homogeneous differential equation 2.10 is the sum of four particular solutions of the form

$$v(x) = e^{kx} \tag{2.20}$$

multiplied by the integration constants. Substitution of the above in 2.10 yields the characteristic equation

$$k^4 EJ - \mu\omega^2 = 0 \tag{2.21}$$

and from there

$$k_{1,2} = \pm \frac{\lambda}{l}$$

$$k_{3,4} = \pm i \frac{\lambda}{l} = ik_{1,2}$$

$$\tag{2.22}$$

where

$$\lambda = l \left(\frac{\mu\omega^2}{EJ} \right)^{1/4} \tag{2.23}$$

The complete integral then takes on the form

$$v(x) = C_1 e^{k_1 x} + C_2 e^{k_2 x} + C_3 e^{k_3 x} + C_4 e^{k_4 x} \tag{2.24}$$

Making use of the familiar formulae for hyperbolic and trigonometric functions — $\cos \lambda x/l = \frac{1}{2}(e^{ik_1 x} + e^{ik_2 x})$, etc. — we may write eq. 2.24 as

$$v(x) = A \cos \frac{\lambda x}{l} + B \sin \frac{\lambda x}{l} + C \cosh \frac{\lambda x}{l} + D \sinh \frac{\lambda x}{l} \tag{2.25}$$

and the relations applying between the two sets of constants as

$$C_1 = \frac{1}{2}(C + D)$$
$$C_2 = \frac{1}{2}(C - D)$$
$$C_3 = \frac{1}{2}(A + iB)$$
$$C_4 = \frac{1}{2}(A + iB)$$

Further, the slope of the tangent is

$$v'(x) = \frac{dv(x)}{dx} = \frac{\lambda}{l} \left(-A \sin \frac{\lambda x}{l} + B \cos \frac{\lambda x}{l} + C \sinh \frac{\lambda x}{l} + D \cosh \frac{\lambda x}{l} \right) \tag{2.26}$$

and the bending moment determined by the second derivative is given by

$$-\frac{M(x)}{EJ} = v''(x) \tag{2.27}$$

where

$$v''(x) = \frac{\lambda^2}{l^2}\left(-A\cos\frac{\lambda x}{l} - B\sin\frac{\lambda x}{l} + C\cosh\frac{\lambda x}{l} + D\sinh\frac{\lambda x}{l}\right) \tag{2.28}$$

The shear force is given by the third derivative, viz.

$$-\frac{T(x)}{EJ} = v'''(x) \tag{2.29}$$

where

$$v'''(x) = \frac{\lambda^3}{l^3}\left(A\sin\frac{\lambda x}{l} - B\cos\frac{\lambda x}{l} + C\sinh\frac{\lambda x}{l} + D\cosh\frac{\lambda x}{l}\right) \tag{2.30}$$

Eqs 2.25 to 2.30 enable us to compute the frequency as well as the modes of free vibration of any uniform beam with any kinds of end support. For the case of a simply supported beam, the applicable boundary conditions will be eqs 2.11. On their introduction in eq. 2.25 and eq. 2.28 we get the following set of four equations:

$$\left.\begin{array}{r} A + C = 0 \\ -A + C = 0 \end{array}\right\} \quad \text{and from there} \quad A = 0, \quad C = 0 \tag{2.31}$$

$$\left.\begin{array}{r} B\sin\lambda + D\sinh\lambda = 0 \\ -B\sin\lambda + D\sinh\lambda = 0 \end{array}\right\} \quad \text{and from there} \quad \begin{array}{l} B\sin\ \lambda = 0 \\ D\sinh\lambda = 0 \end{array}$$

If the solution is to be non-trivial, it must hold that

$$B \neq 0, \quad D = 0, \quad \sin\lambda = 0$$

and

$$\lambda = \pi, 2\pi \ldots j\pi \ldots \tag{2.32}$$

On substituting the above in eq. 2.23 we get for the natural circular frequency

$$\omega_{(j)} = \frac{j^2\pi^2}{l^2}\left(\frac{EJ}{\mu}\right)^{1/2} \tag{2.33}$$

where j is a positive integer.

The shape of the j-th natural mode is given by eq. 2.25 which after substitution takes on the form

$$v(x) = B\sin\frac{j\pi x}{l} = v\left(\frac{l}{2j}\right)\sin\frac{j\pi x}{l} \tag{2.34}$$

where $v(l/2j)$ is the deflection amplitude.

For a beam clamped at both ends the boundary conditions are

$$v(0) = 0 \quad v(l) = 0$$
$$v'(0) = 0 \quad v'(l) = 0$$

(2.35)

and eqs 2.25 and 2.26 yield

$$A + C = 0 \quad C = -A$$
$$B + D = 0 \quad D = -B$$
$$A(\cos \lambda - \cosh \lambda) + B(\sin \lambda - \sinh \lambda) = 0$$
$$-A(\sin \lambda + \sinh \lambda) + B(\cos \lambda - \cosh \lambda) = 0$$

(2.36)

The constants have non-zero values only if the determinant of the coefficients of the last two equations is zero, i.e. if

$$(\cos \lambda - \cosh \lambda)^2 + (\sin^2 \lambda - \sinh^2 \lambda) = 0$$

or after rearrangement

$$\cosh \lambda \cos \lambda - 1 = 0 \qquad (2.37)$$

In the computation of complex systems embodying a number of transversely vibrating bars (continuous beams, rigid frames, etc.) the determination of the integration constants A to D — different for different bars — poses an acute problem. If the system being analysed has n bars, there are $4n$ unknown constants. The equations necessary for their determination are derived from the boundary conditions at the ends of the bars. However, as a solution devised along these lines by Reissner[62,63] suggests, this is a fairly complicated way of obtaining the result. Thus, for example,

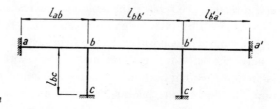

Fig. 2.2 Frame system

even if only transverse vibration is considered, there are twenty unknown constants to be established for the five-bar system shown in Fig. 2.2. This is what is available for the task on hand: two conditions — zero displacement and zero rotation — at each of the outer supports a, a', c, c', i.e. altogether eight conditions; the condition of zero transverse displacements of the ends of all three bars meeting at a joint — three conditions, and that of equal rotations of the three bars in each of the two inner joints — two conditions; and finally, the condition of equilibrium of the end moments of the three bars — two conditions. This gives a total of $8 + 2 \times 5 + 2 = 20$ conditions. As we shall demonstrate presently, the solution may be simplified to a considerable degree through the use of the slope-deflection method[3,4,11,19,32] which works with end forces and moments corresponding to unit amplitudes of the end displacements of the bars.

2.4 Moments and forces at the ends of a laterally vibrating bar

In this section we shall consider systems vibrating with steady harmonic motion. Such motion is apt to arise either in free vibration or in forced vibration produced by an external harmonic load applied for a sufficiently long time.

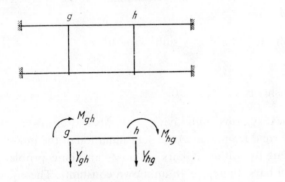
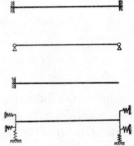

Fig. 2.3 End moments and forces of a bar removed from the system

Fig. 2.4 Various types of bar support

From the system shown in Fig. 2.3 the isolated section of bar gh is vibrating in the transverse direction. To replace the action of the rest of the system on the bar's ends, the ends are loaded by external end moments $M_{gh}(t)$, $M_{hg}(t)$ and external end forces $Y_{gh}(t)$, $Y_{hg}(t)$. Consider as positive the moments acting clockwise at the end of the bar. Accordingly, the left-hand end moment agrees with the bending moment acting at cross-section $x = 0$ in both magnitude and sign, while the right-hand end moment is of opposite sign to the bending moment in section $x = l$. The end forces are considered positive if acting downward at the ends of the bar. Consequently, the right-hand end force $Y_{hg}(t)$ agrees with the shear force acting in section $x = l$ both in magnitude and sign while the left-hand end force $Y_{gh}(t)$ is of opposite sign to the shear force in section $x = 0$. If we know the deformations at the ends of the bar, we may compute all the end moments and forces from eqs 2.25–2.30 written for the end sections of the bar. This applies irrespective of the way the bar is supported at its ends, i.e. whether it is clamped, hinged, free at one end or elastically supported (Fig. 2.4)[33,35,37,38,39].

Let, for example, the right-hand end of a rigidly clamped bar rotate by harmonic motion $\zeta_h \sin \omega t$ with amplitude $\zeta_h = 1$ while the remaining displacements and rotations of the ends are zero (Fig. 2.5).

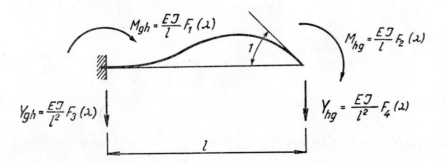

Fig. 2.5 Unit rotation of the right-hand end of bar

If there is no load acting in the intermediate sections of the bar, the applicable equation of motion is eq. 2.10 and its solution eq. 2.25 with derivatives 2.26−2.30. The pertinent boundary conditions are

$$v(0) = 0 \quad v(l) = 0$$
$$v'(0) = 0 \quad v'(l) = 1$$

$$(2.38)$$

On their substitution in eqs 2.25 and 2.26 we get a non-homogeneous set of equations

$$A + C = 0$$
$$B + D = 0$$

$$A \cos \lambda + B \sin \lambda + C \cosh \lambda + D \sinh \lambda = 0$$
$$-A \sin \lambda + B \cos \lambda + C \sinh \lambda + D \cosh \lambda = 1/\lambda$$

$$(2.39)$$

and from it the integration constants

$$A = -C = -\frac{l}{2\lambda} \frac{\sinh \lambda - \sin \lambda}{\cosh \lambda \cos \lambda - 1}$$

$$B = -D = \frac{l}{2\lambda} \frac{\cosh \lambda - \cos \lambda}{\cosh \lambda \cos \lambda - 1}$$

$$(2.40)$$

The amplitudes of the end moments and forces are computed from eqs 2.28 and 2.30. It holds for them that

$$M_{gh} = M(0) = -EJ\,v''(0) = -EJ\frac{\lambda^2}{l^2}\,(-A + C) = \frac{EJ}{l}\,F_1(\lambda) \qquad (2.41)$$

where

$$F_1(\lambda) = -\lambda\,\frac{\sinh\lambda - \sin\lambda}{\cosh\lambda\cos\lambda - 1} \qquad (2.42)$$

further,

$$M_{hg} = -M(l) = EJ\,v''(l) = EJ\frac{\lambda^2}{l^2}\,(-A\cos\lambda - B\sin\lambda + C\cosh\lambda + D\sinh\lambda)$$

$$= \frac{EJ}{l}\,F_2(\lambda) \qquad (2.43)$$

$$Y_{gh} = -T(0) = EJ\,v'''(0) = EJ\frac{\lambda^3}{l^3}\,(-B + D) = \frac{EJ}{l^2}\,F_3(\lambda) \qquad (2.44)$$

$$Y_{hg} = T(l) = -EJ\,v'''(l) = -EJ\frac{\lambda^3}{l^3}\,(A\sin\lambda - B\cos\lambda + C\sinh\lambda + D\cosh\lambda)$$

$$= \frac{EJ}{l^2}\,F_4(\lambda) \qquad (2.45)$$

where

$$F_2(\lambda) = -\lambda\,\frac{\cosh\lambda\sin\lambda - \sinh\lambda\cos\lambda}{\cosh\lambda\cos\lambda - 1}$$

$$F_3(\lambda) = -\lambda^2\,\frac{\cosh\lambda - \cos\lambda}{\cosh\lambda\cos\lambda - 1}$$

$$F_4(\lambda) = \lambda^2\,\frac{\sinh\lambda\sin\lambda}{\cosh\lambda\cos\lambda - 1}$$

$$(2.46)$$

The end forces and moments corresponding to the other unit deformations at the ends of bar gh would be obtained by an analogous procedure. The results of the pertinent computations are set out in Tables 10.1 and 10.2 of the Appendix.

At an arbitrary rotation and displacement of ends g and h

$$v(0) = v_g\,, \quad v(l) = v_h\,, \quad v'(0) = \zeta_g\,, \quad v'(l) = \zeta_h$$

and the integration constants become

$$A = \frac{l}{2\lambda^2}\left[F_1(\lambda)\,\zeta_h + F_2(\lambda)\,\zeta_g - F_3(\lambda)\frac{v_h}{l} - F_4(\lambda)\frac{v_g}{l}\right] + \tfrac{1}{2}v_g$$

$$B = \frac{l}{2\lambda^3}\left[-F_3(\lambda)\,\zeta_h + F_4(\lambda)\,\zeta_g - F_5(\lambda)\frac{v_h}{l} - F_6(\lambda)\frac{v_g}{l}\right] + \frac{l}{2\lambda}\zeta_g$$

$$C = -A + v_g$$

$$D = -B + \frac{l}{\lambda}\zeta_g$$

$$\text{(2.47)}$$

where

$$F_5(\lambda) = \lambda^3 \frac{\sinh\lambda + \sin\lambda}{\cosh\lambda\cos\lambda - 1}$$

$$F_6(\lambda) = -\lambda^3 \frac{\cosh\lambda\sin\lambda + \sinh\lambda\cos\lambda}{\cosh\lambda\cos\lambda - 1}$$

The meaning of the frequency functions $F(\lambda)$ is clear both from formulae 2.41 − 2.47 and from the summary 10.11 in the Appendix which lists them together with their numerical values. See also References 37 and 38.

The amplitude of deflection of an arbitrary cross-section of the bar is then defined by eq. 2.25 in which expressions 2.47 have first been substituted. The end moments and forces are computed from eqs 2.27 and 2.29 or obtained directly by superposition of the values set out in Tables 10.1 and 10.2 in the Appendix. Thus, for example,

$$M_{gh} = \frac{EJ}{l}\left[F_1(\lambda)\,\zeta_h + F_2(\lambda)\,\zeta_g - F_3(\lambda)\frac{v_h}{l} - F_4(\lambda)\frac{v_g}{l}\right] \qquad \text{(2.48)}$$

$$Y_{gh} = \frac{EJ}{l^2}\left[F_3(\lambda)\,\zeta_h - F_4(\lambda)\,\zeta_g + F_5(\lambda)\frac{v_h}{l} + F_6(\lambda)\frac{v_g}{l}\right] \qquad \text{(2.49)}$$

We would proceed similarly when dealing with the end moments and forces of a bar hinged at point g. In this case the boundary conditions are

$$v(0) = v_g \quad v(l) = v_h$$
$$v''(0) = 0 \quad v'(l) = \zeta_h$$

$$\text{(2.50)}$$

and the integration constants are given by

$$A = C = \tfrac{1}{2}v_g$$

$$B = \frac{1}{\cosh \lambda \sin \lambda - \sinh \lambda \cos \lambda} \left[-\frac{\zeta_h l}{\lambda} \sinh \lambda + v_h \cosh \lambda - \tfrac{1}{2}(\sinh \lambda \sin \lambda \right.$$

$$\left. + \cosh \lambda \cos \lambda + 1) v_g \right]$$

$$D = \frac{1}{\cosh \lambda \sin \lambda - \sinh \lambda \cos \lambda} \left[\frac{\zeta_h l}{\lambda} \sin \lambda - v_h \cos \lambda + \tfrac{1}{2}(-\sinh \lambda \sin \lambda \right.$$

$$\left. + \cosh \lambda \cos \lambda + 1) v_g \right]$$

$$(2.51)$$

For a bar clamped at point g and hinged at point h, the boundary conditions are

$$v(0) = v_g \quad v(l) = v_h$$
$$v'(0) = \zeta_g \quad v''(l) = 0$$

$$(2.52)$$

and the integration constants are given by

$$A = \frac{l}{2\lambda^2} \left[F_7(\lambda) \zeta_g - F_8(\lambda) \frac{v_h}{l} - F_9(\lambda) \frac{v_g}{l} \right] + \frac{v_g}{2}$$

$$B = \frac{l}{2\lambda^3} \left[F_9(\lambda) \zeta_g - F_{10}(\lambda) \frac{v_h}{l} - F_{11}(\lambda) \frac{v_g}{l} \right] + \frac{l}{2\lambda} \zeta_g$$

$$C = v_g - A$$

$$D = \frac{l}{\lambda} \zeta_g - B$$

$$(2.53)$$

The end moments and forces are shown in column 6, Table 10.3, 10.4 and 10.5 of the Appendix.

For a bar hinged at both ends, the boundary conditions are

$$v(0) = v_g \quad v(l) = v_h$$
$$v''(0) = 0 \quad v''(l) = 0$$

$$(2.54)$$

and the integration constants are given by

$$A = C = \tfrac{1}{2}v_g$$
$$B = \tfrac{1}{2}(\operatorname{cosec} \lambda \, v_h - \cot \lambda \, v_g)$$
$$D = \tfrac{1}{2}(\operatorname{cosech} \lambda \, v_h - \coth \lambda \, v_g)$$

$$(2.55)$$

The end moments and forces are computed from data listed in column 6, Table 10.6 of the Appendix.

In the case just discussed the dynamic solution differs basically from the static solution. For the case of the static displacement of the supports, the end forces of a bar hinged at both ends are zero.

As the final example consider a cantilever beam with a free left-hand end, whose $M(0) = 0$, $T(0) = 0$. In this case the boundary conditions are

$$v(l) = v_h \quad v'(l) = \zeta_h$$
$$v''(0) = 0 \quad v'''(0) = 0$$

$$(2.56)$$

and the integration constants are

$$A = C = \frac{1}{2} \frac{1}{\cosh \lambda \cos \lambda + 1} \left[-\frac{l}{\lambda}(\sinh \lambda + \sin \lambda)\, \zeta_h + (\cosh \lambda + \cos \lambda)\, v_h \right]$$

$$B = D = \frac{1}{2} \frac{1}{\cosh \lambda \cos \lambda + 1} \left[\frac{l}{\lambda}(\cosh \lambda + \cos \lambda)\, \zeta_h - (\sinh \lambda - \sin \lambda)\, v_h \right]$$

$$(2.57)$$

The end moment and force are computed from the data in column 6, Tables 10.7 and 10.8 of the Appendix.

2.5 Longitudinal vibration

A bar is liable to vibrate not only in the transverse direction but also in the direction of its axis. Longitudinal vibration may take place simultaneously with the transverse vibration; if it does, an exact solution can no longer consider the two motions as independent of one another. At low amplitudes, however, the mutual effect of the two motions may be neglected.

The normal force is defined by the familiar relation

$$N(x, t) = ES \frac{\partial u(x, t)}{\partial x}$$

$$(2.58)$$

where S is the cross-sectional area.

For an increment of length dx

$$dN = ES \frac{\partial^2 u(x, t)}{\partial x^2} dx \qquad (2.59)$$

there is imparted to the bar element an acceleration in the direction of the bar axis, $\partial^2 u(x, t)/\partial t^2$, so that the equation of motion takes on the form

$$ES \frac{\partial^2 u(x, t)}{\partial x^2} - \mu \frac{\partial^2 u(x, t)}{\partial t^2} = 0 \qquad (2.60)$$

If the motion is harmonic

$$u(x, t) = u(x) \sin \omega t \qquad (2.61)$$

eq. 2.60 gives after substitution

$$ES \frac{d^2 u(x)}{dx^2} + \omega^2 \mu\, u(x) = 0 \qquad (2.62)$$

The complete integral of the above equation is

$$u(x) = A \cos \frac{\psi x}{l} + B \sin \frac{\psi x}{l} \qquad (2.63)$$

where

$$\psi = l \left(\frac{\mu \omega^2}{ES} \right)^{1/2} \qquad (2.64)$$

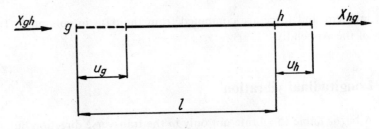

Fig. 2.6 Longitudinal vibration

The amplitude of the normal force is obtained from eq. 2.58 as

$$\frac{N(x)}{ES} = \frac{du(x)}{dx} = u'(x) \qquad (2.65)$$

with

$$\frac{l}{\psi} u'(x) = -A \sin \frac{\psi x}{l} + B \cos \frac{\psi x}{l} \qquad (2.66)$$

Fig. 2.6 shows the longitudinal vibration at which the end points of bar gh vibrate with amplitudes u_g and u_h. Substituting these boundary conditions in eq. 2.63 we get

$$A = u_g$$
$$A \cos \psi + B \sin \psi = u_h$$

$$(2.67)$$

so that

$$B = u_h \cosec \psi - u_g \cot \psi \qquad (2.68)$$

The end forces follow from eq. 2.65 and their magnitudes are (see also Table 10.10 of the Appendix)

$$X_{gh} = -N(0) = \frac{ES}{l} \psi(-u_h \cosec \psi + u_g \cot \psi) \qquad (2.69)$$

$$X_{hg} = N(l) = \frac{ES}{l} \psi(u_h \cot \psi - u_g \cosec \psi) \qquad (2.70)$$

It sometimes happens — e.g. with horizontal bars of rigid frames — that the longitudinal displacements are virtually constant along the whole length of the bar, viz.

$$u(x) = u_g = u_h$$

and the whole bar vibrates as a single unit. This is the case when the bar is rigid $(S \to \infty)$ against longitudinal forces. Although expressions 2.69 and 2.70 then become indefinite, the sum of the two end forces may be computed from the equation

$$(X_{gh} + X_{hg}) + \mu l \omega^2 u_g = 0 \qquad (2.71)$$

which states that the two end forces are in equilibrium with the inertia force of the whole bar.

2.6 Torsional vibration

Equations analogous to those of longitudinal vibration can also be written for rotatory motion of bar elements in torsional vibration, provided, of course, the bar cross-section is either circular or annular. In the case of a rectangular cross-section not greatly different from a square, the equations apply approximately.

The differential equation of rotatory motion is in the form

$$\frac{\partial \left[\frac{\partial \xi(x, t)}{\partial x} GJ_x \right]}{\partial x} - \chi \frac{\partial^2 \xi(x, t)}{\partial x^2} = 0 \qquad (2.72)$$

where $\xi(x, t)$ is the bar rotation about axis X in section x at time t, G the shear modulus, GJ_x the torsional rigidity (for a circular cross-section J_x is the polar moment of inertia of the cross-sectional area with respect to the centre of gravity, i.e. to axis X; for a rectangular cross-section this moment is reduced by a factor), χ is the mass moment of inertia per unit length. The expression in brackets denoting the torsional moment is analogous to the expression of normal force according to eq. 2.58.

If the bar is of constant cross-section and the motion is assumed to be harmonic, i.e.

$$\xi(x, t) = \xi(x) \sin \omega t$$

eq. 2.72 simplifies to

$$GJ_x \frac{d^2\xi(x)}{dx^2} + \chi\omega^2 \, \xi(x) = 0 \tag{2.73}$$

whose complete integral is

$$\xi(x) = A \cos \frac{\vartheta(x)}{l} + B \sin \frac{\vartheta(x)}{l} \tag{2.74}$$

Eq. 2.74 is identical with eq. 2.63 except that $u(x)$ and ψ of the latter are replaced by $\xi(x)$ and ϑ, where

$$\vartheta = l \left(\frac{\mu\omega^2}{GJ_x} \right)^{1/2} \tag{2.75}$$

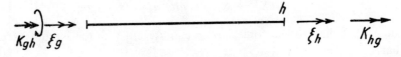

Fig. 2.7 Torsional vibration

This is why the end moments K_{gh}, K_{hg} corresponding to rotations ξ_g and ξ_h of the bar ends (Fig. 2.7) will be computed from formulae similar to those of the end forces X_{gh}, X_{hg} at displacements u_g, u_h (derived in Section 2.5)

$$K_{gh} = \frac{GJ_x}{l} \vartheta(-\xi_h \cosec \vartheta + \xi_g \cot \vartheta)$$

$$K_{hg} = \frac{GJ_x}{l} \vartheta(\xi_h \cot \vartheta - \xi_g \cosec \vartheta)$$

$$\tag{2.76}$$

(see also Table 10.10 of the Appendix).

2.7 Lateral vibration of prismatic bars under an external harmonic load

If a bar is acted upon by a continuous, time variable external load (Fig. 2.8), the differential equation of motion takes on the form

$$EJ \frac{\partial^4 v(x, t)}{\partial x^4} + \mu \frac{\partial^2 v(x, t)}{\partial t^2} - p(x, t) = 0 \qquad (2.77)$$

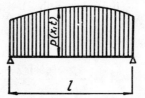

Fig. 2.8 Continuous load varying with time

If the load is harmonic

$$p(x, t) = p(x) \sin \omega t \qquad (2.78)$$

then steady forced vibration is described by the relation

$$v(x, t) = v(x) \sin \omega t \qquad (2.79)$$

On substituting eqs 2.78 and 2.79 in eq. 2.77 and doing some simple rearrangement we get the following equation

$$EJ \frac{d^4 v(x)}{dx^4} - \mu \omega^2 v(x) = p(x) \qquad (2.80)$$

In some instances eq. 2.80 may be solved by direct integration. Thus, for example, if the bar is acted on by a uniform harmonic load, $p(x) = p$ the complete integral of eq. 2.80 is

$$v(x) = A \cos \frac{\lambda x}{l} + B \sin \frac{\lambda x}{l} + C \cosh \frac{\lambda x}{l} + D \sinh \frac{\lambda x}{l} - \frac{p}{\mu \omega^2} \qquad (2.81)$$

Repeated differentiation of the above equation will yield expressions 2.26, 2.28 and 2.30. The integration constants are computed from the boundary conditions.

For a bar clamped at both ends the boundary conditions are

$$v(0) = 0 \quad v(l) = 0$$
$$v'(0) = 0 \quad v'(l) = 0$$

$$(2.82)$$

On substituting them in eqs 2.81 and 2.26 we get

$$A + C = \frac{p}{\mu\omega^2}$$

$$B + D = 0$$

$$A(\cos\lambda - \cosh\lambda) + B(\sin\lambda - \sinh\lambda) = \frac{p}{\mu\omega^2}(1 - \cosh\lambda)$$

$$-A(\sin\lambda + \sinh\lambda) + B(\cos\lambda - \cosh\lambda) = -\frac{p}{\mu\omega^2}\sinh\lambda$$

and from there

$$A = \frac{p}{2\mu\omega^2}\frac{(\cosh\lambda - 1)(\cos\lambda + 1) - \sinh\lambda\sin\lambda}{\cosh\lambda\cos\lambda - 1}$$

$$B = -D = \frac{p}{2\mu\omega^2}\frac{(\cosh\lambda - 1)\sin\lambda + (\cos\lambda - 1)\sinh\lambda}{\cosh\lambda\cos\lambda - 1}$$

$$C = \frac{p}{2\mu\omega^2}\frac{(\cos\lambda - 1)(\cosh\lambda + 1) + \sinh\lambda\sin\lambda}{\cosh\lambda\cos\lambda - 1}$$

$$(2.83)$$

Substituted in eq. 2.81 the above will give the equation of the deflection amplitude. The end moments and forces computed from eqs 2.27, 2.29 turn out as follows:

$$M_{gh} = M(0) = \frac{l^2 p}{\lambda^2}\left(\frac{\cosh\lambda - \cos\lambda - \sinh\lambda\sin\lambda}{\cosh\lambda\cos\lambda - 1}\right) = -\frac{l^2 p}{\lambda^4}[F_3(\lambda) + F_4(\lambda)] \quad (2.84)$$

$$Y_{gh} = -T(0) = \frac{lp}{\lambda}\left(\frac{-\cosh\lambda\sin\lambda - \sinh\lambda\cos\lambda + \sinh\lambda + \sin\lambda}{\cosh\lambda\cos\lambda - 1}\right)$$

$$= \frac{lp}{\lambda^4}[F_6(\lambda) + F_5(\lambda)] \quad (2.85)$$

The procedure as outlined is applicable to all kinds of boundary conditions, i.e. to different types of end support. The pertinent end forces are set out in Table 10.9 of the Appendix.

If the load along the bar axis is of a general character, direct integration is not possible, and other methods, particularly modal analysis (Section 3.5) must be resorted to.

If a bar is loaded with a concentrated force at an intermediate section, then it is advantageous to replace the point of action p by a joint and the original bar gh by a system of two bars, gd and dh (see Fig. 3.7, Section 3.2.2).

2.8   Frequency functions

2.8.1   Functions $F(\lambda)$

The analysis of vibrating beams presented in the preceding sections was facilitated by the introduction of frequency functions $F_1(\lambda)$ to $F_{17}(\lambda)$. The application and significance of these functions will be enlarged upon later on. The numerical tables contained in the Appendix have been set up for $\lambda = 0$ to 6 (functions $F_1(\lambda)$ to $F_{14}(\lambda)$), and for $\lambda = 0$ to 4 (functions $F_{15}(\lambda)$ to $F_{17}(\lambda)$). Values corresponding to λ outside those ranges can be computed by means of approximate formulae to be derived presently. Since for large values of λ it approximately holds that

$$\cosh \lambda \approx \sinh \lambda$$

we may write[41]

$$F_1(\lambda) = -\lambda \frac{\sinh \lambda - \sin \lambda}{\cosh \lambda \cos \lambda - 1} \approx -\lambda \frac{1 - \sin \lambda/\cosh \lambda}{\cos \lambda - 1/\cosh \lambda} \approx -\frac{\lambda}{\cos \lambda} \quad (2.86)$$

and similarly

$$F_2(\lambda) \approx \lambda(1 - \tan \lambda)$$

$$F_3(\lambda) \approx -\frac{\lambda^2}{\cos \lambda}$$

$$F_4(\lambda) \approx \lambda^2 \tan \lambda$$

$$F_5(\lambda) \approx \frac{\lambda^3}{\cos \lambda}$$

$$F_6(\lambda) \approx -\lambda^3(1 + \tan \lambda) \quad (2.87)$$

Analogous formulae for $F_7(\lambda)$ to $F_{17}(\lambda)$ may be found in the Appendix — eqs 10.19 to 10.21.

On the other hand, for low values of λ, the functions are best expanded in a power series. Substituting the familiar relations

$$\cosh \lambda = 1 + \frac{\lambda^2}{2!} + \frac{\lambda^4}{4!} + \cdots$$

$$\sinh \lambda = \frac{\lambda}{1!} + \frac{\lambda^3}{3!} + \frac{\lambda^5}{5!} + \cdots$$

$$\cos \lambda = 1 - \frac{\lambda^2}{2!} + \frac{\lambda^4}{4!} - \frac{\lambda^6}{6!} \cdots$$

$$\sin \lambda = \frac{\lambda}{1!} - \frac{\lambda^3}{3!} + \frac{\lambda^5}{5!} - \frac{\lambda^7}{7!} \cdots$$

$$(2.88)$$

in eq. 2.42 we get

$$F_1(\lambda) = -\lambda \frac{\sinh \lambda - \sin \lambda}{\cosh \lambda \cos \lambda - 1}$$

$$= -\lambda \frac{\lambda/1! + \lambda^3/3! + \lambda^5/5! + \ldots - \lambda/1! + \lambda^3/3! - \lambda^5/5! + \ldots}{(1 + \lambda^2/2! + \lambda^4/4! + \ldots)(1 - \lambda^2/2! + \lambda^4/4! - \ldots) - 1}$$

$$= 2 + 0{\cdot}007\,142\,857\lambda^4 + 0{\cdot}000\,015\,704\lambda^8 + 0{\cdot}000\,000\,032\lambda^{12} + \ldots \quad (2.89)$$

Similarly we find that

$$F_2(\lambda) = \quad 4 - 0{\cdot}009\,523\,810\lambda^4 - 0{\cdot}000\,016\,262\lambda^8 - 0{\cdot}000\,000\,032\lambda^{12} \ldots$$

$$F_3(\lambda) = \quad 6 + 0{\cdot}030\,952\,381\lambda^4 + 0{\cdot}000\,072\,193\lambda^8 + 0{\cdot}000\,000\,148\lambda^{12} \ldots$$

$$F_4(\lambda) = -\ 6 + 0{\cdot}052\,380\,952\lambda^4 + 0{\cdot}000\,076\,617\lambda^8 + 0{\cdot}000\,000\,149\lambda^{12} \ldots$$

$$F_5(\lambda) = -12 - 0{\cdot}128\,571\,429\lambda^4 - 0{\cdot}000\,329\,571\lambda^8 - 0{\cdot}000\,000\,684\lambda^{12} \ldots$$

$$F_6(\lambda) = \quad 12 - 0{\cdot}371\,428\,571\lambda^4 - 0{\cdot}000\,364\,873\lambda^8 - 0{\cdot}000\,000\,693\lambda^{12} \ldots$$

The power series of the remaining functions, established by means of analogous formulae, are reviewed in the Appendix (eqs 10.16–10.17).

2.8.2 Frequency functions $\Phi(\lambda)$

A quantity of considerable importance for dynamic calculations is the value of the integral

$$\int \mu\, v_{(j)}^2(x)\, \mathrm{d}x \quad (2.90)$$

used in the computation of normalised natural modes, in the analysis of damped as well as undamped vibrations, when considering the effect of a moving load, etc.

The mode of harmonic vibration of a prismatic bar, isolated from a frame system, is defined by eq. 2.25, the integration constants being given by formulae 2.47. Eq. 2.25 may therefore be written as

$$v(x) = \zeta_h \frac{l}{2\lambda^3}\left[F_1(\lambda)\,\lambda\left(\cos\frac{\lambda x}{l} - \cosh\frac{\lambda x}{l}\right) - F_3(\lambda)\left(\sin\frac{\lambda x}{l} - \sinh\frac{\lambda x}{l}\right)\right]$$

$$+ \zeta_g \frac{l}{2\lambda^3}\left[F_2(\lambda)\,\lambda\left(\cos\frac{\lambda x}{l} - \cosh\frac{\lambda x}{l}\right) + F_4(\lambda)\left(\sin\frac{\lambda x}{l} - \sinh\frac{\lambda x}{l}\right)\right.$$

$$\left. + \lambda^2\left(\sin\frac{\lambda x}{l} + \sinh\frac{\lambda x}{l}\right)\right] + v_h \frac{1}{2\lambda^3}\left[-F_3(\lambda)\,\lambda\left(\cos\frac{\lambda x}{l} - \cosh\frac{\lambda x}{l}\right)\right.$$

$$- F_5(\lambda)\left(\sin\frac{\lambda x}{l} - \sinh\frac{\lambda x}{l}\right)\right] + v_g\frac{1}{2\lambda^3}\left[-F_4(\lambda)\,\lambda\left(\cos\frac{\lambda x}{l} - \cosh\frac{\lambda x}{l}\right)\right.$$

$$\left.- F_6(\lambda)\left(\sin\frac{\lambda x}{l} - \sinh\frac{\lambda x}{l}\right) + \lambda^3\left(\cos\frac{\lambda x}{l} + \cosh\frac{\lambda x}{l}\right)\right] \tag{2.91}$$

or in the abbreviated form

$$v(x) = \varphi_1(x)\,\zeta_h + \varphi_2(x)\,\zeta_g + \varphi_3(x)\,v_h + \varphi_4(x)\,v_g \tag{2.92}$$

where

$$\varphi_1(x) = \frac{l}{2\lambda^3}\left[F_1(\lambda)\,\lambda\left(\cos\frac{\lambda x}{l} - \cosh\frac{\lambda x}{l}\right) - F_3(\lambda)\left(\sin\frac{\lambda x}{l} - \sinh\frac{\lambda x}{l}\right)\right] \tag{2.93}$$

etc.

On substituting eq. 2.92 in eq. 2.90 we have

$$\mu\int v^2(x)\,\mathrm{d}x = \mu\zeta_h^2\int\varphi_1^2(x)\,\mathrm{d}x + \mu\zeta_g^2\int\varphi_2^2(x)\,\mathrm{d}x + 2\mu\zeta_g\zeta_h\int\varphi_1(x)\,\varphi_2(x)\,\mathrm{d}x$$

$$+ 2\mu\zeta_h v_h\int\varphi_1(x)\,\varphi_3(x)\,\mathrm{d}x + 2\mu\zeta_h v_g\int\varphi_1(x)\,\varphi_4(x)\,\mathrm{d}x$$

$$+ 2\mu\zeta_g v_h\int\varphi_2(x)\,\varphi_3(x)\,\mathrm{d}x + 2\mu\zeta_g v_g\int\varphi_2(x)\,\varphi_4(x)\,\mathrm{d}x$$

$$+ 2\mu v_h v_g\int\varphi_3(x)\,\varphi_4(x)\,\mathrm{d}x + \mu v_h^2\int\varphi_3^2(x)\,\mathrm{d}x + \mu v_g^2\int\varphi_4^2(x)\,\mathrm{d}x$$

and after integration of the right-hand side[37,38]

$$\mu\int v^2(x)\,\mathrm{d}x = \frac{\mu l^3}{\lambda^4}\left\{(\zeta_h^2 + \zeta_g^2)\,\Phi_2(\lambda) + 2\zeta_h\zeta_g\ _1(\lambda)\right.$$

$$+ \frac{2(\zeta_h v_g - \zeta_g v_h)}{l}\ _3(\lambda) + \frac{2(\zeta_h v_h - \zeta_g v_g)}{l}\,\Phi_4(\lambda)$$

$$\left.+ \frac{2v_g v_h}{l^2}\,\Phi_5(\lambda) + \frac{v_h^2 + v_g^2}{l^2}\,\Phi_6(\lambda)\right\} \tag{2.94}$$

where

$$\begin{aligned}
\Phi_1(\lambda) &= \tfrac{1}{4}[F_1(\lambda)\,F_2(\lambda) - F_3(\lambda) - F_1(\lambda)]\\
\Phi_2(\lambda) &= \tfrac{1}{4}[F_1^2(\lambda) - F_2(\lambda)]\\
\Phi_3(\lambda) &= -\tfrac{1}{4}[F_1(\lambda)\,F_4(\lambda) + 2F_3(\lambda)]\\
\Phi_4(\lambda) &= -\tfrac{1}{4}[F_1(\lambda)\,F_3(\lambda) + 2F_4(\lambda)]\\
\Phi_5(\lambda) &= \tfrac{1}{4}[F_3(\lambda)\,F_4(\lambda) - 3F_5(\lambda)]\\
\Phi_6(\lambda) &= \tfrac{1}{4}[F_3^2(\lambda) - 3F_6(\lambda)]
\end{aligned} \tag{2.95}$$

90

If bar gh is hinged at end g, the integration constants are given by eq. 2.51, and integral 2.90 is

$$\int_0^l \mu\, v^2(x)\, dx = \int_0^l \mu \left[A \cos \frac{\lambda x}{l} + B \sin \frac{\lambda x}{l} + C \cosh \frac{\lambda x}{l} + D \sinh \frac{\lambda x}{l} \right]^2 dx$$

$$= \frac{\mu l^3}{\lambda^4} \left\{ \zeta_h^2 \Phi_7(\lambda) + 2 \frac{\zeta_h v_g}{l} \Phi_8(\lambda) + 2 \frac{\zeta_h v_h}{l} \Phi_9(\lambda) + 2 \frac{v_h v_g}{l^2} \Phi_{10}(\lambda) \right.$$

$$\left. + \frac{v_h^2}{l^2} \Phi_{11}(\lambda) + \frac{v_g^2}{l^2} \Phi_{12}(\lambda) \right\} \tag{2.96}$$

where

$$\Phi_7(\lambda) = \tfrac{1}{4}\left[-F_7(\lambda) + F_7^2(\lambda) + 2F_9(\lambda) \right]$$

$$\Phi_8(\lambda) = \tfrac{1}{4}\left[-2F_8(\lambda) + F_7(\lambda) F_8(\lambda) + F_{10}(\lambda) \right]$$

$$\Phi_9(\lambda) = \tfrac{1}{4}\left[-2F_9(\lambda) + F_7(\lambda) F_9(\lambda) + F_{11}(\lambda) \right]$$

$$\Phi_{10}(\lambda) = \tfrac{1}{4}\left[-3F_{10}(\lambda) + F_8(\lambda) F_9(\lambda) \right]$$

$$\Phi_{11}(\lambda) = \tfrac{1}{4}\left[-3F_{11}(\lambda) + 2F_9^2(\lambda) - F_7(\lambda) F_{11}(\lambda) \right]$$

$$\Phi_{12}(\lambda) = \tfrac{1}{4}\left[-3F_{12}(\lambda) + F_8^2(\lambda) \right] \tag{2.97}$$

If a bar is hinged at both ends, the integration constants are given by eq. 2.55 and the integral becomes

$$\int \mu\, v^2(x)\, dx = \frac{\mu l}{\lambda^4} \left\{ 2 v_h v_g\, \Phi_{13}(\lambda) + (v_h^2 + v_g^2)\, \Phi_{14}(\lambda) \right\} \tag{2.98}$$

where

$$\Phi_{13}(\lambda) = -\frac{1}{4}\left[3F_{13}(\lambda) + \frac{F_{10}(\lambda)\, F_{14}(\lambda)}{F_4(\lambda)} \right]$$

$$\Phi_{14}(\lambda) = -\frac{1}{4}\left[3F_{14}(\lambda) + \frac{2F_8(\lambda)\, F_{13}(\lambda)}{F_7(\lambda)} \right] \tag{2.99}$$

And finally, for a cantilever beam with free left end g, the integration constants are those of eq. 2.57 and the integral is

$$\int_0^l \mu\, v^2(x)\, dx = \frac{\mu l^3}{\lambda^4} \left\{ \zeta_h^2\, \Phi_{15}(\lambda) + \frac{2\zeta_h v_h}{l} \Phi_{16}(\lambda) + \frac{v_h^2}{l^2} \Phi_{17}(\lambda) \right\} \tag{2.100}$$

where

$$\Phi_{15}(\lambda) = \tfrac{1}{4}[-F_{15}(\lambda) + F_{15}^2(\lambda) + 2F_{16}(\lambda)]$$
$$\Phi_{16}(\lambda) = \tfrac{1}{4}[-2F_{16}(\lambda) + F_{15}(\lambda)\,F_{16}(\lambda) + F_{17}(\lambda)]$$
$$\Phi_{17}(\lambda) = \tfrac{1}{4}[-3F_{17}(\lambda) + F_{15}(\lambda)\,F_{17}(\lambda) - F_{15}(\lambda)\,F_{11}(\lambda)]$$

$$(2.101)$$

It will be shown in Chapter 6 that the relationship

$$\Phi_i(\lambda) = -\frac{\lambda}{4}\frac{dF_i(\lambda)}{d\lambda} \tag{2.102}$$

can be applied and with its aid the computation of functions $\Phi(\lambda)$ becomes more straightforward than the integration of eq. 2.90.

The numerical tables contained in the Appendix have been set up for $\lambda = 0-6$ (functions $\Phi_1(\lambda) - \Phi_{14}(\lambda)$) and for $\lambda = 0-4$ (functions $\Phi_{15}(\lambda) - \Phi_{17}(\lambda)$). At high λs use can be made of approximate formulae obtained from eqs 2.95, 2.97, 2.99 and 2.101 after substitution from eqs 2.86 and 2.87. Thus, for example,

$$\Phi_1(\lambda) \approx \frac{1}{4}\left(-\frac{\lambda^2}{\cos\lambda}(1-\tan\lambda) + \frac{\lambda^2}{\cos\lambda} - \frac{\lambda}{\cos\lambda}\right) = \frac{\lambda}{4\cos\lambda}(\lambda\tan\lambda - 1) \quad (2.103)$$

At low λs functions $\Phi(\lambda)$ are readily obtained by expansion in a power series. In view of eqs 2.89 and 2.102 it is, for example, found that

$$\Phi_1(\lambda) = -0.007\,142\,857\lambda^4 - 0.000\,031\,408\lambda^8 - 0.000\,000\,095\lambda^{12} - \dots$$

The power series for the other functions $\Phi(\lambda)$ are reviewed in the Appendix, eqs 10.48 – 10.50.

3 FRAME SYSTEMS COMPOSED OF PRISMATIC BARS

3.1 Application of the reciprocal theorem to systems with continuous mass distribution

The equations of motion presented in Chapter 1 were obtained through the use of the reciprocal theorem (Maxwell's extended theorem also called Betti's theorem). The system was considered to be subjected to external as well as inertia forces.

Consider now a system with continuous mass distribution (Fig. 3.1) loaded in the first state with harmonic forces $P \sin \omega t$ acting at sections having coordinates x_p.

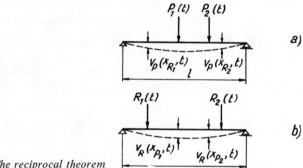

Fig. 3.1 The reciprocal theorem

The system will be set to vibrate, and in steady-state harmonic vibration the deflection at x will vary according to the formula

$$v(x, t) = v_P(x) \sin \omega t \tag{3.1}$$

while the inertia forces will have the magnitude

$$\mu \omega^2 \, v_P(x) \sin \omega t$$

Assume that in the second state the system is loaded with harmonic forces $R \sin \omega t$ acting at sections having coordinates x_R (Fig. 3.1b). The deflections will vary according to the formula

$$v(x, t) = v_R(x) \sin \omega t \tag{3.2}$$

and the inertia forces will have the magnitude

$$\mu\omega^2 \, v_R(x) \sin \omega t$$

The application of the reciprocal theorem to the two states (Figs 3.1a, 3.1b) gives

$$\sum_P P \sin \omega t \, v_R(x_P) \sin \omega t + \int \mu\omega^2 \, v_P(x) \sin \omega t \, v_R(x) \sin \omega t \, dx$$

$$= \sum_R R \sin \omega t \, v_P(x_R) \sin \omega t + \int \mu\omega^2 \, v_R(x) \sin \omega t \, v_P(x) \sin \omega t \, dx \qquad (3.3)$$

or, after cancellation of identical terms,

$$\sum_P P \, v_R(x_P) = \sum_R R \, v_P(x_R) \qquad (3.4)$$

The last equation expresses Rayleigh's reciprocal theorem as applied to two states of vibration varying harmonically at the same frequency ω. Were the frequency of forces R different from the frequency of forces P, the cancellation in eq. 3.3 would not be possible and the simple expression 3.4 never obtained. (In its above form the reciprocal theorem cannot, therefore, be applied to two states one of which is static, the other harmonic.)

It is clear to see from the derivation of eq. 3.3 that the phase of the systems of forces P and R may be shifted at will (instead of dividing the equation by $\sin^2 \omega t$, we could divide it by $\sin \omega t \sin (\omega t + \varphi)$ where φ is the phase shift).

The reciprocal theorem 3.4 can be used to advantage in the solution of some vibrational problems. Consider, for example, the system in Fig. 3.2 consisting

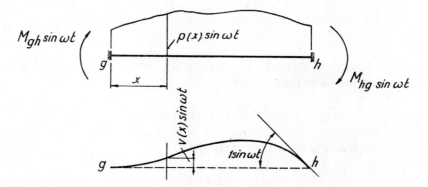

Fig. 3.2 Determination of end moments by means of the reciprocal theorem

of a beam clamped at both ends and subjected to continuous load $p(x) \sin \omega t$, and find the end moment $M_{hg} \sin \omega t$. This problem was already discussed in Section 2.7; now we are going to apply to it Rayleigh's reciprocal theorem. For the second state choose a harmonic motion (Fig. 3.2) at which the right-hand end of the beam rotates harmonically with unit amplitude and circular frequency ω. Denoting by $v(x)$ the deflections of the second state, we may write

$$M_{hg} + \int_0^l p(x)\,v(x)\,\mathrm{d}x = 0 \tag{3.5}$$

If we succeed in evaluating the integral, we get the magnitude of end moment M_{hg} directly. Thus, for example, at a uniform load, $p(x) = p$, and the deflection $v(x)$ is given by eq. 2.25. The integration constants are computed from eq. 2.40, and after substitution in eq. 3.5

$$M_{hg} = \frac{lp}{2\lambda(\cosh \lambda \cos \lambda - 1)} \int_0^l \left[(\sinh \lambda - \sin \lambda)\left(\cos \frac{\lambda x}{l} - \cosh \frac{\lambda x}{l} \right) \right.$$

$$\left. - (\cosh \lambda - \cos \lambda)\left(\sin \frac{\lambda x}{l} - \sinh \frac{\lambda x}{l} \right) \right] \mathrm{d}x$$

$$= \frac{l^2 p}{\lambda^4} \left[F_3(\lambda) + F_4(\lambda) \right] = -M_{gh} \tag{3.6}$$

this result is the same as that given in eq. 2.84 in Section 2.7.

The usefulness of the reciprocal theorem for the analysis of complicated frame systems will be demonstrated in the next paragraphs dealing with the slope-deflection method.

3.2　Analysis of frame systems by the slope-deflection method

3.2.1　The slope-deflection equations

Consider one of the joints, say g, of the vibrating frame system illustrated in Fig. 3.3, subjected to the external moment

$$\mathcal{M}_g \sin \omega t$$

and an external force with the horizontal component

$$\mathcal{X}_g \sin \omega t$$

and the vertical component

$$\mathcal{Y}_g \sin \omega t$$

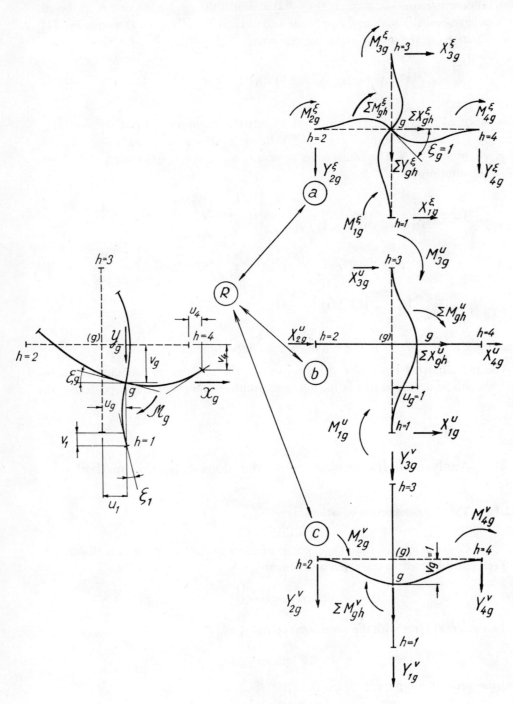

Fig. 3.3 Frame system joint

At steady-state harmonic vibration the joint rotates through an angle

$$\zeta_g \sin \omega t$$

and moves horizontally by

$$u_g \sin \omega t$$

and vertically by

$$v_g \sin \omega t$$

Take the rotation of joint g with unit amplitude (Fig. 3.3a) for the first auxiliary unit state. According to the reciprocal theorem

$$\mathcal{M}_g = \zeta_g \sum_h M_{gh}^\zeta + \sum_h M_{hg}^\zeta \zeta_h + u_g \sum_h X_{gh}^\zeta + \sum_h X_{hg}^\zeta u_h + v_g \sum_h Y_{gh}^\zeta + \sum_h Y_{hg}^\zeta v_h \qquad (3.7a)$$

where

M_{gh}^ζ is the amplitude of the moment at end g of bar gh at $\zeta_g = 1$
X_{hg}^ζ is the amplitude of the horizontal force at end h of bar gh at $\zeta_g = 1$, etc.

Applying analogously the reciprocal theorem to states R and b we get

$$\mathcal{X}_g = \zeta_g \sum_h M_{gh}^u + \sum_h M_{hg}^u \zeta_h + u_g \sum_h X_{gh}^u + \sum_h X_{hg}^u u_h + v_g \sum_h Y_{gh}^u + \sum_h Y_{hg}^u v_h \qquad (3.7b)$$

and to states R and c

$$\mathcal{Y}_g = \zeta_g \sum_h M_{gh}^v + \sum_h M_{hg}^v \zeta_h + u_g \sum_h X_{gh}^v + \sum_h X_{hg}^v u_h + v_g \sum_h Y_{gh}^v + \sum_h Y_{hg}^v v_h \qquad (3.7c)$$

where M_{gh}^u is the moment at end g of bar gh resulting from $u_g = 1$.
Further

$$M_{gh}^\zeta \zeta_g + M_{hg}^\zeta \zeta_h + X_{gh}^\zeta u_g + X_{hg}^\zeta u_h + Y_{gh}^\zeta v_g + Y_{hg}^\zeta v_h = M_{gh} \qquad (3.8)$$

where M_{gh} is the end moment at point g of bar gh for the actual deformation.
Therefore

$$\mathcal{M}_g = \sum_h M_{gh} \qquad (3.9a)$$

and similarly from eqs 3.7b and 3.7c

$$\mathcal{X}_g = \sum_h X_{gh} \qquad (3.9b)$$

$$\mathcal{Y}_g = \sum_h Y_{gh} \qquad (3.9c)$$

X_{gh}, Y_{gh} being respectively the horizontal and vertical components of the end force of bar gh at point g for the actual deformation. Eqs 3.9 express the conditions of equilibrium of external and end forces at joint g.

Eqs 3.9 may be written for all free joints of the system. We may write as many equations 3.9 as there are unknown deformations, i.e. $3s$, s being the number of joints; it is therefore possible to compute the unknown deformations ζ, u, v from the slope-deflection equations 3.9.

Eqs 3.9 can be solved even in the absence of external forces. A non-trivial solution exists only if the determinant of the coefficients of the slope-deflection equations is equal to zero. The circular frequency ω is then no longer arbitrary but one of the natural circular frequencies of the system.

3.2.2   Examples — Dynamic response of frames

Fig. 3.4 shows a simple frame whose bars vibrate in the transverse as well as the longitudinal direction. The two free joints, 1 and 2, can rotate and move both vertically and horizontally. Consequently, there are six unknown components of displacement. We may write for them six slope-deflection equations of the type of eq. 3.7 by applying the reciprocal theorem to states R and a and to states R and f (Fig. 3.5). The first of the equations is of the form

$$\zeta_1 \left[\frac{EJ_{01}}{l_{01}} F_2(\lambda_{01}) + \frac{EJ_{12}}{l_{12}} F_2(\lambda_{12}) \right] + u_1 \frac{EJ_{01}}{l_{01}^2} F_4(\lambda_{01}) - v_1 \frac{EJ_{12}}{l_{12}^2} F_4(\lambda_{12})$$

$$+ \zeta_2 \frac{EJ_{12}}{l_{12}} F_1(\lambda_{12}) - v_2 \frac{EJ_{12}}{l_{12}^2} F_3(\lambda_{12}) = 0 \qquad (3.10)$$

In tabular form the six equations turn out as follows:

Table 3.1

ζ_1	u_1	v_1	ζ_2	u_2	v_2	
a_{11}	a_{12}	a_{13}	a_{14}	a_{15}	a_{16}	$= 0$
a_{12}	a_{22}	a_{23}	a_{24}	a_{25}	a_{26}	$= 0$
a_{13}	a_{23}	a_{33}	a_{34}	a_{35}	a_{36}	$= 0$
a_{14}	a_{24}	a_{34}	a_{44}	a_{45}	a_{46}	$= 0$
a_{15}	a_{25}	a_{35}	a_{45}	a_{55}	a_{56}	$= 0$
a_{16}	a_{26}	a_{36}	a_{46}	a_{56}	a_{66}	$= 0$

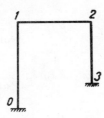

Fig. 3.4 Simple frame

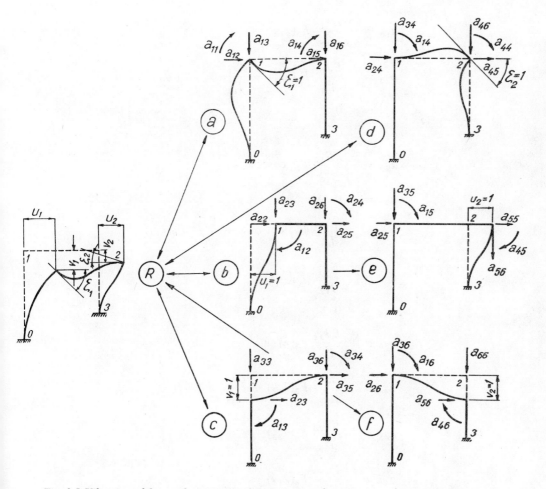

Fig. 3.5 Vibration of frame shown in Fig. 3.4. R — actual motion, a to f — unit states

Their coefficients are:

$$a_{11} = \frac{EJ_{01}}{l_{01}} F_2(\lambda_{01}) + \frac{EJ_{12}}{l_{12}} F_2(\lambda_{12})$$

$$a_{12} = \frac{EJ_{01}}{l_{01}^2} F_4(\lambda_{01})$$

$$a_{13} = - \frac{EJ_{12}}{l_{12}^2} F_4(\lambda_{12})$$

$$a_{14} = \frac{EJ_{12}}{l_{12}} F_1(\lambda_{12})$$

$$a_{15} = 0$$

$$a_{16} = - \frac{EJ_{12}}{l_{12}^2} F_3(\lambda_{12})$$

$$a_{22} = \frac{EJ_{01}}{l_{01}^3} F_6(\lambda_{01}) + \frac{ES_{12}}{l_{12}} \psi_{12} \cot \psi_{12}$$

$$a_{23} = 0$$

$$a_{24} = 0$$

$$a_{25} = - \frac{ES_{12}}{l_{12}} \psi_{12} \operatorname{cosec} \psi_{12}$$

$$a_{26} = 0$$

$$a_{33} = \frac{ES_{01}}{l_{01}} \psi_{01} \cot \psi_{01} + \frac{EJ_{12}}{l_{12}^3} F_6(\lambda_{12})$$

$$a_{34} = \frac{EJ_{12}}{l_{12}^2} F_3(\lambda_{12})$$

$$a_{35} = 0$$

$$a_{36} = \frac{EJ_{12}}{l_{12}^3} F_5(\lambda_{12})$$

$$a_{44} = \frac{EJ_{12}}{l_{12}} F_2(\lambda_{12}) + \frac{EJ_{23}}{l_{23}} F_2(\lambda_{23})$$

$$a_{45} = \frac{EJ_{23}}{l_{23}^2} F_4(\lambda_{23})$$

$$a_{46} = \frac{EJ_{12}}{l_{12}^2} F_4(\lambda_{12})$$

$$a_{55} = \frac{ES_{12}}{l_{12}} \psi_{12} \cot \psi_{12} + \frac{EJ_{23}}{l_{23}^3} F_6(\lambda_{23})$$

$$a_{56} = 0$$

$$a_{66} = \frac{EJ_{12}}{l_{12}^3} F_6(\lambda_{12}) + \frac{ES_{23}}{l_{23}} \psi_{23} \cot \psi_{23}$$

$$(3.11)$$

The natural frequency of the system shown in Fig. 3.5 is sought by trial and error. The procedure consists in choosing ω, computing ψ and λ of the individual bars from formulae 2.64 and 2.23, extracting the numerical values of $F(\lambda)$ from the tables

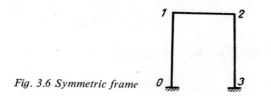

Fig. 3.6 Symmetric frame

(Appendix), computing coefficients a of the equations in Table 3.1, and setting up their determinant. It is repeated until we find an ω for which the determinant is equal to zero.

If the system is symmetric, as shown in Fig. 3.6, the problem is made simple by the fact that the modes of free vibration are either symmetric or antisymmetric (cf. Section 1.5.1). For the symmetric modes

$$\zeta_1 = -\zeta_2, \quad u_1 = -u_2, \quad v_1 = v_2 \qquad (3.12)$$

The slope-deflection equations for the symmetric modes are set out in Table 3.2

Table 3.2

ζ_1	u_1	v_1	
$a_{11} - a_{14}$	a_{12}	$a_{13} + a_{16}$	$= 0$
a_{12}	$a_{22} - a_{25}$	0	$= 0$
$a_{13} + a_{16}$	0	$a_{33} + a_{36}$	$= 0$

and for the antisymmetric modes for which

$$\zeta_1 = \zeta_2, \quad u_1 = u_2, \quad v_1 = -v_2 \qquad (3.13)$$

in Table 3.3.

Table 3.3

ζ_1	u_1	v_1	
$a_{11} + a_{14}$	a_{12}	$a_{13} - a_{16}$	$= 0$
a_{12}	$a_{22} + a_{25}$	0	$= 0$
$a_{13} - a_{16}$	0	$a_{33} - a_{36}$	$= 0$

In the absence of longitudinal vibration $v_1 = v_2 = 0$, $u_1 = u_2$. The symmetric mode is then given by a single equation

$$(a_{11} - a_{14})\,\zeta_1 = 0 \tag{3.14}$$

and the antisymmetric mode by two equations

$$(a_{11} + a_{14})\,\zeta_1 + a_{12}u_1 = 0$$
$$a_{12}\zeta_1 + (a_{22} + a_{25})\,u_1 = 0$$

$$\tag{3.15}$$

where, by eq. 2.71,

$$a_{22} + a_{25} = \frac{EJ_{01}}{l_{01}^3} F_6(\lambda_{01}) - \frac{\mu_{12}l_{12}\omega^2}{2} = \frac{EJ_{01}}{l_{01}^3} F_6(\lambda_{01}) - \frac{EJ_{12}}{l_{12}^3}\frac{\lambda_{12}^4}{2} \tag{3.16}$$

As a further example of the application of the slope-deflection method let us examine the forced vibration of the system shown in Fig. 3.7 consisting of bar gh

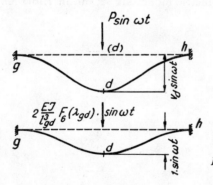

Fig. 3.7 Fixed beam loaded at the midspan
with a harmonic force

clamped at both ends and subjected to a harmonic load $P \sin \omega t$ applied at the midspan. Bar gh may be considered to be composed of two bars, gd and dh. The vibration is symmetric, the only unknown being the vertical displacement v_d. Applying the reciprocal theorem to the two states indicated in Fig. 3.7 we get the slope-deflection equation

$$2 \frac{EJ_{gd}}{l_{gd}^3} F_6(\lambda_{gd}) v_d = P \tag{3.17}$$

and from there — because $l_{gd} = l/2$

$$v_d = \frac{Pl^3}{16EJ \; F_6(\lambda/2)} \tag{3.18}$$

Since in the limiting case of $\lambda = 0$, $F_6(0) = 12$,

$$v_{dst} = \frac{Pl^3}{192EJ} \tag{3.19}$$

which is the static deflection of a beam of length l subjected to time-invariable force P applied midspan.

The moment at end g is given by

$$M_{gh} = M_{gd} = -v_d \frac{EJ}{l_{gd}^2} F_3(\lambda_{gd}) = -\frac{Pl}{4} \frac{F_3(\lambda/2)}{F_6(\lambda/2)} \tag{3.20}$$

and in the limiting case of $\lambda = 0$

$$M_{ghst} = -\frac{Pl}{8}$$

Example 3.1 — Vibration of a multi-storey frame

Consider a multi-storey frame (Fig. 3.8) the top horizontal beam of which carries at the midspan a piece of machinery with unbalanced rotating components. The unbalanced mass m moves clockwise; r is the distance of its centre of gravity from the axis of rotation, n is the number of revolutions per second, $\omega = 2\pi n \; \text{s}^{-1}$ is the circular frequency. The centrifugal force is given by

$$P = mr\omega^2 \tag{3.21}$$

Rotating force P may be resolved in two components: vertically $P \sin \omega t$, and horizontally $P \sin (\omega t + \pi/2)$. The effect of each component may be examined separately, and the resultant obtained through vector addition. To simplify the problem, assume that the cross-sectional area of both columns and beams is so large that longitudinal vibration may be neglected (hence $S \to \infty$), and that the weight of the machinery is small and is included in the uniformly distributed mass of the top beam (μ per unit length).

The following data are specified:

Moments of inertia of horizontal beams:

$$J_{11'} = J_{22'} = J_{33'} = J_2 = 2 \times 10^{-4} \, \text{m}^4$$

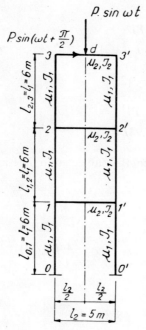

Fig. 3.8 Multi-storey frame

Moments of inertia of columns:

$$J_{01} = J_{12} = J_{23} = J_1 = 1 \times 10^{-4} \, \text{m}^4$$

$$\mu_{11'} = \mu_{22'} = \mu_{33'} = \mu_2 = 0 \cdot 5 \, \text{Mp m}^{-1}/9 \cdot 81 \, \text{m s}^{-2} = 0 \cdot 0510 \, \text{Mp m}^{-2} \, \text{s}^2$$

$$\mu_{01} = \mu_{12} = \mu_{23} = \mu_1 = 0 \cdot 2 \, \text{Mp m}^{-1}/9 \cdot 81 \, \text{m s}^{-2} = 0 \cdot 0204 \, \text{Mp m}^{-2} \, \text{s}^2$$

Modulus of elasticity (steel) $E = 21 \times 10^6 \, \text{Mp m}^{-2}$.

(a) Forced symmetric vibration

Consider first the frame system to be loaded at point d (midspan of the top beam) with the vertical component, $P \sin \omega t$. Base the solution on the principle of superposition. Resolve the load in two parts (Fig. 3.9): one, as shown in Fig. 3.9a, consists of the external load and moments at points 3 and 3', these being so large that no rotation will take place at these points and beam 33' will behave as though perfectly clamped. The moments are given by formula 3.20. The other part of the load, as shown in Fig. 3.9b, consists of the same moments but with opposite signs:

$$\mathscr{M}_3 = \frac{Pl_2}{4} \frac{F_3(\lambda_2/2)}{F_6(\lambda_2/2)} = -\mathscr{M}_{3'} \tag{3.22}$$

The loads according to Figs 3.9a and 3.9b give as their resultant, the load by force $P \sin \omega t$. The rotations of the joints due to the actual load are the same as those due to the moments $\mathscr{M}_3$, $\mathscr{M}_{3'}$. The vertical displacement v_d is obtained

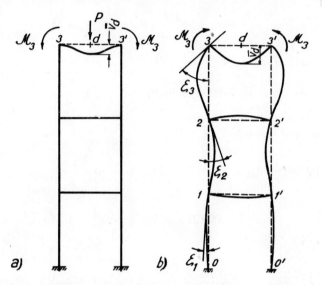

Fig. 3.9 Forced symmetric vibration of a multi-storey frame

by adding up the displacements of Figs 3.9a and b. Since longitudinal vibration has been neglected, the displacements of the joints are zero, and their rotations are the only unknowns. The slope-deflection equations are obtained either by the application of the reciprocal theorem according to eq. 3.7 or from the condition of equilibrium of moments in eq. 3.9. The three slope-deflection equations for the three unknown rotations, $\zeta_1 = -\zeta_{1'}$, $\zeta_2 = -\zeta_{2'}$, $\zeta_3 = -\zeta_{3'}$ are set out in Table 3.4.

Table 3.4

ζ_1	ζ_2	ζ_3	
a_{11}	a_{12}	a_{13}	$= a_{10}$
a_{12}	a_{22}	a_{23}	$= a_{20}$
a_{13}	a_{23}	a_{33}	$= a_{30}$

The coefficients in Table 3.4 are

$$a_{11} = \frac{2EJ_1}{l_1} F_2(\lambda_1) + \frac{EJ_2}{l_2} \left[F_2(\lambda_2) - F_1(\lambda_2) \right] = a_{22}$$

$$a_{12} = \frac{EJ_1}{l_1} F_1(\lambda_1) \qquad\qquad = a_{23}$$

$$a_{13} = 0$$

$$a_{33} = \frac{EJ_1}{l_1} F_2(\lambda_1) + \frac{EJ_2}{l_2} \left[F_2(\lambda_2) - F_1(\lambda_2) \right]$$

$$a_{10} = a_{20} = 0$$

$$a_{30} = \mathcal{M}_3 \quad \text{according to eq. 3.22} \tag{3.23}$$

Numerical example:

Let $m = 0.000\,253$ Mp m^{-1} s^2; $r = 1$ m; $n = 10$ s^{-1}.
Then $\omega = 2\pi n = 62.8$ s^{-1}

$$P = 0.000\,253 \times 1.0 \times 62.8^2 = 1 \text{ Mp}$$

and by eq. 2.23

$$\lambda_1 = 6 \left(\frac{0.0204 \times \omega^2}{21 \times 10^6 \times 1 \times 10^{-4}} \right)^{1/4} = 0.334\omega^{1/2} = 2.654$$

$$\lambda_2 = 5 \left(\frac{0.0510 \times \omega^2}{21 \times 10^6 \times 2 \times 10^{-4}} \right)^{1/4} = 0.296\omega^{1/2} = 2.339$$

The respective functions $F(\lambda)$ are found in the numerical tables in the Appendix:

$$
\begin{aligned}
F_1(\lambda_1) &= 2.397 \\
F_2(\lambda_1) &= 3.483 \\
F_1(\lambda_2) &= 2.229 \\
F_2(\lambda_2) &= 3.699 \\
F_3(\lambda_2/2) &= 6.058 \\
F_6(\lambda_2/2) &= 11.30
\end{aligned}
$$

With the substitution of numerical values, the slope-deflection equations are as indicated in Table 3.5

Table 3.5

ζ_1	ζ_2	ζ_3	
3675	840		$= 0$
840	3675	840	$= 0$
	840	2455	$= 0.670$

and the amplitudes of rotation are:

$$\zeta_1 = \quad 0.1636 \times 10^{-4}$$
$$\zeta_2 = -0.716 \ \times 10^{-4}$$
$$\zeta_3 = \quad 2.97 \ \ \times 10^{-4}$$

(The determinant of the coefficients on the left-hand side of the equations $\Delta = 28.9 \times \times 10^9$ Mp^3 m^3.)

The deflection midspan of the beam is obtained by adding up the deflections of Figs 3.9a and 3.9b. The first is given by formula 3.19, the second is arrived at through the application of the reciprocal theorem to the states corresponding to Figs 3.9a, b

$$-\frac{Pl_2}{4}\frac{F_3(\lambda_2/2)}{F_6(\lambda_2/2)}2\xi_3 + P\bar{v}_d = 0 \tag{3.24}$$

The total deflection midspan is thus

$$v_d = \bar{v}_d + \bar{\bar{v}}_d = \frac{Pl_2^3}{16EJ_2 \times F_6(\lambda_2/2)} + \frac{l_2}{2}\frac{F_3(\lambda_2/2)}{F_6(\lambda_2/2)}\zeta_3 \tag{3.25}$$

and with our numerical values

$$v_d = \frac{5^3}{16 \times 21 \times 10^6 \times 2 \times 10^{-4} \times 11.30} + \frac{5}{2}\frac{6.06}{11.30}2.97 \times 10^{-4} = 5.63 \times 10^{-4} \text{ m}$$

The principle of superposition will also be applied to the computation of the amplitude of end moment $M_{33'}$. The moment of state 3.9a is given by formula 3.20, the moment at deformation corresponding to Fig. 3.9b by formulae 2.41, 2.43 or Table 10.1 in the Appendix:

$$M_{33'} = \frac{EJ_2}{l_2}[F_2(\lambda_2) - F_1(\lambda_2)]\zeta_3 - \frac{Pl_2}{4}\frac{F_3(\lambda_2/2)}{F_6(\lambda_2/2)}$$
$$= 0.1236 \times 2.97 - 0.670 = -0.302 \text{ Mpm}$$

Further

$$M_{32} = \frac{EJ_1}{l_1} \left[F_2(\lambda_1)\,\zeta_3 + F_1(\lambda_1)\,\zeta_2 \right] = 0.302 \text{ Mpm}$$

The remaining moments (in Mpm) are

$$M_{22'} = -0.0890, \quad M_{23} = 0.1630, \quad M_{21} = -0.0739$$
$$M_{11'} = 0.0204, \quad M_{12} = -0.0406, \quad M_{10} = 0.0201$$

Consequently, the check conditions

$$M_{33'} + M_{32} = 0$$
$$M_{23} + M_{22'} + M_{21} = 0$$
$$M_{12} + M_{11'} + M_{10} = 0$$

are satisfied.

For the purpose of comparison, we find below the end moments under a static load. Were the structure subjected to a time-invariable vertical force, $P = 1$ Mp, the moments (in Mpm) would be:

$$M_{33'} = -0.2712, \quad M_{32} = 0.2712$$
$$M_{22'} = -0.0567, \quad M_{23} = 0.1002, \quad M_{21} = -0.0435$$
$$M_{11'} = 0.008\,85, \quad M_{12} = 0.007\,38, \quad M_{10} = -0.0162$$

(The determinant $\Delta = 58.1 \times 10^9$ Mp^3 m^3 in this case.) It is clear to see from the results that the largest dynamic moments differ from the corresponding static ones by about 10%. The angular velocities of forced vibrations are much less in our case than the lowest angular frequency of free vibration which we shall compute later on. With increasing velocity the moments would continually increase and reach infinite values for the velocity equal to the angular frequency of free vibration.

(b)   *Natural frequencies of symmetric vibration*

If in the equations of Table 3.4, force P tends to zero the equations become homogeneous with a solution different from zero only at definite angular velocities. These velocities can be computed from the determinant of all the coefficients with the unknown deformations set to zero. They are the angular frequencies of free vibration, for in their presence the system continues to execute harmonic motion without the action of external forces.

The determinant

$$\Delta = a_{11}a_{22}a_{33} + 2a_{12}a_{23}a_{13} - a_{22}a_{13}^2 - a_{11}a_{23}^2 - a_{33}a_{12}^2$$

$$= \left\{ 2\frac{EJ_1}{l_1}F_2(\lambda_1) + \frac{EJ_2}{l_2}\left[F_2(\lambda_2) - F_1(\lambda_2)\right]\right\}^2$$

$$\times \left\{ \frac{EJ_1}{l_1}F_2(\lambda_1) + \frac{EJ_2}{l_2}\left[F_2(\lambda_2) - F_1(\lambda_2)\right]\right\} - \left[\frac{EJ_1}{l_1}F_1(\lambda_1)\right]^2$$

$$\times \left\{ \frac{3EJ_1}{l_1}F_2(\lambda_1) + \frac{2EJ_2}{l_2}\left[F_2(\lambda_2) - F_1(\lambda_2)\right]\right\}$$

versus ω is plotted in Fig. 3.10. The points at which the curve of Δ intersects the horizontal axis $(\Delta = 0)$ gives the natural circular frequencies (angular velocities)

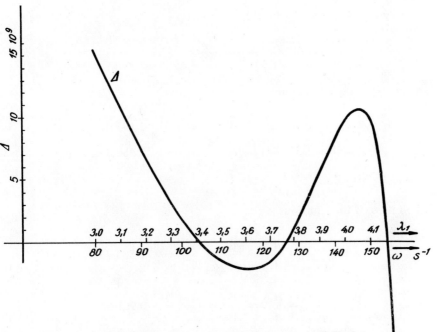

Fig. 3.10 Determinant of the coefficients in the equations set out in Table 3.4

of the system. We read from the diagram that $\omega_{(1)} = 103\cdot703 \text{ s}^{-1}$, $\omega_{(2)} = 126\cdot246 \text{ s}^{-1}$, $\omega_{(3)} = 155\cdot437 \text{ s}^{-1}$, etc.; this corresponds to frequencies $f_{(1)} = \omega_{(1)}/2\pi = 16\cdot5 \text{ Hz}$, $f_{(2)} = 20\cdot1 \text{ Hz}$, $f_{(3)} = 24\cdot8 \text{ Hz}$, etc.

Even though the actual values of the joint rotations are indefinite — the deflection amplitude is likely to have any arbitrary magnitude — the ratios of the rotations are definite (i.e. the modes of free vibration are constant). They are computed from the respective algebraic complements. Table 3.6 shows the slope-deflection equations for the first mode of free vibration $(\omega_{(1)} = 103 \cdot 703 \, \text{s}^{-1}, \, \lambda_1 = 3 \cdot 410 \, 61, \, \lambda_2 = 3 \cdot 005 \, 24)$,

$$F_1(\lambda_1) = 3 \cdot 362 \, 11$$
$$F_2(\lambda_1) = 2 \cdot 305 \, 16$$
$$F_1(\lambda_2) = 2 \cdot 707 \, 74$$
$$F_2(\lambda_2) = 3 \cdot 094 \, 24$$

Table 3.6

ζ_1	ζ_2	ζ_3	
1938·26	1176·74	0	$= 0$
1176·74	1938·26	1176·74	$= 0$
0	1176·74	1131·46	$= 0$

The algebraic complements are

$$A_{11} = 0 \cdot 808 \, 350 \times 10^6, \quad A_{22} = 2 \cdot 193 \, 071 \times 10^6$$
$$A_{12} = -1 \cdot 331 \, 441 \times 10^6, \quad A_{23} = -2 \cdot 280 \, 832 \times 10^6$$
$$A_{13} = 1 \cdot 384 \, 722 \times 10^6, \quad A_{33} = 2 \cdot 372 \, 130 \times 10^6$$

and the ratios of the rotations are

$$\frac{\zeta_{1(1)}}{\zeta_{3(1)}} = \frac{A_{11}}{A_{13}} = \frac{A_{12}}{A_{23}} = \frac{A_{13}}{A_{33}} = 0 \cdot 583 \, 75$$

$$\frac{\zeta_{2(1)}}{\zeta_{3(1)}} = \frac{A_{12}}{A_{13}} = \frac{A_{22}}{A_{23}} = \frac{A_{23}}{A_{33}} = -0 \cdot 961 \, 52$$

The deflection at the midspan of the beam is computed from eq. 3.25 by the substitution of $P = 0$

$$\frac{v_{d(1)}}{\zeta_{3(1)}} = \frac{l_2}{2} \frac{F_3(\lambda_2/2)}{F_6(\lambda_2/2)} = \frac{5}{2} \frac{6 \cdot 16}{10 \cdot 10} = 1 \cdot 525 \, \text{m}$$

At the second mode of natural vibration

$$\omega_{(2)} = 126{\cdot}246 \text{ s}^{-1}$$
$$\lambda_1 = 3{\cdot}763\,10$$
$$\lambda_2 = 3{\cdot}315\,83$$

$$\frac{\zeta_{1(2)}}{\zeta_{3(2)}} = -1{\cdot}012\,55$$

$$\frac{\zeta_{2(2)}}{\zeta_{3(2)}} = 0{\cdot}112\,73$$

$$\frac{v_{d(2)}}{\zeta_{3(2)}} = 1{\cdot}700 \text{ m}$$

The determinant of the coefficients of the equations set out in Table 3.4 again equals zero. Analogously, the third natural circular frequency and the third normal mode are found to be

$$\omega_{(3)} = 155{\cdot}437 \text{ s}^{-1}$$
$$\lambda_1 = 4{\cdot}175\,55$$
$$\lambda_2 = 3{\cdot}679\,26$$

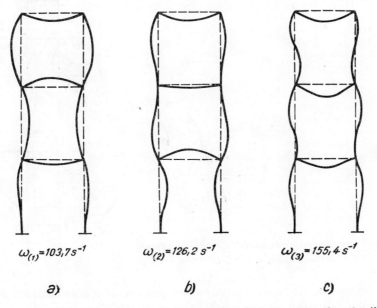

Fig. 3.11 Symmetric free vibration of a multi-storey frame. (a) First natural mode, (b) second natural mode, (c) third natural mode

112

$$\frac{\zeta_{1(3)}}{\zeta_{3(3)}} = 0.775\ 71$$

$$\frac{\zeta_{2(3)}}{\zeta_{3(3)}} = 1.173\ 64$$

$$\frac{v_{d(3)}}{\zeta_{3(3)}} = 2.067$$

The first three modes of free symmetric vibration are illustrated in Fig. 3.11.

(c)   *Forced antisymmetric vibration*

The horizontal component of the external force (cf. eq. 3.21) is $P \sin(\omega t + \pi/2)$. When the symmetric system of Fig. 3.8 is subjected to this antisymmetric load, antisymmetric vibration takes place. There will now be six unknowns, i.e. three rotations

$$\zeta_1 = \zeta_{1'}; \quad \zeta_2 = \zeta_{2'}; \quad \zeta_3 = \zeta_{3'} \tag{3.26a}$$

and three horizontal displacements

$$u_1 = u_{1'}; \quad u_2 = u_{2'}; \quad u_3 = u_{3'} \tag{3.26b}$$

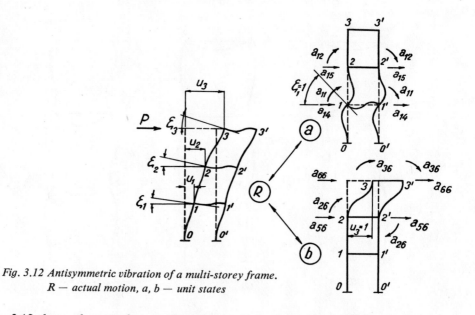

Fig. 3.12 Antisymmetric vibration of a multi-storey frame.
R — actual motion, a, b — unit states

Fig. 3.12 shows the actual state of motion R and two of the six auxiliary unit states. The equations obtained by the application of the reciprocal theorem are given in Table 3.7.

Table 3.7

ζ_1	ζ_2	ζ_3	u_1	u_2	u_3	
a_{11}	a_{12}	0	0	a_{15}	0	$= 0$
a_{12}	a_{22}	a_{23}	a_{24}	0	a_{26}	$= 0$
0	a_{23}	a_{33}	0	a_{35}	a_{36}	$= 0$
0	a_{24}	0	a_{44}	a_{45}	0	$= 0$
a_{15}	0	a_{35}	a_{45}	a_{55}	a_{56}	$= 0$
0	a_{26}	a_{36}	0	a_{56}	a_{66}	$= P/2$

The coefficients of the equations are:

$$a_{11} = \frac{2EJ_1}{l_1} F_2(\lambda_1) + \frac{EJ_2}{l_2} \left[F_1(\lambda_2) + F_2(\lambda_2) \right] = a_{22}$$

$$a_{12} = \frac{EJ_1}{l_1} F_1(\lambda_1) = a_{23}$$

$$a_{15} = -\frac{EJ_1}{l_1^2} F_3(\lambda_1) = -a_{24} = a_{26} = -a_{35}$$

$$a_{33} = \frac{EJ_1}{l_1} F_2(\lambda_1) + \frac{EJ_2}{l_2} \left[F_1(\lambda_2) + F_2(\lambda_2) \right]$$

$$a_{36} = \frac{EJ_1}{l_1^2} F_4(\lambda_1)$$

$$a_{44} = \frac{2EJ_1}{l_1^3} F_6(\lambda_1) - \frac{EJ_2}{l_2^3} \frac{\lambda_2^4}{2} = a_{55}$$

$$a_{45} = \frac{EJ_1}{l_1^3} F_5(\lambda_1) = a_{56}$$

$$a_{66} = \frac{EJ_1}{l_1^3} F_6(\lambda_1) - \frac{EJ_2}{l_2^3} \frac{\lambda_2^4}{2}$$

$$(3.27)$$

Coefficients a_{55} and a_{66} are obtained by referring to eqs 2.71 and 3.16. The first three equations of Table 3.7 express the conditions of equilibrium of the moments, the other three, the conditions of equilibrium of the horizontal forces in joints 1 to 3.

Considering the same numerical values as used in the computation of symmetric vibration in the preceding section ($\lambda_1 = 2 \cdot 654$, $\lambda_2 = 2 \cdot 339$), we obtain the respective functions $F(\lambda)$ from the Tables in the Appendix, and on substituting these in the coefficients of the equations in Table 3.7, get the equations shown in Table 3.8.

Table 3.8

ζ_1	ζ_2	ζ_3	u_1	u_3	u_3	
7420	840	0	0	−451	0	= 0
840	7420	840	+451	0	−451	= 0
0	840	6200	0	+451	−186	= 0
0	+451	0	−648	−187·5	0	= 0
−451	0	+451	−187·5	−648	−187·5	= 0
0	−451	−186	0	−187·5	−576	= 0·50

Solving the equations gives the amplitudes of the deformations

$$u_1 = -1{\cdot}040 \times 10^{-4}\,\mathrm{m}\,; \quad u_2 = 2{\cdot}49 \times 10^{-4}\,\mathrm{m}\,; \quad u_3 = -9{\cdot}00 \times 10^{-4}\,\mathrm{m}$$

$$\zeta_1 = 0{\cdot}204 \times 10^{-4}\,; \quad \zeta_2 = -0{\cdot}463 \times 10^{-4}\,; \quad \zeta_3 = -0{\cdot}388 \times 10^{-4}$$

The computed mode is shown in Fig. 3.13d.

The amplitudes of the moment and force acting at the end of the beam are obtained with the aid of the tables in the Appendix and formula 2.71:

$$X_{33'} = -\frac{\mu_2 l_2 \omega^2}{2} u_3 = \frac{1}{2}\,0{\cdot}0510 \times 5 \times 62{\cdot}8^2 \times 9{\cdot}00 \times 10^{-4} = 0{\cdot}453\ \mathrm{Mp}$$

$$M_{33'} = \frac{EJ_2}{l_2}\left[F_1(\lambda_2) + F_2(\lambda_2)\right]\zeta_3 = -\frac{21 \times 10^6 \times 2 \times 10^{-4}}{5}$$

$$\times \left[2{\cdot}228 + 3{\cdot}700\right] \times 0{\cdot}388 \times 10^{-4} = -0{\cdot}193\ \mathrm{Mpm}$$

and further,

$$X_{32} = 0{\cdot}047\ \mathrm{Mp}$$
$$M_{32} = 0{\cdot}193\ \mathrm{Mpm}$$
$$X_{22'} = -0{\cdot}125\,; \quad X_{23} = 0{\cdot}125\,; \quad X_{21} = 0{\cdot}001\ \mathrm{Mp}$$
$$M_{22'} = -0{\cdot}231\,; \quad M_{23} = 0{\cdot}363\,; \quad M_{21} = -0{\cdot}132\ \mathrm{Mpm}$$
$$X_{11'} = 0{\cdot}0523\,; \quad X_{12} = -0{\cdot}0562\,; \quad X_{10} = 0{\cdot}0039\ \mathrm{Mp}$$
$$M_{11'} = 0{\cdot}101\,; \quad M_{12} = -0{\cdot}146\,; \quad M_{10} = 0{\cdot}044\ \mathrm{Mpm}$$

Were the system acted upon by a horizontal, time-invariable force at bar 33′, $P = 1$ Mp, the resultant deformations would be

$$u_1 = 59{\cdot}5 \times 10^{-4}\,\mathrm{m}\,; \quad u_2 = 136{\cdot}5 \times 10^{-4}\,\mathrm{m}\,; \quad u_3 = 206 \times 10^{-4}\,\mathrm{m}$$

$$\zeta_1 = 5{\cdot}57 \times 10^{-4}\,; \quad \zeta_2 = 5{\cdot}75 \times 10^{-4}\,; \quad \zeta_3 = 3{\cdot}16 \times 10^{-4}$$

and the corresponding moments (in Mpm) would be

$$M_{33'} = 1{\cdot}591 \; ; \quad M_{32} = -1{\cdot}591$$
$$M_{22'} = 2{\cdot}903 \; ; \quad M_{23} = -1{\cdot}409 \; ; \quad M_{21} = -1{\cdot}494$$
$$M_{11'} = 2{\cdot}811 \; ; \quad M_{12} = -1{\cdot}506 \; ; \quad M_{10} = -1{\cdot}305$$

The static deflection of the system is illustrated in Fig. 3.13a. Consulting the figure we see that in this case there is a substantial difference between the moments and deformations under harmonic load and those under static load. The reason for this is that the angular velocity of forced vibration lies between the third and fourth circular frequencies of antisymmetric free vibration; this will be discussed later on.

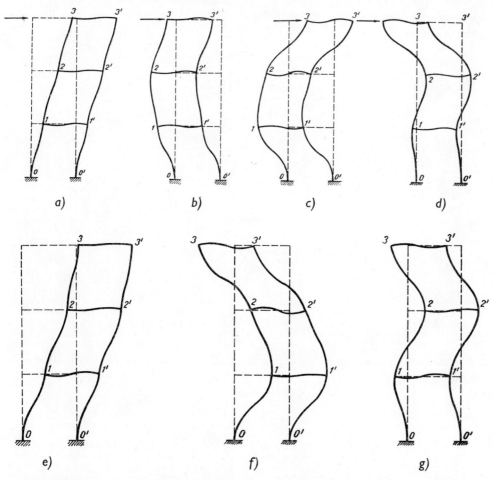

Fig. 3.13 Forced and free antisymmetric vibration of a multi-storey frame (a) $\omega = 0$, (b) $\omega = 17{\cdot}4\,\mathrm{s}^{-1}$, (c) $\omega = 22{\cdot}8\,\mathrm{s}^{-1}$, (d) $\omega = 62{\cdot}8\,\mathrm{s}^{-1}$, (e) $\omega_{(1')} = 8{\cdot}41\,\mathrm{s}^{-1}$, (f) $\omega_{(2')} = 25{\cdot}8\,\mathrm{s}^{-1}$, (g) $\omega_{(3')} = 42{\cdot}4\,\mathrm{s}^{-1}$

Figs 3.13b and 3.13c show the shape of the dynamic deformations at the frequency of forced vibration induced on beam 33′ by a horizontal harmonic force P with circular frequency $\omega = 17\cdot4\ s^{-1}$ or $22\cdot8\ s^{-1}$.

They were computed using the procedure demonstrated in the preceding example.

(d)　　Free vibration

If force P is reduced to zero in the equations shown in Table 3.7, the set becomes homogeneous. The circular frequencies at which a non-zero solution of the set is feasible, are the natural circular frequencies of antisymmetric vibration. As in the previous cases, they are computed from the condition that the determinant of the coefficients on the left-hand sides of the equations should equal zero. In the case under consideration

$$\omega_{(1')} = 8\cdot41\ s^{-1}$$

$$f_{(1')} = 1\cdot34\ \text{Hz}$$

$$\frac{\zeta_1}{u_3} = 0\cdot0335\ \text{m}^{-1}$$

$$\frac{\zeta_2}{u_3} = 0\cdot0237\ \text{m}^{-1}$$

$$\frac{\zeta_3}{u_3} = 0\cdot0081\ \text{m}^{-1}$$

$$\frac{u_1}{u_3} = 0\cdot390$$

$$\frac{u_2}{u_3} = 0\cdot792$$

The ratios of the deformations were again computed from the respective algebraic complements.

Further,

$$\omega_{(2')} = 25\cdot8\ s^{-1}$$

$$f_{(2')} = 4\cdot10\ \text{Hz}$$

and

$$\omega_{(3')} = 42\cdot4\ s^{-1}$$

$$f_{(3')} = 6\cdot75\ \text{Hz}$$

and

$$\omega_{(4')} = 157 \quad s^{-1}$$
$$f_{(4')} = 24.9 \text{ Hz}$$

The first three modes of free antisymmetric vibration are depicted in Figs 3.13e, f, g.

(e) *Estimating the modes of antisymmetric forced and free vibrations*

Fig. 3.13 shows the mode shapes at different frequencies of vibration, Fig. 3.14 the amplitude of horizontal deflection of joint 3 versus the circular frequency of steady forced vibration induced by horizontal force $P \sin \omega t$ with amplitude $P = 1$ Mp. (The amplitude at constant unbalanced mass $m = 0.00249t$, i.e. at $P = 0.000253\omega^2$ Mp is drawn in dotted lines.) The curve enables us to estimate in

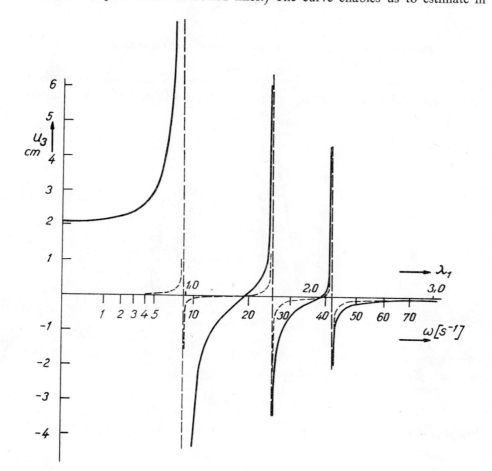

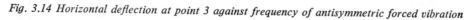

Fig. 3.14 Horizontal deflection at point 3 against frequency of antisymmetric forced vibration

advance the mode of vibration at any frequency of the system being examined. At the points where it intersects the horizontal axis of the coordinates, the deflection changes its sign and a new nodal point appears on the columns. In the system computed in Section (c) (Fig. 3.13d) there are two nodal points on the columns because the curve intersects the horizontal axis twice in the interval $\omega = 0-62{\cdot}8 \text{ s}^{-1}$. Apart from this, u_3 changes its sign together with the remaining deformations (without giving rise to a nodal point) at the natural frequencies at which the determinant of the coefficients of the slope-deflection equations passes through zero and changes its sign. In Fig. 3.14 the natural frequencies are at the points where the curve of deflection u_3 passes through infinity. It is clear from this figure that a new nodal point usually comes in the interval between two natural frequencies.

If, on the other hand, we know the mode of forced vibration at some frequency, we may estimate the natural frequencies between which the given frequency lies, from the number of nodal points and the sign of the deflection u_3.

(f) *Resultant forces and moments under centrifugal force load*

As is well known, the instantaneous values of harmonic quantities may be represented by the projections of rotating vectors (Fig. 3.15). The addition of

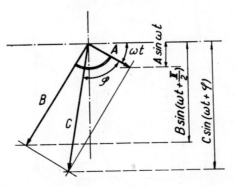

Fig. 3.15 Rotating vector

two harmonic motions with equal frequencies, whose amplitudes A, B differ by $\pi/2$ in phase, gives a harmonic motion with the amplitude

$$C = \pm(A^2 + B^2)^{1/2} \tag{3.28}$$

(the sign of the square root depends on the sign of A), shifted against A by

$$\varphi = \tan^{-1}\frac{B}{A} \tag{3.29}$$

If the unbalance mass of the machinery rotates clockwise, the symmetric vibration lags by $\pi/2$ behind the antisymmetric one. In the example considered, the resultant moments are

$$M_{33'} = -(0.303^2 + 0.193^2)^{1/2} = -0.358 \text{ Mpm}$$

$$\varphi_{33'} = \tan^{-1} \frac{-0.193}{-0.302} = 0.568$$

$$M_{22'} = -(0.0890^2 + 0.231^2)^{1/2} = -0.247 \text{ Mpm}$$

$$\varphi_{22'} = \tan^{-1} \frac{-0.231}{-0.089} = 1.20, \quad \text{etc.}$$

Evidently, the resultant moments do not reach their maximum values at the same time (i.e. there is a phase difference between them).

3.3   Simplified solution

In formulae 2.89 we have expanded the frequency functions in a power series. Consider now only the first two terms of those series. The end moments and forces according to Fig. 3.16a can then be expressed as follows:

$$M_{gh} = \frac{EJ}{l} F_1(\lambda) \approx \frac{2EJ}{l} + \frac{1}{140} \mu l^3 \omega^2 \tag{3.30a}$$

$$M_{hg} = \frac{EJ}{l} F_2(\lambda) \approx \frac{4EJ}{l} - \frac{1}{105} \mu l^3 \omega^2 \tag{3.30b}$$

$$Y_{gh} = \frac{EJ}{l^2} F_3(\lambda) \approx \frac{6EJ}{l^2} + \frac{13}{420} \mu l^2 \omega^2 \tag{3.30c}$$

$$Y_{hg} = \frac{EJ}{l^2} F_4(\lambda) \approx -\frac{6EJ}{l^2} + \frac{11}{210} \mu l^2 \omega^2 \tag{3.30d}$$

and the end forces according to Fig. 3.16b as

$$Y_{gh} = \frac{EJ}{l^3} F_5(\lambda) \approx -\frac{12EJ}{l^3} - \frac{9}{70} \mu l \omega^2 \tag{3.30e}$$

$$Y_{hg} = \frac{EJ}{l^3} F_6(\lambda) \approx \frac{12EJ}{l^3} - \frac{13}{35} \mu l \omega^2 \tag{3.30f}$$

Let us note the physical meaning of the above simplification. Consider the state of motion indicated in Fig. 3.16a and the state of static deformation in Fig. 3.16c, and apply to them the reciprocal theorem. In the former state the system is subjected to end forces and moments, and in addition, to inertia forces with amplitude $\mu\omega^2\,v(x)$, whereas in the latter state the system is subjected to end forces and moments alone. In view of what has been said in Section 3.1, the work of the inertia forces must be taken into account in this case. Therefore

$$M_{gh}\cdot 1 + \omega^2 \int \mu\,v(x)\,v_c(x)\,\mathrm{d}x = M_{hg}^c \tag{3.31}$$

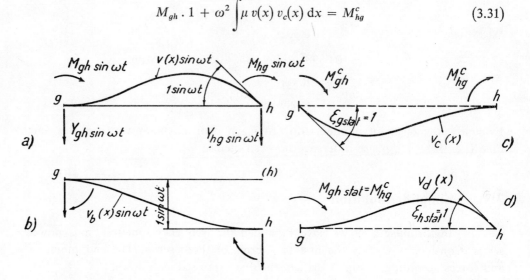

Fig. 3.16 Simplified slope-deflection method. Determination of end forces and moments

Replace now the approximate deflection curve $v(x)$ by the curve of static deflection $v_d(x)$ according to Fig. 3.16d. Since by Maxwell's theorem

$$M_{hg}^c = M_{ghst}$$

then

$$M_{gh} \approx M_{ghst} - \omega^2 \int \mu\,v_c(x)\,v_d(x)\,\mathrm{d}x \tag{3.32}$$

It will be shown in Section 3.4 that the static deflection curves are described by the equations

$$v_c(x) = \frac{x^3}{l^2} - \frac{2x^2}{l} + x$$

$$v_d(x) = \frac{x^3}{l^2} - \frac{x^2}{l}$$

$$\tag{3.33}$$

Since $M_{ghst} = 2EJ/l$, substitution of eq. 3.33 in eq. 3.32 results in

$$M_{gh} = \frac{2EJ}{l} + \frac{l^3}{140}\,\mu\omega^2$$

i.e. expression 3.30a. The remaining end forces and moments are given by analogous relationships. The approximate formulae are thus obtained by replacing, in the computation of inertia forces, the actual deformations by the corresponding static deformations.

Expressions 3.30 substituted in the slope-deflection equations of, for example, Table 3.4 lead to equations of the type shown in Table 3.10 (see below). In the matrix form the latter equations turn out as

$$(B + C\omega^2)\, r = P \tag{3.34}$$

where the matrices are given by

$$B = [b_{ik}]$$
$$C = [c_{ik}] \tag{3.35}$$

and the vectors are

$$r = \begin{bmatrix} \zeta_1 \\ \zeta_2 \\ \zeta_3 \end{bmatrix} \tag{3.36a}$$

$$P = \begin{bmatrix} \mathcal{M}_1 \\ \mathcal{M}_2 \\ \mathcal{M}_3 \end{bmatrix} \tag{3.36b}$$

The problem of finding the natural frequencies thus reduces to that of getting the solution of

$$\det [B + C\omega^2] = 0 \tag{3.37}$$

Since matrix B as well as matrix C have constant terms, this can be done more easily than finding the natural frequencies through the exact solution discussed in Section 3.2.

The natural frequencies computed from the approximate relations 3.30 are, of course, only approximate. They may, however, be corrected to the exact values. The approximate values of $F(\lambda)$ computed from the first two terms of series 2.89 correspond to exact values of $F(\lambda)$ for somewhat lower arguments λ. Thus, e.g. for $\lambda = 2$, the computed $F_1(\lambda)$ is

$$F_1(\lambda) \approx 2 + \frac{2^4}{140} = 2 \cdot 114$$

Consulting the numerical tables in the Appendix we see that the argument appertaining to this value of $F_1(\lambda)$ is $\lambda = 1 \cdot 98$, i.e. about 1% less. It turns out that this difference is the same for function

$$F_2(\lambda) \approx 4 - \frac{2^4}{105} = 3 \cdot 847$$

and nearly the same for functions $F_3(\lambda)$ to $F_6(\lambda)$.

For $\lambda = 3$, the computed $F_1(\lambda)$ is

$$F_1(\lambda) \approx 2 + \frac{3^4}{140} = 2 \cdot 58$$

and the argument read from the tables is $\lambda = 2 \cdot 88$, i.e. 4% less. For $\lambda = 4$, the difference increases to 10%. Since λ is proportional to the square root of ω, the differences for frequencies are about double. The frequencies of free vibration computed from the slope-deflection equations with approximate coefficients are, therefore, higher than those obtained through the exact solution. If, however, the λs of all the bars of a system are approximately the same, the exact values may be found by applying the corrections shown in Table 3.9.

Table 3.9

λ (simplified solution)	2·00	2·50	3·00	3·20	3·40	3·60	3·80	4·00	4·20	4·40	4·60
λ (exact solution)	1·98	2·45	2·88	3·03	3·19	3·34	3·48	3·60	3·70	3·80	3·90
difference λ	1%	2%	4%	5%	6%	7%	8%	10%	12%	13·5%	15%
difference ω	2%	4%	8%	10%	12%	14%	15%	19%	22%	25%	28%

It is recommended that only functions $F_1(\lambda)$ to $F_6(\lambda)$ be used in the simplified solution, since computations involving the remaining functions are even less accurate. If some of the bars have hinged or free ends, the number of slope-deflection equations becomes, of course, higher. In the next example we shall show how to proceed in such a case.

Example 3.2

Fig. 3.17 shows a continuous beam with two spans. The only unknown deformation is rotation ζ_1 of joint 1. For free vibration the slope-deflection equation is

$$\left[\frac{EJ_{01}}{l_{01}} F_7(\lambda_{01}) + \frac{EJ_{12}}{l_{12}} F_7(\lambda_{12}) \right] \zeta_1 = 0 \tag{3.38}$$

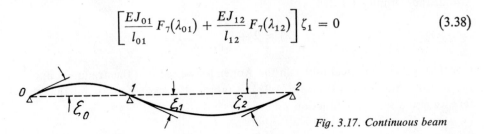

Fig. 3.17. Continuous beam

If we wish to avoid using function $F_7(\lambda)$, we must write three equations expressing the condition of equilibrium of the moments in joints 0, 1, 2 as follows:

$$\frac{EJ_{01}}{l_{01}} \left[F_2(\lambda_{01}) \, \zeta_0 + F_1(\lambda_{01}) \, \zeta_1 \right] = 0$$

$$\frac{EJ_{01}}{l_{01}} F_1(\lambda_{01}) \, \zeta_0 + \left[\frac{EJ_{01}}{l_{01}} F_2(\lambda_{01}) + \frac{EJ_{12}}{l_{12}} F_2(\lambda_{12}) \right] \zeta_1 + \frac{EJ_{12}}{l_{12}} F_1(\lambda_{12}) \, \zeta_2 = 0$$

$$\frac{EJ_{12}}{l_{12}} \left[F_1(\lambda_{12}) \, \zeta_1 + F_2(\lambda_{12}) \, \zeta_2 \right] = 0$$

$$(3.39)$$

In the exact solution, eqs 3.39 are equivalent to eqs 3.38. In the approximate solution, eqs 3.39 are more accurate than eq. 3.38. To simplify the discussion, consider the case of $l_{01} = l_{12}$, $J_{01} = J_{12}$, $\mu_{01} = \mu_{12}$. Then the exact lowest value of $\lambda = \pi$. From eq. 3.38 and eq. 10.16 in the Appendix

$$F_7(\lambda) \approx 3 - 0 \cdot 019 \, 048 \lambda^4 = 0$$

and from there

$$\lambda \approx \left(\frac{3}{0 \cdot 019 \, 048} \right)^{1/4} = 3 \cdot 53$$

The approximate value is thus 12% higher than the exact one. More exact results are obtained through the use of eqs 3.39. Then

$$F_2(\lambda) \, \zeta_0 + F_1(\lambda) \, \zeta_1 = 0$$
$$F_1(\lambda) \, \zeta_0 + F_2(\lambda) \, \zeta_1 = 0$$

and for the first natural mode, $\zeta_0 = -\zeta_1$, giving the frequency equation

$$F_2(\lambda) - F_1(\lambda) = 0 \, .$$

Substituting the approximate values corresponding to eq. 2.89 in the above yields

$$4 - 0 \cdot 009 \, 524 \lambda^4 - 2 - 0 \cdot 007 \, 143 \lambda^4 = 0$$

and from there

$$\lambda = \left(\frac{2}{0 \cdot 016 \, 667} \right)^{1/4} = 3 \cdot 31$$

With the correction read from Table 3.9

$$\lambda = 3 \cdot 31 \times 0 \cdot 95 = 3 \cdot 14 \, ,$$

and this agrees very well with the exact value, i.e. π.

Example 3.3

Let us apply the simplified method to the computation of the symmetric vibration of the frame shown in Fig. 3.8. (Example 3.1.) On substituting the approximate values the slope-deflection equations are as shown in Table 3.10.

Table 3.10

ζ_1	ζ_2	ζ_3	
$b_{11} - c_{11}\omega^2$	$b_{12} - c_{12}\omega^2$	0	$= 0$
$b_{12} - c_{12}\omega^2$	$b_{22} - c_{22}\omega^2$	$b_{13} - c_{23}\omega^2$	$= 0$
0	$b_{23} - c_{23}\omega^2$	$b_{33} - c_{33}\omega^2$	$= a_{03}$

The coefficients have the following values

$$b_{11} = \frac{8EJ_1}{l_1} + \frac{2EJ_2}{l_2} = \frac{8 \times 21 \times 10^6 \times 1 \times 10^{-4}}{6} + \frac{2 \times 21 \times 10^6 \times 2 \times 10^{-4}}{5}$$

$$= 2800 + 1680 = 4480 = b_{22}$$

$$b_{12} = \frac{2EJ_1}{l_1} = 700 = b_{23}$$

$$b_{33} = \frac{4EJ_1}{l_1} + \frac{2EJ_2}{l_2} = 1400 + 1680 = 3080$$

$$c_{11} = \frac{2}{105}\mu_1 l_1^3 + \left(\frac{1}{105} + \frac{1}{140}\right)\mu_2 l_2^3 = \frac{2 \times 0.0204 \times 6^3}{105} + \frac{0.0510 \times 5^3}{60}$$

$$= 0.0839 + 0.1062 = 0.1901 = c_{22}$$

$$c_{12} = -\frac{1}{140}\mu_1 l_1^3 = -\frac{0.0204 \times 6^3}{140} = -0.0315 = c_{23}$$

$$c_{33} = \frac{1}{105}\mu_1 l_1^3 + \left(\frac{1}{105} + \frac{1}{140}\right)\mu_2 l_2^3 = 0.0420 + 0.1062 = 0.1482$$

For free vibration $a_{03} = 0$.

With the substitution of these coefficients the equations of Table 3.10 are as shown in Table 3.11.

Table 3.11

ζ_1	ζ_2	ζ_3	
$4480 - 0{\cdot}1901\omega^2$	$700 + 0{\cdot}0315\omega^2$	0	$= 0$
$700 + 0{\cdot}0315\omega^2$	$4480 - 0{\cdot}1901\omega^2$	$700 + 0{\cdot}0315\omega^2$	$= 0$
0	$700 + 0{\cdot}0315\omega^2$	$3080 - 0{\cdot}1482\omega^2$	$= 0$

The first non-trivial solution of the above equations is for $\omega = 117 \text{ s}^{-1}$ which gives $\lambda_1 = 0{\cdot}328 \sqrt{117} \approx 3{\cdot}55$, $\lambda_2 = 0{\cdot}296 \sqrt{117} \approx 3{\cdot}20$, $\lambda_{av} \approx 3{\cdot}38$.

According to Table 3.9, the exact value of ω is about 12% less, i.e. $\omega_{(1)} \approx 0{\cdot}88 \times$ $\times 117 = 103 \text{ s}^{-1}$. This agrees within the limits of computation accuracy with $\omega_{(1)} = 103{\cdot}7 \text{ s}^{-1}$ obtained in Example 3.1.

In the second non-trivial solution, $\omega = 150 \text{ s}^{-1}$, giving $\lambda_1 = 0{\cdot}328 \sqrt{150} \approx 4{\cdot}0$, $\lambda_2 = 0{\cdot}296 \sqrt{150} \approx 3{\cdot}60$, $\lambda_{av} \approx 3{\cdot}80$. In this case the difference read in Table 3.9 is about 15%; hence

$$\omega_{(2)} = 0{\cdot}85 \times 150 = 127 \text{ s}^{-1}$$

again close enough to the exact value, $\omega_{(2)} = 126 \text{ s}^{-1}$.

In the third, $\omega = 194 \text{ s}^{-1}$, i.e. $\lambda_1 = 4{\cdot}56$, $\lambda_2 = 4{\cdot}12$, $\lambda_{av} = 4{\cdot}34$; the difference is about 28%, hence

$$\omega_{(3)} = 0{\cdot}76 \times 194 = 148 \text{ s}^{-1}$$

Since in this case the difference is already large, the data in Table 3.9 are evidently not reliable enough at high λs.

When computing the parameters of antisymmetric vibration of our system, we should bear in mind that coefficients a_{44}, a_{55} and a_{66} in formulae 3.27 contain the terms $EJ_2/l_2^3 \times \lambda^4/2$ which do not follow from the approximate relations. This is why Table 3.9 should not be used for correcting the results of computation in this case.

In order to reduce the λs that are either too high, or differ too much one from another, and thus refine the computation, the bars of the system may be divided into shorter elements by means of secondary joints.

Higher accuracy of computation is also achieved by taking a larger number of terms of power series 2.89 which expands the frequency functions.

3.4   The energy method

In either form — exact or simplified — the slope-deflection method has been proved to be very efficient for the analysis of frame systems. This is why we shall consider only two of the remaining methods applicable to dynamic computations, i.e. the energy method for solving free vibration, and the method of modal analysis for solving forced vibration. When applying the energy method one can use to advantage some of the results obtained in the preceding sections.

3.4.1   Computing the lowest frequency from the static deflection curves

The principles of the energy method have been explained in Section 1,6. Referring to it we write for systems composed of bars of constant cross-section that

$$\omega_{(1)}^2 = \frac{\sum \int_0^l EJ\, v''^2(x)\, dx}{\sum \int_0^l \mu\, v^2(x)\, dx} \tag{3.40}$$

where the numerator expresses double the maximum potential energy in the first natural mode. The summation sign $\sum$ includes all the bars of the system. If $v(x)$ is not the first natural mode, formula 3.40 yields but an approximate value of $\omega_{(1)}$. Assume that the first natural mode shape of the system is represented by the deformation produced by a suitable static rotation and displacement of the system joints.

The static deflection of section x of bar gh is defined by the equation

$$M(x) = -EJ \frac{d^2 v(x)}{dx^2} \tag{3.41}$$

If there is no load applied between the end points of the bar, the moment in section x is

$$M(x) = M_{gh} - Y_{gh}x \tag{3.42}$$

Substitution of eq. 3.42 in eq. 3.41 and two integrations of eq. 3.41 give the general solution of the latter, i.e. the equation of the static deflection curve

$$v(x) = C_1 x^3 + C_2 x^2 + C_3 x + C_4 \tag{3.43}$$

The integration constants are computed from the four conditions of the bar supports.

For a bar with clamped ends the conditions of support are expressed in terms of displacement and rotation of the ends. From eq. 3.43

$$v_g = v(0) = C_4$$
$$\zeta_g = v'(0) = C_3$$
$$v_h = v(l) = C_1 l^3 + C_2 l^2 + v'(0)\, l + v(0)$$
$$\zeta_h = v'(l) = 3C_1 l^2 + 2C_2 l + v'(0)$$

and from there

$$C_1 = \frac{1}{l^2}\left(\zeta_g + \zeta_h - 2\frac{v_h - v_g}{l}\right)$$

$$C_2 = \frac{1}{l}\left(-2\zeta_g - \zeta_h + 3\frac{v_h - v_g}{l}\right)$$

$$(3.44)$$

With the constants substituted in eq. 3.43, we compute the denominator of expression 3.40 as follows

$$\int_0^l \mu\, v^2(x)\,\mathrm{d}x = \mu\left\{\frac{l^3}{105}(\zeta_g^2 + \zeta_h^2) - \frac{l^3}{70}\zeta_g\zeta_h + \frac{13l^2}{210}(\zeta_g v_h - \zeta_h v_g)\right.$$

$$\left. + \frac{11l^2}{105}(\zeta_g v_g - \zeta_h v_h) + \frac{13l}{35}(v_g^2 + v_h^2) + \frac{9l}{35}v_g v_h\right\}$$

$$(3.45)$$

and the numerator of eq. 3.40

$$\int_0^l EJ\, v''^2(x)\,\mathrm{d}x = \frac{4EJ}{l}\left\{\zeta_g^2 + \zeta_h^2 + \zeta_g\zeta_h + 3(\zeta_g + \zeta_h)\frac{v_g - v_h}{l} + 3\left(\frac{v_g - v_h}{l}\right)^2\right\}\quad (3.46)$$

For a bar hinged at end g, the conditions of support are given by the equalities $M_{gh} = -EJ\, v''(0) = 0$; $v(0) = v_g$; $v(l) = v_h$; $v'(l) = \zeta_h$. In this case we get

$$C_1 = \frac{1}{2l^2}\left(\zeta_h - \frac{v_h - v_g}{l}\right)$$

$$C_2 = 0$$

$$C_3 = \frac{1}{2}\left(-\zeta_h + 3\frac{v_h - v_g}{l}\right)$$

$$C_4 = v_g$$

$$(3.47)$$

and

$$\int_0^l \mu\, v^2(x)\,\mathrm{d}x = \mu\left\{\frac{2l^3}{105}\zeta_h^2 - \frac{11l^2}{140}\zeta_h v_g - \frac{6l^2}{35}\zeta_h v_h\right.$$

$$\left. + \frac{33l}{140}v_g^2 + \frac{17l}{35}v_h^2 + \frac{39l}{140}v_g v_h\right\}$$

$$(3.48)$$

$$\int_0^l EJ\, v''^2(x)\,\mathrm{d}x = \frac{3EJ}{l}\left\{\zeta_h^2 + 2\zeta_h\frac{v_g - v_h}{l} + \left(\frac{v_g - v_h}{l}\right)^2\right\}$$

$$(3.49)$$

For a bar hinged at both ends, the conditions of support are $M_{gh} = -EJ\,v''(0) = 0$; $M_{hg} = EJ\,v''(l) = 0$; $v(0) = v_g$; $v(l) = v_h$, and the integration constants are

$$C_1 = C_2 = 0$$

$$C_3 = \frac{v_h - v_g}{l}$$

$$C_4 = v_g$$

$$(3.50)$$

and

$$\int_0^l \mu\,v^2(x)\,\mathrm{d}x = \frac{\mu l}{3}\left(v_g^2 + v_h^2 + v_g v_h\right) \qquad (3.51)$$

$$\int_0^l EJ\,v''^2(x)\,\mathrm{d}x = 0 \qquad (3.52)$$

And finally, for a cantilever beam, the conditions of support are $M_{gh} = = -EJ\,v''(0) = 0$; $Y_{gh} = EJ\,v'''(0) = 0$; $v(l) = v_h$; $v'(l) = \zeta_h$, and the integration constants are

$$C_1 = C_2 = 0$$
$$C_3 = \zeta_h$$
$$C_4 = v_h - \zeta_h l$$

$$(3.53)$$

and

$$\int_0^l \mu\,v^2(x)\,\mathrm{d}x = \mu l\left(\frac{\zeta_h^2 l^2}{3} - \zeta_h v_h l + v_h^2\right) \qquad (3.54)$$

$$\int_0^l EJ\,v''^2(x)\,\mathrm{d}x = 0 \qquad (3.55)$$

Example 3.4

Consider the frame shown in Fig. 3.18. The system was analysed in Reference 26 using the method of dividing the columns and the horizontal bar into equal elements and replacing the integral in the denominator of eq. 3.40 by a summation. What we plan to do here is to determine the approximate value of the fundamental circular frequency by direct computation of the integrals. The specification of the frame is as follows: $l_{01} = l$, $J_{01} = J$, $\mu_{01} = \mu$, $l_{11'} = 2l/3$, $J_{11'} = J$, $\mu_{11'} = 3\mu$. If a vertical force $P = 1$ is applied at the midspan of bar $11'$, the slope-deflection equation is

$$\zeta_1 \frac{EJ}{l}\left(4 + \frac{3}{2}\,2\right) = \frac{2}{3}\frac{l}{8}$$

and from there

$$\zeta_1 = \frac{l^2}{84EJ}$$

The static deflection at the midspan of bar $11'$ is then

$$v_p = \frac{l}{6}\zeta_1 + \frac{l^3}{2 \times 12EJ \times 3^3} = \frac{2l^3}{567EJ}$$

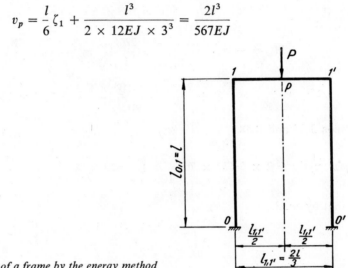

Fig. 3.18 Solution of a frame by the energy method

On substituting the computed values in formula 3.45 for all bars (assuming bar $11'$ to be composed of two bars: $1-p$ and $p-1'$) and then in formula 3.40 in which the numerator is expressed as the work done by the force $P = 1$ on displacement v_p we get

$$\omega_{(1)} \approx \frac{(2l^3/567EJ)^{1/2}}{\left\{2\mu l\left[\dfrac{l^2(1 + 3/3^3)}{105}(l^2/84EJ)^2 + \dfrac{13}{35}\dfrac{3}{3}(2l^3/567EJ)^2 + \right.\right.} \approx \frac{15\cdot9}{l^2}\left(\frac{EJ}{\mu}\right)^{1/2}$$

$$\left.\left. + \dfrac{13}{210}\dfrac{3l}{3^2}(l^2/84EJ)(2l^3/567EJ)\right]\right\}^{1/2}$$

This value agrees with that reported in Reference 26 but differs by about 11% from the exact one (obtained by following the solution of eq. 3.14, in Section 3.2.2) viz:

$$\omega_{(1)} = \frac{14\cdot29}{l^2}\left(\frac{EJ}{\mu}\right)^{1/2}$$

The numerical computation just presented — although shorter by far than the evaluation by means of summation — gives clear evidence about the shortcomings of this method against those of the slope-deflection method.

Example 3.5

Consider a three-span continuous beam (the notation is the same as in Fig. 7.5) with $J_{01} = 3\cdot6\,\text{m}^4$, $\mu_{01} = 0\cdot477\,\text{Mpm}^{-2}\,\text{s}^2$ in the first and third spans, and $J_{11'} = 6\cdot0\,\text{m}^4$, $\mu_{11'} = 0\cdot573\,\text{Mpm}^{-2}\,\text{s}^2$ in the second span; $l_{01} = l_{1'0'} = 70\,\text{m}$, $l_{11'} = 90\,\text{m}$; $E\,(\text{steel}) = 21 \times 10^6\,\text{Mp/m}^2$. At points 1 and 1' apply symmetric moments so large that the rotation of the joints $\zeta_1 = -\zeta_{1'} = 1$. The moments are

$$M_1 = -M_{1'} = \frac{3EJ_{01}}{l_{01}} + \frac{2EJ_{11'}}{l_{11'}} = \frac{3 \times 21 \times 10^6 \times 3\cdot6}{70}$$

$$+ \frac{2 \times 21 \times 10^6 \times 6\cdot0}{90} = 6\cdot04 \times 10^6\,\text{Mpm}$$

By eqs 3.45 and 3.48

$$\int_0^l \mu\,v^2(x)\,\text{d}x = 2 \times 0\cdot477 \times 70^3\,\frac{2}{105} + 0\cdot573 \times 90^3\left(\frac{2}{105} + \frac{1}{70}\right) = 20\,200\,\text{Mpms}^2$$

so that by eq. 3.40

$$\omega_{(1)} \approx \left(\frac{2 \times 6\cdot04 \times 10^2}{2\cdot02}\right)^{1/2} = 24\cdot4\,\text{s}^{-1}$$

The exact value computed by the slope-deflection method from the equation

$$\frac{EJ_{01}}{l_{01}}F_7(\lambda_{01}) + \frac{EJ_{11'}}{l_{11'}}\left[F_2(\lambda_{11'}) - F_1(\lambda_{11'})\right] = 0 \qquad (3.56)$$

is

$$\omega_{(1)} = 21\cdot0\,\text{s}^{-1};$$

consequently, the error of the energy method is about 16%.

3.4.2 Computing the fundamental frequency from the dynamic deflection curves

The energy method yields more accurate results if the dynamic deflection curve appertaining to a harmonic load is used in place of the hitherto employed static deflection curve.

This, of course, can be done only on the condition that the frequency of the harmonic load is estimated close enough to the first natural frequency. For bars of constant cross-section and constant mass, the integral $\int \mu\,v^2(x)\,\text{d}x$ was evaluated in Section 2.8.2 for different types of end supports. What remains is to determine the integral in the denominator of eq. 3.40.

At the instant a bar is in its limit position (maximum deflection) the potential energy is equal to the work done by the amplitude of the loads on half of the displacements. The load consists partly of forces acting at the ends of the bar and partly of the product (negative) of mass and acceleration, $-\mu\omega^2 v(x)$. For bars clamped at both ends, the end moments and forces are defined by eqs 2.48 and 2.49; hence the double quantity of potential energy is given by

$$\int_0^l EJ\, v''^2(x)\, dx = \frac{EJ}{l}\left[F_2(\lambda)\,(\zeta_g^2 + \zeta_h^2) + 2F_1(\lambda)\,\zeta_g\zeta_h + 2F_3(\lambda)\frac{\zeta_h v_g - \zeta_g v_h}{l} \right.$$

$$\left. + 2F_4(\lambda)\frac{\zeta_h v_h - \zeta_g v_g}{l} + 2F_5(\lambda)\frac{v_g v_h}{l^2} + F_6(\lambda)\frac{v_g^2 + v_h^2}{l^2} \right]$$

$$+ \omega^2 \int_0^l \mu\, v^2(x)\, dx = \mathscr{A}_1 + \omega^2 \int_0^l \mu\, v^2(x)\, dx \qquad (3.57)$$

where $\mathscr{A}_1$ is shorthand notation of the first part of the expression.

For a bar hinged at end g, it is similarly found that

$$\int_0^l EJ\, v''^2(x)\, dx = \frac{EJ}{l}\left[F_7(\lambda)\,\zeta_h^2 + 2F_8(\lambda)\frac{\zeta_h v_g}{l} + 2F_9(\lambda)\frac{\zeta_h v_h}{l} + 2F_{10}(\lambda)\frac{v_g v_h}{l^2} \right.$$

$$\left. + F_{11}(\lambda)\frac{v_h^2}{l^2} + F_{12}(\lambda)\frac{v_g^2}{l^2} \right] + \omega_0^2 \int_0^l \mu\, v^2(x)\, dx$$

$$= \mathscr{A}_2 + \omega^2 \int_0^l \mu\, v^2(x)\, dx \qquad (3.58)$$

For a bar hinged at both ends

$$\int_0^l EJ\, v''^2(x)\, dx = \frac{EJ}{l^3}\left[F_{14}(\lambda)\,(v_g^2 + v_h^2) + 2F_{13}(\lambda)\,v_g v_h \right] + \omega^2 \int_0^l \mu\, v^2(x)\, dx$$

$$= \mathscr{A}_3 + \omega^2 \int_0^l \mu\, v^2(x)\, dx \qquad (3.59)$$

and finally, for a cantilever beam

$$\int_0^l EJ\, v''^2(x)\, dx = \frac{EJ}{l}\left[F_{15}(\lambda)\,\zeta_h^2 + 2F_{16}(\lambda)\frac{\zeta_h v_h}{l} + F_{17}(\lambda)\frac{v_h^2}{l^2} \right] + \omega^2 \int_0^l \mu\, v^2(x)\, dx$$

$$= \mathscr{A}_4 + \omega^2 \int_0^l \mu\, v^2(x)\, dx \qquad (3.60)$$

The natural circular frequency is obtained from eq. 3.40 with the substitution of the pertinent of expressions 3.57−3.60:

$$\omega_{(1)}^2 \approx \omega^2 + \frac{\sum \mathscr{A}_k}{\sum \int_0^l \mu\, v^2(x)\, \mathrm{d}x}\,,\quad (k = 1, 2, 3, 4) \tag{3.61}$$

where the summation indicated by $\sum$ includes all the bars of the structure, and the subscript of $\mathscr{A}$ depends on the type of support of the bar in question.

Example 3.6

Consider again the continuous beam of Example 3.5. Starting out from the approximate value $\omega_{(1)} = 24 \cdot 4 \text{ s}^{-1}$ we get

$$\lambda_{01} = 70 \left(\frac{0 \cdot 477 \times 24 \cdot 4^2}{21 \times 10^6 \times 3 \cdot 6} \right)^{1/4} = 3 \cdot 08\,, \quad \lambda_{12} = 90 \left(\frac{0 \cdot 573 \times 24 \cdot 4^2}{21 \times 10^6 \times 6 \cdot 0} \right)^{1/4} = 3 \cdot 65$$

and by eqs 2.94 and 2.96

$$\int_0^l \mu\, v^2(x)\, \mathrm{d}x = 47\,700 \text{ Mpm s}^{-2}$$

By eqs 3.57 and 3.58

$$\omega_{(1)}^2 \approx 24 \cdot 4^2 + \frac{\dfrac{2 \times 21 \times 10^6 \times 3 \cdot 6}{70}\, 0 \cdot 358 + 2\, \dfrac{21 \times 10^6 \times 6}{90}(1 \cdot 52 - 4 \cdot 04)}{47\,700} = 464 \text{ s}^{-2}$$

$$\omega_{(1)} \approx 21 \cdot 5 \text{ s}^{-1}$$

which differs by about 2% from the exact value $\omega_{(1)} = 21 \cdot 0 \text{ s}^{-1}$. Even though the computation could be repeated at will (always starting out from the approximate value of $\omega_{(1)}$ last obtained) to get the result to the desired degree of accuracy, the simple slope-deflection method as described in Sections 3.2 and 3.3 is far more straightforward.

3.5   Modal analysis

The principles of this method were explained in Section 1.2.2. If our considerations are restricted to transverse motion, the equation applying to systems composed of bars with constant cross-section executing forced vibration, is

$$\frac{\partial^2 v(x, t)}{\partial t^2} + \frac{EJ}{\mu} \frac{\partial^4 v(x, t)}{\partial x^4} = \frac{p(x, t)}{\mu} \tag{3.62}$$

where $p(x, t)$ is the continuous load varying with respect to both x and t.

Expand the load and the deflections in series

$$v(x, t) = \sum_{j=1}^{\infty} q_{(j)}(t)\, v_{(j)}(x) \tag{3.63}$$

$$\frac{p(x, t)}{\mu} = \sum_{j=1}^{\infty} p_{(j)}(t)\, v_{(j)}(x) \tag{3.64}$$

Coefficients $p_{(j)}(t)$ are computed as shown in Section 1.2.2, formula 1.61: Equation 3.64 is multiplied by ordinate $v_{(i)}(x)$ of the i-th natural mode and by the mass per unit length μ and integrated over all of the bars of the system. Since the natural modes are orthogonal,

$$\sum \int_0^l \mu\, v_{(j)}(x)\, v_{(i)}(x)\, dx = 0 \quad \text{for} \quad i \neq j \tag{3.65}$$

where the summation includes all of the bars of the system. From eq. 3.64

$$p_{(j)}(t) = \frac{\displaystyle\sum \int_0^l p(x, t)\, v_{(j)}(x)\, dx}{\displaystyle\sum \int_0^l \mu\, v_{(j)}^2(x)\, dx} \tag{3.66}$$

Substitution of eqs 3.63 and 3.64 in eq. 3.62 gives

$$\sum_j \frac{d^2 q_{(j)}(t)}{dt^2}\, v_{(j)}(x) + \sum_j \frac{EJ}{\mu} \frac{d^4 v_{(j)}(x)}{dx^4}\, q_{(j)}(t) = \sum_j p_{(j)}(t)\, v_{(j)}(x) \tag{3.67}$$

Since by eq. 2.10

$$\frac{EJ}{\mu} \frac{d^4 v_{(j)}(x)}{dx^4} = \omega_{(j)}^2\, v_{(j)}(x) \tag{3.68}$$

eq. 3.67 changes to the equation

$$\sum_j \left[\frac{d^2 q_{(j)}(t)}{dt^2} + \omega_{(j)}^2\, q_{(j)}(t) - p_{(j)}(t) \right] v_{(j)}(x) = 0 \tag{3.69}$$

which can be satisfied for all xs only on the condition that

$$\frac{d^2 q_{(j)}(t)}{dt^2} + \omega_{(j)}^2\, q_{(j)}(t) = p_{(j)}(t) \tag{3.70}$$

which we recall is the same as eq. 1.64.

If the load is harmonic, $p_{(j)} \sin \omega t$, then eq. 1.67 can be applied analogously to give

$$q_{(j)}(t) = q_{(j)} \sin \omega t + a_{(j)} \cos \omega_{(j)} t + b_{(j)} \sin \omega_{(j)} t \qquad (3.71)$$

where

$$q_{(j)} = \frac{p_{(j)}}{\omega_{(j)}^2 - \omega^2} \qquad (3.72)$$

and $a_{(j)}$, $b_{(j)}$ are the integration constants which depend on the initial conditions of motion.

N o t e: With the normalised natural modes formula 3.66 simplifies to

$$p_{(j)}(t) = \sum \int_0^l p(x, t)\, V_{(j)}(x)\, \mathrm{d}x \qquad (3.73)$$

where — analogously to eq. 1.17

$$V_{(j)}(x) = \frac{v_{(j)}(x)}{\left(\sum \int_0^l \mu v_{(j)}^2(x)\, \mathrm{d}x\right)^{1/2}} \qquad (3.74)$$

represents the normalised j-th natural mode. The integral in the denominator is computed as shown in Section 2.8.2. However, more often than not it is better not to normalise the natural modes.

Section 2.8.2 has indicated the way of evaluating the denominator in formula 3.66. What remains is to evaluate the integral in the numerator for some typical loads. At a uniform load $p(x, t) = p(t)$ the deflection is given by eq. 2.25. For a bar clamped at both ends the constants are given by formula 2.47 and

$$\int_0^l p(t)\, v(x)\, \mathrm{d}x = p(t) \int_0^l \left[A \cos \frac{\lambda x}{l} + B \sin \frac{\lambda x}{l} + C \cosh \frac{\lambda x}{l} + D \sinh \frac{\lambda x}{l} \right] \mathrm{d}x$$

$$= \left\{ (\zeta_g - \zeta_h) \frac{l^2}{\lambda^4} \left[F_3(\lambda) + F_4(\lambda) \right] \right.$$

$$\left. - (v_g + v_h) \frac{l}{\lambda^4} \left[F_5(\lambda) + F_6(\lambda) \right] \right\} p(t) \qquad (3.75)$$

For a bar hinged at end g

$$\int_0^l p(t)\, v(x)\, \mathrm{d}x = - \left\{ \zeta_h \frac{l^2}{\lambda^4} \left[F_8(\lambda) + F_9(\lambda) \right] + v_g \frac{l}{\lambda^4} \left[F_{10}(\lambda) + F_{12}(\lambda) \right] \right.$$

$$\left. + v_h \frac{l}{\lambda^4} \left[F_{10}(\lambda) + F_{11}(\lambda) \right] \right\} p(t) \qquad (3.76)$$

and for a bar hinged at both ends

$$\int_0^l p(t)\, v(x)\, \mathrm{d}x = -(v_g + v_h)\frac{l}{\lambda^4}\left[F_{13}(\lambda) + F_{14}(\lambda)\right] p(t) \qquad (3.77)$$

Finally, for a uniformly loaded cantilever bar

$$\int_0^l p(t)\, v(x)\, \mathrm{d}x = -\left[\zeta_h \frac{l^2}{\lambda^4} F_{16}(\lambda) + v_h \frac{l}{\lambda^4} F_{17}(\lambda)\right] p(t) \qquad (3.78)$$

All we need to know to evaluate expressions 2.94−2.101 and 3.75−3.78 and by their help find $p_{(j)}(t)$ (eq. 3.66) and $q_{(j)}$ (eq. 3.72) are the natural frequencies, and the deformations of the joints in free vibration. If we know the modes of free vibration, $v(x, t)$ is obtained from eq. 3.63.

If a concentrated load $P(t)$ is applied to section $x = a$, the integral in the numerator of eq. 3.66 is computed as

$$\lim \int_a^{a+\Delta a} \frac{P(t)}{\Delta a} v(x)\, \mathrm{d}x = P(t)\, v(a) \qquad (3.79)$$

$$\Delta a \to 0.$$

Example 3.7

Consider again the multi-storey frame shown in Fig. 3.8 to which the slope-deflection method was applied in Example 3.1. Let us now examine its symmetric forced vibration using the method of modal analysis. From Example 3.1 we know the first three frequencies and modes of free vibration

$$\omega_{(1)} = 103\cdot703\ \mathrm{s}^{-1}, \quad \zeta_{1(1)} = \quad 0\cdot584, \quad \zeta_{2(1)} = -0\cdot962, \quad \zeta_{3(1)} = 1$$

$$v_{p(1)} = 1\cdot525\ \mathrm{m}$$

$$\omega_{(2)} = 126\cdot246\ \mathrm{s}^{-1}, \quad \zeta_{1(2)} = -1\cdot013, \quad \zeta_{2(2)} = \quad 0\cdot113, \quad \zeta_{3(2)} = 1$$

$$v_{p(2)} = 1\cdot700\ \mathrm{m}$$

$$\omega_{(3)} = 155\cdot437\ \mathrm{s}^{-1}, \quad \zeta_{1(3)} = \quad 0\cdot776, \quad \zeta_{2(3)} = \quad 1\cdot174, \quad \zeta_{3(3)} = 1$$

$$v_{p(3)} = 2\cdot067\ \mathrm{m}$$

By eq. 2.94 for the first mode of free vibration (at $\lambda_1 = 3\cdot410\,61$, $\lambda_2 = 3\cdot005\,24$)

$$\sum \int_0^l \mu\, v_{(1)}^2(x)\, \mathrm{d}x = \frac{4\mu_1 l_1^3}{\lambda_1^4}\left[(\zeta_{1(1)}^2 + \zeta_{2(1)}^2 + \tfrac{1}{2}\zeta_{3(1)}^2)\, \Phi_2(\lambda_1) + (\zeta_{1(1)}\zeta_{2(1)}\right.$$

$$+ \zeta_{2(1)}\zeta_{3(1)}) \, \Phi_1(\lambda_1)] + \frac{2\mu_2 l_2^3}{\lambda_2^4} (\zeta_{1(1)}^2 + \zeta_{2(1)}^2 + \zeta_{3(1)}^2)$$

$$\times [\Phi_2(\lambda_2) - \Phi_1(\lambda_2)] = \frac{4 \times 0{\cdot}0204 \times 6^3}{3{\cdot}410\ 61^4}$$

$$\times [(0{\cdot}584^2 + 0{\cdot}962^2 + 0{\cdot}5)\, 2{\cdot}248$$

$$+ (-0{\cdot}584 \times 0{\cdot}962 - 0{\cdot}962 \times 1)\,(-1{\cdot}904)]$$

$$+ \frac{2 \times 0{\cdot}0510 \times 5^3}{3{\cdot}005\ 24^4} (0{\cdot}584^2 + 0{\cdot}962^2 + 1^2)\, 1{\cdot}916 = 1{\cdot}573 \text{ Mpms}^2$$

If a vertical harmonic force with amplitude $P = 1$ Mp and circular frequency $\omega = 62{\cdot}8 \text{ s}^{-1}$ acts at the mid-span of bar $33'$, then by eq. 3.79

$$\sum \int_0^l p(x, t)\, v_{(1)}(x)\, \mathrm{d}x = P \sin \omega t \, v_{p(1)} = 1{\cdot}525 \sin \omega t \text{ Mpm}$$

$$p_{(1)} = \frac{1{\cdot}525}{1{\cdot}573} = 0{\cdot}969 \text{ s}^{-2}$$

If only steady-state vibration is considered, amplitude $q_{(1)}$ is given by eq. 3.72

$$q_{(1)} = \frac{0{\cdot}969}{(103{\cdot}7^2 - 62{\cdot}8^2)} = 1{\cdot}424 \times 10^{-4}$$

Similarly, for the second natural mode $(\lambda_1 = 3{\cdot}763\ 10, \; \lambda_2 = 3{\cdot}315\ 83)$

$$\int_0^l \mu \, v_{(2)}^2(x)\, \mathrm{d}x = \frac{4 \times 0{\cdot}0204 \times 6^3}{3{\cdot}763\ 10^4} [(1{\cdot}013^2 + 0{\cdot}113^2 + 0{\cdot}5)\, 4{\cdot}831$$

$$+ (-1{\cdot}013 \times 0{\cdot}113 + 0{\cdot}113 \times 1)\,(-4{\cdot}268)]$$

$$+ \frac{2 \times 0{\cdot}0510 \times 5^3}{3{\cdot}315\ 83^4} (1{\cdot}013^2 + 0{\cdot}113^2 + 1^2)\, 3{\cdot}442 = 1{\cdot}388 \text{ Mpms}^2$$

$$\int_0^l p(x, t)\, v_{(2)}(x)\, \mathrm{d}x = P \sin \omega t \, v_{p(2)} = 1{\cdot}700 \sin \omega t \text{ Mpm}$$

$$p_{(2)} = \frac{1{\cdot}700}{1{\cdot}388} = 1{\cdot}225 \text{ s}^{-2}$$

$$q_{(2)} = \frac{1{\cdot}225}{(126{\cdot}2^2 - 62{\cdot}8^2)} = 1{\cdot}021 \times 10^{-4}$$

And finally, for the third natural mode ($\lambda_1 = 4\cdot17555$, $\lambda_2 = 3\cdot67926$)

$$\int_0^l \mu \, v_{(3)}^2(x) \, dx = \frac{4 \times 0\cdot0204 \times 6^3}{4\cdot175\,55^4} \left[(0\cdot776^2 + 1\cdot174^2 + 0\cdot5) \, 16\cdot290 \right.$$

$$+ (0\cdot776 \times 1\cdot174 + 1\cdot174 \times 1)\,(-15\cdot450)]$$

$$+ \frac{2 \times 0\cdot0510 \times 5^3}{3\cdot679\,26^4} (0\cdot776^2 + 1\cdot174^2 + 1^2)\, 7\cdot417 = 2\cdot011 \text{ Mpms}^2$$

$$\int_0^l p(x, t)\, v_{(3)}(x) \, dx = P \sin \omega t \, v_{p(3)} = 2\cdot067 \sin \omega t \text{ Mpm}$$

$$P_{(3)} = \frac{2\cdot067}{2\cdot011} = 1\cdot028 \text{ s}^{-2}$$

$$q_{(3)} = \frac{1\cdot028}{(155\cdot4^2 - 62\cdot8^2)} = 0\cdot509 \times 10^{-4}$$

By eq. 3.63 the resultant amplitudes of rotation of joints 1, 2, 3 are

$$\zeta_1 \approx \sum_{j=1}^{3} q_{(j)}\zeta_{1(j)} = (1\cdot424 \times 0\cdot584 - 1\cdot021 \times 1\cdot013 + 0\cdot509 \times 0\cdot776) \times 10^{-4}$$

$$= 0\cdot192 \times 10^{-4}$$

$$\zeta_2 \approx \sum_{j=1}^{3} q_{(j)}\zeta_{2(j)} = (-1\cdot424 \times 0\cdot962 + 1\cdot021 \times 0\cdot113 + 0\cdot509 \times 1\cdot174) \times 10^{-4}$$

$$= -0\cdot657 \times 10^{-4}$$

$$\zeta_3 \approx \sum_{j=1}^{3} q_{(j)}\zeta_{3(j)} = (1\cdot424 + 1\cdot021 + 0\cdot509) \times 10^{-4} = 2\cdot954 \times 10^{-4}$$

and the amplitude of deflection at the midspan of bar 33′ is given by

$$v_p \approx \sum_{j=1}^{3} q_{(j)}v_{p(j)} = (1\cdot424 \times 1\cdot525 + 1\cdot021 \times 1\cdot700 + 0\cdot509 \times 2\cdot067) \times 10^{-4}$$

$$= 4\cdot96 \times 10^{-4} \text{ m}$$

The above results (arrived at through the use of the first three terms of the series) are in fairly good agreement with the exact results of the slope-deflection method. The only significant difference is at the point of application v_p of the load. The method of modal analysis is best suited to cases involving the effect of a force at various exciting frequencies, e.g. the computation of resonance curves, etc. It is not so well suited for problems similar to the one just discussed — load at a given exciting frequency; those are solved more readily and less laboriously by the slope-deflection method.

3.6   Three-dimensional frame systems

In a three-dimensional frame system every joint is capable of sustaining six deformations, i.e. displacements and rotations in the direction of three coordinate axes. The unknowns can be found from the six slope-deflection equations that — ana-

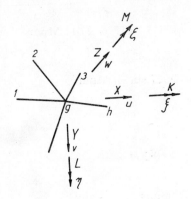

Fig. 3.19 Forces, moments, displacements and rotations at a joint of a three-dimensional frame. (Double-point arrows denote rotation and moment about axis Z (X, Y), i.e. rotation and moment in plane XY (YZ, ZX)).

logous to eq. 3.9 — express the conditions of equilibrium at the joint. It holds for the joint shown in Fig. 3.19 that

$$\sum_h K_{gh} = \mathcal{K}_g$$

$$\sum_h L_{gh} = \mathcal{L}_g$$

$$\sum_h M_{gh} = \mathcal{M}_g$$

$$\sum_h X_{gh} = \mathcal{X}_g$$

$$\sum_h Y_{gh} = \mathcal{Y}_g$$

$$\sum_h Z_{gh} = \mathcal{Z}_g$$

$$(3.80)$$

where K_{gh} is the amplitude of the moment about axis X acting at end g of bar gh, $\mathcal{K}_g$ the amplitude of the moment of external load about axis X; and similarly, $\mathcal{L}_g$ the amplitude of the moment of external load about axis Y.

The slope-deflection equations may also be obtained directly, by the application of the reciprocal theorem to the actual state of deformation and the auxiliary unit states described by the relations

$$\xi_g = 1 , \quad \eta_g = 0 , \quad \zeta_g = 0 , \quad u_g = 0 , \quad v_g = 0 , \quad w_g = 0 , \quad \text{etc.}$$

The notation used is as indicated in Table 3.12:

Table 3.12

Axis	X	Y	Z
Rotation about the respective axis	ξ	η	ζ
Moment	K	L	M
Moment of external load	$\mathcal{K}$	$\mathcal{L}$	$\mathcal{M}$
Displacement	u	v	w
Force	X	Y	Z
External force	$\mathcal{X}$	$\mathcal{Y}$	$\mathcal{Z}$

The procedure will be illustrated in the next example.

Example 3.8

The columns of the three-dimensional system shown in Fig. 3.20 have elastically restrained lower ends requiring the application of forces C_x, C_y, C_z to achieve a compression of unit length in the direction of axes X, Y, Z and moments C_ξ, C_η and C_ζ

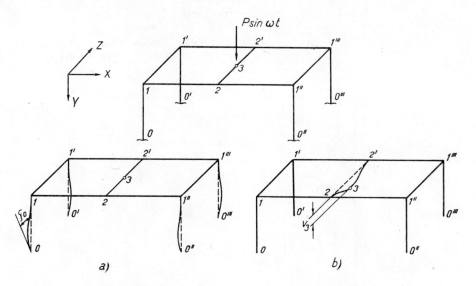

Fig. 3.20 Three-dimensional frame. a, b — auxiliary unit states of deformation

for the rotation by a unit angle about axes X, Y, Z. (In a more general case, the relation between displacements and constants C may be more general, too — cf. Section 1.3.) The bars of the system are of constant cross-section and mass per unit length; and only at point 3 is there also a concentrated mass m and a harmonic force acting in the vertical direction. The longitudinal vibration is neglected for the horizontal bars but considered for the columns. On the strength of this simplification and the symmetry of the system, the number of equations reduces to 11 (for an asymmetric system with 11 joints the number would be $11 \times 6 = 66$ equations). The equations are set out in Table 3.13. Two of the auxiliary unit states are shown in Figs 3.20a and b.

Table 3.13

ξ_0	ζ_0	u_0	v_0	w_0	ξ_1	ζ_1	v_1	ξ_2	v_2	v_3	
a_{11}				a_{15}	a_{16}						$=0$
	a_{22}	a_{23}				a_{27}					$=0$
	a_{23}	a_{33}				a_{37}					$=0$
			a_{44}				a_{48}				$=0$
a_{15}				a_{55}	a_{56}						$=0$
a_{16}				a_{56}	a_{66}		a_{68}	a_{69}			$=0$
	a_{27}	a_{37}				a_{77}	a_{78}		$a_{7,10}$		$=0$
			a_{48}		a_{68}	a_{78}	a_{88}		$a_{8,10}$		$=0$
					a_{69}			a_{99}	$a_{9,10}$	$a_{9,11}$	$=0$
						$a_{7,10}$	$a_{8,10}$	$a_{9,10}$	$a_{10,10}$	$a_{10,11}$	$=0$
								$a_{9,11}$	$a_{10,11}$	$a_{11,11}$	$=P/2$

$$a_{11} = C_\xi + \left[\frac{EJ_x}{l} F_2(\lambda_x)\right]_{01}$$

$$a_{15} = \left[\frac{EJ_x}{l^2} F_4(\lambda_x)\right]_{01}$$

$$a_{16} = \left[\frac{EJ_x}{l} F_1(\lambda_x)\right]_{01}$$

$$a_{22} = C_\zeta + \left[\frac{EJ_z}{l} F_2(\lambda_z)\right]_{01}$$

$$a_{23} = -\left[\frac{EJ_z}{l^2} F_4(\lambda_z)\right]_{01}$$

$$a_{27} = \left[\frac{EJ_z}{l} F_1(\lambda_z)\right]_{01}$$

$$a_{33} = C_x + \left[\frac{EJ_z}{l^3} F_6(\lambda_z)\right]_{01}$$

$$a_{37} = \left[\frac{EJ_z}{l^2} F_3(\lambda_z)\right]_{01}$$

$$a_{44} = C_y + \left[\frac{ES}{l} \psi \cot \psi\right]_{01}$$

$$a_{48} = -\left[\frac{ES}{l} \psi \operatorname{cosec} \psi\right]_{01}$$

$$a_{55} = C_z + \left[\frac{EJ_x}{l^3} F_6(\lambda_x)\right]_{01}$$

$$a_{56} = -\left[\frac{EJ_x}{l^2} F_3(\lambda_x)\right]_{01}$$

$$a_{66} = \left[\frac{EJ_x}{l} F_2(\lambda_x)\right]_{01} + \left[\frac{EJ_x}{l}(F_2(\lambda_x) - F_1(\lambda_x))\right]_{11'} + \left[\frac{GJ_x}{l} \vartheta \cot \vartheta\right]_{12}$$

$$a_{68} = \left[\frac{EJ_x}{l^2}(F_3(\lambda_x) + F_4(\lambda_x))\right]_{11'}$$

$$a_{69} = -\left[\frac{GJ_x}{l} \vartheta \operatorname{cosec} \vartheta\right]_{12}$$

$$a_{77} = \left[\frac{EJ_z}{l} F_2(\lambda_z)\right]_{01} + \left[\frac{EJ_z}{l} F_2(\lambda_z)\right]_{12} + \left[\frac{GJ_z}{l} \vartheta(\cot \vartheta - \operatorname{cosec} \vartheta)\right]_{11'}$$

$$a_{78} = -\left[\frac{EJ_z}{l^2} F_4(\lambda_z)\right]_{12}$$

$$a_{7,10} = -\left[\frac{EJ_z}{l^2} F_3(\lambda_z)\right]_{12}$$

$$a_{88} = \left[\frac{ES}{l} \psi \cot \psi\right]_{01} + \left[\frac{EJ_x}{l^3}[F_5(\lambda_x) + F_6(\lambda_x)]\right]_{11'} + \left[\frac{EJ_z}{l^3} F_6(\lambda_z)\right]_{12}$$

$$a_{8,10} = \left[\frac{EJ_z}{l^3} F_5(\lambda_z)\right]_{12}$$

$$a_{99} = \left[\frac{GJ_x}{l} \vartheta \cot \vartheta\right]_{12} + \frac{1}{2}\left[\frac{EJ_x}{l} F_2(\lambda_x)\right]_{23}$$

$$a_{9,10} = \frac{1}{2}\left[\frac{EJ_x}{l^2} F_4(\lambda_x)\right]_{23}$$

$$a_{9,11} = \frac{1}{2}\left[\frac{EJ_x}{l^3} F_3(\lambda_x)\right]_{23}$$

$$a_{10,10} = \left[\frac{EJ_z}{l^2}F_6(\lambda_z)\right]_{12} + \frac{1}{2}\left[\frac{EJ_x}{l^3}F_6(\lambda_x)\right]_{23}$$

$$a_{10,11} = \frac{1}{2}\left[\frac{EJ_x}{l^3}F_5(\lambda_x)\right]_{23}$$

$$a_{11,11} = \frac{1}{2}\left[\frac{EJ_x}{l^3}F_6(\lambda_x)\right]_{23} - \frac{m\omega^2}{4}$$

Coefficients a were computed with the aid of Tables 10.1, 10.2 and 10.10 in the Appendix, and formulae 2.69, 2.70 and 2.76. The subscripts of the brackets denote the respective bars. In cases when the bars are of deep cross-section, it is necessary to also consider the secondary effects (to be discussed in Chapter 8) and replace functions $F(\lambda)$ by functions $F(a, b)$. If, on the other hand, the cross-sections are small, one can neglect the inertia forces arising in torsional vibration about the longitudinal axis of the bars. Then $\vartheta \cot \vartheta = \vartheta \operatorname{cosec} \vartheta \approx 1$.

4 SYSTEMS WITH NON-PRISMATIC MEMBERS

4.1 Straight bars of varying cross-section

The equation of motion of a straight non-prismatic bar is obtained in the same way as that for a prismatic bar (Chapter 2). Because the bending moment is described by expression 2.14 and the shear force by

$$T(x, t) = \frac{\partial M(x, t)}{\partial x} = -\frac{\partial}{\partial x}\left(EJ(x)\frac{\partial^2 v(x, t)}{\partial x^2}\right) \tag{4.1}$$

the equation of equilibrium of an element, not subjected to external forces, gives

$$\frac{\partial^2}{\partial x^2}\left[EJ(x)\frac{\partial^2 v(x, t)}{\partial x^2}\right] + \mu(x)\frac{\partial^2 v(x, t)}{\partial t^2} = 0 \tag{4.2}$$

For harmonic vibration

$$v(x, t) = v(x) \sin \omega t$$

eq. 4.2 takes on the form

$$\frac{d^2}{dx^2}\left[EJ(x)\frac{d^2 v(x)}{dx^2}\right] - \mu(x) \omega^2 v(x) = 0 \tag{4.3}$$

Eq. 4.3 may be solved by direct integration only in the case where $\mu(x)$ and $J(x)$ are expressed by suitable functions (cf. Section 4.1.1 below). Otherwise one must resort to one of the approximate or iterative methods, e.g. the energy, Galerkin's, or Ritz's methods, etc. It will be shown in Section 4.1.2 that the slope-deflection method leads to accurate results when applied to bars with a step-wise varying cross-section, and to approximate results when applied to bars with continuously varying cross-sections. A simplified slope-deflection method will be outlined in Section 4.1.3. The method of discrete mass points explained in Chapter 1 can of course also be used for the analysis of systems with non-prismatic bars.

4.1.1 Direct solution*

4.1.1.1 Equation of motion and its solution

As mentioned above a direct solution of eq. 4.2 is possible only if the mass per unit length, $\mu(x)$, and the cross-sectional moment of inertia $J(x)$ are expressed by some suitable functions, such as[48]

$$\mu(x) = \mu_g \left(1 + c\,\frac{x}{l}\right)^n \tag{4.4}$$

$$J(x) = J_g \left(1 + c\,\frac{x}{l}\right)^{n+2} \tag{4.5}$$

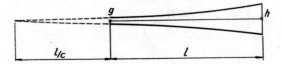

Fig. 4.1 Bar of varying cross-section

where constants c and n are computed from the relations

$$c = \left(\frac{\mu_g}{\mu_h}\frac{J_h}{J_g}\right)^{1/2} - 1$$

$$\frac{n}{2} = \frac{\log \mu_h - \log \mu_g}{\log J_h - \log J_g - \log \mu_h + \log \mu_g} \tag{4.6}$$

and J_g, J_h, μ_g, μ_h are respectively the moments of inertia and the masses per unit length at ends g and h.

Note: For bars of specified dimensions, the above formulae approximately describe the actual conditions. It is found that the actual variation of cross-section is approximated far better by the somewhat more general formulae

$$\mu(x) = a \left(1 + c\,\frac{x}{l}\right)^n$$

$$J(x) = b \left(1 + c\,\frac{x}{l}\right)^{n+2} \tag{4.7}$$

---

* The solution presented in this section has been worked out by Z. Němeček and P. Řeřicha, members of the author's staff.

where constants a, b, c and number n should be chosen to fit the case being considered.

Eq. 4.3 may be given the form

$$J(x) v''''(x) + 2J'(x) v'''(x) + J''(x) v''(x) - \frac{\mu(x)}{E} \omega^2 v(x) = 0 \qquad (4.8)$$

where

$$v''(x) = \frac{d^2 v(x)}{dx^2}, \quad \text{etc.}$$

Set

$$\xi = 1 + c \frac{x}{l}$$

$$(4.9)$$

Substitution of eqs 4.4, 4.5 and 4.9 in eq. 4.8 gives

$$\xi^{n+2} \frac{d^4 v(\xi)}{d\xi^4} + 2(n + 2) \xi^{n+1} \frac{d^3 v(\xi)}{d\xi^3} + (n + 2)(n + 1) \xi^n \frac{d^2 v(\xi)}{d\xi^2}$$

$$- \frac{\mu_g \omega^2}{EJ_g} \frac{l^4}{c^4} \xi^n v(\xi) = 0$$

or after rearrangement

$$\xi^2 \frac{d^4 v(\xi)}{d\xi^4} + 2(n + 2) \xi \frac{d^3 v(\xi)}{d\xi^3} + (n + 2)(n + 1) \frac{d^2 v(\xi)}{d\xi^2} - \frac{\lambda_g^4}{c^4} v(\xi) = 0 \qquad (4.10)$$

where

$$\lambda_g = l \left(\frac{\mu_g \omega^2}{EJ_g} \right)^{1/4} \qquad (4.11)$$

Writing simply v instead of $v(\xi)$ and introducing new variables β and γ according to the relations

$$\beta = \frac{2\lambda_g}{c} \xi^{1/2} \qquad (4.12)$$

and

$$v = \frac{\gamma}{\beta^n} \qquad (4.13)$$

in place of variables v and ξ, we get (if $c \neq 0$)

$$\frac{d\beta}{d\xi} = \frac{2\lambda_g^2}{c^2 \beta}$$

$$(4.14)$$

$$\frac{d^2\beta}{d\xi^2} = -\frac{4\lambda_g^4}{c^4\beta^3}$$

$$\frac{d^3\beta}{d\xi^3} = \frac{24\lambda_g^6}{c^6\beta^5}$$

$$\frac{d^4\beta}{d\xi^4} = -\frac{240\lambda_g^8}{c^8\beta^7}$$

$$(4.14)$$

and

$$\frac{dv}{d\xi} = \frac{d\beta}{d\xi}\frac{dv}{d\beta} = \frac{2\lambda_g^2}{c^2\beta}\frac{dv}{d\beta}$$

$$\frac{d^2v}{d\xi^2} = \frac{4\lambda_g^4}{c^4\beta^2}\left(\frac{d^2v}{d\beta^2} - \frac{1}{\beta}\frac{dv}{d\beta}\right)$$

$$\frac{d^3v}{d\xi^3} = \frac{8\lambda_g^6}{c^6\beta_8^3}\left(\frac{d^3v}{d\beta^3} - \frac{3}{\beta}\frac{d^2v}{d\beta^2} + \frac{3}{\beta^2}\frac{dv}{d\beta}\right)$$

$$\frac{d^4v}{d\xi^4} = \frac{16\lambda_g^8}{c^8\beta^4}\left(\frac{d^4v}{d\beta^4} - \frac{6}{\beta}\frac{d^3v}{d\beta^3} + \frac{15}{\beta^2}\frac{d^2v}{d\beta^2} - \frac{15}{\beta^3}\frac{dv}{d\beta}\right)$$

$$(4.15)$$

and on differentiating with respect to β

$$\frac{dv}{d\beta} = \frac{1}{\beta^n}\frac{d\gamma}{d\beta} - \frac{n}{\beta^{n+1}}\gamma$$

$$\frac{d^2v}{d\beta^2} = \frac{1}{\beta^n}\frac{d^2\gamma}{d\beta^2} - \frac{2n}{\beta^{n+1}}\frac{d\gamma}{d\beta} + \frac{n(n+1)}{\beta^{n+2}}\gamma$$

$$\frac{d^3v}{d\beta^3} = \frac{1}{\beta^n}\frac{d^3\gamma}{d\beta^3} - \frac{3n}{\beta^{n+1}}\frac{d^2\gamma}{d\beta^2} + \frac{3n(n+1)}{\beta^{n+2}}\frac{d\gamma}{d\beta} + \frac{n(n+1)(n+2)}{\beta^{n+3}}\gamma$$

$$\frac{d^4v}{d\beta^4} = \frac{1}{\beta^n}\frac{d^4\gamma}{d\beta^4} - \frac{4n}{\beta^{n+1}}\frac{d^3\gamma}{d\beta^3} + \frac{6n(n+1)}{\beta^{n+2}}\frac{d^2\gamma}{d\beta^2} - \frac{4n(n+1)(n+2)}{\beta^{n+3}}\frac{d\gamma}{d\beta}$$

$$+ \frac{n(n+1)(n+2)(n+3)}{\beta^{n+4}}\gamma$$

$$(4.16)$$

Substitution of eqs 4.12–4.16 in eq. 4.10 and rearrangement gives

$$\frac{d^4\gamma}{d\beta^4} + \frac{2}{\beta}\frac{d^3\gamma}{d\beta^3} - \frac{2n^2+1}{\beta^2}\frac{d^2\gamma}{d\beta^2} + \frac{2n^2+1}{\beta^3}\frac{d\gamma}{d\beta} + \left[\frac{n^2(n^2-4)}{\beta^4} - 1\right]\gamma = 0 \quad (4.17)$$

The last equation can also be written as

$$\left(\frac{d^2}{d\beta^2} + \frac{1}{\beta}\frac{d}{d\beta} - \frac{n^2}{\beta^2} + 1\right)\left(\frac{d^2}{d\beta^2} + \frac{1}{\beta}\frac{d}{d\beta} - \frac{n^2}{\beta^2} - 1\right)\gamma = 0 \qquad (4.18)$$

and is satisfied if either

$$\frac{d^2\gamma}{d\beta^2} + \frac{1}{\beta}\frac{d\gamma}{d\beta} + \left(1 - \frac{n^2}{\beta^2}\right)\gamma = 0$$

or

$$\frac{d^2\gamma}{d\beta^2} + \frac{1}{\beta}\frac{d\gamma}{d\beta} - \left(1 + \frac{n^2}{\beta^2}\right)\gamma = 0$$

$$(4.19)$$

Eqs 4.19 are Bessel's equations whose solution may be expressed in terms of cylindrical functions. The general integral of eq. 4.18 is

$$\gamma = C_1 J_n(\beta) + C_2 N_n(\beta) + C_3 I_n(\beta) + C_4 K_n(\beta) \qquad (4.20)$$

where J_n, N_n, I_n, K_n are the cylindrical functions of the n-th order. By formula 4.13 the deflection $v = v(\beta)$ is

$$v(\beta) = \frac{1}{\beta^n}[C_1 J_n(\beta) + C_2 N_n(\beta) + C_3 I_n(\beta) + C_4 K_n(\beta)] \qquad (4.21)$$

Further, the amplitudes of the tangent rotation $\zeta(x)$, bending moment $M(x)$ and shear force $T(x)$ at section x are

$$\zeta(x) = \frac{dv(x)}{dx} \qquad (4.22)$$

$$M(x) = -EJ(x)\frac{d^2v(x)}{dx^2} \qquad (4.23)$$

$$T(x) = -\frac{d}{dx}\left[EJ(x)\frac{d^2v(x)}{dx^2}\right] \qquad (4.24)$$

By the familiar relations, the derivatives of the cylindrical functions are

$$\frac{d}{d\beta}\left[\frac{J_n(\beta)}{\beta^n}\right] = -\frac{J_{n+1}(\beta)}{\beta^n} \qquad \frac{d}{d\beta}\left[\frac{I_n(\beta)}{\beta^n}\right] = \frac{I_{n+1}(\beta)}{\beta^n}$$

$$\frac{d}{d\beta}\left[\frac{N_n(\beta)}{\beta^n}\right] = -\frac{N_{n+1}(\beta)}{\beta^n} \qquad \frac{d}{d\beta}\left[\frac{K_n(\beta)}{\beta^n}\right] = -\frac{K_{n+1}(\beta)}{\beta^n}$$

$$(4.25)$$

$$\frac{d^2}{d\beta^2}\left[\frac{J_n(\beta)}{\beta^n}\right] = \frac{J_{n+2}(\beta)}{\beta^n} - \frac{J_{n+1}(\beta)}{\beta^{n+1}} \qquad \frac{d^2}{d\beta^2}\left[\frac{I_n(\beta)}{\beta^n}\right] = \frac{I_{n+2}(\beta)}{\beta^n} + \frac{I_{n+1}(\beta)}{\beta^{n+1}}$$

$$\frac{d^2}{d\beta^2}\left[\frac{N_n(\beta)}{\beta^n}\right] = \frac{N_{n+2}(\beta)}{\beta^n} - \frac{N_{n+1}(\beta)}{\beta^{n+1}} \qquad \frac{d^2}{d\beta^2}\left[\frac{K_n(\beta)}{\beta^n}\right] = \frac{K_{n+2}(\beta)}{\beta^n} - \frac{K_{n+1}(\beta)}{\beta^{n+1}}$$

$$(4.26)$$

$$\frac{d^3}{d\beta^3}\left[\frac{J_n(\beta)}{\beta^n}\right] = -\frac{J_{n+3}(\beta)}{\beta^n} + 3\frac{J_{n+2}(\beta)}{\beta^{n+1}} \qquad \frac{d^3}{d\beta^3}\left[\frac{I_n(\beta)}{\beta^n}\right] = \frac{I_{n+3}(\beta)}{\beta^n} + 3\frac{I_{n+2}(\beta)}{\beta^{n+1}}$$

$$\frac{d^3}{d\beta^3}\left[\frac{N_n(\beta)}{\beta^n}\right] = -\frac{N_{n+3}(\beta)}{\beta^n} + 3\frac{N_{n+2}(\beta)}{\beta^{n+1}} \qquad \frac{d^3}{d\beta^3}\left[\frac{K_n(\beta)}{\beta^n}\right] = -\frac{K_{n+3}(\beta)}{\beta^n} + 3\frac{K_{n+2}(\beta)}{\beta^{n+1}}$$

$$(4.27)$$

Differentiation of eq. 4.21 and relations 4.25, 4.26 and 4.27 leads to

$$\frac{dv}{d\beta} = -\frac{1}{\beta^n}\left[C_1 J_{n+1}(\beta) + C_2 N_{n+1}(\beta) - C_3 I_{n+1}(\beta) + C_4 K_{n+1}(\beta)\right] \qquad (4.28)$$

$$\frac{d^2v}{d\beta^2} = \frac{1}{\beta^n}\left[C_1 J_{n+2}(\beta) + C_2 N_{n+2}(\beta) + C_3 I_{n+2}(\beta) + C_4 K_{n+2}(\beta)\right]$$

$$- \frac{1}{\beta^{n+1}}\left[C_1 J_{n+1}(\beta) + C_2 N_{n+1}(\beta) - C_3 I_{n+1}(\beta) + C_4 K_{n+1}(\beta)\right] \qquad (4.29)$$

$$\frac{d^3v}{d\beta^3} = -\frac{1}{\beta^n}\left[C_1 J_{n+3}(\beta) + C_2 N_{n+3}(\beta) - C_3 I_{n+3}(\beta) + C_4 K_{n+3}(\beta)\right.$$

$$\left. + \frac{3}{n+1} C_1 J_{n+2}(\beta) + C_2 N_{n+2}(\beta) + C_3 I_{n+2}(\beta) + C_4 K_{n+2}(\beta)\right] \qquad (4.30)$$

With regard to eqs 4.9, 4.15 and 4.28−4.30, formulae 4.22−4.24 appear as follows:

$$\zeta(\beta) = -\frac{2\lambda_g^2}{cl\beta^{n+1}}\left[C_1 J_{n+1}(\beta) + C_2 N_{n+1}(\beta) - C_3 I_{n+1}(\beta) + C_4 K_{n+1}(\beta)\right] \quad (4.31)$$

$$M(\beta) = -\frac{4EJ\lambda_g^4}{l^2 c^2 \beta^{n+2}}\left[C_1 J_{n+2}(\beta) + C_2 N_{n+2}(\beta) + C_3 I_{n+2}(\beta) + C_4 K_{n+2}(\beta)\right] \quad (4.32)$$

$$T(\beta) = -\frac{8EJ\lambda_g^6}{l^3 c^3 \beta^{n+3}}\left[C_1 J_{n+1}(\beta) + C_2 N_{n+1}(\beta) + C_3 I_{n+1}(\beta) - C_4 K_{n+1}(\beta)\right] \quad (4.33)$$

In the last of these expressions use was made of the recurrent relations

$$\frac{2n}{\beta} J_n(\beta) - J_{n+1}(\beta) = J_{n-1}(\beta)$$

$$\frac{2n}{\beta} N_n(\beta) - N_{n+1}(\beta) = N_{n-1}(\beta)$$

$$\frac{2n}{\beta} I_n(\beta) + I_{n+1}(\beta) = I_{n-1}(\beta)$$

$$\frac{2n}{\beta} K_n(\beta) - K_{n+1}(\beta) = -K_{n-1}(\beta)$$

$$(4.34)$$

To express variable β in terms of x, we refer to eqs 4.7, 4.9, 4.11 and 4.12 and write

$$\lambda_g^4 = l^4 \left(1 + c\,\frac{x}{l}\right)^2 \frac{\mu(x)\,\omega^2}{E\,J(x)} \tag{4.35}$$

$$\beta(x) = \frac{2l}{c}\left(1 + c\,\frac{x}{l}\right)\left(\frac{\mu(x)\,\omega^2}{E\,J(x)}\right)^{1/4} \tag{4.36}$$

and on substituting these in eqs 4.31−4.33

$$\zeta(x) = -\frac{\lambda(x)}{l\,\beta''(x)}\left[C_1 J_{n+1} + C_2 N_{n+1} - C_3 I_{n+1} + C_4 K_{n+1}\right] \tag{4.37}$$

$$M(x) = -\frac{E\,J(x)\,\lambda^2(x)}{l^2\,\beta''(x)}\left[C_1 J_{n+2} + C_2 N_{n+2} + C_3 I_{n+2} + C_4 K_{n+2}\right] \tag{4.38}$$

$$T(x) = -\frac{E\,J(x)\,\lambda^3(x)}{l^3\,\beta''(x)}\left[C_1 J_{n+1} + C_2 N_{n+1} + C_3 I_{n+1} - C_4 K_{n+1}\right] \tag{4.39}$$

where

$$\lambda(x) = l\left(\frac{\mu(x)\,\omega^2}{E\,J(x)}\right)^{1/4} \tag{4.40}$$

150

4.1.1.2 Beam clamped at both ends

(a)   Free vibration

Substituted in eqs 4.21 and 4.37 the boundary conditions

$$v(0) = v_g = 0, \quad v(l) = v_h = 0$$
$$v'(0) = \zeta_g = 0, \quad v'(l) = \zeta_h = 0$$

$$(4.41)$$

lead to the equations set out in Table 4.1:

Table 4.1

C_1	C_2	C_3	C_4	
$J_n(\beta_g)$	$N_n(\beta_g)$	$I_n(\beta_g)$	$K_n(\beta_g)$	$= 0$
$J_{n+1}(\beta_g)$	$N_{n+1}(\beta_g)$	$-I_{n+1}(\beta_g)$	$K_{n+1}(\beta_g)$	$= 0$
$J_n(\beta_h)$	$N_n(\beta_h)$	$I_n(\beta_h)$	$K_n(\beta_h)$	$= 0$
$J_{n+1}(\beta_h)$	$N_{n+1}(\beta_h)$	$-I_{n+1}(\beta_h)$	$K_{n+1}(\beta_h)$	$= 0$

In these equations β_g and β_h are given by

$$\beta_g = \frac{2\lambda_g}{c}$$

$$\beta_h = \frac{2\lambda_g}{c}(1 + c)^{1/2} = 2\lambda_h \frac{1 + c}{c} \qquad (4.42)$$

For the solution to be non-trivial, the determinant of the coefficients in Table 4.1 must be zero,

$$\text{i.e. } \Delta = 0 \qquad (4.43)$$

Solving this equation gives the values of λ_g and in turn, (eq. 4.11) the natural frequencies

$$\omega_{(j)} = \frac{\lambda_g^2}{l^2}\left(\frac{EJ_g}{\mu_g}\right)^{1/2} \qquad (4.44)$$

(b)   Forced vibration — Forces and moments acting at the ends of a beam

Consider the case of the right-hand end of the beam executing forced harmonic motion as shown in Fig. 4.2, and find the end moments and forces by using the procedure described in Chapter 2. The boundary conditions are

$$v(0) = v_g = 0, \quad v(l) = v_h \neq 0$$
$$v'(0) = \zeta_h = 0, \quad v'(l) = \zeta_h \neq 0 \tag{4.45}$$

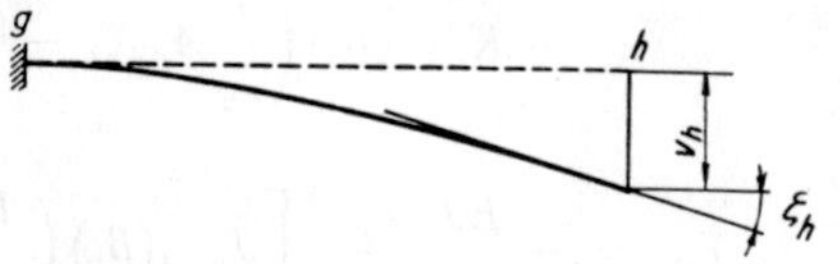

Fig. 4.2 Deformation of a bar clamped at both ends

On substituting in eqs 4.21 and 4.37 we get the pertinent equations as set out in Table 4.2.

Table 4.2

C_1	C_2	C_3	C_4	
$J_n(\beta_g)$	$N_n(\beta_g)$	$I_n(\beta_g)$	$K_n(\beta_g)$	$= 0$
$J_{n+1}(\beta_g)$	$N_{n+1}(\beta_g)$	$-I_{n+1}(\beta_g)$	$K_{n+1}(\beta_g)$	$= 0$
$J_n(\beta_h)$	$N_n(\beta_h)$	$I_n(\beta_h)$	$K_n(\beta_h)$	$= \beta_h^n v_h$
$J_{n+1}(\beta_h)$	$N_{n+1}(\beta_h)$	$-I_{n+1}(\beta_h)$	$K_{n+1}(\beta_h)$	$= -(l\beta_h^n/\lambda_h)\,\zeta_h$

Their solution yields the integration constants

$$C_1 = \frac{\beta_h^n}{\Delta}\left(-\frac{lA_{14}}{\lambda_h}\zeta_h + A_{13}v_h\right)$$

$$C_2 = \frac{\beta_h^n}{\Delta}\left(-\frac{lA_{24}}{\lambda_h}\zeta_h + A_{23}v_h\right)$$

$$C_3 = \frac{\beta_h^n}{\Delta}\left(-\frac{lA_{34}}{\lambda_h}\zeta_h + A_{33}v_h\right)$$

$$C_4 = \frac{\beta_h^n}{\Delta}\left(-\frac{lA_{44}}{\lambda_h}\zeta_h + A_{43}v_h\right)$$

$$\tag{4.46}$$

where Δ is the determinant of the coefficients on the left-hand side of the equations in Table 4.2, and A_{ik} the algebraic complement appertaining to the element of the i-th column and k-th row. On substituting the constants in eqs 4.38 and 4.39 we have

$$M_{gh} = \frac{EJ_g}{l^2} \frac{\lambda_g^{n+2}}{\lambda_h^n \Delta} \left[J_{n+2}(\beta_g) \left(\frac{l}{\lambda_h} A_{14}\zeta_h - A_{13}v_h \right) \right.$$

$$+ N_{n+2}(\beta_g) \left(\frac{l}{\lambda_h} A_{24}\zeta_h - A_{23}v_h \right) + I_{n+2}(\beta_g) \left(\frac{l}{\lambda_h} A_{34}\zeta_h - A_{33}v_h \right)$$

$$\left. + K_{n+2}(\beta_g) \left(\frac{l}{\lambda_h} A_{44}\zeta_h - A_{43}v_h \right) \right]$$

$$Y_{gh} = -\frac{EJ_g}{l^3} \frac{\lambda_g^{n+3}}{\lambda_h^n \Delta} \left[J_{n+1}(\beta_g) \left(\frac{l}{\lambda_h} A_{14}\zeta_h - A_{13}v_h \right) \right.$$

$$+ N_{n+1}(\beta_g) \left(\frac{l}{\lambda_h} A_{24}\zeta_h - A_{23}v_h \right) + I_{n+1}(\beta_g) \left(\frac{l}{\lambda_h} A_{34}\zeta_h - A_{33}v_h \right)$$

$$\left. - K_{n+1}(\beta_g) \left(\frac{l}{\lambda_h} A_{44}\zeta_h - A_{43}v_h \right) \right]$$

$$M_{hg} = -\frac{EJ_h}{l^2} \frac{\lambda_h^2}{\Delta} \left[J_{n+2}(\beta_h) \left(\frac{l}{\lambda_h} A_{14}\zeta_h - A_{13}v_h \right) \right.$$

$$+ N_{n+2}(\beta_h) \left(\frac{l}{\lambda_h} A_{24}\zeta_h - A_{23}v_h \right) + I_{n+2}(\beta_h) \left(\frac{l}{\lambda_h} A_{34}\zeta_h - A_{33}v_h \right)$$

$$\left. + K_{n+2}(\beta_h) \left(\frac{l}{\lambda_h} A_{44}\zeta_h - A_{43}v_h \right) \right]$$

$$Y_{hg} = \frac{EJ_h}{l^3} \frac{\lambda_h^3}{\Delta} \left[J_{n+1}(\beta_h) \left(\frac{l}{\lambda_h} A_{14}\zeta_h - A_{13}v_h \right) \right.$$

$$+ N_{n+1}(\beta_h) \left(\frac{l}{\lambda_h} A_{24}\zeta_h - A_{23}v_h \right) + I_{n+1}(\beta_h) \left(\frac{l}{\lambda_h} A_{34}\zeta_h - A_{33}v_h \right)$$

$$\left. - K_{n+1}(\beta_h) \left(\frac{l}{\lambda_h} A_{44}\zeta_h - A_{43}v_h \right) \right]$$

$$(4.47)$$

Because of eq. 4.34 and the fact that for cylindrical functions

$$\begin{vmatrix} J_n(z) & N_n(z) \\ J_n'(z) & N_n'(z) \end{vmatrix} = -\begin{vmatrix} J_n(z) & N_n(z) \\ J_{n+1}(z) & N_{n+1}(z) \end{vmatrix} = \frac{2}{\pi z}$$

$$\begin{vmatrix} I_n(z) & K_n(z) \\ I_n'(z) & K_n'(z) \end{vmatrix} = \begin{vmatrix} I_n(z) & K_n(z) \\ I_{n+1}(z) & -K_{n+1}(z) \end{vmatrix} = -\frac{1}{z}$$

$$(4.48)$$

expressions 4.47 may be given the form

$$M_{gh} = \frac{EJ_g}{l} \left[F_1(\lambda_g, \lambda_h)\, \zeta_h - \frac{J_h}{J_g} F_3(\lambda_h, \lambda_g)\, \frac{v_h}{l} \right]^*$$

$$Y_{gh} = \frac{EJ_g}{l^2} \left[F_3(\lambda_g, \lambda_h)\, \zeta_h + F_5(\lambda_g, \lambda_h)\, \frac{v_h}{l} \right]$$

$$M_{hg} = \frac{EJ_h}{l} \left[F_2(\lambda_g, \lambda_h)\, \zeta_h + F_4(\lambda_g, \lambda_h)\, \frac{v_h}{l} \right]$$

$$Y_{hg} = \frac{EJ_h}{l^2} \left[F_4(\lambda_g, \lambda_h)\, \zeta_h + F_6(\lambda_g, \lambda_h)\, \frac{v_h}{l} \right]$$

$$(4.49)$$

where

$$F_1(\lambda_g, \lambda_h) = \frac{2\lambda_g^{n+2}}{\Delta \lambda_h^{n+1}} \left[A_{34} I_n(\beta_g) + A_{44} K_n(\beta_g) \right] \tag{4.49a}$$

$$F_2(\lambda_g, \lambda_h) = -\lambda_h \left\{ \frac{2(n+1)}{\beta_h} + \frac{2}{\Delta} \left[A_{34} I_n(\beta_h) + A_{44} K_n(\beta_h) \right] \right\} \tag{4.49b}$$

$$F_3(\lambda_g, \lambda_h) = \frac{2\lambda_g^{n+3}}{\Delta \lambda_h^{n+1}} \left[A_{34} I_{n+1}(\beta_g) - A_{44} K_{n+1}(\beta_g) \right] \tag{4.49c}$$

$$F_4(\lambda_g, \lambda_h) = \lambda_h^2 \left\{ 1 + \frac{2}{\Delta} \left[A_{34} I_{n+1}(\beta_h) - A_{44} K_{n+1}(\beta_h) \right] \right\} \tag{4.49d}$$

$$F_5(\lambda_g, \lambda_h) = \frac{2\lambda_g^{n+3}}{\Delta \lambda_h^{n}} \left[A_{33} I_{n+1}(\beta_g) - A_{43} K_{n+1}(\beta_g) \right] \tag{4.49e}$$

$$F_6(\lambda_g, \lambda_h) = \frac{2\lambda_h^3}{\Delta} \left[A_{33} I_{n+1}(\beta_h) + A_{43} K_{n+1}(\beta_h) \right] \tag{4.49f}$$

4.1.1.3 Beam hinged at one end

One proceeds analogously to the above when dealing with a bar clamped at end h and hinged at end g. For the deformations as shown in Fig. 4.3 the boundary conditions are

$$v(0) = v_g \quad v(l) = v_h$$
$$v''(0) = 0 \quad v'(l) = \zeta_h$$

$$(4.50)$$

---

* $F_3(\lambda_h, \lambda_g)$ is obtained from $F_3(\lambda_g, \lambda_h)$ by interchanging the subscripts in the respective formula (as well as in Δ and in complements A_{33}, A_{44}).

Combined with the boundary conditions, eqs 4.21, 4.37 and 4.38 give the equations shown in Table 4.3

Table 4.3

C_1	C_2	C_3	C_4	
$J_n(\beta_g)$	$N_n(\beta_g)$	$I_n(\beta_g)$	$K_n(\beta_g)$	$= \beta_g^n v_g$
$J_{n+2}(\beta_g)$	$N_{n+2}(\beta_g)$	$I_{n+2}(\beta_g)$	$K_{n+2}(\beta_g)$	$= 0$
$J_n(\beta_h)$	$N_n(\beta_h)$	$I_n(\beta_h)$	$K_n(\beta_h)$	$= \beta_h^n v_h$
$J_{n+1}(\beta_h)$	$N_{n+1}(\beta_h)$	$-I_{n+1}(\beta_h)$	$K_{n+1}(\beta_h)$	$= -(l\beta_h^n/\lambda_h)\,\zeta_h$

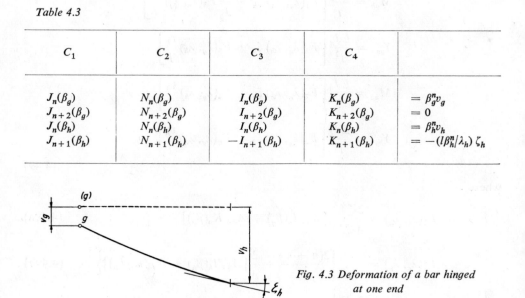

Fig. 4.3 Deformation of a bar hinged at one end

The above equations give the integration constants, and when these constants are substituted in eqs 4.38 and 4.39, the end forces and moments can be obtained:

$$Y_{gh} = \frac{EJ_g}{l^2}\left[F_{12}(\lambda_g, \lambda_h)\frac{v_g}{l} + F_8(\lambda_g, \lambda_h)\,\zeta_h + F_{10}(\lambda_g, \lambda_h)\frac{v_h}{l}\right]$$

$$M_{hg} = \frac{EJ_h}{l}\left[\frac{J_g}{J_h}F_8(\lambda_g, \lambda_h)\frac{v_g}{l} + F_7(\lambda_g, \lambda_h)\,\zeta_h + F_9(\lambda_g, \lambda_h)\frac{v_h}{l}\right]$$

$$Y_{hg} = \frac{EJ_h}{l^2}\left[\frac{J_g}{J_h}F_{10}(\lambda_g, \lambda_h)\frac{v_g}{l} + F_9(\lambda_g, \lambda_h)\,\zeta_h + F_{11}(\lambda_g, \lambda_h)\frac{v_h}{l}\right]$$

$$(4.51)$$

where

$$F_7(\lambda_g, \lambda_h) = -\frac{\lambda_h}{\Delta}\left[A_{14}\,J_{n+2}(\beta_h) + A_{24}\,N_{n+2}(\beta_h)\right.$$

$$\left. + A_{34}\,I_{n+2}(\beta_h) + A_{44}\,K_{n+2}(\beta_h)\right] \qquad (4.52)$$

$$F_8(\lambda_g, \lambda_h) = -\frac{\lambda_g^{n+3}}{\Delta\lambda_h^{n+1}}\left[A_{14}\,J_{n+1}(\beta_g) + A_{24}\,N_{n+1}(\beta_g)\right.$$

$$\left. + A_{34}\,I_{n+1}(\beta_g) - A_{44}\,K_{n+1}(\beta_g)\right] \qquad (4.53)$$

$$F_9(\lambda_g, \lambda_h) = \frac{\lambda_h^2}{\Delta}\left[A_{14}\,J_{n+1}(\beta_h) + A_{24}\,N_{n+1}(\beta_h)\right.$$

$$\left. + A_{34}\,I_{n+1}(\beta_h) - A_{44}\,K_{n+1}(\beta_h)\right] \tag{4.54}$$

$$F_{10}(\lambda_g, \lambda_h) = \frac{\lambda_g^{n+3}}{\Delta\lambda_h^n}\left[A_{13}\,J_{n+1}(\beta_g) + A_{23}\,N_{n+1}(\beta_g)\right.$$

$$\left. + A_{33}\,I_{n+1}(\beta_g) - A_{43}\,K_{n+1}(\beta_g)\right] \tag{4.55}$$

$$F_{11}(\lambda_g, \lambda_h) = -\frac{\lambda_h^3}{\Delta}\left[A_{13}\,J_{n+1}(\beta_h) + A_{23}\,N_{n+1}(\beta_h)\right.$$

$$\left. + A_{33}\,I_{n+1}(\beta_h) - A_{43}\,K_{n+1}(\beta_h)\right] \tag{4.56}$$

$$F_{12}(\lambda_g, \lambda_h) = \frac{\lambda_g^3}{\Delta}\left[A_{11}\,J_{n+1}(\beta_g) + A_{21}\,N_{n+1}(\beta_g)\right.$$

$$\left. + A_{31}\,I_{n+1}(\beta_g) - A_{41}\,K_{n+1}(\beta_g)\right] \tag{4.57}$$

Symbols Δ and A_{ik} denote respectively the determinant and the algebraic complements of the coefficients in Table 4.3.

4.1.1.4 Beam hinged at both ends

According to Fig. 4.4 the boundary conditions are

$$v(0) = 0 \quad v(l) = v_h$$
$$v''(0) = 0 \quad v''(l) = 0$$
$$\tag{4.58}$$

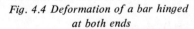

Fig. 4.4 Deformation of a bar hinged at both ends

Proceeding in the same way as before, we first get from the boundary conditions, (eqs 4.58), and eqs 4.21, 4.38 the equations (Table 4.4) for computing the integration constants C, and with them substituted into eq. 4.39, the end forces Y_{gh}, Y_{hg} are obtained.

156

Table 4.4

C_1	C_2	C_3	C_4	
$J_n(\beta_g)$	$N_n(\beta_g)$	$I_n(\beta_g)$	$K_n(\beta_g)$	$= 0$
$J_{n+2}(\beta_g)$	$N_{n+2}(\beta_g)$	$I_{n+2}(\beta_g)$	$K_{n+2}(\beta_g)$	$= 0$
$J_n(\beta_h)$	$N_n(\beta_h)$	$I_n(\beta_h)$	$K_n(\beta_h)$	$= v_h\beta_h^n$
$J_{n+2}(\beta_h)$	$N_{n+2}(\beta_h)$	$I_{n+2}(\beta_h)$	$K_{n+2}(\beta_h)$	$= 0$

$$Y_{gh} = \frac{EJ_g}{l^3} F_{13}(\lambda_g, \lambda_h)\, v_h$$

$$Y_{hg} = \frac{EJ_h}{l^3} F_{14}(\lambda_g, \lambda_h)\, v_h$$

(4.59)

where

$$F_{13}(\lambda_g, \lambda_h) = \frac{\lambda_g^{n+3}}{\Delta \lambda_h^n}\left[A_{13}\, J_{n+1}(\beta_g) + A_{23}\, N_{n+1}(\beta_g) + A_{33}\, I_{n+1}(\beta_g)\right.$$
$$\left. - A_{43}\, K_{n+1}(\beta_g)\right]$$

(4.60)

$$F_{14}(\lambda_g, \lambda_h) = -\frac{\lambda_h^3}{\Delta}\left[A_{13}\, J_{n+1}(\beta_h) + A_{23}\, N_{n+1}(\beta_h) + A_{33}\, I_{n+1}(\beta_h)\right.$$
$$\left. - A_{43}\, K_{n+1}(\beta_h)\right]$$

(4.61)

and symbols Δ, A_{ik} refer to Table 4.4.

The natural frequencies of a simply supported beam are computed from the condition that determinant Δ be made equal to zero.

4.1.1.5 Cantilever beam

For the deformation shown in Fig. 4.5 the boundary conditions are

$$v''(0) = 0 \quad v(l) = v_h$$
$$T(0) = 0 \quad v'(l) = \zeta_h$$

(4.62)

On substituting these conditions into eqs 4.21, 4.37, 4.38, 4.39 we get the equations as shown in Table 4.5

Table 4.5

C_1	C_2	C_3	C_4	
$J_{n+2}(\beta_g)$	$N_{n+2}(\beta_g)$	$I_{n+2}(\beta_g)$	$K_{n+2}(\beta_g)$	$= 0$
$J_{n+1}(\beta_g)$	$N_{n+1}(\beta_g)$	$I_{n+1}(\beta_g)$	$-K_{n+1}(\beta_g)$	$= 0$
$J_n(\beta_h)$	$N_n(\beta_h)$	$I_n(\beta_h)$	$K_n(\beta_h)$	$= v_h \beta_h^n$
$J_{n+1}(\beta_h)$	$N_{n+1}(\beta_h)$	$-I_{n+1}(\beta_h)$	$K_{n+1}(\beta_h)$	$= -(l\beta_h^n/\lambda_h)\,\zeta_h$

Fig. 4.5 Deformation of a cantilever beam

and with their aid the end moment and force can be expressed as

$$M_{hg} = \frac{EJ_h}{l}\left[F_{15}(\lambda_g, \lambda_h)\,\zeta_h + F_{16}(\lambda_g, \lambda_h)\frac{v_h}{l}\right]$$

$$Y_{hg} = \frac{EJ_h}{l^2}\left[F_{16}(\lambda_g, \lambda_h)\,\zeta_h + F_{17}(\lambda_g, \lambda_h)\frac{v_h}{l}\right]$$

(4.63)

where

$$F_{15}(\lambda_g, \lambda_h) = -\lambda_h\left\{\frac{2(n+1)}{\beta_h} + \frac{2}{\Delta}\left[A_{34}\,I_n(\beta_h) + A_{44}\,K_n(\beta_h)\right]\right\} \qquad (4.64)$$

$$F_{16}(\lambda_g, \lambda_h) = \frac{\lambda_h^2}{\Delta}\left[A_{14}\,J_{n+1}(\beta_h) + A_{24}\,N_{n+1}(\beta_h) + A_{34}\,I_{n+1}(\beta_h)\right.$$
$$\left. - A_{44}\,K_{n+1}(\beta_h)\right] \qquad (4.65)$$

$$F_{17}(\lambda_g, \lambda_h) = -\frac{2\lambda_h^3}{\Delta}\left[A_{33}\,I_{n+1}(\beta_h) - A_{43}\,K_{n+1}(\beta_h)\right] \qquad (4.66)$$

Symbols Δ and A_{ik} refer to the equations of Table 4.5.

The natural frequencies of a cantilever beam with the right-hand end restrained as shown are computed from the condition that determinant Δ be made equal to zero.

158

4.1.2 Solution by the slope-deflection method

The slope-deflection method as outlined in Chapters 2 and 3 leads to exact results in the case of a step-wise varying cross-section[37,38] (Fig. 4.6), for then the bar can be subdivided into segments of constant cross-section and treated as a system

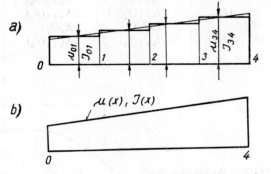

Fig. 4.6 (a) Bar with stepwise variation in cross-section; (b) bar with continuous variation in cross-section

composed of prismatic elements. The procedure will be illustrated by way of a numerical example.

Example 4.1

Compute the moments and forces acting at the ends of the bar shown in Fig. 4.6a if the bar vibrates in a steady-state forced vibration produced by a harmonic motion of its ends which simultaneously perform vertical motion with amplitudes v_0 and v_4 and rotation with amplitudes ζ_0 and ζ_4. Consider the bar to be a system composed of four prismatic elements and three intermediate joints with altogether six unknown deformations, i.e. three rotations $\zeta_1, \zeta_2, \zeta_3$ and three displacements v_1, v_2, v_3. The conditions of equilibrium lead to the equations set out in Table 4.6.

Table 4.6

ζ_1	ζ_2	ζ_3	v_1	v_2	v_3	
a_1	a_{12}		b_1	$-b_{12}$		$= d_1$
a_{12}	a_2	a_{23}	b_{12}	b_2	$-b_{23}$	$= d_2$
	a_{23}	a_3		b_{23}	b_3	$= d_3$
b_1	b_{12}		c_1	c_{12}		$= d_4$
$-b_{12}$	b_2	b_{23}	c_{12}	c_2	c_{23}	$= d_5$
	$-b_{23}$	b_3		c_{23}	c_3	$= d_6$

The coefficients of these equations are as follows:

$$a_{i,i+1} = \frac{EJ_{i,i+1}}{l_{i,i+1}} F_1(\lambda_{i,i+1})$$

or in abbreviated notation

$$a_{i,i+1} = \left[\frac{EJ}{l} F_1(\lambda)\right]_{i,i+1} \qquad \text{for} \quad i = 1, 2$$

$$a_i = \left[\frac{EJ}{l} F_2(\lambda)\right]_{i,i-1} + \left[\frac{EJ}{l} F_2(\lambda)\right]_{i,i+1} \qquad \text{for} \quad i = 1, 2, 3$$

$$b_{i,i+1} = \left[\frac{EJ}{l^2} F_3(\lambda)\right]_{i,i+1} \qquad i = 1, 2$$

$$b_i = \left[\frac{EJ}{l^2} F_4(\lambda)\right]_{i,i-1} - \left[\frac{EJ}{l^2} F_4(\lambda)\right]_{i,i+1} \qquad \text{for} \quad i = 1, 2, 3$$

$$c_{i,i+1} = \left[\frac{EJ}{l^3} F_5(\lambda)\right]_{i,i+1} \qquad i = 1, 2$$

$$c_i = \left[\frac{EJ}{l^3} F_6(\lambda)\right]_{i,i-1} + \left[\frac{EJ}{l^3} F_6(\lambda)\right]_{i,i+1} \qquad i = 1, 2, 3$$

$$(4.67)$$

$$d_1 = -\left[\frac{EJ}{l} F_1(\lambda)\right]_{01} \zeta_0 - \left[\frac{EJ}{l^2} F_3(\lambda)\right]_{01} v_0$$

$$d_3 = -\left[\frac{EJ}{l} F_1(\lambda)\right]_{34} \zeta_4 + \left[\frac{EJ}{l^2} F_3(\lambda)\right]_{34} v_4$$

$$d_4 = \left[\frac{EJ}{l^2} F_3(\lambda)\right]_{01} \zeta_0 - \left[\frac{EJ}{l^3} F_5(\lambda)\right]_{01} v_0$$

$$d_6 = -\left[\frac{EJ}{l^2} F_3(\lambda)\right]_{34} \zeta_4 - \left[\frac{EJ}{l^3} F_5(\lambda)\right]_{34} v_4$$

$$d_2 = d_5 = 0$$

$$(4.68)$$

If the bar is simultaneously subjected to an external harmonic load, the loading terms would be

$$\bar{d}_1 = d_1 + \mathcal{M}_1$$
$$\bar{d}_2 = \mathcal{M}_2$$
$$\bar{d}_3 = d_3 + \mathcal{M}_3$$
$$\bar{d}_4 = d_4 + \mathcal{Y}_1, \quad \text{etc.}$$

where $\mathcal{M}_1, \mathcal{M}_2 \dots \mathcal{Y}_1$ are the respective components of external load applied to the beam.

The equations in Table 4.6 give the unknown deformations and Tables 10.1 and 10.2 of the Appendix the end moments and forces, e.g.

$$M_{01} = M_{04} = \left[\frac{EJ}{l} F_2(\lambda)\right]_{01} \zeta_0 + \left[\frac{EJ}{l} F_1(\lambda)\right]_{01} \zeta_1 - \left[\frac{EJ}{l^2} F_4(\lambda)\right]_{01} v_0$$

$$- \left[\frac{EJ}{l^2} F_3(\lambda)\right]_{01} v_1 \tag{4.69}$$

etc.

Numerical example

The dimensions of the bar according to Fig. 4.6a are:

$$l_{04} = 10 \text{ m} ; \quad l_{01} = l_{12} = l_{23} = l_{34} = l/4$$

$$\mu_{01} = 0 \cdot 1125 \text{ t/m} ; \qquad \mu_{12} = 0 \cdot 1375 \text{ t/m} ; \qquad \mu_{23} = 0 \cdot 1625 \text{ t/m} ;$$

$$\mu_{34} = 0 \cdot 1875 \text{ t/m}$$

$$J_{01} = 1 \cdot 125 \times 10^{-4} \text{ m}^4 ; \quad J_{12} = 1 \cdot 375 \times 10^{-4} \text{ m}^4 ; \quad J_{23} = 1 \cdot 625 \times 10^{-4} \text{ m}^4 ;$$

$$J_{34} = 1 \cdot 875 \times 10^{-4} \text{ m}^4$$

Find the end moments and forces in forced harmonic motion with frequency $\omega = 18 \cdot 16 \text{ s}^{-1}$ if $E = 21 \times 10^6 \text{ Mpm}^{-2}$.

The dimensionless quantity λ is the same for all four segments, viz.

$$\lambda = l \left(\frac{\mu\omega^2}{EJ}\right)^{1/4} = \frac{10}{4} \left(\frac{0 \cdot 1 \times 18 \cdot 16^2}{10^{-4} \times 21 \times 10^6 \times 9 \cdot 81}\right)^{1/4} = 0 \cdot 50$$

With the above data the coefficients of Table 4.6 can be computed to get the equations shown in Table 4.7

Table 4.7

ζ_1	ζ_2	ζ_3	v_1	v_2	v_3	
3·99940	1·10025		0·23987	−1·32043		$= -0 \cdot 90020 \zeta_0 - 1 \cdot 08035 v_0$
1·10025	4·79928	1·30029	1·32043	0·23987	−1·56050	$= 0$
	1·30029	5·59916		1·56050	0·23987	$= -1 \cdot 50034 \zeta_4 + 1 \cdot 80058 v_4$
0·23987	1·32043		1·91629	−1·05670		$= 1 \cdot 08035 \zeta_0 + 0 \cdot 86458 v_0$
−1·32043	0·23987	1·56050	−1·05670	2·29955	1·24884	$= 0$
	−1·56050	0·23987		1·24884	2·68280	$= -1 \cdot 80058 \zeta_4 + 1 \cdot 44097 v_4$

As there is no external load acting at points 1 to 3, $\mathcal{M}_1 = \mathcal{M}_2 = \ldots \mathcal{Y}_3 = 0$. All the coefficients of the equations are divided by $E \times 10^{-4} = 21 \times 10^2$.

The solution of the equations of Table 4.7 gives the deformations

$$
\begin{aligned}
\zeta_1 &= 0\cdot129\,63\zeta_0 - 0\cdot390\,32\zeta_4 - 0\cdot119\,53v_0 + 0\cdot132\,74v_4 \\
\zeta_2 &= -0\cdot251\,04\zeta_0 - 0\cdot237\,55\zeta_4 - 0\cdot150\,28v_0 + 0\cdot149\,08v_4 \\
\zeta_3 &= -0\cdot267\,06\zeta_0 + 0\cdot268\,16\zeta_4 - 0\cdot115\,94v_0 + 0\cdot093\,13v_4 \\
v_1 &= 1\cdot330\,94\zeta_0 - 0\cdot509\,60\zeta_4 + 0\cdot835\,91v_0 + 0\cdot189\,88v_4 \\
v_2 &= 1\cdot106\,97\zeta_0 - 1\cdot490\,92\zeta_4 + 0\cdot482\,78v_0 + 0\cdot560\,75v_4 \\
v_3 &= 0\cdot393\,15\zeta_0 - 1\cdot527\,33\zeta_4 + 0\cdot146\,79v_0 + 0\cdot876\,53v_4
\end{aligned}
$$

$$(4.70)$$

The end moments and forces are obtained from formula 4.69 and Tables 10.1, 10.2 in the Appendix:

$$
\begin{aligned}
M_{04} &= 1015\cdot4\,\zeta_0 + 645\cdot15\zeta_4 + 144\cdot36\,v_0 - 179\cdot87\,v_4 \\
M_{40} &= 645\cdot15\zeta_0 + 1368\cdot8\,\zeta_4 + 221\cdot27\,v_0 - 170\cdot19\,v_4 \\
Y_{04} &= 144\cdot36\zeta_0 + 221\cdot27\zeta_4 + 22\cdot027v_0 - 43\cdot612\,v_4 \\
Y_{40} &= -179\cdot87\zeta_0 - 170\cdot19\zeta_4 - 43\cdot612v_0 + 13\cdot622\,v_4
\end{aligned}
$$

$$(4.71)$$

If the bar cross-section varies continuously, the slope-deflection method yields approximate results. In such a case the procedure consists of subdividing the bar into several segments whose cross-section and mass per unit length, μ, are assumed to be constant. For example the bar in Fig. 4.6b has the characteristics

$$l = 10 \text{ m}$$

$$J(x) = \left(1 + \frac{x}{l}\right) 10^{-4} \text{ m}^4$$

$$\mu(x) = 0\cdot1 \left(1 + \frac{x}{l}\right) \text{ t/m}$$

On subdividing it in four segments and assuming the segments to be approximately prismatic, we get a bar with a step-wise varying cross-section which we have solved in the preceding example.

4.1.3   Simplified solution

In Section 3.3 we have evolved a simplified solution for the end forces in prismatic bars. The results obtained there can, in part, be applied also to systems with non-prismatic bars[46]. If a variable $\mu(x)$ is substituted for constant μ in eq. 3.32, the formula also holds good for a non-prismatic bar. The same goes for Table 3.9 if λ is computed for a suitably chosen cross-section of the bar. The procedure for computing a system, composed of non-prismatic bars is thus as follows: first, the end moments and forces are determined from formula 3.32 using the static deflection curves which are usually available from the preceding static analysis. As is customary with the slope-deflection method, this is followed by the setting-up of the slope-deflection equations expressing the conditions of equilibrium at the individual joints. We shall illustrate the procedure by way of the following example.

Example 4.2

Fig. 4.7 shows the load-carrying structure of a bridge. The structure is symmetric and divided at joints marked in the figure into non-prismatic bars ab, bs, sb', $b'a'$, and columns cb of constant cross-section. The longitudinal vibration of the bars being neglected, the joints b, b' execute rotation but not translation.

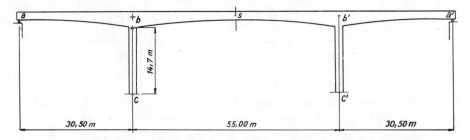

Fig. 4.7 Frame system with non-prismatic members

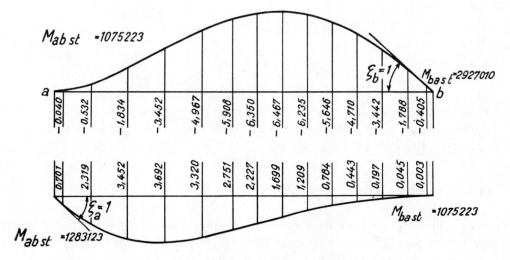

Fig. 4.8 Static deflection curves of bar $a—b$

The moments at the ends of the horizontal bars are to be computed by the simplified method outlined above. The static deflection curves, which are also the influence lines of static end moments and forces are shown in Fig. 4.8 for bar *ab* and

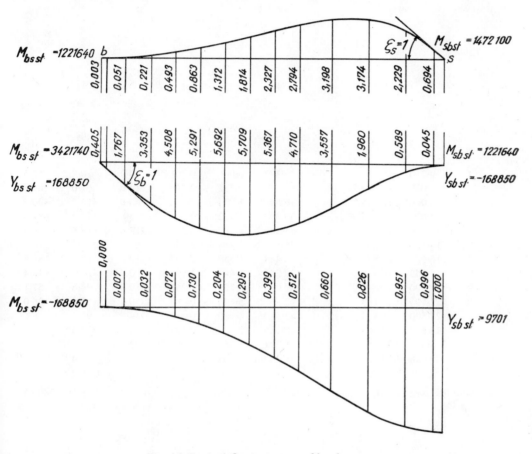

Fig. 4.9 Static deflection curves of bar b—s

in Fig. 4.9 for bar *bs*. The end moments are computed from formula 3.32 with the integral on the right-hand side being evaluated by numerical integration:

$$M_{ab(\zeta_a=1)} = M_{abst} - \omega^2 \sum_i m_i v_{1i}^2$$

$$M_{ba(\zeta_a=1)} = M_{bast} - \omega^2 \sum_i m_i v_{1i} v_{2i}$$

(4.72)

where m_i is the mass of the *i*-th segment of the bar (Table 4.8)

$\quad v_{1i}, v_{2i}$ are the ordinates of the influence lines at the centre of gravity of the *i*-th segment

Figs 4.8 and 4.9 also show the static end moments and forces corresponding to the deflection curves. The numerical computation for bar ab is given in Table 4.8; the remaining bars were computed in an analogous manner and the results summarised in Table 4.9.

Table 4.8

Point	1	2	3	4	5	6	7	8	9
	x_i	μ_i	Δx	$m_i = \mu_i \Delta x$	$-v_{1i}$	v_{2i}	$m_i \cdot v_{1i}^2$	$-m_i \times v_{1i}v_{2i}$	$m_i \cdot v_{2i}^2$
1	0·750	16·931	1·6	25·397	0·040	0·701	0·0	0·7	12·5
2	3·000	16·479	3	49·439	0·532	2·319	14·0	61·0	265·8
3	6·000	16·516	3	49·549	1·834	3·452	166·6	313·7	590·5
4	9·000	16·669	3	50·009	3·452	3·692	595·8	637·2	681·6
5	12·000	17·800	3	53·400	4·967	3·320	1317·6	880·6	588·5
6	14·500	18·366	2	36·733	5·908	2·751	1282·0	597·1	278·1
7	16·500	18·725	2	37·448	6·350	2·227	1510·0	529·5	185·7
8	18·500	19·128	2	38·257	6·467	1·699	1600·0	420·4	110·4
9	20·500	19·547	2	39·094	6·235	1·209	1519·9	294·6	57·1
10	22·500	21·183	2	42·367	5·646	0·784	1350·4	187·5	26·0
11	24·500	21·830	2	43·660	4·710	0·443	968·6	91·2	8·6
12	26·500	22·563	2	45·126	3·442	0·197	534·8	30·5	1·7
13	28·588	23·352	2·175	50·791	1·788	0·045	162·3	4·1	0·1
14	30·088	37·924	0·825	31·287	0·405	0·003	5·1	0·0	0·0

The units used are Mp, m, s.

$$\sum = \quad 11027·0 \quad 4048·1 \quad 2806·6$$
$$\sum/9·81 = \quad 1124·46 \quad 412·80 \quad 286·20$$

Table 4.9

	ζ_a	ζ_b	ζ_s	v_s
M_{ab}	$1\ 283\ 123 - 286·20\omega^2$	$1\ 075\ 223 + \quad 412·80\omega^2$		
M_{ba}	$1\ 075\ 223 + 412·80\omega^2$	$2\ 927\ 010 - 1124·46\omega^2$		
M_{bs}		$3\ 421\ 740 - \quad 819·45\omega^2$	$1\ 221\ 640 + 301·02\omega^2$	$-168\ 850 - 57·35\omega^2$
M_{bc}		$2\ 143\ 961 - \quad 301·15\omega^2$		
$\sum$	$1\ 075\ 223 + 412·80\omega^2$	$8\ 492\ 711 - 2245·06\omega^2$	$1\ 221\ 640 + 301·02\omega^2$	$-168\ 850 - 57·35\omega^2$
M_{sb}		$1\ 221\ 640 + \quad 301·02\omega^2$	$1\ 472\ 100 - 207·56\omega^2$	
Y_{sb}		$-168\ 850 - \quad\ \ 57·35\omega^2$		$9\ 701 - 15·111\omega^2$

Between c and b the columns are of constant cross-section; above point d they intersect the vertical beams and their rigidity greatly increases as a result. For the purposes of computation, part db was assumed to be infinitely rigid (Fig. 4.10).

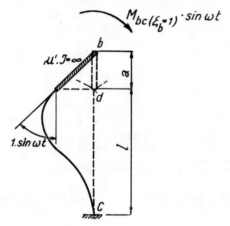

Fig. 4.10 Haunched bar

At their lower ends, point c, the columns are assumed to be fixed; this assumption is, of course, only satisfied for the case of the foundation being fixed on very strong rock.

For vibration according to Fig. 4.10 the amplitude of end moment M_{bc} is given by eq. 4.73 and obtained from Tables 10.1, 10.2 of the Appendix. It is

$$M_{bc(\zeta_b = 1)} = \frac{EJ}{l}\left[F_2(\lambda) - \frac{2a}{l}F_4(\lambda) + \frac{a^2}{l^2}F_6(\lambda)\right] - \tfrac{1}{3}\mu'\omega^2 a^3 \qquad (4.73)$$

where the last term expresses the moment of inertia forces and μ' the assumed mass per unit length of the rigid part of the column of length a.

Note: The end moments of bar ab were computed on the assumption that the bar is clamped at point a. Actually, however, that end is hinged. To satisfy the actual conditions of support, we must take rotation ζ_a as another unknown and complement our system of slope-deflection equations by the equation

$$M_{ab} = M_{ab(\zeta_a = 1)}\zeta_a + M_{ab(\zeta_b = 1)}\zeta_b = 0 \qquad (4.74)$$

This procedure is analogous to that used in Section 3.3, Example 3.2.

(a) Symmetric vibration

The mode shape for symmetric vibration is shown in Fig. 4.11. The slope-deflection equations are of the form

$$M_{ab} = 0 \tag{4.75}$$

$$M_{ba} + M_{bs} + M_{bc} = 0 \tag{4.76}$$

$$Y_{sb} = 0 \tag{4.77}$$

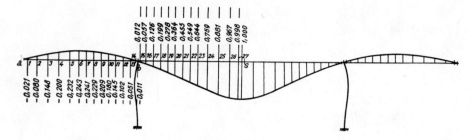

Fig. 4.11 Symmetric mode of free vibration of system shown in Fig. 4.7

where

$$M_{ab} = M_{ab(\zeta_a=1)}\zeta_a + M_{ab(\zeta_b=1)}\zeta_b$$

$$M_{ba} = M_{ba(\zeta_a=1)}\zeta_a + M_{ba(\zeta_b=1)}\zeta_b$$

$$M_{bs} = M_{bs(\zeta_b=1)}\zeta_b + M_{bs(v_s=1)}v_s$$

$$M_{bc} = M_{bc(\zeta_b=1)}\zeta_b$$

$$Y_{sb} = Y_{sb(\zeta_b=1)}\zeta_b + Y_{sb(v_s=1)}v_s$$

$$\tag{4.78}$$

ζ_a, ζ_b denoting the rotations of joints a and b, and v_s the vertical displacement of point s.

Table 4.10

ζ_a	ζ_b	v_s	
$1\,283\,123 - 286{\cdot}20\omega^2$	$1\,075\,223 + 412{\cdot}80\omega^2$		$= 0$
$1\,075\,223 + 412{\cdot}80\omega^2$	$8\,492\,711 - 2245{\cdot}06\omega^2$	$-168\,850 - 57{\cdot}350\omega^2$	$= 0$
	$-168\,850 - 57{\cdot}35\omega^2$	$9\,701 - 15{\cdot}111\omega^2$	$= 0$

With the numerical values substituted in eq. 4.78 and the latter in eqs 4.75, 4.76 and 4.77 we get the slope-deflection equations set out in Table 4.10. The condition that the determinant of the coefficients should be zero leads to the following cubic equation in ω^2

$$\omega^6 - 16\ 078{\cdot}807\omega^4 + 36\ 175\ 074{\cdot}800\omega^2 - 9\ 353\ 401\ 530{\cdot}0 = 0$$

whose roots are

$$\omega_{(1)} = 17{\cdot}2 \ \ s^{-1}$$
$$\omega_{(2)} = 48{\cdot}4 \ \ s^{-1}$$
$$\omega_{(3)} = 115{\cdot}9 \ \ s^{-1}$$

The exact value of the first natural frequency is lower than that obtained by this simplified method. Going by the criteria stated earlier (see Table 3.9) the difference will be about 2%. For higher frequencies the difference will be appreciably higher.

The mode of free symmetric vibration is given by deformations ζ_a, ζ_b, v_s one of which may be chosen at will and the remaining computed from the respective equations. Assuming that $v_s = 1$, we get

$$\zeta_a = -0{\cdot}02803$$
$$\zeta_b = \ \ \ 0{\cdot}02804$$

The deformations of the individual bars in free vibration are then obtained by multiplying the ordinates of the influence lines (Figs 4.8, 4.9) by the respective end deformations, and superposing the results. The resultant mode shape for circular frequency $\omega_{(1)} = 17{\cdot}2\,s^{-1}$ is shown in Fig. 4.11.

(b) Antisymmetric vibration

The mode shape of antisymmetric vibration is depicted in Fig. 4.12. There are again three slope-deflection equations

$$M_{ab} = 0$$
$$M_{ba} + M_{bs} + M_{bc} = 0$$
$$M_{sb} = 0$$

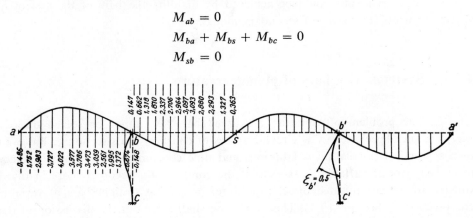

Fig. 4.12 Antisymmetric mode of free vibration of system shown in Fig. 4.7

Table 4.11

ζ_a		ζ_b	ζ_s	
$1\,283\,123 - 286{\cdot}20\omega^2$		$1\,075\,223 + 412{\cdot}80\omega^2$		$= 0$
$1\,075\,223 + 412{\cdot}80\omega^2$		$8\,492\,711 - 2245{\cdot}06\omega^2$	$1\,221\,640 + 301{\cdot}02\omega^2$	$= 0$
		$1\,221\,640 + 301{\cdot}02\omega^2$	$1\,472\,100 - 207{\cdot}56\omega^2$	$= 0$

Proceeding analogously to the case of symmetric vibration we get the equations set out in Table 4.11. The condition that the determinant of the coefficients should be zero leads to the following cubic equation in ω^2

$$72{\cdot}060\,547\omega^6 - 2\,075\,875{\cdot}447\omega^4 + 11\,663\,891\,301\omega^2 - 12\,424\,917\,638\,041 = 0$$

the roots of which are

$$\omega_{(1)} = 37{\cdot}3\ \text{s}^{-1}$$
$$\omega_{(2)} = 75{\cdot}4\ \text{s}^{-1}$$
$$\omega_{(3)} = 147{\cdot}4\ \text{s}^{-1}$$

As to the difference between the exact and simplified method: for the first natural frequency it will be about 6%; for higher natural frequencies the values computed by the simplified method should be considered for reference only.

The mode of free vibration is given by the ratios of ζ_a, ζ_b and ζ_s. The natural mode at $\omega_{(1)}$ computed for $\zeta_s = -0{\cdot}5$ is shown in Fig. 4.12. The other two deformations are $\zeta_b = 0{\cdot}3602$ and $\zeta_a = -0{\cdot}6728$.

More accurate results could be achieved by dividing the bars of the system in shorter elements (by means of secondary joints).

4.2 Systems with bars of plane curvature

The problem of vibration of curved bars[9,30] cannot be solved directly except in such special cases as circular arches of constant cross-section[75]. But even then a practical calculation is laborious and time consuming. And when it comes to curved bars of variable cross-section, the computation becomes more involved still (recall the complicated direct procedure for solving straight bars of variable cross-section − Section 4.1.1). This is why we shall leave out the discussion of the direct method and turn to the indirect or approximate solutions.

4.2.1 Method of discrete mass points

Fig. 4.13 shows a bar with a generally curved centre line and a generally varying cross-section. On dividing the bar lengthwise into short segments and concentrating the masses of the segments at their centres of gravity, we obtain an elastic system

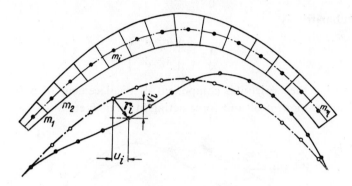

Fig. 4.13 Vibration of a bar of plane curvature

with discrete mass points. A system with n such points moving unrestricted in either
direction has $2n$ degrees of freedom, for each point can be displaced in the vertical
as well as in the horizontal direction. The solution of a system with a finite number of
degrees of freedom was discussed in Chapter 1. Using the method of influence
coefficients we write the equation of motion of the system as eq. 1.100

$$r(t) = -\Delta m\, \ddot{r}(t) \tag{4.79}$$

where r is the vector of displacement

$$r(t) = \begin{bmatrix} r_1(t) \\ r_2(t) \\ r_i(t) \\ r_n(t) \end{bmatrix} \tag{4.80}$$

where

$$r_i(t) = \begin{bmatrix} u_i(t) \\ v_i(t) \end{bmatrix} \tag{4.81}$$

The diagonal matrix of mass is given by

$$m = \begin{bmatrix} m_1 & & & & \\ & m_2 & & & \\ & & m_3 & & \\ & & & \ddots & \\ & & & & m_n \end{bmatrix} \tag{4.82}$$

and the matrix of influence coefficients is

$$\Delta = \begin{bmatrix} \Delta_{11} & \Delta_{12} & \cdots & \Delta_{1n} \\ \Delta_{21} & \Delta_{22} & \cdots & \Delta_{2n} \\ \vdots & \vdots & & \vdots \\ \Delta_{n1} & \Delta_{n2} & \cdots & \Delta_{nn} \end{bmatrix} \qquad (4.83)$$

where the submatrix is

$$\Delta_{ik} = \begin{bmatrix} \delta_{ik}^{uu} & \delta_{ik}^{uv} \\ \delta_{ik}^{vu} & \delta_{ik}^{vv} \end{bmatrix} \qquad (4.84)$$

with δ_{ik}^{uv} denoting the horizontal component of displacement of point k produced by the vertical component of unit force acting at point i, etc.

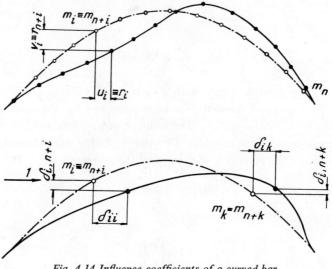

Fig. 4.14 Influence coefficients of a curved bar

Since matrix Δ is symmetric, $\Delta_{ik} = \Delta'_{ki}$ where Δ' is the transpose of the matrix. Using the symbols of Fig. 4.14, we may write the equations of motion in the form

$$u_i(t) = -(m_1 \delta_{i1} \ddot{u}_1(t) + m_2 \delta_{i2} \ddot{u}_2(t) + \ldots m_1 \delta_{i,n+1} \ddot{v}_2(t) + m_2 \delta_{i,n+2} \ddot{v}_2(t) + \ldots)$$

which, with the notation

$$u_i(t) = r_i(t)$$
$$v_i(t) = r_{n+i}(t)$$
$$m_i \;\; = m_{n+i}$$

$$(4.85)$$

becomes

$$r_i(t) = - \sum_{k=1}^{2n} m_k \delta_{ik} \ddot{r}_k(t) \tag{4.86}$$

where n is the total number of discrete mass points.

In harmonic vibration where

$$r_i(t) = r_i \sin \omega t$$

we have

$$r_i = \omega^2 \sum_{k=1}^{2n} m_k \delta_{ik} r_k$$

$$\tag{4.87}$$

Eq. 4.87 may be solved by any one of the methods discussed in Chapter 1, e.g. by the method of successive approximations. An illustrative example will be given in Chapter 7.

4.2.2 The slope-deflection method

In a system with curved bars each bar is again subdivided into short segments assumed to be approximately prismatic (Fig. 4.15). If there are n segments of the bar, there are $n - 1$ intermediate joints. In each joint there are three unknown deformations, rotation ζ_i and horizontal and vertical displacements, u_i and v_i. To determine

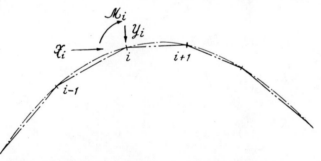

Fig. 4.15 Curved bar divided into prismatic elements

them we need three slope-deflection equations to express the conditions of equilibrium of the end forces and moments in the joint. The procedure is a generalisation of the solution evolved in Section 4.1.2. For joint i of the system shown in Fig. 4.15

$$M_{i,i-1} + M_{i,i+1} = \mathcal{M}_i$$
$$X_{i,i-1} + X_{i,i+1} = \mathcal{X}_i$$
$$Y_{i,i-1} + Y_{i,i+1} = \mathcal{Y}_i$$

$$\tag{4.88}$$

Referring to Fig. 4.15 we see that X and Y are respectively the horizontal and the vertical axes. Apart from this each bar segment has its own system of coordinate axes A, B, C (Fig. 4.16), A being the longitudinal axis and C the axis at right angles to the plane of the drawing. $\mathcal{M}_i$, $\mathcal{X}_i$, $\mathcal{Y}_i$ are the components of external load. The end moments of bar $i - k$ are given by the formula

$$M_{ik} = a_{zz}\zeta_i + b_{zx}u_i + b_{zy}v_i + c_{zz}\zeta_k + d_{zx}u_k + d_{zy}v_k \tag{4.89}$$

and the end forces are

$$X_{ik} = b_{xz}\zeta_i + e_{xx}u_i + e_{xy}v_i - d_{zx}\zeta_k + f_{xx}u_k + f_{xy}v_k \tag{4.90}$$

$$Y_{ik} = b_{yz}\zeta_i + e_{yx}u_i + e_{yy}v_i - d_{zy}\zeta_k + f_{yx}u_k + f_{yy}v_k \tag{4.91}$$

where

$$a_{zz} = \frac{EJ}{l} F_2(\lambda)$$

$$b_{zx} = -\frac{EJ}{l^2} F_4(\lambda) \cos \beta_x = b_{xz}$$

$$b_{zy} = -\frac{EJ}{l^2} F_4(\lambda) \cos \beta_y = b_{yz}$$

$$c_{zz} = \frac{EJ}{l} F_1(\lambda)$$

$$d_{zx} = -\frac{EJ}{l^2} F_3(\lambda) \cos \beta_x$$

$$d_{zy} = -\frac{EJ}{l^2} F_3(\lambda) \cos \beta_y$$

$$e_{xx} = \frac{EJ}{l^3} F_6(\lambda) \cos^2 \beta_x + \frac{ES}{l} \psi \cot \psi \cos^2 \alpha_x$$

$$e_{xy} = \frac{EJ}{l^3} F_6(\lambda) \cos \beta_x \cos \beta_y + \frac{ES}{l} \psi \cot \psi \cos \alpha_x \cos \alpha_y = e_{yx}$$

$$f_{xy} = \frac{EJ}{l^3} F_5(\lambda) \cos \beta_x \cos \beta_y - \frac{ES}{l} \psi \, \text{cosec} \, \psi \cos \alpha_x \cos \alpha_y = f_{yx}$$

$$\tag{4.92}$$

The remaining coefficients e_{yy}, f_{xx}, f_{yy} are obtained from the above formulae through the interchange of the subscripts. The notation is as indicated in Fig. 4.16: angles α_x, β_x are the angles formed by axes A and B with axis X, α_y the angle between axes A and Y.

Formulae 4.89−4.92 follow from the application of Tables 10.1 and 10.2 in the Appendix. When using formulae 4.92 one should consult Fig. 4.16 for the correct sense of directional angles α, β, γ. Angle $\alpha_{ik} = \alpha_{ki} + 180°$, etc.

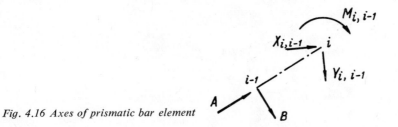

Fig. 4.16 Axes of prismatic bar element

The slope-deflection method has both advantages and shortcomings when compared with the method of discrete mass points. Advantages are that no matrices of influence coefficients need to be computed and that the segments assumed to be prismatic may be fairly long. A shortcoming is the necessity to compute three unknowns ζ_i, u_i, v_i (against only two (u_i, v_i) with of the method of discrete mass points) at each joint.

4.2.3 Simplified solution

The procedure is analogous to that described in Sections 3.3 and 4.1.3 except for the obvious difference that now there are three components of the end forces, i.e. a moment and a horizontal and a vertical component of force. They are again computed from the static influence lines[44]. To determine the forces acting at the ends of a bar, we must know six influence lines, each having horizontal and vertical components of displacement. The lines formed at the unit displacements of the left- and right-hand ends are, however, bound by the following simple relations:

$$u(s)_{(u_g=1)} = 1 - u(s)_{(u_h=1)}$$
$$v(s)_{(u_g=1)} = - v(s)_{(u_h=1)}$$
$$u(s)_{(v_g=1)} = - u(s)_{(v_h=1)}$$
$$v(s)_{(v_g=1)} = 1 - v(s)_{(v_h=1)} \tag{4.93}$$

where $v(s)_{(u_g=1)}$ denotes the vertical displacement at point s at unit end displacement $u_g = 1$, etc. Because of the above relationships, four deflection lines are sufficient for our purpose. The end moments and forces are computed from the formulae

$$M_{gh(u_h=1)} \approx M_{ghst} - \omega^2 \int \mu(s) \left[u(s)_{(\zeta_g=1)} u(s)_{(u_h=1)} + v(s)_{(\zeta_g=1)} v(s)_{(u_h=1)} \right] ds$$

$$X_{hg(v_h=1)} \approx X_{hgst} - \omega^2 \int \mu(s) \left[u(s)_{(u_h=1)} u(s)_{(v_h=1)} + v(s)_{(u_h=1)} v_i(s)_{(v_h=1)} \right] ds$$

etc. or − after the integral has been replaced by a summation − from

$$M_{gh(u_h=1)} \approx M_{ghst} - \omega^2 \sum_i m_i u_{i(\zeta_g=1)} u_{i(u_h=1)} - \omega^2 \sum_i m_i v_{i(\zeta_g=1)} v_{i(u_h=1)} \quad (4.94)$$

etc.

Because of eq. 4.93

$$X_{gh(\zeta_h=1)} = -X_{hg(\zeta_h=1)} - \omega^2 \sum m_i \cdot u_{i(\zeta_h=1)}$$

$$Y_{gh(\zeta_h=1)} = -Y_{hg(\zeta_h=1)} - \omega^2 \sum m_i \cdot v_{i(\zeta_h=1)}$$

$$X_{gh(u_h=1)} = -X_{hg(u_h=1)} - \omega^2 \sum m_i \cdot u_{i(u_h=1)}$$

$$X_{gh(u_g=1)} = \quad X_{hg(u_g=1)} - \omega^2 \sum m_i + \omega^2 \sum m_i \cdot u_{i(u_h=1)}$$

$$Y_{gh(u_g=1)} = -Y_{hg(u_g=1)} + \omega^2 \sum m_i \cdot v_{i(u_h=1)}$$

$$Y_{gh(v_g=1)} = \quad Y_{hg(v_g=1)} - \omega^2 \sum m_i + \omega^2 \sum m_i \cdot v_{i(v_h=1)}$$

$$(4.95)$$

etc. The procedure will be illustrated by way of the following example.

Example 4.3

Consider the arch shown in Fig. 4.17. The system is symmetric, with the dimensions as indicated in Table 4.12. Applying the simplified slope-deflection method, we first divide the system into two bars, gp and ph. The static deflection

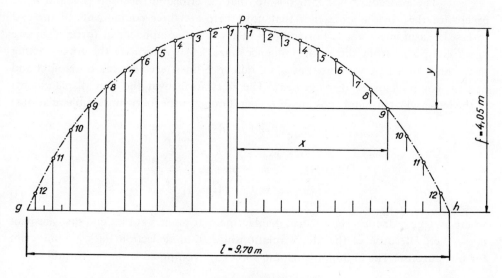

Fig. 4.17 Arch

curves for unit end deformations are shown in Fig. 4.18. The horizontal components of displacement, u_i, are set out in Table 4.13, the vertical ones, v_i, in Table 4.14.

In abbreviated notation the components of displacement are

$$u_{ai} = u_{i(\zeta_g=1)} \quad v_{ai} = v_{i(\zeta_g=1)}$$
$$u_{bi} = u_{i(\zeta_p=1)} \quad v_{bi} = v_{i(\zeta_p=1)}$$
$$u_{ci} = u_{i(u_p=1)} \quad v_{ci} = v_{i(u_p=1)}$$
$$u_{di} = u_{i(v_p=1)} \quad v_{di} = v_{i(v_p=1)}$$

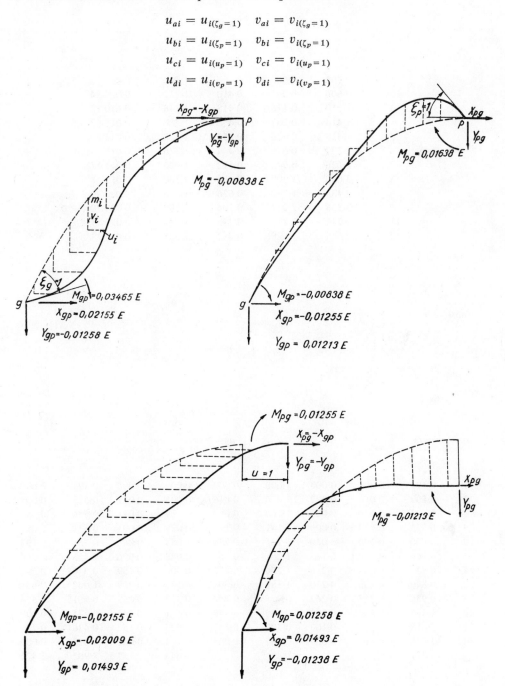

Fig. 4.18 Unit static deformations of arch (a) $\xi_g = 1$, (b) $\xi_p = 1$, (c) $u_p = 1$, (d) $v_p = 1$

Table 4.12

Segment	x m	y m	Δs m	S m^2	J m^4
1	0·202	0·008	0·406	1·216	0·009 36
2	0·606	0·045	0·410	1·232	0·009 74
3	1·010	0·128	0·418	1·248	0·010 12
4	1·414	0·252	0·432	1·280	0·010 92
5	1·818	0·430	0·452	1·328	0·012 20
6	2·222	0·655	0·484	1·376	0·013 57
7	2·626	0·945	0·512	1·440	0·015 55
8	3·030	1·300	0·566	1·500	0·017 58
9	3·434	1·730	0·626	1·576	0·020 39
10	3·838	2·264	0·704	1·680	0·024 70
11	4·242	2·872	0·795	1·784	0·029 57
12	4·647	3·628	0·912	1·928	0·037 33

Table 4.13

1	2	3	4	5	6	7	8	9	10
Point	Mass (volume)	Ordinates							Dynamic
i	m_i/ϱ	u_a	u_b	u_c	u_d	$\dfrac{m_i u_a}{\varrho}$	$\dfrac{m_i u_b}{\varrho}$	$\dfrac{m_i u_c}{\varrho}$	$\dfrac{m_i u_d}{\varrho}$
1	0·494	−0·008	−0·008	1·000	−0·000	−0·005	−0·005	0·494	0·000
2	0·505	−0·010	−0·022	1·016	−0·016	−0·005	−0·011	0·513	−0·008
3	0·522	−0·036	−0·022	1·066	−0·069	−0·019	−0·011	0·556	−0·036
4	0·553	−0·070	0·009	1·141	−0·157	−0·039	0·005	0·631	−0·087
5	0·600	−0·090	0·070	1·225	−0·272	−0·054	0·042	0·735	−0·163
6	0·666	−0·072	0·144	1·281	−0·385	−0·048	0·096	0·853	−0·256
7	0·737	0·008	0·216	1·276	−0·472	0·006	0·159	0·940	−0·347
8	0·849	0·160	0·270	1·181	−0·508	0·136	0·229	1·002	−0·431
9	0·986	0·357	0·262	0·975	−0·464	0·352	0·259	0·961	−0·456
10	1·183	0·567	0·203	0·655	−0·338	0·671	0·241	0·775	−0·400
11	1·418	0·652	0·107	0·303	−0·197	0·925	0·152	0·430	−0·279
12	1·758	0·412	0·004	0·007	−0·005	0·723	0·007	0·012	−0·009
Σ	10·271					2·643	1·163	7·902	−2·472

Units: Mp, m.

By formulae 4.94, 4.95 and Tables 4.13, 4.14 we get for the end forces and moments

$\zeta_p = 1$

$$M_{pg} = 0.016\,38E - \varrho\omega^2(\ 0.246 + 0.955) = 0.016\,38E - 1.201\varrho\omega^2$$
$$X_{pg} = 0.012\,55E - \varrho\omega^2(\ 1.082 - 0.306) = 0.012\,55E - 0.776\varrho\omega^2$$
$$Y_{pg} = -0.012\,13E + \varrho\omega^2(\ 0.470 + 1.159) = -0.012\,13E + 1.629\varrho\omega^2$$
$$M_{gp} = -0.008\,38E + \varrho\omega^2(-0.358 + 0.792) = -0.008\,38E + 0.434\varrho\omega^2$$
$$X_{gp} = -0.012\,55E - \varrho\omega^2(\ 1.163 - 0.776) = -0.012\,55E - 0.387\varrho\omega^2$$
$$Y_{gp} = 0.012\,13E + \varrho\omega^2(\ 2.303 - 1.629) = 0.012\,13E + 0.674\varrho\omega^2$$

$u_p = 1$

$$M_{pg} = 0.012\,55E - 0.776\varrho\omega^2$$
$$X_{pg} = 0.020\,09E - \varrho\omega^2(\ 8.274 + 0.138) = 0.020\,09E - 8.412\varrho\omega^2$$
$$Y_{pg} = -0.014\,93E + \varrho\omega^2(\ 2.416 - 0.277) = -0.014\,93E + 2.139\varrho\omega^2$$
$$M_{gp} = -0.021\,55E - \varrho\omega^2(\ 1.036 + 0.448) = -0.021\,55E - 1.484\varrho\omega^2$$
$$X_{gp} = -0.020\,09E - \varrho\omega^2(\ 7.902 - 8.412) = -0.020\,09E + 0.510\varrho\omega^2$$
$$Y_{gp} = 0.014\,93E - \varrho\omega^2(\ 0.929 + 2.139) = 0.014\,93E - 3.068\varrho\omega^2$$

11	12	13	14	15	16	17	18	19	20

complements

$m_i u_a^2/\varrho$	$m_i u_a u_b/\varrho$	$m_i u_a u_c/\varrho$	$m_i u_a u_d/\varrho$	$m_i u_b^2/\varrho$	$m_i u_b u_c/\varrho$	$m_i u_b u_d/\varrho$	$m_i u_c^2/\varrho$	$m_i u_c u_d/\varrho$	$m_i u_d^2/\varrho$
0.000	0.000	−0.004	0.000	0.000	−0.004	0.000	0.494	0.000	0.000
0.000	0.000	−0.005	0.000	0.000	−0.011	0.000	0.522	−0.008	0.000
0.001	0.000	−0.020	0.000	0.000	−0.012	0.001	0.592	−0.038	0.003
0.003	0.000	−0.044	0.006	0.000	0.006	−0.001	0.720	−0.099	0.014
0.005	−0.004	−0.066	0.015	0.003	0.051	−0.011	0.900	−0.199	0.044
0.003	−0.007	−0.061	0.018	0.014	0.123	−0.037	1.091	−0.328	0.099
0.000	0.001	0.008	−0.003	0.034	0.203	−0.075	1.200	−0.443	0.164
0.022	0.037	0.161	−0.069	0.062	0.270	−0.116	1.182	−0.509	0.219
0.125	0.092	0.343	−0.163	0.068	0.252	−0.120	0.936	−0.445	0.212
0.381	0.137	0.439	−0.227	0.049	0.158	−0.081	0.507	−0.262	0.135
0.602	0.099	0.280	−0.182	0.016	0.046	−0.030	0.130	−0.085	0.055
0.298	0.003	0.005	−0.004	0.000	0.000	0.000	0.000	0.000	0.000
1.440	0.358	1.036	−0.609	0.246	1.082	−0.470	8.274	−2.416	0.945

$v_p = 1$

$$M_{pg} \qquad\qquad\qquad\qquad = -0{\cdot}012\,13E + 1{\cdot}629\varrho\omega^2$$

$$X_{pg} \qquad\qquad\qquad\qquad = -0{\cdot}014\,93E + 2{\cdot}139\varrho\omega^2$$

$$Y_{pg} = \quad 0{\cdot}012\,38E - \varrho\omega^2(\quad 0{\cdot}945 + 1{\cdot}860) \quad = \quad 0{\cdot}012\,38E - 2{\cdot}805\varrho\omega^2$$

$$M_{gp} = \quad 0{\cdot}012\,58E - \varrho\omega^2(-0{\cdot}609 + 0{\cdot}626) \quad = \quad 0{\cdot}012\,58E - 0{\cdot}017\varrho\omega^2$$

$$X_{gp} = \quad 0{\cdot}014\,93E + \varrho\omega^2(\quad 2{\cdot}472 - 2{\cdot}139) \quad = \quad 0{\cdot}014\,93E + 0{\cdot}333\varrho\omega^2$$

$$Y_{gp} = -0{\cdot}012\,38E - \varrho\omega^2(\quad 2{\cdot}769 - 2{\cdot}805) \quad = -0{\cdot}012\,38E + 0{\cdot}036\varrho\omega^2$$

$\zeta_g = 1$

$$M_{pg} \qquad\qquad\qquad\qquad = -0{\cdot}008\,38E + 0{\cdot}434\varrho\omega^2$$

$$X_{pg} \qquad\qquad\qquad\qquad = -0{\cdot}021\,55E - 1{\cdot}484\varrho\omega^2$$

$$Y_{pg} \qquad\qquad\qquad\qquad = \quad 0{\cdot}012\,58E - 0{\cdot}017\varrho\omega^2$$

$$M_{gp} = \quad 0{\cdot}003\,465E - \varrho\omega^2(\quad 1{\cdot}440 + 2{\cdot}029) \quad = \quad 0{\cdot}003\,465E - 3{\cdot}469\varrho\omega^2$$

$$X_{gp} = \quad 0{\cdot}021\,55E \; - \varrho\omega^2(\quad 2{\cdot}643 - 1{\cdot}484) \quad = \quad 0{\cdot}021\,55E - 1{\cdot}159\varrho\omega^2$$

$$Y_{gp} = -0{\cdot}012\,58E \; - \varrho\omega^2(\quad 4{\cdot}014 - 0{\cdot}017) \quad = -0{\cdot}012\,58E - 3{\cdot}997\varrho\omega^2$$

Table 4.14

1	2	3	4	5	6	7	8	9	10
Point	Mass	Ordinates							Dynamic
i	m_i/ϱ	v_a	v_b	v_c	v_d	$\dfrac{m_i v_a}{\varrho}$	$\dfrac{m_i v_b}{\varrho}$	$\dfrac{m_i v_c}{\varrho}$	$\dfrac{m_i v_d}{\varrho}$
1	0·494	0·000	−0·202	0·000	1·000	0·000	−0·100	0·000	0·494
2	0·505	0·040	−0·444	0·035	0·929	0·020	−0·224	0·017	0·469
3	0·522	0·113	−0·563	0·085	0·814	0·059	−0·294	0·044	0·426
4	0·553	0·213	−0·583	0·136	0·671	0·118	−0·322	0·075	0·371
5	0·600	0·329	−0·539	0·175	0·522	0·197	−0·324	0·105	0·313
6	0·666	0·447	−0·454	0·195	0·380	0·298	−0·302	0·130	0·253
7	0·737	0·552	−0·352	0·194	0·255	0·407	−0·259	0·142	0·188
8	0·849	0·629	−0·245	0·171	0·153	0·534	−0·208	0·145	0·130
9	0·986	0·654	−0·150	0·131	0·079	0·645	−0·148	0·129	0·078
10	1·183	0·607	−0·075	0·081	0·031	0·718	−0·088	0·096	0·037
11	1·418	0·466	−0·024	0·033	0·007	0·661	−0·034	0·046	0·010
12	1·758	0·203	0·000	0·000	0·000	0·357	0·000	0·000	0·000
Σ	10·271					4·014	−2·303	0·929	2·769

Units: Mp, m.

$u_g = 1$

$$M_{pg} \qquad\qquad\qquad\qquad\qquad = -0.012\,55E - 0.387\varrho\omega^2$$
$$X_{pg} \qquad\qquad\qquad\qquad\qquad = -0.020\,09E + 0.510\varrho\omega^2$$
$$Y_{pg} \qquad\qquad\qquad\qquad\qquad = 0.014\,93E + 0.333\varrho\omega^2$$
$$M_{gp} \qquad\qquad\qquad\qquad\qquad = 0.021\,55E - 1.159\varrho\omega^2$$
$$X_{gp} = 0.020\,09E - \varrho\omega^2(\ 10.271 - 7.902 + 0.510) = 0.020\,09E - 2.880\varrho\omega^2$$
$$Y_{gp} = -0.014\,93E + \varrho\omega^2(\ 0.929 - 0.333) \qquad = -0.014\,93E + 0.596\varrho\omega^2$$

$v_g = 1$

$$M_{pg} \qquad\qquad\qquad\qquad\qquad = 0.012\,13E + 0.674\varrho\omega^2$$
$$X_{pg} \qquad\qquad\qquad\qquad\qquad = 0.014\,93E - 3.068\varrho\omega^2$$
$$Y_{pg} \qquad\qquad\qquad\qquad\qquad = -0.012\,38E + 0.036\varrho\omega^2$$
$$M_{gp} \qquad\qquad\qquad\qquad\qquad = -0.012\,58E - 3.997\varrho\omega^2$$
$$X_{gp} \qquad\qquad\qquad\qquad\qquad = -0.014\,93E + 0.596\varrho\omega^2$$
$$Y_{gp} = 0.012\,38E - \varrho\omega^2(\ 10.271 - 2.769 + 0.036) = 0.012\,38E - 7.538\varrho\omega^2$$

11	12	13	14	15	16	17	18	19	20

complements

$m_i v_a^2/\varrho$	$m_i v_a v_b/\varrho$	$m_i v_a v_c/\varrho$	$m_i v_a v_d/\varrho$	$m_i v_b^2/\varrho$	$m_i v_b v_c/\varrho$	$m_i v_b v_d/\varrho$	$m_i v_c^2/\varrho$	$m_i v_c v_d/\varrho$	$m_i v_d^2/\varrho$
0·000	0·000	0·000	0·000	0·020	0·000	−0·099	0·000	0·000	0·494
0·001	−0·009	0·001	0·019	0·099	−0·008	−0·208	0·001	0·016	0·436
0·007	−0·033	0·005	0·048	0·165	−0·025	−0·239	0·004	0·036	0·346
0·025	−0·069	0·016	0·079	0·188	−0·044	−0·216	0·010	0·050	0·249
0·065	−0·106	0·034	0·103	0·174	−0·057	−0·169	0·018	0·055	0·164
0·133	−0·135	0·058	0·113	0·137	−0·059	−0·115	0·025	0·049	0·096
0·224	−0·143	0·079	0·104	0·091	−0·050	−0·066	0·028	0·036	0·048
0·336	−0·130	0·091	0·082	0·051	−0·036	−0·032	0·025	0·022	0·020
0·422	−0·097	0·084	0·051	0·022	−0·019	−0·012	0·017	0·010	0·006
0·436	−0·054	0·058	0·022	0·007	−0·007	−0·003	0·008	0·003	0·001
0·308	−0·016	0·022	0·005	0·001	−0·001	0·000	0·002	0·000	0·000
0·072	0·000	0·000	0·000	0·000	0·000	0·000	0·000	0·000	0·000
2·029	−0·792	0·448	0·626	0·955	−0·306	−1·159	0·138	0·277	1·860

For the symmetric vibration of the system shown in Fig. 4.17 there is just one slope-deflection equation for the vertical forces at the point p

$$2(0 \cdot 012\,38E - 2 \cdot 805 \varrho \omega^2)\, v_p = 0$$

from which $\left(\text{with } E/\varrho = 9 \cdot 81 \times 10^6 \text{ m}^2 \text{ s}^{-2}\right)$

$$\omega_{(1)} = \left(\frac{0 \cdot 012\,38E}{2 \cdot 805\varrho}\right)^{1/2} = 0 \cdot 0665 \left(\frac{E}{\varrho}\right)^{1/2} = 208 \text{ s}^{-1}$$

The two slope-deflection equations for the antisymmetric vibration are

$$(0 \cdot 016\,38E - 1 \cdot 201\varrho\omega^2)\,\zeta_p + (0 \cdot 012\,55E - 0 \cdot 776\varrho\omega^2)\,u_p = 0$$
$$(0 \cdot 012\,55E - 0 \cdot 776\varrho\omega^2)\,\zeta_p + (0 \cdot 020\,09E - 8 \cdot 412\varrho\omega^2)\,u_p = 0$$

The natural circular frequency is obtained from the condition that the determinant of the coefficients be made equal to zero; hence $\omega_{(1')} = 0 \cdot 0363(E/\varrho)^{1/2}$, $\omega_{(2')} = 0 \cdot 117(E/\varrho)^{1/2}$, and with $E/\varrho = 9 \cdot 8 \times 10^6 \text{ m}^2 \text{ s}^{-2}$, $\omega_{(1')} = 113 \text{ s}^{-1}$, $\omega_{(2')} = 367 \text{ s}^{-1}$, the last value being for reference only.

4.2.4   Modal analysis

The equations of motion of a system executing forced vibration which is caused by a continuous, time varying load $p(s, t)$ in an arbitrary direction (Fig. 4.19) are $(\delta(s, a)$ is the matrix of influence functions)

$$r(a, t) = \int \delta(s, a) \left[-\mu(s)\,\ddot{r}(s, t) + p(s, t)\right] \mathrm{d}s = 0 \tag{4.96}$$

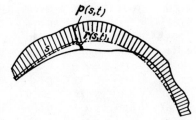

Fig. 4.19 Forced vibration of bar with curved centre line

Both the deflections and the loads can be expanded in a series of natural modes

$$r(s, t) = \sum_j q_{(j)}(t)\, r_{(j)}(s) \tag{4.97}$$

$$p(s, t) = \sum_j \mu(s)\, p_{(j)}(t)\, r_{(j)}(s) \tag{4.98}$$

For free harmonic vibration, $r(s, t) = r_{(j)}(s) \sin \omega t$, we get from (4.96)

$$r_{(j)}(a) = \omega_{(j)}^2 \int \mu(s) \, \delta(s, a) \, r_{(j)}(s) \, ds \tag{4.99}$$

The coefficients of the series 4.98 are computed from eq. 4.98 by multiplying the equation by $r_{(j)}(s) \, ds$ and integrating the result over the whole system:

$$p_{(j)}(t) = \frac{\int p(s, t) \, r_{(j)}(s) \, ds}{\int \mu(s) \, r_{(j)}^2(s) \, ds} \tag{4.100}$$

Eqs 4.96−4.99 then give

$$\ddot{q}_{(j)}(t) + \omega_{(j)}^2 \, \dot{q}_{(j)}(t) = p_{(j)}(t) \tag{4.101}$$

For forced damped vibration the above equation takes on the form (cf. Chapter 6)

$$\ddot{q}_{(j)}(t) + 2\omega_b \, \dot{q}_{(j)}(t) + \omega_{(j)}^2(t) = p_{(j)}(t) \tag{4.102}$$

4.3 Systems with bars of skew curvature

These systems are analysed by essentially the same methods as systems with bars of plane curvature (discussed in Section 4.2). No direct, exact solution is possible except in some special cases (cf. References 9, 17 and 30).

4.3.1 Method of discrete mass points

The solution evolved in Section 4.2.1 can readily be generalised to bars of skew curvature (Fig. 4.20). Eqs 4.79, 4.80, 4.82 and 4.83 continue to hold but the displacement vector r_i at point i has three components

$$r_i(t) = \begin{bmatrix} u_i(t) \\ v_i(t) \\ w_i(t) \end{bmatrix} \tag{4.103}$$

and the sub-matrix of the influence coefficients is

$$\Delta_{ik} = \begin{bmatrix} \delta_{ik}^{uu} & \delta_{ik}^{uv} & \delta_{ik}^{uw} \\ \delta_{ik}^{vu} & \delta_{ik}^{vv} & \delta_{ik}^{vw} \\ \delta_{ik}^{wu} & \delta_{ik}^{wv} & \delta_{ik}^{ww} \end{bmatrix} \tag{4.104}$$

The equations of the form of eq. 4.85 now are (Fig. 4.21)

$$u_i(t) = r_i(t)$$
$$v_i(t) = r_{n+1}(t)$$
$$w_i(t) = r_{2n+i}(t)$$
$$m_i = m_{n+1} = m_{n+2}$$

$$(4.105)$$

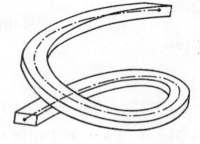

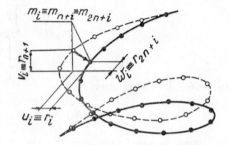

Fig. 4.20 Bar of skew curvature   Fig. 4.21 Concentration of mass at discrete points

where n is the total number of discrete points, and eqs 4.86, 4.87 are replaced by

$$r_i(t) = -\sum_{k=1}^{3n} m_k \delta_{ik} \ddot{r}_k(t)$$

$$(4.106)$$

$$r_i(t) = \omega^2 \sum_{k=1}^{3n} m_k \delta_{ik} r_k$$

$$(4.107)$$

4.3.2 The slope-deflection method

The computational procedure is analogous to that in Section 4.2.2. The bar is divided by intermediate joints into short prismatic segments (Fig. 4.22). For each joint there are six conditions of equilibrium expressed by six slope-deflection equations (cf. References 37 and 38)

$$K_{i,i-1} + K_{i,i+1} = \mathcal{K}_i$$
$$L_{i,i-1} + L_{i,i+1} = \mathcal{L}_i$$
$$M_{i,i-1} + M_{i,i+1} = \mathcal{M}_i$$
$$X_{i,i-1} + X_{i,i+1} = \mathcal{X}_i$$
$$Y_{i,i-1} + Y_{i,i+1} = \mathcal{Y}_i$$
$$Z_{i,i-1} + Z_{i,i+1} = \mathcal{Z}_i$$

$$(4.108)$$

where $K_{i,i-1}$, $L_{i,i-1}$, $M_{i,i-1}$ are the amplitudes of the end moments of bar i, $i-1$

acting at point i about axes X, Y, Z; $X_{i,i-1}$, $Y_{i,i-1}$, $Z_{i,i-1}$ are the components of the end forces. The symbols to the right of eq. 4.108 denote the respective components of the external load.

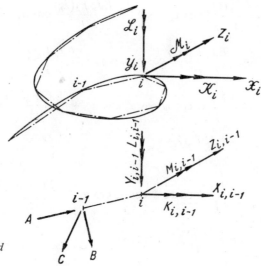

Fig. 4.22 Bar of skew curvature divided into prismatic elements

The end forces and moments of an arbitrary bar, say ik, depend on the rotations ξ_i, η_i, ζ_i, ξ_k, η_k, ζ_k and the displacements u_i, v_i, w_i, u_k, v_k, w_k of the bar ends. Tables 10.1, 10.2 in the Appendix, and eq. 2.69, lead to the relations set out in Table 4.15.

Table 4.15

	ξ_i	η_i	ζ_i	u_i	v_i	w_i	ξ_k	η_k	ζ_k	u_k	v_k	w_k
$K_{ik} =$	a_{xx}	a_{xy}	a_{xz}	b_{xx}	b_{xy}	b_{xz}	c_{xx}	c_{xy}	c_{xz}	d_{xx}	d_{xy}	d_{xz}
$L_{ik} =$	a_{xy}	a_{yy}	a_{yz}	b_{yx}	b_{yy}	b_{yz}	c_{xy}	c_{yy}	c_{yz}	d_{yx}	d_{yy}	d_{yz}
$M_{ik} =$	a_{xz}	a_{yz}	a_{zz}	b_{zx}	b_{zy}	b_{zz}	c_{xz}	c_{yz}	c_{zz}	d_{zx}	d_{zy}	d_{zz}
$X_{ik} =$	b_{xx}	b_{yx}	b_{zx}	e_{xx}	e_{xy}	e_{xz}	$-d_{xx}$	$-d_{yx}$	$-d_{zx}$	f_{xx}	f_{xy}	f_{xz}
$Y_{ik} =$	b_{xy}	b_{yy}	b_{zy}	e_{xy}	e_{yy}	e_{yz}	$-d_{xy}$	$-d_{yy}$	$-d_{zy}$	f_{xy}	f_{yy}	f_{yz}
$Z_{ik} =$	b_{xz}	b_{yz}	b_{zz}	e_{xz}	e_{yz}	e_{zz}	$-d_{xz}$	$-d_{yz}$	$-d_{zz}$	f_{xz}	f_{yz}	f_{zz}

In the table

$$a_{xx} = \frac{GJ_a}{l}\cos^2\alpha_x + \frac{EJ_b\,F_2(\lambda_b)}{l}\cos^2\beta_x + \frac{EJ_c\,F_2(\lambda_c)}{l}\cos^2\gamma_x$$

$$a_{xy} = \frac{GJ_a}{l}\cos\alpha_x\cos\alpha_y + \frac{EJ_b\,F_2(\lambda_b)}{l}\cos\beta_x\cos\beta_y + \frac{EJ_c\,F_2(\lambda_c)}{l}\cos\gamma_x\cos\gamma_y$$

$$b_{xy} = \frac{EJ_b\,F_4(\lambda_b)}{l^2}\cos\beta_x\cos\gamma_y - \frac{EJ_c\,F_4(\lambda_c)}{l^2}\cos\gamma_x\cos\beta_y$$

$$c_{xy} = -\frac{GJ_a}{l}\cos\alpha_x\cos\alpha_y + \frac{EJ_b\,F_1(\lambda_b)}{l}\cos\beta_x\cos\beta_y + \frac{EJ_c\,F_1(\lambda_c)}{l}\cos\gamma_x\cos\gamma_y$$

$$d_{xy} = \frac{EJ_b\,F_3(\lambda_b)}{l^2}\cos\beta_x\cos\gamma_y - \frac{EJ_c\,F_3(\lambda_c)}{l^2}\cos\gamma_x\cos\beta_y$$

$$e_{xy} = \frac{ES}{l}\psi\cot\psi\cos\alpha_x\cos\alpha_y + \frac{EJ_c\,F_6(\lambda_c)}{l^3}\cos\beta_x\cos\beta_y$$

$$+ \frac{EJ_b\,F_6(\lambda_b)}{l^3}\cos\gamma_x\cos\gamma_y$$

$$f_{xy} = -\frac{ES}{l}\psi\operatorname{cosec}\psi\cos\alpha_y\cos\alpha_z + \frac{EJ_c\,F_5(\lambda_c)}{l^3}\cos\beta_x\cos\beta_y$$

$$+ \frac{EJ_b\,F_5(\lambda_b)}{l^3}\cos\gamma_x\cos\gamma_y$$

$$(4.109)$$

The remaining coefficients, e.g. a_{zx}, etc. are obtained from formulae 4.109 by a simple interchange of the subscripts. The symbols used in the formulae have the following meaning: l — the span of the segment, J_b — the moment of inertia about the bar's axis B, $\lambda_b = l(\mu\omega^2/EJ_b)^{1/4}$, etc.; ψ is computed from eq. 2.64; $\alpha_x, \beta_x, \gamma_x$ are the angles between the positive direction of axis X and axes A, B, C. $F(\lambda)$ are the functions tabulated in the Appendix.

The expressions from eqs 4.108 are substituted in Table 4.15 to obtain the unknown deformations.

Note: In the computation of torques GJ_a/l we have neglected to include the effect of the inertia forces of the rotating bar. A more exact computation using eq. 2.76 would not be difficult; however, we have neglected the effects of rotatory inertia also in the computation of bending moments and shall not consider them until Chapter 8. They are in fact of importance only in massive bars.

4.3.3 Simplified solution

The method described in Section 4.2.3 is also suitable for systems with bars of skew curvature. The forces acting at end g, corresponding to an arbitrary end deformation of bar gh are given by six components, three components of moments about axes X, Y, Z and three components of forces in the direction of those axes.

They are computed analogously to eq. 4.72. Thus, for example, the moment about axis X at point g, with point h undergoing displacement $w_h = 1$ along axis Z is approximately

$$K_{gh(w_h = 1)} \approx K_{ghst} - \omega^2 \sum_i m_i \mathbf{r}_{1i} \mathbf{r}_{2i}$$

$$= K_{ghst} - \omega^2 \sum_i m_i (u_{1i} u_{2i} + v_{1i} v_{2i} + w_{1i} w_{2i}) \qquad (4.110)$$

where $\mathbf{r}_{1i}$ is the displacement vector at point i in the case where end g of bar gh rotates about axis X with $\xi_g = 1$. The components of displacement $\mathbf{r}_{1i}$ in the direction of axes X, Y, Z are u_{1i}, v_{1i}, w_{1i}. Vector $\mathbf{r}_{2i}$ denotes the displacement at point i when end h is displaced in the direction of axis Z with $w_h = 1$. K_{ghst} denotes the static value of the moment.

5   CYCLOSYMMETRIC SYSTEMS. SYSTEMS WITH REPEATED ELEMENTS

5.1   Three-dimensional cyclosymmetric systems*

Three-dimensional cyclosymmetric frame systems are symmetric about a bundle of vertical planes dividing the space in equal parts. The planes of symmetry intersect one another on the vertical axis of the system. The systems are composed of horizontal bars forming regular polygons (rings) and oblique − or possibly also vertical, straight or curved − bars (ribs) lying in the planes of symmetry. Multiple symmetry of the systems makes for a greatly simplified solution (cf. Section 1.5.2).

An exemplary system of this kind is shown in Fig. 5.1. Each joint is denoted by the symbol of the ring (generally J) and the symbol of the rib (generally k). Thus the fraction J/k refers to the joint in which the J-th ring intersects the k-th rib, fraction $J/(k-1)$ to the joint at the point of intersection of the J-th ring and $(k-1)$-th rib, etc. A separate system of orthogonal axes X, Y, Z is chosen for each joint: Z is the horizontal axis in the plane of the rib, Y the vertical axis, and X the axis at right angles to the plane of the rib. If N is the total number of rings and h that of ribs, there are $6Nh$ unknown joint deformations, and we need $6Nh$ slope-deflection equations to solve for them.

Symbols used for denoting the forces and deformations are reviewed in Table 3.12, Section 3.6.

The unknown deformation components of joint J/k may be divided into components in the planes of symmetry − $w_{J/k}$, $\xi_{J/k}$, $v_{J/k}$ (termed symmetric components), and into components $\eta_{J/k}$, $\zeta_{J/k}$, $u_{J/k}$ (antisymmetric components). The bars of the ribs lie in the planes of symmetry; their symmetric (antisymmetric) deformation components are therefore caused only by symmetric (antisymmetric) forces.

---

* A static solution of these systems is presented in Reference 65; for a dynamic analysis made by the author refer to References 37 and 38.

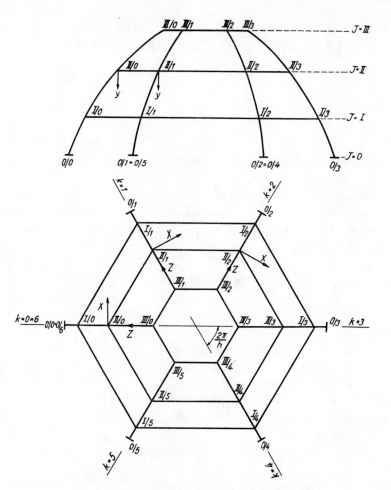

Fig. 5.1 Cyclosymmetric frame structure of a dome

5.1.1 Steady-state forced vibration*

The slope-deflection equations

The slope-deflection equations are obtained in a manner similar to that used in Chapter 3. They represent the conditions of equilibrium of forces and moments acting at the various joints of the system:

* To avoid the tedious task of writing the equations of cyclosymmetric systems twice, we have reversed our usual procedure and shall examine forced vibration first. In this section the symbols of forces and deformations denote the amplitudes in simple, steady harmonic motion.

$$K_{J/k,k-1} + K_{J/k,k+1} + K_{J,J-1/k} + K_{J,J+1/k} = \mathcal{K}_{J/k} \tag{5.1}$$

$$Y_{J/k,k-1} + Y_{J/k,k+1} + Y_{J,J-1/k} + Y_{J,J+1/k} = \mathcal{Y}_{J/k} \tag{5.2}$$

$$Z_{J/k,k-1} + Z_{J/k,k+1} + Z_{J,J-1/k} + Z_{J,J+1/k} = \mathcal{Z}_{J/k} \tag{5.3}$$

$$L_{J/k,k-1} + L_{J/k,k+1} + L_{J,J-1/k} + L_{J,J+1/k} = \mathcal{L}_{J/k} \tag{5.4}$$

$$M_{J/k,k-1} + M_{J/k,k+1} + M_{J,J-1/k} + M_{J,J+1/k} = \mathcal{M}_{J/k} \tag{5.5}$$

$$X_{J/k,k-1} + X_{J/k,k+1} + X_{J,J-1/k} + X_{J,J+1/k} = \mathcal{X}_{J/k} \tag{5.6}$$

The subscripts refer to the bar and the point of action. Thus, for example, $K_{J,J+1/k}$ denotes the end moment about axis X at point J/k of bar $J/k - (J+1)/k$; $Z_{J/k,k-1}$ the end force in the direction of axis Z acting at point J/k of bar $J/k - J/(k-1)$, etc. Written letters $\mathcal{K}_{J/k}$, $\mathcal{X}_{J/k}$, etc. denote the respective components of external load acting at point J/k. The end moments and forces can be expressed in terms of the end deformations of the bars, and the equations of unknown deformations obtained from eqs 5.1 to 5.6. The latter can also be arrived at directly, by the application of the reciprocal theorem. Thus, e.g., for the state of actual deformation (Fig. 5.2) and the unit auxiliary state a (Fig. 5.2a) in which $w_{J/k} = 1$ and all the remaining deformations are zero, there holds the equation

$$
\begin{aligned}
&K_J^w \xi_{J/k} + K_{2J}^w(\xi_{J/k+1} + \xi_{J/k-1}) + K_{J+1,J}^w \xi_{J+1/k} + K_{J-1,J}^w \xi_{J-1/k} + Y_J^w v_{J/k} \\
&+ Y_{2J}^w(v_{J/k+1} + v_{J/k-1}) + Y_{J+1,J}^w v_{J+1/k} + Y_{J-1,J}^w v_{J-1/k} + Z_J^w w_{J/k} \\
&+ Z_{2J}^w(w_{J/k+1} + w_{J/k-1}) + Z_{J+1,J}^w w_{J+1,J} + Z_{J-1,J}^w w_{J-1/k} + L_{2J}^w(\eta_{J/k+1} - \eta_{J/k-1}) \\
&+ M_{2J}^w(\zeta_{J/k+1} - \zeta_{J/k-1}) + X_{2J}^w(u_{J/k+1} - u_{J/k-1}) = \mathcal{Z}_{J/k} \tag{5.7}
\end{aligned}
$$

which is identical to eq. 5.3. The meaning of the symbols used in the above is evident from Fig. 5.2a: L_{2J}^w is the amplitude of the end moment about axis Y at point $J/(k+1)$ of bar $J/k - J/(k+1)$, if point J/k is undergoing displacement with amplitude $w_{J/k} = 1$, etc. Analogous equations would be obtained by using the auxiliary state $\xi_{J/k} = 1$. Let us denote the moments and forces in this state by M_J^ξ, K_{2J}^ξ, etc. All the superscripts on the left-hand side of eq. 5.7 will now change to ξ, and the loading term on the right-hand side will be $\mathcal{K}_{J/k}$ (eq. 5.1). For the state $v_{J/k} = 1$ the procedure is wholly analogous.

For the auxiliary state $u_{J/k} = 1$ (Fig. 5.2b) eq. 5.6 takes on the form

$$
\begin{aligned}
&K_{2J}^u(\xi_{J/k+1} - \xi_{J/k-1}) + Y_{2J}^u(v_{J/k+1} - v_{J/k-1}) + Z_{2J}^u(w_{J/k+1} - w_{J/k-1}) + L_J^u \eta_{J/k} \\
&+ L_{2J}^u(\eta_{J/k-1} + \eta_{J/k+1}) + L_{J+1,J}^u \eta_{J+1/k} + L_{J-1,J}^u \eta_{J-1/k} + M_J^u \zeta_{J/k} \\
&+ M_{2J}^u(\zeta_{J/k-1} + \zeta_{J/k+1}) + M_{J+1,J}^u \zeta_{J+1/k} + M_{J-1,J}^u \zeta_{J-1/k} + X_J^u u_{J/k} \\
&+ X_{2J}^u(u_{J/k-1} + u_{J/k+1}) + X_{J+1,J}^u u_{J+1/k} + X_{J-1,J}^u u_{J-1/k} = \mathcal{X}_{J/k} \tag{5.8}
\end{aligned}
$$

The equations for states $\eta_{J/k} = 1$ and $\zeta_{J/k} = 1$ are obtained by changing the superscripts.

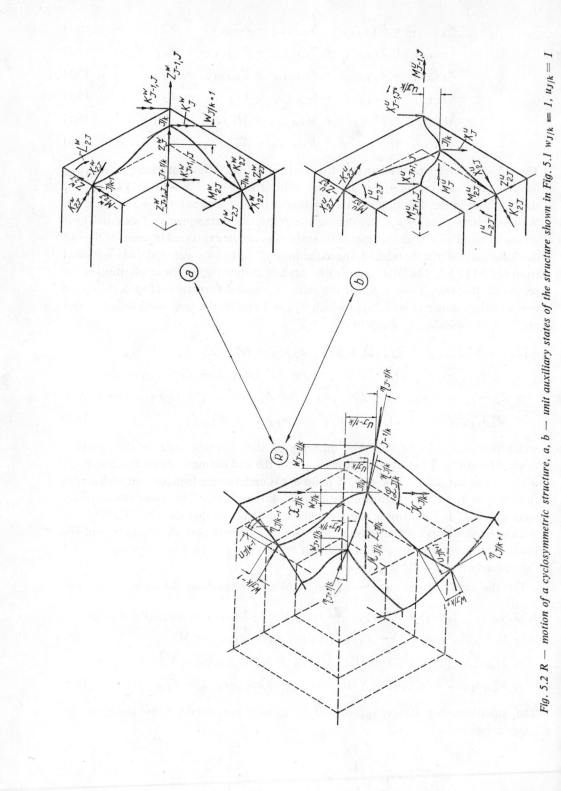

Fig. 5.2 R — motion of a cyclosymmetric structure, a, b — unit auxiliary states of the structure shown in Fig. 5.1 $w_{J/k} = 1$, $u_{J/k} = 1$

Eq. 5.7 may be given the form

$$a_\xi \Delta_k^2 \xi + a_v \Delta_k^2 v + a_w \Delta_k^2 w + b_\xi \Delta_J^2 \xi + b_v \Delta_J^2 v + b_w \Delta_J^2 w + c_\xi \Delta_J \xi + c_v \Delta_J v$$

$$+ c_w \Delta_J w + d_\eta \Delta_k \eta + d_\zeta \Delta_k \zeta + d_u \Delta_k u + e_\xi \xi_{J/k} + e_v v_{J/k} + e_w w_{J/k} = \mathscr{L}_{J/k} \qquad (5.9)$$

where

$$\Delta_k^2 \xi = \xi_{J/k-1} - 2\xi_{J/k} + \xi_{J/k+1}$$

$$\Delta_k \xi = \xi_{J/k+1} - \xi_{J/k-1}$$

$$\Delta_J^2 \xi = \xi_{J-1/k} - 2\xi_{J/k} + \xi_{J+1/k}$$

$$\Delta_J \xi = \xi_{J+1/k} - \xi_{J-1/k}$$

$$(5.10)$$

The coefficients of the equations are constant for all ks, and their values are as follows:

$$a_\xi = K_{2J}^w, \quad a_v = Y_{2J}^w, \quad a_w = Z_{2J}^w$$

$$b_\xi = \tfrac{1}{2}(K_{J+1,J}^w + K_{J-1,J}^w) \quad \text{etc.}$$

$$c_\xi = \tfrac{1}{2}(K_{J+1,J}^w - K_{J-1,J}^w) \quad \text{etc.}$$

$$d_\eta = L_{2J}^w, \quad d_\zeta = M_{2J}^w, \quad d_u = X_{2J}^w$$

$$e_\xi = K_J^w + 2K_{2J}^w + K_{J+1,J}^w + K_{J-1,J}^w$$

$$(5.11)$$

Transformation of deformations

Eqs 5.7 and 5.8 are difference equations of the second order (strictly speaking, partial difference equations, for they contain differences Δ_k and Δ_J), solved by procedures analogous to those used for similar differential equations. The unknown deformations can be expressed as* (cf. Section 1.5.2.1, formulae 1.115 and 1.116)

$$\xi_{J/k} = \sum_{j=0}^{h/2} \left[s_j'(\xi_J) \sin \frac{2\pi jk}{h} + s_j''(\xi_J) \cos \frac{2\pi jk}{h} \right]$$

$$v_{J/k} = \sum_{j=0}^{h/2} \left[s_j'(v_J) \sin \frac{2\pi jk}{h} + s_j''(v_J) \cos \frac{2\pi jk}{h} \right]$$

$$w_{J/k} = \sum_{j=0}^{h/2} \left[s_j'(w_J) \sin \frac{2\pi jk}{h} + s_j''(w_J) \cos \frac{2\pi jk}{h} \right]$$

$$(5.12)$$

* Applies to an even number of ribs. For an odd number of ribs $\sum$ is the sum from $j = 0$ to $(h-1)/2$.

$$\eta_{J/k} = \sum_{j=0}^{h/2} \left[r'_j(\eta_J) \cos \frac{2\pi jk}{h} - r''_j(\eta_J) \sin \frac{2\pi jk}{h} \right]$$

$$\zeta_{J/k} = \sum_{j=0}^{h/2} \left[r'_j(\zeta_J) \cos \frac{2\pi jk}{h} - r''_j(\zeta_J) \sin \frac{2\pi jk}{h} \right]$$

$$u_{J/k} = \sum_{j=0}^{h/2} \left[r'_j(u_J) \cos \frac{2\pi jk}{h} - r''_j(u_J) \sin \frac{2\pi jk}{h} \right]$$

(5.13)

And similarly so, the forces of external load are given by

$$\mathscr{K}_{j/k} = \sum_{j=0}^{h/2} \left[S'_j(\xi_J) \sin \frac{2\pi jk}{h} + S''_j(\xi_J) \cos \frac{2\pi jk}{h} \right]$$

$$\mathscr{X}_{J/k} = \sum_{j=0}^{h/2} \left[R'_j(u_J) \cos \frac{2\pi jk}{h} - R''_j(u_J) \sin \frac{2\pi jk}{h} \right], \quad \text{etc.}$$

(5.14)

The new unknowns s and r, on the other hand, can be expressed by means of the original ones.

Multiply eqs 5.12, 5.13 and 5.14 by $\sin 2\pi ik/h$ (or $\cos 2\pi ik/h$) and take the sum for all $k = 0$ to $h - 1$. Recall that the following relations hold good:

$$\sum_{k=1}^{h-1} \sin \frac{2\pi ik}{h} \sin \frac{2\pi jk}{h} \begin{cases} = 0 & \text{for } i \neq j \text{ and for } i = j = 0 \\ = h/2 & \text{for } i = j \neq 0 \end{cases}$$

$$\sum_{k=1}^{h-1} \cos \frac{2\pi ik}{h} \sin \frac{2\pi jk}{h} = 0$$

$$\sum_{k=0}^{h-1} \cos \frac{2\pi ik}{h} \cos \frac{2\pi jk}{h} \begin{cases} = 0 & \text{for } i \neq j \\ = h & \text{for } i = j = 0 \\ = h/2 & \text{for } i = j \neq 0 \end{cases}$$

(5.15)

For $j = 0$

$$s'_0(\xi_J) = 0$$

$$s''_0(\xi_J) = \frac{1}{h} \sum_{k=0}^{h-1} \xi_{J/k}$$

$$S'_0(w_j) = 0$$

$$S''_0(w_j) = \frac{1}{h} \sum_{k=0}^{h-1} \mathscr{L}_{J/k}$$

(5.16)

and so on for all the symmetric components of deformations and load. Also

$$r_0'(\eta_J) = \frac{1}{h}\sum_{k=0}^{h-1}\eta_{J/k}$$

$$r_0''(\eta_J) = 0$$

$$(5.17)$$

and so on for the antisymmetric components.

For $j = h/2$

$$s_{h/2}'(\xi_J) = 0$$

$$s_{h/2}''(\xi_J) = \frac{1}{h}\sum_{k=0}^{h-1}(-1)^k\,\xi_{J/k}$$

$$(5.18)$$

$$r_{h/2}'(\eta_J) = \frac{1}{h}\sum_{k=0}^{h-1}(-1)^k\,\eta_{J/k}$$

$$r_{h/2}''(\eta_J) = 0$$

$$R_{h/2}'(\eta_J) = \frac{1}{h}\sum_{k=0}^{h-1}(-1)^k\,\mathscr{L}_{J/k}$$

$$R_{h/2}''(\eta_J) = 0$$

$$(5.19)$$

For general j we then get

$$s_j'(\xi_J) = \frac{2}{h}\sum_{k=0}^{h-1}\xi_{J/k}\sin\frac{2\pi jk}{h}$$

$$s_j''(\xi_J) = \frac{2}{h}\sum_{k=0}^{h-1}\xi_{J/k}\cos\frac{2\pi jk}{h}$$

$$(5.20)$$

and similarly so for all the remaining symmetric deformations and loads, e.g.

$$S_j'(w_J) = \frac{2}{h}\sum_{k=1}^{h-1}\mathscr{L}_{J/k}\sin\frac{2\pi jk}{h}, \quad \text{etc.}$$

$$(5.21)$$

For the antisymmetric components

$$r_j'(u_J) = \frac{2}{h}\sum_{k=0}^{h-1}u_{J/k}\cos\frac{2\pi jk}{h}$$

$$r_j''(u_J) = -\frac{2}{h}\sum_{k=0}^{h-1}u_{J/k}\sin\frac{2\pi jk}{h}$$

$$R_j'(\zeta_J) = \frac{2}{h}\sum_{k=1}^{h-1}\mathscr{M}_{J/k}\cos\frac{2\pi jk}{h}$$

$$(5.22)$$

Eqs 5.12, 5.13 represent the expansion of an arbitrary deformation state of forced harmonic vibration in n fundamental modes characterised by coordinates r'_j, s'_j and r''_j, s''_j (for $j = 0$ to $h/2$, or for $j = 0$ to $(h - 1)/2$) which may be called the generalised coordinates of displacement. These fundamental modes are mutually orthogonal (see eq. 5.15).

Transformed equations

Substitute now expressions 5.12, 5.13, 5.21 in the slope-deflection equation 5.7. Because (cf. eqs 1.112 and 1.125)

$$\cos \frac{2\pi j(k - 1)}{h} \pm \cos \frac{2\pi j(k + 1)}{h} = \begin{cases} 2 \cos \dfrac{2\pi j}{h} \cos \dfrac{2\pi jk}{h} \\[2mm] 2 \sin \dfrac{2\pi j}{h} \sin \dfrac{2\pi jk}{h} \end{cases}$$

$$\sin \frac{2\pi j(k - 1)}{h} \pm \sin \frac{2\pi j(k + 1)}{h} = \begin{cases} 2 \cos \dfrac{2\pi j}{h} \sin \dfrac{2\pi jk}{h} \\[2mm] -2 \sin \dfrac{2\pi j}{h} \cos \dfrac{2\pi jk}{h} \end{cases}$$

$$(5.23)$$

and at identical $\sin 2\pi jk/h$ (or $\cos 2\pi jk/h$) the terms on both sides of the equations must equal one another, we get ordinary difference equations which for $j \neq 0$, $h/2$ may be written in the following form

$$a_{Jj}^{w\xi} s'_j(\xi_J) + a_{Jj}^{wv} s'_j(v_J) + a_{Jj}^{ww} s'_j(w_J) + b_{Jj}^{w\eta} r'_j(\eta_J)$$
$$+ b_{Jj}^{w\zeta} r'_j(\zeta_J) + b_{Jj}^{wu} r'_j(u_J) + c_J^{w\xi} s'_j(\xi_{J+1})$$
$$+ c_J^{wv} s'_j(v_{J+1}) + c_J^{ww} s'_j(w_{J+1}) + d_J^{w\xi} s'_j(\xi_{J-1})$$
$$+ d_J^{wv} s'_j(v_{J-1}) + d_J^{ww} s'_j(w_{j-1})$$
$$= \frac{2}{h} \sum_{k=0}^{h-1} \mathscr{L}_{J/k} \sin \frac{2\pi jk}{h}$$

$$(5.24)$$

The coefficients of the equations (the first superscript w denotes the displacement, the second ξ, v, etc. determines the force quantity on the right-hand side in the sense of Table 3.12, the first subscript J denotes the number of the ring, the second, j, the number of the fundamental mode) are:

$$a_{Jj}^{w\xi} = K_J^w + 2K_{2J}^w \cos \frac{2\pi j}{h}$$

$$a_{Jj}^{wv} = Y_J^w + 2Y_{2J}^w \cos \frac{2\pi j}{h}$$

$$a_{Jj}^{ww} = Z_J^w + 2Z_{2J}^w \cos \frac{2\pi j}{h}$$

$$b_{Jj}^{w\eta} = -2L_{2J}^w \sin \frac{2\pi j}{h}$$

$$b_{Jj}^{w\zeta} = -2M_{2J}^w \sin \frac{2\pi j}{h}$$

$$b_{Jj}^{wu} = -2X_{2J}^w \sin \frac{2\pi j}{h}$$

$$c_J^{w\xi} = K_{J+1,J}^w$$
$$c_J^{wy} = Y_{J+1,J}^w$$
$$c_J^{ww} = Z_{J+1,J}^w$$
$$d_J^{w\xi} = K_{J-1,J}^w$$
$$d_J^{wv} = Y_{J-1,J}^w$$
$$d_J^{ww} = Z_{J-1,J}^w$$

$$(5.25)$$

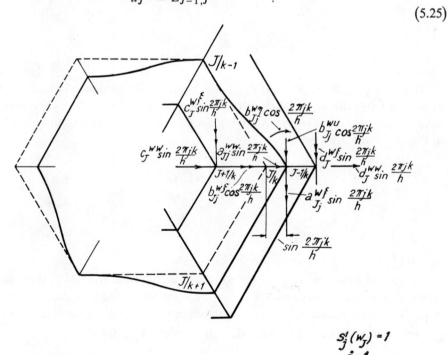

$$s_j^i(w_J) = 1$$
$$j = 1$$

Fig. 5.3 Generalised symmetric unit state $s'(w_j) = 1$

The equation of s_j'', r_j'' has the same coefficients as eq. 5.24 and its right-hand side is

$$\frac{2}{h} \sum_{k=0}^{h=1} \mathscr{Z}_{J/k} \cos \frac{2\pi jk}{h} = S''(w_J)$$

Note: Eqs 5.24 are obtained directly by the application of the reciprocal theorem to the actual state of motion R (Fig. 5.2) and the unit auxiliary state of the j-th fundamental mode (Fig. 5.3) characterised by the condition that

$$s_j'(w_J) = 1 \tag{5.26}$$

and that all the remaining s_j and r_j are zero.

Since by eqs 5.12 and 5.13 the actual state of motion may be expanded in the fundamental modes which are orthogonal because of eq. 5.15, the only component of motion that will come into play in the working equation is the one characterised by coordinates r_j', s_j' with the same subscript j as in eq. 5.26.

When the system is deformed as shown in Fig. 5.3, the force acting at point k in the direction of axis Y is (eq. 5.23)

$$Y_J^w \sin \frac{2\pi jk}{h} + Y_{2J}^w \left(\sin \frac{2\pi j(k-1)}{h} + \sin \frac{2\pi j(k+1)}{h} \right)$$

$$= \left(Y_J^w + 2Y_{2J}^w \cos \frac{2\pi j}{h} \right) \sin \frac{2\pi jk}{h} = a_{jJ}^{wv} \sin \frac{2\pi jk}{h}$$

Referring to eq. 5.20 we see that a_{jJ}^{ww} is the generalised coordinate of force Y_j for the generalised coordinate of displacement $s_j(w_j)$ equal to one. The other coefficients have an analogous meaning. Applying the reciprocal theorem to the states shown in Fig. 5.2 and 5.3 we get

$$\sum_{k=1}^{h-1} \sin^2 \frac{2\pi jk}{h} \sum_{\beta=\xi,v,w} \left[a_{jJ}^{w\beta} s_j'(\beta_J) + c_J^{w\beta} s_j'(\beta_{J+1}) + d_J^{w\beta} s_j'(\beta_{J-1}) \right]$$

$$+ \sum_{\beta=\eta,\zeta,u} \cos^2 \frac{2\pi jk}{h} b_{jJ}^{w\beta} r_j'(\beta_J) = \sum_{k=1}^{h-1} \mathscr{Z}_{J/k} \sin \frac{2\pi jk}{h}$$

and in view of eq. 1.137 this is eq. 5.24.

The second equation obtained for $s_j'(\xi_J) = 1$ differs from eq. 5.24 in that the superscript w is replaced by ξ, and $\mathscr{Z}_{J/k}$ on the right-hand side is replaced by moments $\mathscr{K}_{J/k}$. The third equation is written with superscript v and the loading term is $\mathscr{Y}_{J/k}$.

Turning now to the antisymmetric deformation (Fig. 5.4)

$$r_j'(u_j) = 1 \tag{5.27}$$

for $j \neq 0,\ h/2$ we get analogously

$$a_{Jj}^{u\eta}\, r'_j(\eta_J) + a_{Jj}^{u\zeta}\, r'_j(\zeta_J) + a_{Jj}^{uu}\, r'_j(u_J) + b_{Jj}^{u\xi}\, s'_j(\xi_J) + b_{Jj}^{uv}\, s'_j(v_J) + b_{Jj}^{uw}\, s'_j(w_J)$$

$$+ c_J^{u\eta}\, r'_j(\eta_{J+1}) + c_J^{u\zeta}\, r'_j(\zeta_{J+1}) + c_J^{uu}\, r'_j(u_{J+1}) + d_J^{u\eta}\, r'_j(\eta_{J-1}) + d_J^{u\xi}\, r'_j(\zeta_{J-1})$$

$$+ d_J^{uu}\, r'_j(u_{J-1}) = \frac{2}{h}\sum_{k=0}^{h-1} \mathscr{X}_{J/k}\cos\frac{2\pi jk}{h} \tag{5.28}$$

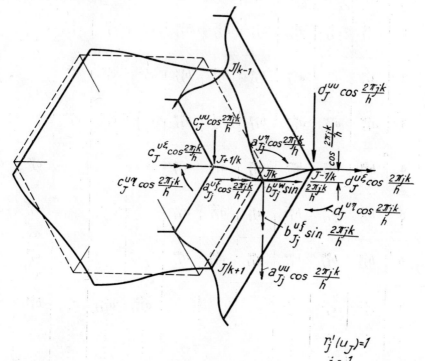

Fig. 5.4 Generalised antisymmetric unit state $r'(u_j) = 1$

where

$$a_{Jj}^{u\eta} = L_J^u + 2L_{2J}^u \cos\frac{2\pi j}{h}$$

$$a_{Jj}^{u\zeta} = M_J^u + 2M_{2J}^u \cos\frac{2\pi j}{h}$$

$$b_{Jj}^{uw} = 2Z_J^u \sin\frac{2\pi j}{h}$$

$$c_J^{u\eta} = L_{J+1,J}^u$$

$$d_J^{u\zeta} = M_{J-1,J}^u,\quad \text{etc.} \tag{5.29}$$

198

Table 5.1

Eye		I						II				
Deformation		1	2	3	4	5	6	1	2	3	4	5
		s_j			r_j			s_j			r_j	
Equation		(ξ_I)	(v_I)	(w_I)	(η_I)	(ζ_I)	(u_I)	(ξ_{II})	(v_{II})	(w_{II})	(η_{II})	(ζ_{II})
I	1	$a^{\xi\xi}_{Ij}$	$a^{\xi v}_{Ij}$	$a^{\xi w}_{Ij}$	$b^{\xi\eta}_{Ij}$	$b^{\xi\zeta}_{Ij}$	$b^{\xi u}_{Ij}$	$c^{\xi\xi}_{I}$	$c^{\xi v}_{I}$	$c^{\xi w}_{I}$		
	2	$a^{\xi v}_{Ij}$	a^{vv}_{Ij}	a^{vw}_{Ij}	$b^{v\eta}_{Ij}$	$b^{v\zeta}_{Ij}$	b^{vu}_{Ij}	$c^{v\xi}_{I}$	c^{vv}_{I}	c^{vw}_{I}		
	3	$a^{\xi w}_{Ij}$	a^{vw}_{Ij}	a^{ww}_{Ij}	$b^{w\eta}_{Ij}$	$b^{w\zeta}_{Ij}$	b^{wu}_{Ij}	$c^{w\xi}_{I}$	c^{wv}_{I}	c^{ww}_{I}		
	4	$b^{\xi\eta}_{Ij}$	$b^{v\eta}_{Ij}$	$b^{w\eta}_{Ij}$	$a^{\eta\eta}_{Ij}$	$a^{\eta\zeta}_{Ij}$	$a^{\eta u}_{Ij}$				$c^{\eta\eta}_{I}$	$c^{\eta\zeta}_{I}$
	5	$b^{\xi\zeta}_{Ij}$	$b^{v\zeta}_{Ij}$	$b^{w\zeta}_{Ij}$	$a^{\eta\zeta}_{Ij}$	$a^{\zeta\zeta}_{Ij}$	$a^{\zeta u}_{Ij}$				$c^{\zeta\eta}_{I}$	$c^{\zeta\zeta}_{I}$
	6	$b^{\xi u}_{Ij}$	b^{vu}_{Ij}	b^{wu}_{Ij}	$a^{\eta u}_{Ij}$	$a^{\zeta u}_{Ij}$	a^{uu}_{Ij}				$c^{u\eta}_{I}$	$c^{u\zeta}_{I}$
II	1	$c^{\xi\xi}_{I}$	$c^{v\xi}_{I}$	$c^{w\xi}_{I}$				$a^{\xi\xi}_{IIj}$	$a^{\xi v}_{IIj}$	$a^{\xi w}_{IIj}$	$b^{\xi\eta}_{IIj}$	$b^{\xi\zeta}_{IIj}$
to N	⋮	⋮	⋮	⋮	⋮			⋮	⋮	⋮	⋮	⋮

Further equations are obtained by changing superscript u to η (or ζ), and forces $\mathscr{X}_{J/K}$ to moments $\mathscr{L}_{J/K}$ (or $\mathscr{M}_{J/K}$). The equations defining r'_j have the same coefficients on their left-hand sides, and the sum $2/h \sum\limits_{k=0}^{h-1} \mathscr{X}_{J/K} \sin 2\pi jk/h = S''_j(u_J)$ on their right-hand sides, etc. In abridged notation all the equations may be written as follows

$$\sum_{\beta=\xi,v,w} \left[a^{\alpha\beta}_{Jj} s_j(\beta_J) + c^{\alpha\beta}_{J} s_j(\beta_{J+1}) + d^{\alpha\beta}_{J} s_j(\beta_{J-1}) \right] + \sum_{\beta=\eta,\zeta,u} b^{\alpha\beta}_{Jj} r_j(\beta_J) = S_j(\alpha_J)$$

where

$$\alpha = \xi, v, w$$

$$\sum_{\beta=\eta,\zeta,u} \left[a^{\alpha\beta}_{Jj} r_j(\beta_J) + c^{\alpha\beta}_{J} r_j(\beta_{J+1}) + d^{\alpha\beta}_{J} r_j(\beta_{J-1}) \right] + \sum_{\beta=\xi,v,w} b^{\alpha\beta}_{Jj} s_j(\beta_J) = R_j(\alpha_J) \qquad (5.30)$$

6	III to N 1	...	A	B
		s_j		
(u_{II})	(ξ_{III})	$(v_{III})\ldots$		
			$= \dfrac{2}{h}\displaystyle\sum_{k=0}^{h-1}\mathscr{X}_{I/k}\sin\dfrac{2\pi jk}{h}$	$= \dfrac{2}{h}\displaystyle\sum_{k=0}^{h-1}\mathscr{X}_{I/k}\cos\dfrac{2\pi jk}{h}$
			$= \dfrac{2}{h}\displaystyle\sum_{k=0}^{h-1}\mathscr{Y}_{I/k}\sin\dfrac{2\pi jk}{h}$	$= \dfrac{2}{h}\displaystyle\sum_{k=0}^{h-1}\mathscr{Y}_{I/k}\cos\dfrac{2\pi jk}{h}$
			$= \dfrac{2}{h}\displaystyle\sum_{k=0}^{h-1}\mathscr{Z}_{I/k}\sin\dfrac{2\pi jk}{h}$	$= \dfrac{2}{h}\displaystyle\sum_{k=0}^{h-1}\mathscr{Z}_{I/k}\cos\dfrac{2\pi jk}{h}$
$c_I^{\eta u}$			$= \dfrac{2}{h}\displaystyle\sum_{k=0}^{h-1}\mathscr{L}_{I/k}\cos\dfrac{2\pi jk}{h}$	$= -\dfrac{2}{h}\displaystyle\sum_{k=0}^{h-1}\mathscr{L}_{I/k}\sin\dfrac{2\pi jk}{h}$
$c_I^{\zeta u}$			$= \dfrac{2}{h}\displaystyle\sum_{k=0}^{h-1}\mathscr{M}_{I/k}\cos\dfrac{2\pi jk}{h}$	$= -\dfrac{2}{h}\displaystyle\sum_{k=0}^{h-1}\mathscr{M}_{I/k}\sin\dfrac{2\pi jk}{h}$
c_I^{uu}			$= \dfrac{2}{h}\displaystyle\sum_{k=0}^{h-1}\mathscr{X}_{I/k}\cos\dfrac{2\pi jk}{h}$	$= -\dfrac{2}{h}\displaystyle\sum_{k=0}^{h-1}\mathscr{X}_{I/k}\sin\dfrac{2\pi jk}{h}$
$b_{IIj}^{\xi u}$	$c_{II}^{\xi\xi}$	$c_{II}^{\xi v}$	$= \dfrac{2}{h}\displaystyle\sum_{k=0}^{h-1}\mathscr{X}_{II/k}\sin\dfrac{2\pi jk}{h}$	$= \dfrac{2}{h}\displaystyle\sum_{k=0}^{h-1}\mathscr{X}_{II/k}\cos\dfrac{2\pi jk}{h}$
$\vdots$	$\vdots$	$\vdots$	$\vdots$	$\vdots$

where

$$\alpha = \eta,\ \zeta,\ u\,.$$

Eqs 5.30 apply to all single and double prime values and to all j and J, i.e. also to $j = 0$ and $j = h/2$ provided due regard is given to relations 5.16 to 5.19. Their total is thus $6Nh$, that is to say, the same as that of equations of the type of 5.7 and 5.8. But while the untransformed equations are mutually dependent, the transformed equations for an even (odd) number of ribs decompose into $h - 2\ (h - 1)$ independent systems of $6N$ equations and four (two) systems of $3N$ equations. This makes for considerable simplification of the solution. Thus, for example, if a cyclosymmetric system has 12 ribs and 6 rings, i.e. 72 joints, there are altogether $6 \times 72 = 432$ unknown components of deformation. However, it is not necessary to solve all 432 equations at once, for they can be decomposed into 10 independent systems each with 36 equations, every pair of which has, moreover, identical coefficients, and into 4 systems of 18 equations each (for $j = 0$ and $j = 6$).

We have proved by eqs 5.24 and 5.28 that a load of fundamental mode j produces deformation of the same mode j. Table 5.1 shows the equations written in the tabular form for $j \neq 0 \neq n/2$. Because of the validity of the reciprocal theorem, the coefficients of the equations must be diagonally symmetric. From this it follows that for coefficients a and b we may interchange the superscripts, and for coefficients c and d use the relation

$$d^{\alpha\beta}_{J+1} = c^{\beta\alpha}_J \qquad (5.31)$$

The primes of unknowns s, r are left out: the single-prime values s', r' are computed from the loading terms in column A, the double-prime values s'', r'' from those in column B.

Example 5.1

Fig. 5.5 shows a three-dimensional cyclosymmetric system with a single ring and the axes of the bars forming an equilateral triangle; the ribs are represented by vertical columns. All the bars of the system are straight, and of constant cross-section.

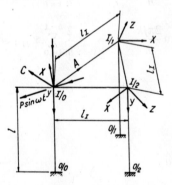

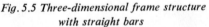

Fig. 5.5 Three-dimensional frame structure
with straight bars

The principal axes of the columns lie in axes X, Y, Z. The bars of the ring have longitudinal axis A, vertical axis B and the horizontal axis C which is at right angles to plane AB. At joint $I/0$ the system is subjected to a harmonic force, $P \sin \omega t$ acting in the direction of axis Z. The amplitude is

$$\mathscr{Z}_{I/0} = P$$

and all the remaining load components are zero.

First, expand the external load in the fundamental modes. It is given by

$$S_0'(w_I) = 0$$

$$S_0''(w_I) = \frac{1}{3} \sum_{k=0}^{2} \mathscr{L}_{J/k} = \frac{P}{3}$$

(5.32)

and the remaining components S_0, R_0 are zero.

Further

$$S_1'(w_I) = 0$$

$$S_1''(w_I) = \frac{2}{3} \sum_{k=0}^{2} \mathscr{L}_{I/k} \cos \frac{2\pi k}{3} = \frac{2}{3} P$$

(5.33)

The slope-deflection equations are established by applying eqs 5.30. For $j = 0$, $s_0' = r_0' = 0$. For s_0'' the equations are as shown in Table 5.2

Table 5.2

$s_0''(\xi_I)$	$s_0''(v_I)$	$s_0''(w_I)$	
$a_{I0}^{\xi\xi}$	$a_{I0}^{\xi v}$	$a_{I0}^{\xi w}$	$= 0$
$a_{I0}^{\xi v}$	a_{I0}^{vv}		$= 0$
$a_{I0}^{\xi w}$		a_{I0}^{ww}	$= \frac{1}{3}P$

Coefficients a are computed by referring to eq. 5.25 and using Tables 10.1 and 10.2 in the Appendix. They are

$$a_{I0}^{\xi\xi} = \frac{EJ_x}{l} F_2(\lambda_x) + \frac{3EJ_c}{2l_I} [F_2(\lambda_c) - F_1(\lambda_c)]$$

$$a_{I0}^{\xi v} = - \frac{\sqrt{(3)}\, EJ_c}{l_I^2} [F_3(\lambda_c) + F_4(\lambda_c)]$$

$$a_{I0}^{\xi w} = - \frac{EJ_x}{l^2} F_4(\lambda_x)$$

$$a_{I0}^{vv} = \frac{ES}{l} \psi \cot \psi + \frac{2EJ_c}{l_I^3} [F_6(\lambda_c) + F_5(\lambda_c)]$$

$$a_{I0}^{ww} = \frac{EJ_x}{l^3} F_6(\lambda_x) + \frac{EJ_b}{2l_I^3} [F_6(\lambda_b) + F_5(\lambda_b)] + \frac{3}{2} \frac{ES_I}{l_I} \psi_I(\cot \psi_I + \operatorname{cosec} \psi_I)$$

(5.34)

In the above l is the span, S — the cross-sectional area, J_x — the moment of inertia about axis X of the column; the data for the horizontal bars are: l_I — the span, S_I — the cross-sectional area, J_b, J_c — the moments of inertia about axes B and C, respectively.

For $j = 1$, the application of Table 5.1 results in the equations set out in Table 5.3. Because of eq. 5.33, $r_1' = s_1' = 0$.

Table 5.3

$s_1''(\xi_I)$	$s_1''(v_I)$	$s_1''(w_I)$	$r_1''(\eta_I)$	$r_1''(\zeta_I)$	$r_1''(u_I)$	
$a_{I1}^{\xi\xi}$	$a_{I1}^{\xi v}$	$a_{I1}^{\xi w}$		$b_{I1}^{\xi\zeta}$		$= 0$
$a_{I1}^{\xi v}$	a_{I1}^{vv}			$b_{I1}^{v\zeta}$		$= 0$
$a_{I1}^{\xi w}$		a_{I1}^{ww}	$b_{I1}^{w\eta}$		b_{I1}^{wu}	$= \frac{2}{3}P$
		$b_{I1}^{w\eta}$	$a_{I1}^{\eta\eta}$		$a_{I1}^{\eta u}$	$= 0$
$b_{I1}^{\xi\zeta}$	$b_{I1}^{v\zeta}$			$a_{I1}^{\zeta\zeta}$	$a_{I1}^{\zeta u}$	$= 0$
		b_{I1}^{wu}	$a_{I1}^{\eta u}$	$a_{I1}^{\zeta u}$	a_{I1}^{uu}	$= 0$

In the equations

$$a_{I1}^{\xi\xi} = \frac{EJ_x}{l} F_2(\lambda_x) + \frac{3EJ_c}{2l_I} \left[F_2(\lambda_c) + \tfrac{1}{2}F_1(\lambda_c) \right] + \frac{3GJ_a}{4l_I}$$

$$a_{I1}^{\xi v} = -\frac{\sqrt{(3)}\, EJ_c}{l_I^2} \left[F_4(\lambda_c) - \tfrac{1}{2}F_3(\lambda_c) \right]$$

$$a_{I1}^{\xi w} = -\frac{EJ_x}{l^2} F_4(\lambda_x)$$

$$a_{I1}^{vv} = \frac{ES}{l} \psi \cot\psi + \frac{2EJ_c}{l_I^3} \left[F_6(\lambda_c) - \tfrac{1}{2}F_5(\lambda_c) \right]$$

$$a_{I1}^{ww} = \frac{EJ_x}{l} F_6(\lambda_x) + \frac{EJ_b}{l_I^3} \left[F_6(\lambda_b) - \tfrac{1}{2}F_5(\lambda_b) \right] + \frac{3ES_I}{2l_I} \psi_I(\cot\psi_I - \tfrac{1}{2}\operatorname{cosec}\psi_I)$$

$$a_{I1}^{\eta\eta} = \frac{GJ_y}{l} + \frac{EJ_b}{l_I} \left[2F_2(\lambda_b) - F_1(\lambda_b) \right]$$

$$a_{I1}^{\eta u} = \frac{\sqrt{(3)}\, EJ_b}{l_I^2} \left[F_4(\lambda_b) + \tfrac{1}{2}F_3(\lambda_b) \right]$$

$$a_{I1}^{\zeta\zeta} = \frac{EJ_z}{l} F_2(\lambda_z) + \frac{EJ_c}{4l_I} \left[2F_2(\lambda_c) - F_1(\lambda_c) \right] + \frac{3}{4}\frac{GJ_a}{l_I}$$

$$a_{I1}^{\zeta u} = \frac{EJ_z}{l} F_4(\lambda_z)$$

$$a_{I1}^{uu} = \frac{EJ_z}{l^3} F_6(\lambda_z) + \frac{3EJ_b}{2l_I^3} \left[F_6(\lambda_b) + \tfrac{1}{2}F_5(\lambda_b) \right] + \frac{ES_I}{4l_I} \psi_I(2 \cot \psi_I + \operatorname{cosec} \psi_I)$$

$$b_{I1}^{\xi r} = -\frac{3}{4} \frac{EJ_c}{l_I} F_1(\lambda_c) + \frac{3}{4} \frac{GJ_a}{l_I}$$

$$b_{I1}^{v\zeta} = -\frac{\sqrt{(3)}\,EJ_c}{2l_I^2} F_3(\lambda_c)$$

$$b_{I1}^{w\eta} = \frac{\sqrt{(3)}\,EJ_b}{2l_I^2} F_3(\lambda_b)$$

$$b_{I1}^{wu} = \frac{3EJ_b}{4l_I^2} F_5(\lambda_b) - \frac{3ES_I}{4l_I} \psi_I \operatorname{cosec} \psi_I$$

$$(5.35)$$

The torques were computed without regard to inertia forces arising in rotation about axis A. The equations in Tables 5.2 and 5.3 give the values of s'' and r'', and eqs 5.12 and 5.13 give the actual deformations at points $I/1$, $I/2$, $I/3$.

In the foregoing analysis of forced vibration no use was made of the expansion in natural modes of which there is an infinite number in systems with continuously distributed mass. The load was expanded in the fundamental cyclosymmetric modes whose number is the same as the number of ribs in the system. This method works well for forced vibration with a single, given frequency. But if we wish to solve forced non-harmonic vibration, we must adopt the method of modal analysis, i.e. the expansion in natural modes (Sections 1.2.2 and 3.5), the modes to be determined by a procedure which we shall outline in the next section and in Example 5.2.

5.1.2 Free vibration

Free vibration of three-dimensional cyclosymmetric systems is described by the equations of Table 5.1 and eqs 5.30 with the right-hand sides set to zero. For a non-trivial solution to exist in such a case, the determinant of the coefficients in one of the j-th systems of equations must be equal to zero. It is therefore clear that free harmonic vibration will always occur in some fundamental mode. Since a system composed of bars with continuously distributed mass has an infinite number of degrees of freedom, infinitely many natural frequencies $\omega_{j(1)}, \omega_{j(2)}, \omega_{j(3)} \cdots \omega_{j(\infty)}$ are associated with each fundamental mode j, and if $j \neq 0, h/2$, two natural mode shapes with each of these frequencies exist (cf. Section 1.5.2.1, formulae 1.115 and

1.116). The fundamental mode j decides the deformation conditions in the rings. The deformation conditions in the ribs are different for different is, however.

Example 5.2

Find the first natural frequency of the frame shown in Fig. 5.5, vibrating in the fundamental mode s_0''. The specification of the frame is as follows:

columns:

$$l = 5\,\text{m}\,,\quad J_x = 5 \times 10^{-4}\,\text{m}^4\,,\quad \mu = 0\!\cdot\!5\,\text{t/m}\,;$$

longitudinal vibration is neglected;

horizontal bars:

$$l_I = 4\,\text{m}\,;\quad J_a = 1 \times 10^{-4}\,\text{m}^4\,,\quad J_b = 1 \times 10^{-5}\,\text{m}^4\,,\quad J_c = 2 \times 10^{-4}\,\text{m}^4\,,$$

$$S_I = 1 \times 10^{-2}\,\text{m}^2\,,\quad \mu_I = 0\!\cdot\!1\,\text{t/m}\,,$$

$$E = 21 \times 10^6\,\text{Mp/m}^2\,,\quad G = 8\!\cdot\!1 \times 10^6\,\text{Mp/m}^2$$

Further,

$$\lambda_x = l\left(\frac{\mu}{EJ_x}\right)^{1/4}\sqrt{\omega} = 5\left(\frac{0\!\cdot\!5}{9\!\cdot\!81 \times 21 \times 10^6 \times 5 \times 10^{-4}}\right)^{1/4}\sqrt{\omega} = 0\!\cdot\!235\sqrt{\omega}$$

$$\lambda_b = \frac{l_I}{l}\left(\frac{\mu_I J_x}{\mu J_b}\right)^{1/4}\lambda_x = \frac{4}{5}\left(\frac{0\!\cdot\!1 \times 5 \times 10^{-4}}{0\!\cdot\!5 \times 1 \times 10^{-5}}\right)^{1/4}\lambda_x = 1\!\cdot\!42\lambda_x = 0\!\cdot\!334\sqrt{\omega}$$

$$\lambda_c = \frac{l_I}{l}\left(\frac{\mu_I J_x}{\mu J_c}\right)^{1/4}\lambda_x = \frac{4}{5}\left(\frac{0\!\cdot\!1 \times 5 \times 10^{-4}}{0\!\cdot\!5 \times 2 \times 10^{-4}}\right)^{1/4}\lambda_x = 0\!\cdot\!672\lambda_x = 0\!\cdot\!158\sqrt{\omega}$$

$$\psi_I = l_I\left(\frac{\mu_I}{ES_I}\right)^{1/2}\omega = 4\left(\frac{0\!\cdot\!1}{9\!\cdot\!81 \times 21 \times 10^6 \times 1 \times 10^{-2}}\right)^{1/2}\omega = 0\!\cdot\!882 \times 10^{-3}\omega$$

$$= 0\!\cdot\!0160\lambda_x^2$$

Substitution of the above values in the equations of Table 5.2 gives the equations set out in Table 5.4 (Because the columns are incompressible, deformation v drops out). The equations have a non-zero solution if

$$\lambda_x = 3\!\cdot\!33$$

i.e.

$$\omega_{0(1)} = 201\,\text{s}^{-1}$$

$$f_{0(1)} = 32\,\text{Hz}$$

Table 5.4

$s_0''(\xi_I)$	$s_0''(w_I)$	
$2100F_2(\lambda_x) + 1575[F_2(\lambda_c) - F_1(\lambda_c)]$	$-420F_4(\lambda_x)$	$= 0$
$-420F_4(\lambda_x)$	$84F_6(\lambda_x) + 1{\cdot}643[F_6(\lambda_b) + F_5(\lambda_b)] +$ $+\ 78\,800\psi_I\,(\cot\psi_I + \operatorname{cosec}\psi_I)$	$= 0$

Note: Non-harmonic vibration of cyclosymmetric systems may be solved by their expansion in natural modes (cf. Sections 1.2.2 and 1.5).

5.2 Continuous systems with repeated elements

5.2.1 Introductory notes

Continuous systems with repeated elements have been used by designers of all periods. In course of time they have changed in appearance but not in principle, as borne out by Figs 5.6a, b, c showing the development from early viaducts to modern bridges. What has remained unchanged are the economic advantages of the systems; combined with recent advances in precasting and standardisation

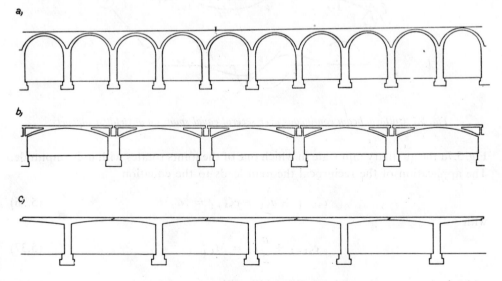

Fig. 5.6 Systems with repeated elements

they have been the main reason for structures composed of repeated elements finding a steadily broadening field of application. Though they tend to be more slender than before, and consequently more sensitive to dynamic effects, they are expected to meet ever more stringent dynamic demands.

In the analysis that follows we shall make full use of all the simplifications in numerical procedures which repetition of identical elements affords. The solutions at which we shall arrive will resemble those obtained in Section 5.1 for three-dimensional cyclosymmetric systems. This is as it should be, for cyclosymmetric systems are but a special case of systems with repeated elements. The difference between them and the two-dimensional systems is that their elements are repeated on the periphery of a circle.

In the sections that follow we shall examine vibrations of different kinds of systems with repeated elements, starting with uniform beams continuous over several equal spans and ending with continuous arches.

5.2.2 Uniform beam continuous over several equal spans on unyielding supports

5.2.2.1 *The slope-deflection method*

Consider a continuous beam with an arbitrary number of spans vibrating in harmonic motion. Fig. 5.7R shows several spans of such a beam in actual motion,

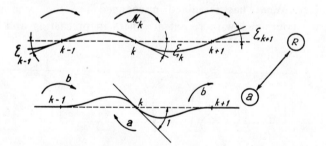

Fig. 5.7 Uniform beam continuous over several equal spans on unyielding supports

Fig. 5.7a the auxiliary unit state in which one of the joints rotates with unit amplitude. The application of the reciprocal theorem leads to the equation

$$c\zeta_{k-1} + a\zeta_k + c\zeta_{k+1} = \mathcal{M}_k \tag{5.36}$$

with

$$c\zeta_{k-1} + \frac{a}{2}\,\zeta_k = M_{k,k-1} \tag{5.37}$$

which is the moment at point k of bar $k(k-1)$ at actual rotations ζ_k, ζ_{k-1}. It holds, therefore, that (cf. eq. 3.9a)

$$M_{k,k-1} + M_{k,k+1} = \mathcal{M}_k \tag{5.38}$$

According to Table 10.1 in the Appendix, coefficients a and c are

$$a = \frac{2EJ}{l} F_2(\lambda)$$

$$c = \frac{EJ}{l} F_1(\lambda) \tag{5.39}$$

Eq. 5.36, with no change in coefficients a and c can be written for all joints of the system. It is therefore a difference equation of the second order with constant coefficients, and can be given the following form

$$c\, \Delta^2 \zeta_{k-1} + (a + 2c)\, \zeta_k = \mathcal{M}_k \tag{5.40}$$

where

$$\Delta^2 \zeta_{k-1} = \zeta_{k+1} - 2\zeta_k + \zeta_{k-1} \tag{5.41}$$

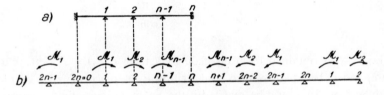
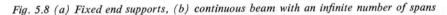

Fig. 5.8 (a) Fixed end supports, (b) continuous beam with an infinite number of spans

The solution is analogous to that of three-dimensional cyclosymmetric systems (cf. eq. 5.13)

$$\zeta_k = \sum_{j=0}^{h/2} \left(r'_j \cos\frac{2\pi jk}{h} - r''_j \sin\frac{2\pi jk}{h} \right) \tag{5.42}$$

where j and h are integers. Integer h and coefficient r_j are determined from the boundary conditions.

5.2.2.2 Fixed end supports

Steady-state forced vibration

Denote the number of spans by n. The boundary conditions (Fig. 5.8a) are satisfied if the rotations of the end supports are zero:

$$\zeta_0 = \zeta_n = 0$$

Now

$$h = 2n \tag{5.43}$$

and

$$r'_j = 0 \tag{5.44}$$

Therefore

$$\zeta_k = -\sum_{j=1}^{n-1} r''_j \sin \frac{\pi jk}{n} \tag{5.45}$$

Substitute expression 5.45 in eq. 5.36 written for all joints of the system, and multiply the result by $[-\sin(\pi jk/n)]$. Because of eqs 5.15 and 5.23 the sum of all the equations for $k = 1$ to $n - 1$ then gives

$$a_j r''_j = -\frac{2}{n} \sum_{k=1}^{n-1} \mathcal{M}_k \sin \frac{\pi jk}{n} \tag{5.46}$$

where

$$a_j = a + 2c \cos \frac{\pi j}{n} \tag{5.47}$$

Note: Eq. 5.46 is obtained from eq. 5.30 by setting $r''_j(\zeta_j) = r''_j$, letting all the remaining r_j and s_j be equal to zero, and writing a_j in place of a_{jj}. Complete agreement with Section 5.1 would be reached if the beam had an infinite number of spans (Fig. 5.8b) and were subjected to a load repeated periodically after every $h = 2n$ spans.

Free vibration

Eq. 5.46 has a non-zero solution also in the case where the external load is not present. Then

$$a_j = 0 \tag{5.48}$$

and substituting eqs 5.39 and 5.47 gives

$$F_2(\lambda) + F_1(\lambda) \cos \frac{\pi j}{n} = 0 \quad \text{for} \quad j = 1 \text{ to } n - 1 \tag{5.48a}$$

The values for which the equation is satisfied, determine the natural frequencies of the system. Thus, for example, for the natural frequencies of the 5-span system shown in Fig. 5.10

$$F_2(\lambda) + 0\cdot809F_1(\lambda) = 0$$
$$F_2(\lambda) + 0\cdot309F_1(\lambda) = 0$$
$$F_2(\lambda) - 0\cdot309F_1(\lambda) = 0$$
$$F_2(\lambda) - 0\cdot809F_1(\lambda) = 0$$

$$(5.49)$$

The numerical values of λ_j are taken from the tables of functions $F(\lambda)$ in the Appendix:

$$\lambda_1 = 4\cdot55$$
$$\lambda_2 = 4\cdot16$$
$$\lambda_3 = 3\cdot70$$
$$\lambda_4 = 3\cdot31$$

$$(5.50)$$

Since

$$\omega = \frac{\lambda^2}{l^2}\left(\frac{EJ}{\mu}\right)^{1/2}$$

$$(5.51)$$

the natural frequencies arranged in order of their magnitudes are as follows:

$$f_4 = \frac{\omega}{2\pi} = \frac{\lambda_4^2}{2\pi l^2}\left(\frac{EJ}{\mu}\right)^{1/2} = \frac{\lambda_4^2}{\pi^2}f^*$$

$$f_4 = 1\cdot11f^* = f_{(1)}$$
$$f_3 = 1\cdot39f^* = f_{(2)}$$
$$f_2 = 1\cdot75f^* = f_{(3)}$$
$$f_1 = 2\cdot10f^* = f_{(4)}$$

$$(5.52)$$

where

$$f^* = \frac{\pi}{2l^2}\left(\frac{EJ}{\mu}\right)^{1/2}$$

$$(5.53)$$

is the lowest natural frequency of a simply supported beam. Numerals in parentheses denote the position of the frequencies in the sequence of their magnitudes. The fifth natural frequency equals the natural frequency of a beam clamped at both ends, for which $r_0'' = 0$ and

$$\lambda_0 = 4\cdot73$$
$$f_{(5)} = 2\cdot26f^*$$

$$(5.54)$$

The values specified by eqs 5.52 and 5.54 represent the lowest natural frequencies of the system. In each fundamental mode there exists an unlimited number of higher natural frequencies. The next higher values of λ are

$$\lambda_1 = 6\cdot46$$
$$\lambda_2 = 6\cdot84$$
$$\lambda_3 = 7\cdot28$$
$$\lambda_4 = 7\cdot67$$
$$\lambda_5 = 7\cdot85$$

$$(5.55)$$

and the corresponding frequencies are

$$f_{(6)} = 4\cdot23f^* \qquad (5.56a)$$
$$f_{(7)} = 4\cdot75f^* \qquad (5.56b)$$
$$f_{(8)} = 5\cdot36f^* \qquad (5.56c)$$
$$f_{(9)} = 5\cdot96f^* \qquad (5\cdot56d)$$
$$f_{(10)} = 6\cdot25f^* \qquad (5\cdot56e)$$

Figs. 5.10a to 5.10e show the first five natural modes of vibration.

The values given by eq. 5.55 and higher values of λ may be computed from the simplified formulae 2.87 for $F(\lambda)$

$$F_1(\lambda) \approx -\frac{\lambda}{\cos \lambda}$$

$$F_2(\lambda) \approx \lambda(1 - \tan \lambda)$$

Substituting in eq. 5.48 gives the simple equation

$$\cos \lambda_j - \sin \lambda_j = \cos \frac{\pi j}{n}$$

or

$$\sin\left(\lambda_j - \frac{\pi}{4}\right) = -\frac{\sqrt{2}}{2} \cos \frac{\pi j}{n} \quad \text{for} \quad j = 0 \text{ to } n$$

$$(5.57)$$

It is evident that frequencies $f_{(1)}$ to $f_{(5)}$ lie in the range of the first frequency of a simply supported beam, f^*, and of a beam clamped at both ends; the second group of frequencies lies between $4f^*$ — the second frequency of a simply supported beam — and the second frequency of a beam clamped at both ends, $f_{(10)}$.

Formula 5.57 makes it possible to calculate the higher natural frequencies to adequate accuracy in a very simple way. It approximately holds that

$$\sin\left(\lambda_j - \frac{\pi}{4}\right) \approx \left(\lambda_j - \frac{\pi}{4}\right)$$

and therefore by formula 5.57

$$\frac{\lambda_j}{\pi} \approx \frac{1}{4} - \frac{\sqrt{2}}{2\pi} \cos \frac{\pi j}{n}$$

In the second group of natural modes it is then

$$\frac{\lambda_j}{\pi} \approx 2 + \frac{1}{4} - 0 \cdot 225 \cos \frac{\pi j}{n}$$

Thus, for example, for $n = 5$

$$\frac{\lambda_1}{\pi} \approx 2 \cdot 25 - 0 \cdot 225 \times 0 \cdot 809 = 2 \cdot 07$$

$$\lambda_1 \approx 6 \cdot 50 \text{ against } \lambda_1 = 6 \cdot 46 \text{ according to } 5.55$$

$$\frac{\lambda_2}{\pi} \approx 2 \cdot 25 - 0 \cdot 225 \times 0 \cdot 309 = 2 \cdot 18$$

$$\lambda_2 \approx 6 \cdot 84, \quad \text{which value agrees with that of } 5.55$$

5.2.2.3 *Hinged end supports*

Eqs 5.36—5.42 remain valid in this case, too. The only things that change are the boundary conditions which are satisfied if

$$\zeta_{-1} = \zeta_1 = \zeta_{2n-1}$$
$$\zeta_{n+1} = \zeta_{n-1}$$

$$(5.58)$$

and the span $-1, 0$ (or $n, n + 1$) is loaded antisymmetrically to span $0, 1$ (or $n - 1, n$). The result is arrived at by considering a beam with an unlimited number of spans loaded as indicated in Fig. 5.9b.

Under these conditions

$$n = \frac{h}{2}$$

$$r_j'' = 0$$

$$(5.59)$$

and therefore

$$\zeta_k = \sum_{j=0}^{n} r'_j \cos \frac{\pi j k}{n} \qquad (5.60)$$

Proceeding analogously as in Section 5.2.2.2 we get

$$a_j r'_j = \frac{2}{n} \sum_{k=0}^{n} \mathcal{M}_k \cos \frac{\pi j k}{n} \quad \text{for} \quad j \neq 0 \neq n \qquad (5.61)$$

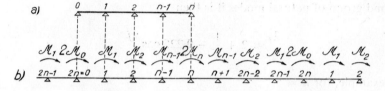

Fig. 5.9 (a) Hinged end supports, (b) continuous beam with an infinite number of spans

According to eq. 5.17, for $j = 0$

$$a_0 r'_0 = \frac{1}{n} \sum_{k=0}^{n} \mathcal{M}_k \qquad (5.62)$$

and by eq. 5.19 for $j = n$

$$a_n r'_n = \frac{1}{n} \sum_{k=0}^{n} (-1)^k \mathcal{M}_k \qquad (5.63)$$

Eq. 5.47 continues to apply.

Free vibration

For free vibration eq. 5.61 again leads to the relation $a_j = 0$ so that eqs 5.48 to 5.52 apply. Eq. 5.54 becomes invalid, and eq. 5.63 that replaces it gives for $\mathcal{M}_k = 0$

$$a_n = 0$$

from there by eq. 5.47

$$F_2(\lambda) - F_1(\lambda) = 0 \qquad (5.64)$$

and

$$\lambda = \pi$$

Accordingly, the lowest frequency of the beam in Fig. 5.11 is the same as that of a simply supported beam

$$f_5 = f^* = f_{(1)}$$

The sixth natural frequency is by eq. 5.62 obtained from the equation

$$a_0 = 0$$

from which by eq. 5.47

$$F_2(\lambda) + F_1(\lambda) = 0 \tag{5.65}$$

and

$$\lambda = 2\pi$$

so that

$$f_{(6)} = 4f^*$$

i.e. the second natural frequency of a simply supported beam. The other natural frequencies are as defined by formulae 5.56a−d; formula 5.56e does not apply here.

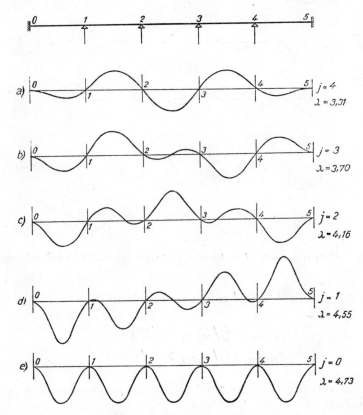

Fig. 5.10. First five natural modes of a continuous beam with fixed end supports

It follows from the foregoing that for $j = 1$ to $n - 1$ the natural frequencies of a uniform beam continuous over several equal spans are the same for fixed as for hinged supports. The corresponding mode shapes are different, however (cf. Figs. 5.10 and 5.11).

The natural frequencies of continuous beams with n spans are comprehensively set out in Table 5.5.

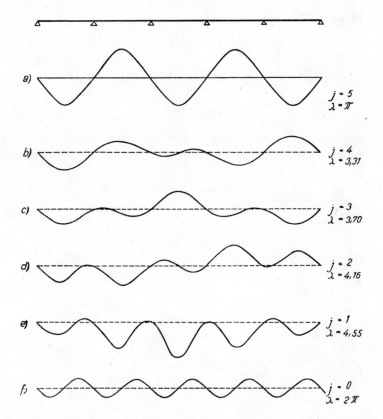

Fig. 5.11 First five natural modes of a continuous beam with hinged end supports

Note: The frequencies of further groups are calculated from the formulae

$$f_{(2n+i)} = \left[\sqrt{f_{(n+i)}} + \sqrt{f^*}\right]^2$$
$$f_{(3n+i)} = \left[\sqrt{f_{(n+i)}} + 2\sqrt{f^*}\right]^2, \quad \text{etc.}$$

Thus, for example, the twelfth natural frequency of a five span $(n = 5)$ beam with clamped ends is

$$f_{(12)} = \left(\sqrt{4\cdot75} + 1\right)^2 f^* = 3\cdot18^2 f^* = 10\cdot1 f^*$$

and the seventeenth natural frequency is

$$f_{(17)} = \left(3\cdot18 + 1\right)^2 f^* = 17\cdot5 f^*$$

Table 5.5

Number of spans	Continuous beam hinged at both ends											
	first group of natural frequencies						second group of natural frequencies					
1	f^*						$4f^*$					
2	f^*			$1.56f^*$			$4f^*$			$5.06f^*$		
3	f^*		$1.28f^*$		$1.87f^*$		$4f^*$		$4.55f^*$		$5.58f^*$	
4	f^*	$1.17f^*$		$1.56f^*$		$2.01f^*$	$4f^*$	$4.33f^*$		$5.06f^*$		$5.84f^*$
5	f^*	$1.11f^*$	$1.39f^*$	$1.75f^*$		$2.10f^*$	$4f^*$	$4.23f^*$	$4.75f^*$	$5.36f^*$		$5.96f^*$
6	f^*	$1.08f^*$	$1.28f^*$	$1.56f^*$	$1.87f^*$	$2.15f^*$	$4f^*$	$4.16f^*$	$4.55f^*$	$5.06f^*$	$5.58f^*$	$6.05f^*$

Number of spans	Continuous beam clamped at both ends											
	first group of natural frequencies						second group of natural frequencies					
1	$2.36f^*$						$6.25f^*$					
2	$1.56f^*$			$2.36f^*$			$5.06f^*$			$6.25f^*$		
3	$1.28f^*$		$1.87f^*$		$2.36f^*$		$4.55f^*$		$5.58f^*$		$6.25f^*$	
4	$1.17f^*$	$1.56f^*$		$2.01f^*$		$2.36f^*$	$4.33f^*$	$5.06f^*$		$5.84f^*$		$6.25f^*$
5	$1.11f^*$	$1.39f^*$	$1.75f^*$	$2.10f^*$		$2.36f^*$	$4.23f^*$	$4.75f^*$	$5.36f^*$	$5.96f^*$		$6.25f^*$
6	$1.08f^*$	$1.28f^*$	$1.56f^*$	$1.87f^*$	$2.15f^*$	$2.36f^*$	$4.16f^*$	$4.55f^*$	$5.06f^*$	$5.58f^*$	$6.05f^*$	$6.25f^*$

5.2.3 Uniform beam continuous over several equal spans on elastic supports

There are two unknown deformations in each of the elastic supports (Fig. 5.12): vertical displacement v_k which is symmetric, and rotation ζ_k which is antisymmetric. The pertinent unit auxiliary states are shown in Figs 5.12a and b. The application of the reciprocal theorem leads to the following equations

$$ev_{k-1} + dv_k + ev_{k+1} - b\zeta_{k-1} + b\zeta_{k+1} = \mathcal{Y}_k$$
$$bv_{k-1} - bv_{k+1} + c\zeta_{k-1} + a\zeta_k + c\zeta_{k+1} = \mathcal{M}_k$$

$$(5.66)$$

216

Coefficients a and c are given by formulae 5.39 and

$$b = \frac{EJ}{l^2} F_3(\lambda)$$

$$d = \frac{2EJ}{l^3} F_6(\lambda) + C$$

$$e = \frac{EJ}{l^3} F_5(\lambda)$$

(5.67)

where C is the spring constant of the elastic support.

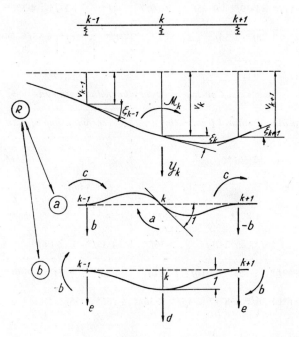

Fig. 5.12 Beam on elastic supports. R — actual state, a, b — unit states

The equations are solved by substitution (cf. eqs 5.12 and 5.13)

$$v_k = \sum_{j=0}^{h/2} \left(s_j' \sin \frac{2\pi jk}{h} + s_j'' \cos \frac{2\pi jk}{h} \right)$$

$$\zeta_k = \sum_{j=0}^{h/2} \left(r_j' \cos \frac{2\pi jk}{h} - r_j'' \sin \frac{2\pi jk}{h} \right)$$

(5.68)

Computation of natural modes by the method of successive approximations ($j = 4$)

Table 5.6

Symmetric load

1	2	3	4	5	6	7	8	9	10	11	12	13	14	15	16	17	18	19	20
		Influence coefficients of:							First approximation					Second approximation				Third approximation	
Point i	Mass m_i (Mp m^{-1} s^2)	deflection; $\delta_{ik}^s \times 10^5$ (m Mp^{-1})					moment; δ_{M_s} (m)	σ_i	2×9 $m_i\sigma_i$ (Mp m^{-1} s^2)	8×10 $m_i\sigma_i\delta_{M_s}$ (Mp s^2)	$\sum m_k\sigma_k\delta_{ik}^s \times 10^5$ (s^2)	$r^4 \sin\frac{2\pi}{5} \times 10^5 \delta_{M_s}$ (s^2)	$12+13$ $\sigma_i \times 10^5$ (s^2)	2×14 $m_i\sigma_i \times 10^3$ (Mpm^{-1}s^4)	8×15 $m_i\sigma_i\delta_{M_s} \times 10^3$ (Mp s^4)	$\sum m_k\sigma_k\delta_{ik}^s \times 10^8$ (s^4)	$r_4 \sin\frac{2\pi}{5} \times 10^8 \delta_{M_s}$ (s^4)	$\sigma_i \times 10^8$ (s^4)	$14:19$ $\omega_{(1)}^2$
		1	2	3	4	5													
1	0·373	0·058	0·112	0·262	0·335	0·378	0·745	1·0	0·373	0·278	1·0	90·6	91·6	0·342	0·255	1·4	131·6	133·0	689
2	0·301	0·112	1·530	3·280	4·525	5·222	2·168	1·8	0·542	1·175	12·6	263·8	276·4	0·832	1·803	17·9	382·9	400·8	690
3	0·247	0·262	3·280	9·395	4·290	17·035	3·427	3·2	0·791	2·711	39·5	416·9	456·4	1·128	3·865	54·9	605·2	660·1	690
4	0·212	0·335	4·525	14·290	26·275	33·370	4·417	4·4	0·933	4·121	69·6	537·4	607·0	1·287	5·684	99·0	780·1	879·1	690
5	0·194	0·378	5·222	17·035	33·370	47·540	4·972	4·8	0·933	4·629	92·0	604·9	696·9	1·352	6·722	131·2	878·1	1009·3	690

$2\overline{M}_{k,k+1(\zeta_k=1)} = 0.454 \times 10^5 \text{ Mp m}, \quad \overline{M}_{k+1,k(\zeta_k=1)} = 0.162 \times 10^5 \text{ Mp m}$

$\sum = 12.914 = M_s$

$\sum = 18.329 = M_s \times 10^3$

$\sin\frac{2\pi}{5} = 0.951, \qquad \cos\frac{2\pi}{5} = 0.309, \qquad \cos\frac{4\pi}{5} = -0.809$

$a_4 = [0.454 + 2 \times 0.162(-0.809)]10^5 = 0.192 \times 10^5 \text{ Mp m} \text{ (by eq. 5.92)}$

$R_4 = 2M_s \sin\frac{2\pi}{5} = 2 \times 12.914 \times 0.951 =$

$= 24.56 \text{ Mp s}^2 \text{ (by eq. 5.89)}$

$r_4 = \frac{R_4}{a_4} = \frac{24.56}{0.192 \times 10^5} =$

$127.9 \times 10^{-5} \text{ m}^{-1}\text{s}^2 \text{ (by eq. 5.91)}$

$10^5 r_4 \sin\frac{2\pi}{5} = 127.9 \times 0.951 = 121.7 \text{ m}^{-1}\text{s}^2$

$10^5 r_4 \cos\frac{2\pi}{5} = 127.9 \times 0.309 = 39.5 \text{ m}^{-1}\text{s}^2$

$R_4 = 2\left(M_s \sin\frac{2\pi}{5} + M_r \cos\frac{2\pi}{5}\right) = 2(18.329 \times 0.951 + 1.287 \times$

$\times 0.309)10^{-3} = 35.66 \times 10^{-3} \text{ Mp s}^4$

$r_4 = \frac{R_4}{a_4} = \frac{35.66 \times 10^{-3}}{0.192 \times 10^5} = 185.7 \times 10^{-8} \text{m}^{-1}\text{s}^4$

$10^8 r_4 \sin\frac{2\pi}{5} = 185.7 \times 0.951 = 176.6 \text{ m}^{-1} \text{s}^4$

$10^8 r_4 \cos\frac{2\pi}{5} = 185.7 \times 0.309 = 57.4 \text{ m}^{-1} \text{s}^4$

Antisymmetric load

1	2	3	4	5	6	7	8	9	10	11	12	13	14	15	16	17	18	19	20
		Influence coefficients of:							First approximation					Second approximation				Third approximation	
Point i	Mass m_i (Mp m^{-1} s^2)	deflection; $\delta_{ik}^r \times 10^5$ (m Mp^{-1})					moment δ_{M_r} (m)	ϱ_i	2×9 $m_i\varrho_i$	8×10 $m_i\varrho_i\delta_{M_r}$	$\sum m_k\varrho_k\delta_{ik}^r \times 10^5$ (s^2)	$r_4 \cos\frac{2\pi}{5} \times 10^5 \delta_{M_r}$ (s^2)	$12+13$ $\varrho_i \times 10^5$ (s^2)	2×14 $m_i\varrho_i \times 10^3$ (Mpm^{-1}s^4)	8×15 $m_i\varrho_i\delta_{M_r} \times 10^3$ (Mp s^4)	$\sum m_k\varrho_k\delta_{ik}^r \times 10^8$ (s^4)	$r_4 \cos\frac{2\pi}{5} \times 10^8 \delta_{M_r}$ (s^4)	$\varrho_i \times 10^8$ (s^4)	$14:19$ $\omega_{(1)}^2$
		1	2	3	4	5													
1	0·373	—	0·061	0·062	0·054	0·023	0·735	—	—	—	—	29·1	29·1	0·109	0·079	0·0	42·2	42·2	689
2	0·301	0·061	1·324	1·994	1·791	0·726	1·795	—	—	—	—	71·0	71·0	0·214	0·384	1·0	103·0	104·0	682
3	0·247	0·062	1·994	4·828	4·922	2·080	2·240	—	—	—	—	88·5	88·5	0·219	0·492	2·4	128·5	130·9	675
4	0·212	0·054	1·791	4·922	7·091	3·320	1·865	—	—	—	—	73·7	73·7	0·156	0·292	2·8	107·0	109·8	670
5	0·194	0·023	0·726	2·080	3·320	2·484	0·725	—	—	—	—	28·7	28·7	0·056	0·040	1·3	41·6	42·9	668

$\sum = 1.287$

Eqs (5.68) are not a general solution of eq. 5.66, and, as will be shown later on, cannot satisfy arbitrary boundary conditions.

5.2.3.1 *End supports permitting vertical displacement*

Fig. 5.13 shows a beam resting on elastic supports of which the end ones permit vertical displacement but no rotation. The spring constant of the outer supports is $C/2$, and of the inner ones C.

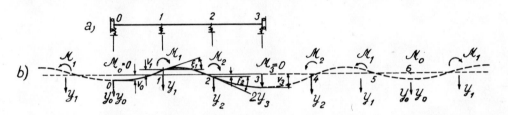

Fig. 5.13 *Uniform beam continuous over several equal spans on elastic supports, end supports permitting vertical displacement*

The boundary conditions are

$$\zeta_0 = \zeta_n = 0$$

and for them

$$r_j' = s_j' = 0 \quad \text{and} \quad h = 2n$$

As shown in Fig. 5.13b, the beam may be made into one of unlimited length subjected to a periodic load. Then eq. 5.68 gives

$$v_k = \sum_{j=0}^{n} s_j'' \cos \frac{\pi j k}{n}$$

$$\zeta_k = - \sum_{j=0}^{n} r_j'' \sin \frac{\pi j k}{n}$$

$$(5.69)$$

Substituting in eq. 5.66 and using the procedure outlined in Section 5.1 gives the equations (cf. eq. 5.30)

$$d_j s_j'' + b_j r_j'' = \frac{2}{n} \sum_{k=0}^{n} \mathscr{Y}_k \cos \frac{\pi j k}{n}$$

$$b_j s_j'' + a_j r_j'' = - \frac{2}{n} \sum_{k=1}^{n-1} \mathscr{M}_k \sin \frac{\pi j k}{n}$$

$$(5.70)$$

where d_j is written in place of a_j^{vv}, and a_j in place of $a_i^{\zeta\zeta}$, and

$$d_j = d + 2e \cos \frac{\pi j}{n}$$

$$b_j = - 2b \sin \frac{\pi j}{n}$$

$$(5.71)$$

a_j is defined by eq. 5.47.

Eq. 5.70 applies to the case of $j \neq 0 \neq n$. For $j = 0$

$$d_0 s_j'' = \frac{1}{n} \sum_{k=0}^{n} \mathscr{Y}_k \qquad (5.72)$$

and for $j = n$

$$d_n s_j'' = \frac{1}{n} \sum_{k=0}^{n} (-1)^k \mathscr{Y}_k \qquad (5.73)$$

The free vibration of the system is described by eqs 5.70, 5.72 and 5.73 with the right-hand sides set to zero. From the condition that the determinant of the coefficients should be equal to zero we get by eq. 5.70

$$d_j a_j - b_j^2 = 0$$

or after substitution

$$\left[F_2(\lambda) + F_1(\lambda) \cos \frac{\pi j}{n} \right] \left[F_6(\lambda) + F_5(\lambda) \cos \frac{\pi j}{n} + \frac{Cl^3}{2EJ} \right] - F_3^2(\lambda) \sin^2 \frac{\pi j}{n} = 0 \quad (5.74)$$

For $j = 0$ $(j = n)$

$$d_0 = 0 \quad (d_n = 0)$$

or

$$\frac{2EJ}{l^3} F_6(\lambda) + C \pm \frac{2EJ}{l^3} F_5(\lambda) = 0 \qquad (5.75)$$

5.3.3.2 *End supports permitting rotation*

The particular solution 5.67 can also be used in the case of end supports permitting rotation but no vertical displacement (Fig. 5.14). Then the boundary conditions are (cf. eq. 1.170)

$$
\begin{aligned}
v_0 &= v_n = 0 \\
v_{-1} &= -v_1 \\
v_{n+1} &= -v_{n-1} \\
\zeta_{-1} &= \zeta_1 \\
\zeta_{n+1} &= \zeta_{n-1}
\end{aligned}
$$

$$(5.76)$$

and the load applied to spans $-1,0$ and $n, n+1$ of the beam vibrating in forced vibration, acts as shown in Fig. 5.14b. At $h = 2n$ these conditions are satisfied by the equations

$$v_k = \sum_{j=0}^{n} s'_j \sin \frac{\pi j}{n}$$

$$\zeta_k = \sum_{j=0}^{n} r'_j \cos \frac{\pi j}{n}$$

$$(5.77)$$

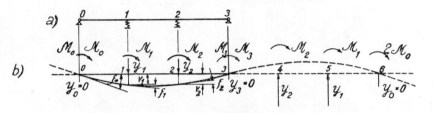

Fig. 5.14 Uniform beam continuous over several equal spans on elastic supports, end supports permitting rotation

Substituting them in eq. 5.66 gives

$$d_j s'_j + b_j r'_j = \frac{2}{n} \sum_{k=1}^{n-1} \mathcal{Y}_k \sin \frac{\pi j k}{n}$$

$$b_j s'_j + a_j r'_j = \frac{2}{n} \sum_{k=0}^{n} \mathcal{M}_k \cos \frac{\pi j k}{n} \quad \text{for} \quad j \neq 0, n$$

$$(5.78)$$

$$a_0 r'_0 = \frac{1}{n} \sum_{k=0}^{n} \mathcal{M}_k \tag{5.79}$$

$$a_n r'_n = \frac{1}{n} \sum_{k=0}^{n} (-1)^k \mathcal{M}_k \tag{5.80}$$

The natural frequencies of the system are defined by eq. 5.74, with eqs 5.64 and 5.65 applying in place of eqs 5.75. The modes of free vibration are different from those in the preceding case.

Example 5.3

Consider a uniform beam continuous over 5 spans resting on elastic supports and end supports permitting only rotation (Fig. 5.15), with the following specification: $l = 15$ m, $\mu = 0.165$ Mpm^{-2} s^2, $J = 0.0223$ m^4, $E = 2.4 \times 10^6$ Mpm^{-2}. $C = 761$ Mp/m is the force required to compress the elastic support by 1 m.

The lowest natural frequency of the beam is at $j = 1$. It is determined from eq. 5.74 which with the numerical values substituted in, turns out as follows:

$$\left[F_2(\lambda) + \cos \frac{\pi}{5} F_1(\lambda) \right]\left[F_6(\lambda) + \cos \frac{\pi}{5} F_5(\lambda) + 24 \right] - \left[\sin \frac{\pi}{5} F_3(\lambda) \right]^2 = 0$$

The lowest value of λ that satisfies the equation is 2.587, i.e.

$$5\cdot44 \times 3\cdot60 - 4\cdot43^2 = 0$$

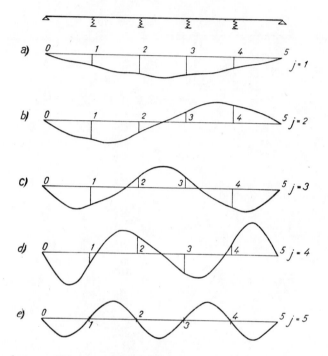

Fig. 5.15 Uniform beam continuous over five equal spans on elastic supports, end supports permitting rotation. First five modes of free vibration

The natural mode is given by the ratio

$$\frac{r_1}{s_1} l = -\frac{d_1}{b_1} l = -\frac{b_1}{a_1} l = \frac{3\cdot60}{4\cdot43} = \frac{4\cdot43}{5\cdot44} = 0\cdot813$$

By eq. 5.77 the deformations of the joints then are

$$v_1 = \sin \frac{\pi}{5} = 0\cdot588 ; \quad v_2 = \sin \frac{2\pi}{5} = 0\cdot951 ; \quad \text{etc.}$$

$$\zeta_0 = \frac{0\cdot813}{l} ; \quad \zeta_1 = \frac{0\cdot813}{l} \cos \frac{\pi}{5} = \frac{0\cdot658}{l} ; \quad \zeta_2 = \frac{0\cdot813}{l} \cos \frac{2\pi}{5} = \frac{0\cdot251}{l} ; \quad \text{etc.}$$

To these deformations one can compute the variations of the deflections of the spans from eqs 2.25 and 2.47. For the deflection midspan of $k, k + 1$ the computation yields

$$v\left(\frac{l}{2}\right) = -\frac{1}{2}(v_k + v_{k+1})\frac{F_5(\frac{1}{2}\lambda)}{F_6(\frac{1}{2}\lambda)} + (\zeta_k - \zeta_{k+1})\frac{l\,F_3(\frac{1}{2}\lambda)}{4F_6(\lambda\frac{1}{2})}$$

$$= \left[-\frac{F_5(\frac{1}{2}\lambda)}{F_6(\frac{1}{2}\lambda)}s'_j\cos\frac{\pi}{2n}j + \frac{l}{2}\frac{F_3(\frac{1}{2}\lambda)}{F_6(\frac{1}{2}\lambda)}r'_j\sin\frac{\pi}{2n}j\right]\sin\frac{\pi}{n}(k+\tfrac{1}{2})j$$

In the case considered, the deflection midspan of 2, 3 $(\lambda/2 \approx 1{\cdot}30)$ is

$$v\left(\frac{l_{2,3}}{2}\right) = \frac{12{\cdot}37}{10{\cdot}94}\cos\frac{\pi}{10} + \frac{6{\cdot}08}{10{\cdot}94}0{\cdot}813\sin\frac{\pi}{10} = 1{\cdot}14$$

and of 1, 2

$$v\left(\frac{l_{1,2}}{2}\right) = 1{\cdot}14\sin\frac{3\pi}{10} = 0{\cdot}922, \quad \text{etc.}$$

The deflections are shown in Fig. 5.15a. The natural frequency corresponding to this mode is

$$f_{(1)} = \frac{\lambda_1^2}{2\pi l^2}\left(\frac{EJ}{\mu}\right)^{1/2} = \frac{2{\cdot}587^2}{2\pi}\,2{\cdot}53 = 2{\cdot}70\text{ s}^{-1}$$

At $j = 2$, eq. 5.74 becomes

$$\left[F_2(\lambda) + \cos\frac{2\pi}{5}F_1(\lambda)\right]\left[F_6(\lambda) + \cos\frac{2\pi}{5}F_5(\lambda) + 24\right] - \left[\sin\frac{2\pi}{5}F_3(\lambda)\right]^2 = 0$$

The equation is satisfied by $\lambda = 2{\cdot}605$. The natural mode is given by the ratio

$$\frac{r'_2}{s'_2}l = \frac{12{\cdot}28}{7{\cdot}22} = \frac{7{\cdot}22}{4{\cdot}26} = 1{\cdot}70$$

and from there the deformations of the joints are

$$v_1 = \sin\frac{2\pi}{5} = 0{\cdot}951; \quad v_2 = \sin\frac{4\pi}{5} = 0{\cdot}588; \quad \text{etc.}$$

$$\zeta_0 = \frac{1{\cdot}70}{l}; \quad \zeta_1 = \frac{1{\cdot}70}{l}\cos\frac{2\pi}{5} = \frac{0{\cdot}525}{l}; \quad \zeta_2 = \frac{1{\cdot}70}{l}\cos\frac{4\pi}{5} = -\frac{1{\cdot}375}{l}; \quad \text{etc.}$$

The deflections are shown in Fig. 5.15b. The natural frequency is

$$f_{(2)} = \frac{2{\cdot}605^2}{2\pi}\,2{\cdot}53 = 2{\cdot}73\text{ s}^{-1}$$

Similarly, at $j = 3$, eq. 5.74 is satisfied by $\lambda = 2.708$; the corresponding ratio is

$$\frac{r'_3}{s'_3} l = \frac{21.00}{7.52} = \frac{7.52}{2.68} = 2.80$$

and the deformations of the joints (Fig. 5.15c) are

$$v_1 = \sin \frac{3\pi}{5} = 0.951 ; \quad v_2 = \sin \frac{6\pi}{5} = -0.588 ; \quad \text{etc.}$$

$$\zeta_0 = \frac{2.80}{l} ; \quad \zeta_1 = \frac{2.80}{l} \cos \frac{3\pi}{5} = -\frac{0.865}{l} ; \quad \zeta_2 = \frac{2.80}{l} \cos \frac{6\pi}{5} = -\frac{2.26}{l} ; \quad \text{etc.}$$

The corresponding natural frequency is

$$f_{(3)} = \frac{2.708^2}{2} 2.53 = 2.95 \text{ s}^{-1}$$

At $j = 4$ (Fig. 5.15d)

$$\lambda = 2.937 , \quad \frac{r'_4}{s'_4} l = \frac{25.20}{5.18} = \frac{5.18}{1.060} = 4.87$$

$$v_1 = \sin \frac{4\pi}{5} = 0.588 , \quad v_2 = \sin \frac{8\pi}{5} = -0.951 , \quad \text{etc.}$$

$$\zeta_0 = \frac{4.87}{l} , \quad \zeta_1 = \frac{4.87}{l} \cos \frac{4\pi}{5} = -\frac{3.94}{l} , \quad \zeta_2 = \frac{4.87}{l} \cos \frac{8\pi}{5} = \frac{1.50}{l} , \quad \text{etc.}$$

and the frequency

$$f_{(4)} = \frac{2.937^2}{2\pi} 2.53 = 3.47 \text{ s}^{-1}$$

And finally, at $j = 5$, there holds the condition 5.64 and $\lambda = \pi$. The corresponding mode shape is identical with the first mode shape of a simply supported beam (Fig. 5.15e). All the vertical displacements v_k are zero. The natural frequency is

$$f_{(5)} = \frac{\pi}{2} 2.53 = 3.97 \text{ s}^{-1}$$

The higher frequencies and natural modes are obtained by the same procedure using the higher values of λ that satisfy eq. 5.74. The results are

$$\text{for} \quad j = 4 , \quad \lambda = 4.02 , \quad f_{(6)} = \frac{4.02^2}{2\pi} 2.53 = 6.52 \text{ s}^{-1}$$

for $j = 3$, $\lambda = 4\cdot55$, $f_{(7)} = \dfrac{4\cdot55^2}{2\pi} \, 2\cdot53 = 8\cdot34 \text{ s}^{-1}$

for $j = 2$, $\lambda = 5\cdot13$, $f_{(8)} = \dfrac{5\cdot13^2}{2\pi} \, 2\cdot53 = 10\cdot6 \text{ s}^{-1}$

for $j = 1$, $\lambda = 5\cdot72$, $f_{(9)} = \dfrac{5\cdot72^2}{2\pi} \, 2\cdot53 = 13\cdot2 \text{ s}^{-1}$

for $j = 0$, $\lambda = 2$, $f_{(10)} = 2\pi \, . \, 2\cdot53 = 15\cdot9 \text{ s}^{-1}$

For the purposes of comparison we list below the natural frequencies of a simply supported beam of identical dimensions (without intermediate supports) and span of $5l$, for which $1/5^2 l^2 \, . \, (EJ/\mu)^{1/2} = 2\cdot53/5^2 = 0\cdot1013 \text{ s}^{-1}$

$$f_{(1)} = 0\cdot159 \text{ s}^{-1}, \quad f_{(6)} = 5\cdot72 \text{ s}^{-1}$$
$$f_{(2)} = 0\cdot637 \text{ s}^{-1}, \quad f_{(7)} = 7\cdot78 \text{ s}^{-1}$$
$$f_{(3)} = 1\cdot43 \text{ s}^{-1}. \quad f_{(8)} = 10\cdot18 \text{ s}^{-1}$$
$$f_{(4)} = 2\cdot54 \text{ s}^{-1}, \quad f_{(9)} = 12\cdot87 \text{ s}^{-1}$$
$$f_{(5)} = 3\cdot97 \text{ s}^{-1}, \quad f_{(10)} = 15\cdot9 \text{ s}^{-1}$$

The natural frequency of a simply supported beam with an identical span of $5l$ but resting on an elastic foundation with the specific stiffness of the subsoil $C/l = 761/15 = 50\cdot7 \text{ Mp m}^{-1}$ would be

$$f_{(k)} = \frac{1}{2\pi} \left(\frac{k^4\pi^4 EJ + Cl^3}{\mu l^4} \right)^{1/2} \tag{5.81}$$

and for the given numerical values

$$f_{(1)} = 2\cdot80 \text{ s}^{-1}, \quad f_{(6)} = 6\cdot36 \text{ s}^{-1}$$
$$f_{(2)} = 2\cdot86 \text{ s}^{-1}, \quad f_{(7)} = 8\cdot27 \text{ s}^{-1}$$
$$f_{(3)} = 3\cdot14 \text{ s}^{-1}, \quad f_{(8)} = 10\cdot56 \text{ s}^{-1}$$
$$f_{(4)} = 3\cdot78 \text{ s}^{-1}, \quad f_{(9)} = 13\cdot2 \text{ s}^{-1}$$
$$f_{(5)} = 4\cdot87 \text{ s}^{-1}, \quad f_{(10)} = 16\cdot1 \text{ s}^{-1}$$

5.2.4 Beam continuous over several spans on elastic supports — cross-section varying within the spans

Consider now the more general case of a beam with a cross-section which varies within the spans (Fig. 5.21) resting on elastic supports (Fig. 5.15). Though the equations deduced in Sections 5.2.2 and 5.2.3 continue to apply, the coefficients a, b, c, d, e can no longer be determined from the simple expressions 5.39 and 5.67.

Given certain assumptions they can be found using the more complicated relations 4.51−4.57; but even so, the solution is far from easy because there are no tables of functions $F(\lambda_g, \lambda_h)$. This is why we shall examine the problem by the method described in the preceding sections in combination with the methods of discrete mass points and successive approximations. In addition we shall also apply the simplified slope-deflection method (Section 5.2.4.2).

5.2.4.1 Solution of free vibration by the method of successive approximations

The method of successive approximations is based (cf. Section 1.2.1) on repeated computation of the static deflection curve appertaining to the static load. It will clearly be to advantage when dealing with systems with repeated elements, to choose the initial load in the fundamental mode j; as we shall show presently, the deflection produced by it will also be of the fundamental mode j. Another circumstance we shall make use of in the computations is that the individual spans of the system are symmetric about the axis passing through the centre of the span and perpendicular to the beam centre line.

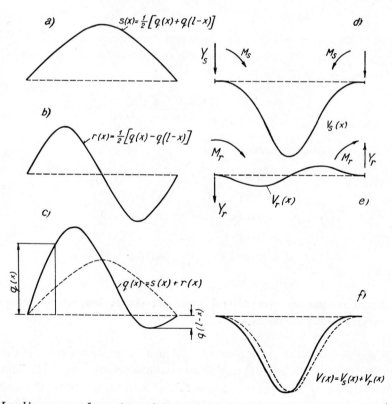

Fig. 5.16 Load in one span of a continuous beam — (a) Symmetric and (b) antisymmetric components

Deformation of a single span

Consider first a single-span beam clamped at both ends and subjected to a continuous static load $q(x)$ (Fig. 5.16c). The load may be resolved into the symmetric component $s(x)$ and the antisymmetric component $r(x)$ (Fig. 5.16a, b)

$$q(x) = s(x) + r(x)$$
$$s(x) = \tfrac{1}{2}[q(x) + q(l-x)] = s(l-x)$$
$$r(x) = \tfrac{1}{2}[q(x) - q(l-x)] = -r(l-x)$$

$$(5.82)$$

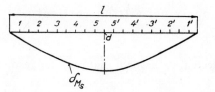

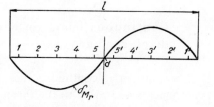

Fig. 5.17 (a) Influence line M_s Fig. 5.17 (b) Influence line M_r

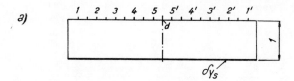

a)

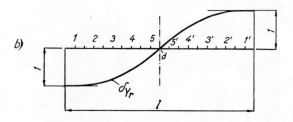

b)

Fig. 5.18 (a) Influence line Y_s, (b) influence line Y_r

The symmetric load component, $s(x)$, produces symmetric deformation $V_s(x)$ of the beam, the left support being subjected to force Y_s and moment M_s, the right one to Y_s and $-M_s$ (Fig. 5.16d). The antisymmetric load component $r(x)$ produces antisymmetric deformation $V_r(x)$, and forces and moments $\pm Y_r$, M_r (Fig. 5.16e). At point x the total deformation is thus (Fig. 5.16f)

$$V(x) = V_s(x) + V_r(x)$$

$$(5.83)$$

Consider the same beam and assume that its ends are simultaneously rotated through unit angle in opposite directions. The rotation of the ends will cause the deflection line — which is also the influence line of moment M_s produced by the symmetric load — to be symmetric. The influence line ordinate δ_{M_s} (Fig. 5.17a) at point x defines the value of moment M_s acting on the left support of the clamped beam if two unit forces $P = 1$ are applied to sections x and $l - x$.

Figs 5.17b, 5.18a, b show the other end deformations and the lines δ_{M_r}, δ_{Y_s}, δ_{Y_r} appertaining to them. Ordinates δ_{M_r}, δ_{Y_r} give respectively the moment and force acting on the left support if two antisymmetric unit forces are applied to sections x and $l - x$. Ordinate δ_{Y_s} is of unit value in all cross-sections.

Continuous beam on end supports permitting vertical displacement

Consider a continuous beam subjected to a load that can be resolved in components according to the formula

$$q_{k,k+1}(x) = \sum_{j=0}^{n} \left[s_j''(x) \cos \frac{\pi j(k + \frac{1}{2})}{n} - r_j''(x) \sin \frac{\pi j(k + \frac{1}{2})}{n} \right] \tag{5.84}$$

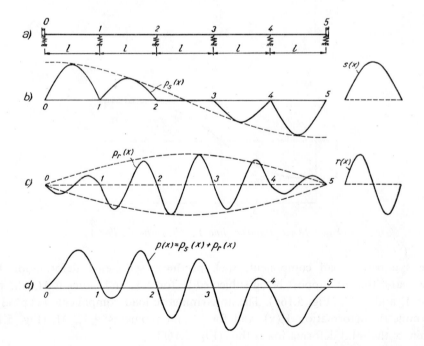

Fig. 5.19 *Uniform beam continuous over five equal spans with end supports permitting displacement. Load of type $j = 1$. (b) Symmetric load component, (c) antisymmetric load component*

where $q_{k,k+1}(x)$ is the load in section x of span k, $k + 1$. We shall first examine the static deflection produced by a load of mode j with the symmetric component

$$p_s(x) = s(x) \cos \frac{\pi j(k + \frac{1}{2})}{n} \tag{5.85}$$

and the antisymmetric component

$$p_r(x) = -r(x) \sin \frac{\pi j(k + \frac{1}{2})}{n} \tag{5.86}$$

where j is an arbitrary number from 0 to n. Fig. 5.19b, c shows a beam with five spans $(n = 5)$ in the case of $j = 1$.

Were we in a position to prevent all the joints from undergoing displacement and rotation, the deflection produced by the load defined by eqs 5.85 and 5.86 (cf. eq. 5.83) would be

$$V_s(x) \cos \frac{\pi j(k + \frac{1}{2})}{n} - V_r(x) \sin \frac{\pi j(k + \frac{1}{2})}{n}$$

Consider a continuous beam on elastic supports; the spring constant of the end supports is $C/2$, and of the intermediate supports, C. A beam on unyielding supports is a special case of this more general solution. The loading force and moment at joint k are

$$\mathcal{Y}_k = Y_s \left[\cos \frac{\pi j(k + \frac{1}{2})}{n} + \cos \frac{\pi j(k - \frac{1}{2})}{n} \right]$$

$$- Y_r \left[\sin \frac{\pi j(k + \frac{1}{2})}{n} - \sin \frac{\pi j(k - \frac{1}{2})}{n} \right] = S_j \cos \frac{\pi j k}{n} \tag{5.87}$$

$$\mathcal{M}_k = M_s \left[\cos \frac{\pi j(k + \frac{1}{2})}{n} - \cos \frac{\pi j(k - \frac{1}{2})}{n} \right]$$

$$- M_r \left[\sin \frac{\pi j(k + \frac{1}{2})}{n} + \sin \frac{\pi j(k - \frac{1}{2})}{n} \right] = -R_j \sin \frac{\pi j k}{n} \tag{5.88}$$

where

$$S_j = 2 \left(Y_s \cos \frac{\pi j}{2n} - Y_r \sin \frac{\pi j}{2n} \right)$$

$$R_j = 2 \left(M_s \sin \frac{\pi j}{2n} + M_r \cos \frac{\pi j}{2n} \right)$$

$$\tag{5.89}$$

These forces and moments produce generalised deformations r_j, s_j so that the displacements and rotations arising at the various joints are

$$v_k = s_j \cos \frac{\pi jk}{n}$$

$$\zeta_k = -r_j \sin \frac{\pi jk}{n}$$

(5.90)

s_j, r_j are defined by the equations

$$d_j s_j + b_j r_j = S_j$$
$$b_j s_j + a_j r_j = R_j$$

(5.91)

which differ from eq. 5.70 in that

$$a_j = 2\overline{M}_{k,k+1(\zeta_k=1)} + 2\overline{M}_{k+1,k(\zeta_k=1)} \cos \frac{\pi j}{n}$$

$$b_j = 2\overline{Y}_{k+1,k(\zeta_k=1)} \sin \frac{\pi j}{n} = -2\overline{M}_{k+1,k(v_k=1)} \sin \frac{\pi j}{n}$$

$$d_j = 2\overline{Y}_{k,k+1(v_k=1)} + C + 2\overline{Y}_{k+1,k(v_k=1)} \cos \frac{\pi j}{n}$$

(5.92)

where $\overline{M}_{k,k+1(\zeta_k=1)}$ is the static moment acting at end k of bar k, $k+1$ if end k is rotated through $\zeta_k = 1$, etc.

If the joints undergo displacement and rotation according to eq. 5.90, the bar deforms in span k, $k+1$. This deformation has a symmetric component which — because $\delta_{Y_s} = 1$ — is

$$\tfrac{1}{2}(v_k + v_{k+1}) + \tfrac{1}{2}(\zeta_k - \zeta_{k+1}) \delta_{M_s}$$

and an antisymmetric component

$$\tfrac{1}{2}(v_k - v_{k+1}) \delta_{Y_r} + \tfrac{1}{2}(\zeta_k + \zeta_{k+1}) \delta_{M_r}$$

Further

$$\frac{1}{2}(v_k \pm v_{k+1}) = s_j \left[\cos \frac{\pi jk}{n} \pm \cos \frac{\pi j(k+1)}{n} \right] = \left\{ \begin{array}{l} s_j \cos \dfrac{\pi j}{2n} \cos \dfrac{\pi j(k+\frac{1}{2})}{n} \\[2mm] s_j \sin \dfrac{\pi j}{2n} \sin \dfrac{\pi j(k+\frac{1}{2})}{n} \end{array} \right.$$

(5.93)

$$\frac{1}{2}(\zeta_k \pm \zeta_{k+1}) = r_j\left[-\sin\frac{\pi jk}{n} \pm \sin\frac{\pi j(k+1)}{n}\right] = \begin{cases} r_j \sin\dfrac{\pi j}{2n}\cos\dfrac{\pi j(k+\frac{1}{2})}{n} \\[3mm] -r_j\cos\dfrac{\pi j}{2n}\sin\dfrac{\pi j(k+\frac{1}{2})}{n} \end{cases}$$

$$(5.93)$$

The total deflection produced by load $p(x)$ defined by eqs 5.85 and 5.86 in span k, $k+1$ is the sum of the symmetric and antisymmetric components

$$v(x) = \sigma_j(x)\cos\frac{\pi j(k+\frac{1}{2})}{n} - \varrho_j(x)\sin\frac{\pi j(k+\frac{1}{2})}{n} \qquad (5.94)$$

where

$$\sigma_j(x) = V_s(x) + \delta_{Y_s}s_j\cos\frac{\pi j}{2n} + \delta_{M_s}r_j\sin\frac{\pi j}{2n}$$

$$\varrho_j(x) = V_r(x) - \delta_{Y_r}s_j\sin\frac{\pi j}{2n} + \delta_{M_r}r_j\cos\frac{\pi j}{2n}$$

$$(5.95)$$

Expression $\sigma_j(x)$ arises from the superposition of

(1) the deflection $V_s(x)$ of the clamped beam resulting from load $s(x)$,
(2) the deflection resulting from the symmetric displacement $s_j\cos(\pi j/2n)$ of the ends of − and hence also of the whole of − the beam,
(3) the deflection resulting from symmetric rotations of the ends of the beam, the right end rotating through angle $+r_j\sin(\pi j/2n)$, the left one through angle $-r_j\sin(\pi j/2n)$.

Expression $\varrho_j(x)$ arises from the superposition of

(1) the deflection $V_r(x)$ of the clamped beam resulting from load $r(x)$,
(2) the deflection resulting from the left end undergoing displacement $-s_j\sin(\pi j/2n)$, the right one displacement $+s_j\sin(\pi j/2n)$,
(3) the deflection resulting from the ends rotating through $r_j\cos(\pi j/2n)$.

The load curves are presented in Fig. 5.19. It is evident from formulae 5.85, 5.86 and 5.94 that the expressions of load and deformation are of the same fundamental mode.

The results of the analysis carried out in the preceding paragraphs can be used in the computation of frequencies and modes of free vibration by the method of successive approximations. The procedure is as follows: For the first approximation choose any deflection curve ${}^1v(x)$ that can be resolved in components according to eqs 5.85 and 5.86, and compute the respective symmetric and antisymmetric load components $\mu(x)\,^1v(x)$ [$\mu(x)$ denoting the mass per unit length at point x]. These load components give rise to deflection $^2v(x)$ as the second approximation to the free vibration mode. Deflection $^2v(x)$ has a symmetric and an antisymmetric component according to eq. 5.94. It is clear that the computation may be carried out for merely one span since the deflection in any span can easily be determined with the aid of $\sigma(x)$ and $\varrho(x)$ defined by eq. 5.95. The process may be repeated until a sufficient agreement is reached between the forms of the last-but-one approximation $^{k-1}v(x)$ and the last approximation $^kv(x)$.

Then

$$\omega_j^2 = \frac{^{k-1}v(x)}{^kv(x)}$$

It is evident from what has just been said that the ratio of the symmetric and antisymmetric components, too, will become stabilised in the last approximations, so that the natural circular frequency may be computed from the formula

$$\omega_j^2 = \frac{^{k-1}\sigma_j(x)}{^k\sigma_j(x)} = \frac{^{k-1}\varrho_j(x)}{^k\varrho_j(x)} \tag{5.96}$$

Formula 5.95 applies to $j \neq 0, n$. For $j = 0$, the antisymmetric component of eq. 5.94 drops out and therefore

$$\omega_0^2 = \frac{^{k-1}\sigma_0(x)}{^k\sigma_0(x)} \tag{5.97}$$

For $j = n$ it is the symmetric component that drops out and

$$\omega_n^2 = \frac{^{k-1}\varrho_n(x)}{^k\varrho_n(x)} \tag{5.98}$$

Formulae 5.97 and 5.98 express the natural circular frequency of a single-span beam.

Practising engineers usually do not consider the mass as continuously distributed but divide the beam into a finite number of segments and assume that the segment masses are concentrated at their centres of gravity.

Continuous beam on end supports permitting rotation

The solution remains nearly the same as in the preceding case. However, the load of the j-th mode is divided according to the law

$$p(x) = p_s(x) + p_r(x) = s(x)\sin\frac{\pi j(k + \tfrac{1}{2})}{n} + r(x)\cos\frac{\pi j(k + \tfrac{1}{2})}{n} \tag{5.99}$$

Analogous to eqs 5.87 and 5.88 the force and moment at joint k now are

$$\mathscr{Y}_k = Y_s \left[\sin \frac{\pi j(k + \frac{1}{2})}{n} + \sin \frac{\pi j(k - \frac{1}{2})}{n} \right]$$

$$Y_r \left[\cos \frac{\pi j(k + \frac{1}{2})}{n} - \cos \frac{\pi j(k - \frac{1}{2})}{n} \right] = S_j \sin \frac{\pi j k}{n} \qquad (5.100)$$

$$\mathscr{M}_k = M_s \left[\sin \frac{\pi j(k + \frac{1}{2})}{n} - \sin \frac{\pi j(k - \frac{1}{2})}{n} \right]$$

$$M_r \left[\cos \frac{\pi j(k + \frac{1}{2})}{n} + \cos \frac{\pi j(k - \frac{1}{2})}{n} \right] = R_j \cos \frac{\pi j k}{n} \qquad (5.101)$$

Eqs 5.89, 5.91 and 5.92 continue to apply but eqs 5.90 are replaced by

$$v_k = s_j \sin \frac{\pi j k}{n}$$

$$\zeta_k = r_j \cos \frac{\pi j k}{n}$$

$$(5.102)$$

On the strength of this, eq. 5.93 is replaced by

$$\frac{1}{2}(v_k \pm v_{k+1}) = \left\{ \begin{array}{l} s_j \cos \dfrac{\pi j}{2n} \sin \dfrac{\pi j(k + \frac{1}{2})}{n} \\[3mm] -s_j \sin \dfrac{\pi j}{2n} \cos \dfrac{\pi j(k + \frac{1}{2})}{n} \end{array} \right.$$

$$\frac{1}{2}(\zeta_k \pm \zeta_{k+1}) = \left\{ \begin{array}{l} r_j \cos \dfrac{\pi j}{2n} \cos \dfrac{\pi j(k + \frac{1}{2})}{n} \\[3mm] r_j \sin \dfrac{\pi j}{2n} \sin \dfrac{\pi j(k + \frac{1}{2})}{n} \end{array} \right.$$

$$(5.103)$$

and eq. 5.94 by

$$v(x) = \sigma_j \sin \frac{\pi j(k + \frac{1}{2})}{n} + \varrho_j \cos \frac{\pi j(k + \frac{1}{2})}{n} \qquad (5.104)$$

with eq. 5.95 applying.

Eq. 5.96 continues to hold for $j \neq 0, n$. For $j = 0$

$$\omega_0^2 = \frac{{}^{k-1}\varrho_0(x)}{{}^k \varrho_0(x)}$$

for $j = n$

$$\omega_n^2 = \frac{^{k-1}\sigma_n(x)}{^k\sigma_n(x)}$$

Example 5.4

Consider the continuous beam in Fig. 5.20 with cross-section varying within the spans and dimensions as shown in Fig. 5.21. The axis of the beam is assumed to be straight, and Young's modulus $E = 2 \cdot 4 \times 10^6$ Mp/m². The influence

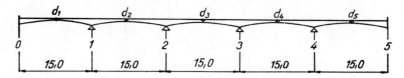

Fig. 5.20 Beam continuous over five spans — cross-section varying within the spans

lines of deflection $\delta_i^s(x)$ for a symmetric load consisting of two unit forces acting on the clamped beam are shown in Fig. 5.22a, the influence line of moment $M_s(x)$ resulting from the clamping, is shown in Fig. 5.17a. The influence lines $\delta_i^r(x)$, $\delta_{M_r}(x)$ shown in Figs 5.22b and 5.17b belong to an antisymmetric load. The influence coefficients are set out in matrix form in Table 5.6.

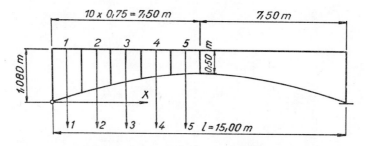

Fig. 5.21 One span of beam shown in Fig. 5.20

A detailed computation by the method of successive approximations was made for $j = 4$, $j = 1$ and $j = 0$. Table 5.6 reviews the computation for $j = 4$ at which the natural frequency is the lowest of all. In auxiliary computations carried out in the preparation of the table, use was made of formulae 5.91 and 5.92 with $s_j = 0$, and of eqs 5.89; column 14 was computed from formula 5.95. From the ratio of the last two approximations we get

$$\omega_4^2 = \omega_{(1)}^2 = \frac{696 \cdot 9 \times 10^3}{1\,009 \cdot 3} = 690 \text{ s}^{-2} \; ; \; f_{(1)} = 4 \cdot 18 \text{ s}^{-1}$$

The natural mode in span k, $k + 1$ is given by the formula

$$v_{i(1)} = \sigma_i \cos \frac{\pi}{n} j\left(k + \tfrac{1}{2}\right) - \varrho_i \sin \frac{\pi}{n} j\left(k + \tfrac{1}{2}\right)$$

where $j = 4$, and σ_i, ϱ_i are the last approximations.

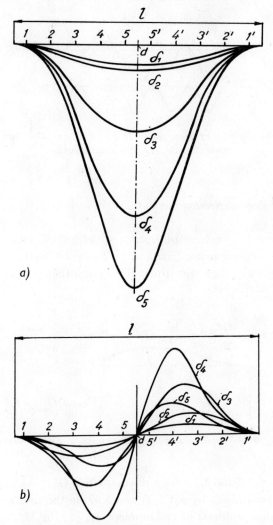

Fig. 5.22 *Influence lines of deflection of a clamped span. (a) Symmetric load, (b) antisymmetric load*

Proceeding analogously we obtain

$$\omega_1^2 = \omega_{(4)}^2 = 4\ 530\ \text{s}^{-2}, \quad f_{(4)} = 10 \cdot 73\ \text{s}^{-1}$$

and finally

$$\omega_0^2 = \omega_{(5)}^2 = 6\ 080\ \text{s}^{-2}, \quad f_{(5)} = 12 \cdot 43\ \text{s}^{-1}$$

the last value also being the fundamental frequency of a clamped beam. The natural mode shapes (first, fourth and fifth) are shown in Fig. 5.23.

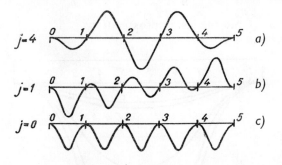

Fig. 5.23 *Modes of free vibration of beam shown in Fig. 5.20. (a) First natural mode, (b) fourth natural mode, (c) fifth natural mode*

5.2.4.2 *Simplified slope-deflection method*

In Section 4.1.3 we have outlined a simplified computation of end forces and moments of a non-prismatic bar, and in Table 4.8 showed an example of a numerical computation of these quantities. If we know them we can use the equations

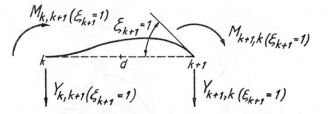

Fig. 5.24 *Simplified slope-deflection method. End moments*

derived in Sections 5.2.2 and 5.2.3 for a prismatic bar. Of course, coefficients a and c in them are no longer defined by eqs 5.39 and 5.67; in the case we are dealing with, $a/2$ (or c) denotes the amplitude of end moment $M_{k,k+1}$ (or $M_{k+1,k}$) if end k rotates with unit amplitude (cf. Fig. 5.24). Coefficients b, d, e have an analogous meaning.

The solution is convenient whenever the computation is carried out under the assumptions mentioned in Section 4.1.3. We shall illustrate it by way of the following example:

Example 5.5

Consider the beam shown in Fig. 5.20 which we have examined in Example 5.4. To enhance the accuracy of the solution, divide the individual spans of the beam by means of secondary joints d_1 to d_5. The end forces and moments of bar kd at various end deformations may be solved by the approximate method described at length in Section 4.1.3. Their numerical values (in Mp and m) are as follows:

$$Y_{dk(v_d=1)} = 1{\cdot}916 . 10^3 - 0{\cdot}294\omega^2$$
$$Y_{dk(\zeta_k=1)} = -9{\cdot}68 . 10^3 - 0{\cdot}386\omega^2 = M_{kd(v_d=1)}$$
$$M_{kd(\zeta_k=1)} = 55{\cdot}3 . 10^3 - 2{\cdot}025\omega^2$$
$$M_{dk(\zeta_k=1)} = 17{\cdot}1 . 10^3 + 0{\cdot}508\omega^2 = M_{kd(\zeta_d=1)}$$
$$M_{dk(\zeta_d=1)} = 17{\cdot}9 . 10^3 - 0{\cdot}256\omega^2$$

$$(5.105)$$

The end moments of the whole span k, $k + 1$ are obtained by the slope-deflection method from the equations

$$2M_{dk(\zeta_d=1)}\zeta_d + M_{dk(\zeta_k=1)} = 0$$
$$2\,Y_{dk(v_d=1)}v_d + Y_{dk(\zeta_k=1)} = 0$$

$$(5.106)$$

and from there

$$\zeta_d = -\frac{M_{dk(\zeta_k=1)}}{2M_{dk(\zeta_d=1)}}$$

$$v_d = -\frac{Y_{dk(\zeta_k=1)}}{2Y_{dk(v_d=1)}}$$

$$(5.107)$$

The moments acting at ends k and $k + 1$ are then (see Fig. 5.12)

$$\frac{a}{2} = M_{kd(\zeta_k=1)} + M_{kd(\zeta_d=1)}\zeta_d + M_{kd(v_d=1)}v_d = M_{kd(\zeta_k=1)} - \frac{M^2_{kd(\zeta_d=1)}}{2M_{dk(\zeta_d=1)}}$$
$$- \frac{M^2_{kd(v_d=1)}}{2Y_{dk(v_d=1)}} \quad (5.108)$$

$$c = M_{kd(\zeta_d=1)}\zeta_d - M_{kd(v_d=1)}v_d = -\frac{M^2_{kd(\zeta_d=1)}}{2M_{dk(\zeta_d=1)}} + \frac{M^2_{kd(v_d=1)}}{2Y_{dk(v_d=1)}} \quad (5.109)$$

Using eq. 5.48 and the above values we can now compute the frequencies and modes of free vibration of the continuous beam. We find that the lowest frequency is at $j = 4$, with

$$a_4 = a + 2c \cos\frac{4\pi}{5} = 0 \quad (5.110)$$

Substituting from eqs 5.108, 5.109, 5.105 and $\cos (4\pi/5) = -0.809$ gives

$$2(55.3 \times 10^3 - 2.025\omega^2) - \frac{(17.1 \times 10^3 + 0.508\omega^2)^2}{17.9 \times 10^3 - 0.256\omega^2} (1 - 0.809)$$

$$- \frac{(9.68 \times 10^3 + 0.386\omega^2)^2}{1.916 \times 10^3 - 0.294\omega^2} (1 + 0.809) = 0 \qquad (5.111)$$

and from there

$$\omega_4^2 = \omega_{(1)}^2 = 690 \text{ s}^{-2}$$

$$f_{(1)} = \frac{\omega_{(1)}}{2\pi} = 4.18 \text{ Hz}$$

This value is the same as that computed in Example 5.4. No correction (Table 3.9) is necessary because the average value of λ is

$$\lambda = l \left(\frac{\mu\omega^2}{EJ}\right)^{1/4} = 7.5 \left(\frac{0.247 \times 690}{2.4 \times 10^6 \times 0.0223 \times 1.5}\right)^{1/4} = 1.61$$

(for the mean values of μ and J).

The second natural frequency corresponds to $j = 3$ where $\cos (3\pi/5) = -0.309$. The equation

$$a - 2 \times 0.309c = 0$$

results in

$$\omega_{(2)}^2 = 1450 \text{ s}^{-2}$$

$$f_{(2)} = 6.07 \text{ Hz}$$

For the third natural frequency

$$j = 2, \quad \cos \frac{2\pi}{5} = 0.309$$

$$a + 2 \times 0.309c = 0$$

$$\omega_{(3)}^2 = 2785 \text{ s}^{-2}$$

$$f_{(3)} = 8.41 \text{ Hz}$$

For the fourth natural frequency

$$j = 1, \quad \cos \frac{\pi}{5} = 0.809$$

$$a + 2 \times 0.809c = 0$$

$$\omega_{(4)}^2 = 4790 \text{ s}^{-2}$$

$$f_{(4)} = 11.0 \text{ Hz}$$

Note: The average λ is about 2·6; by Table 3.9 the corrected value of $f_{(4)}$ is about 4% less than the one just obtained, i.e. $0·96 \times 11·0 = 10·55$ Hz. The frequency computed in the preceding Section is 10·73 Hz, i.e. about 2% more than the corrected, and about 2% less than the uncorrected value. In this case Table 3.9 fails to afford absolutely exact results for the reason that λ varies too much with varying cross-section.

The fifth natural frequency is the same as that of a clamped beam. The applicable relations are

$$a + 2c = 0$$

or

$$Y_{dk(v_d=1)} = 0$$

so that

$$1·916 \times 10^3 - 0·294\omega_{(5)}^2 = 0$$

and from there

$$\omega_{(5)}^2 = 6520 \text{ s}^{-2}$$
$$f_{(5)} = 12·85 \text{ Hz}$$

The exact value of $f_{(5)}$, 12.43 Hz, obtained in Example 5.4 is about 3% less. The natural modes are defined by formula 5.45.

5.2.5   Frames with repeated elements

Repetition of elements is quite a common practice in continuous and multi-storey frames; in these structures, too, it makes for considerable simplification of the calculation[45].

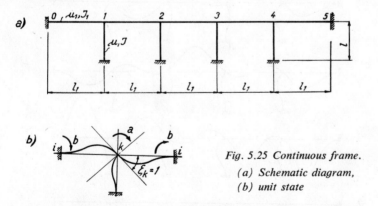

Fig. 5.25 Continuous frame.
(a) Schematic diagram,
(b) unit state

Consider the continuous frame shown in Fig. 5.25a. The application of the reciprocal theorem to the actual state of motion and to the unit auxiliary state indicated in Fig. 5.25b results in the equations set out in Table 5.7. ($b = c$, cf. eq. 5.39).

Table 5.7

ζ_1	ζ_2	ζ_3	ζ_4	
a	c			$= M_1$
c	a	c		$= M_2$
	c	a	c	$= M_3$
		c	a	$= M_4$

These equations are the same as eqs 5.36 except for moment a of the auxiliary state which is defined by the relation

$$a = \frac{2EJ_1}{l_1} F_2(\lambda_1) + \frac{EJ}{l} F_2(\lambda) \tag{5.112}$$

where subscript 1 refers to the horizontal bars; the data of the columns are without subscripts. The other relation that applies is

$$c = \frac{EJ_1}{l_1} F_1(\lambda_1)$$

so that the formulae derived in Section 5.2.2.2 can be used in an analogous way.

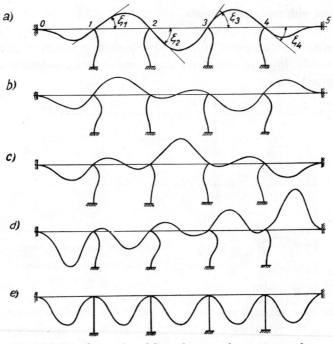

Fig. 5.26 First five modes of free vibration of a continuous frame

Free vibration is described by the frequency equation

$$a_j = 0$$

i.e.

$$a + 2c \cos \frac{\pi j}{n} = 0$$

or after substitution

$$2F_2(\lambda_1) + 2F_1(\lambda_1) \cos \frac{\pi j}{n} + \frac{J l_1}{J_1 l} F_2(\lambda) = 0 \quad \text{for} \quad j = 1 \dots n - 1 \quad (5.113)$$

Apart from this, the horizontal bars are apt to vibrate like clamped beams so that by eq. 2.37

$$\cosh \lambda_1 \cos \lambda_1 - 1 = 0$$

The natural modes of vibration are shown in Fig. 5.26.

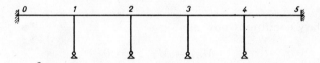

Fig. 5.27 Continuous frame with hinged column bases

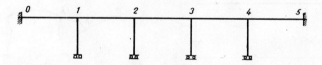

Fig. 5.28 Continuous frame with column bases permitting horizontal displacement but no rotation

The analysis of the frame illustrated in Fig. 5.27 proceeds along the same lines except that the frequency equation is written as

$$2F_2(\lambda_1) + 2F_1(\lambda_1) \cos \frac{\pi j}{n} + \frac{J l_1}{J_1 l} F_7(\lambda) = 0 \qquad (5.114)$$

The same applies to the frame in Fig. 5.28 whose frequency equation is

$$F_2(\lambda_1) + F_1(\lambda_1) \cos \frac{\pi j}{n} + \frac{J l_1}{2 J_1 l} [F_2(2\lambda) - F_1(2\lambda)] = 0 \qquad (5.115)$$

As the next example consider the frame in Fig. 5.29 with the end columns having half the cross-section of the central columns. In this case we must also include the rotation of joints 0 and n. The slope-deflection equations are again eqs 5.36 but their

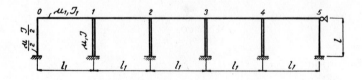

Fig. 5.29 Other types of continuous frame

solution is now expression 5.60. For free vibration the frequency equation 5.113 applies to $j = 0$ to n. The modes of free vibration generally differ from those of the frame shown in Fig. 5.25.

The slope-deflection equations of the frame in Fig. 5.30 are also given by eq. 5.36 with

$$a = \frac{2EJ_1}{l_1} F_2(\lambda_1) + \frac{EJ}{l} F_2(\lambda) \qquad (5.116)$$

but the boundary conditions now are

$$M_{01} = M_{n,n+1} = 0 \qquad (5.117)$$

In this case the solution of eq. 5.36 is that defined by expression 5.42.

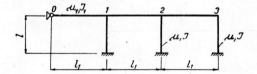

Fig. 5.30 Continuous frame with hinged end cross bar

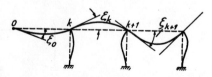

Fig. 5.31 First natural mode of system shown in Fig. 5.30

Because of condition 5.117 it must be the case that

$$\zeta_k F_2(\lambda_1) + \zeta_{k+1} F_1(\lambda_1) = 0 \quad \text{for} \quad k = 0, k = n \qquad (5.118)$$

and substitution from eq. 5.42 gives

$$\left[F_2(\lambda_1) + F_1(\lambda_1) \cos \frac{\pi j}{n} \right] r_j' - F_1(\lambda_1) \sin \frac{\pi j}{n} r_j'' = 0 \qquad (5.119)$$

For $j \neq 0$, n there apply simultaneously eq. 5.119 stating the ratio r_j''/r_j' and eq. 5.113 for computing the natural frequencies. For $j = 0$, n, simultaneous fulfilment of eqs 5.113 and 5.119 is not possible, however. Instead, the natural frequency is computed from the equation

$$F_7(\lambda_1) + \frac{Jl_1}{J_1 l} F_2(\lambda) = 0 \qquad (5.120)$$

and the natural mode defined by the relation

$$. M_{k,k+1} = 0$$

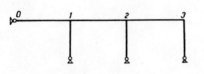

Fig. 5.32 Frame shown in Fig. 5.30 with hinged bases

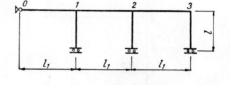

Fig. 5.33 Frame shown in Fig. 5.30 with column bases permitting displacement but no rotation

so that eq. 5.118 may be satisfied for any k. From there

$$\frac{\zeta_{k+1}}{\zeta_k} = - \frac{F_2(\lambda_1)}{F_1(\lambda_1)} = A \qquad (5.121)$$

or

$$\zeta_k = A^k \zeta_0$$

where ζ_0 is of arbitrary magnitude. The mode of free vibration is shown in Fig. 5.31.

The frames shown in Figs 5.32 and 5.33 (differing as to the kind of column bases) can be solved in an analogous manner. The multi-storey frame according to Fig. 5.34 with horizontal bars all of the same cross-section can be divided in two frames of the type shown in Fig. 5.33, and its symmetric vibration found accordingly.

Example 5.6

The frame in Fig. 5.34 has the following dimensions:

$$l_1 = 6 \text{ m}, \quad \mu_1 = 0 \cdot 2 \text{ t/m}, \quad J_1 = 1 \times 10^{-4} \text{ m}^4$$
$$l_2 = 2l = 5 \text{ m}, \quad \mu = \mu_2 = 0 \cdot 5 \text{ t/m}, \quad J = J_2 = 2 \times 10^{-4} \text{ m}^4$$

Therefore

$$\frac{Jl_1}{2J_1 l} = 2 \cdot 4$$

and

$$\frac{2\lambda}{\lambda_1} = 0 \cdot 882$$

Find the natural frequencies and the corresponding modes of free vibration restricting your considerations to symmetric vibration.

The lowest value of parameter λ_1 is obtained from the equation

$$F_7(\lambda_1) + \frac{Jl_1}{2J_1 l}[F_2(2\lambda) - F_1(2\lambda)] = 0$$

which is analogous to eq. 5.120.

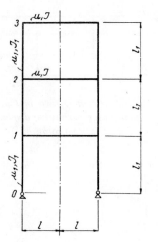

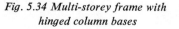

Fig. 5.34 Multi-storey frame with hinged column bases

With the given numerical values substituted, the equation is

$$F_7(\lambda_1) + 2\cdot4[F_2(0\cdot882\lambda_1) - F_1(0\cdot882\lambda_1)] = 0$$

the lowest value of parameter λ_1 satisfying it is 3·32 so that

$$\omega_{(1)} = \frac{\lambda_1^2}{l^2}\left(\frac{EJ_1}{\mu_1}\right)^{1/2} = \frac{3\cdot32^2}{6^2}\left(\frac{21 \times 10^6 \times 1 \times 10^{-4} \times 9\cdot81}{0\cdot2}\right)^{1/2} = 98\cdot3 \text{ s}^{-1}$$

$$f_{(1)} = 15\cdot6 \text{ Hz}$$

The mode of free vibration is defined by the ratio 5.121

$$\frac{\zeta_3}{\zeta_2} = \frac{\zeta_2}{\zeta_1} = \frac{\zeta_1}{\zeta_0} = -\frac{F_2(\lambda_1)}{F_1(\lambda_1)} = -0\cdot793$$

The amplitude of the deflection midway on the frame is computed from the formula

$$v = \frac{F_3(\lambda)}{F_6(\lambda)}\zeta_1 l \approx 0\cdot6\zeta_1 l$$

The frequency equation for the second and third modes of free vibration is analogous to eq. 5.113

$$2F_2(\lambda_1) + 2F_1(\lambda_1) \cos \frac{\pi j}{n} + \frac{Jl_1}{2J_1 l}[F_2(2\lambda) - F_1(2\lambda)] = 0$$

On substituting the numerical values we get for $j = 2$

$$2F_2(\lambda_1) - F_1(\lambda_1) + 2 \cdot 4[F_2(0 \cdot 882\lambda_1) - F_1(0 \cdot 882\lambda_1)] = 0$$

In this case the lowest value of parameter λ_1 is

$$\lambda_1 = 3 \cdot 56$$

so that

$$\omega_{(2)} = 113 \quad \text{s}^{-1}$$
$$f_{(2)} = 18 \cdot 0 \text{ Hz}$$

Eq. 5.119 gives the ratio

$$\frac{r_2''}{r_2'} = \frac{F_2(\lambda_1) + F_1(\lambda_1) \cos \dfrac{2\pi}{3}}{F_1(\lambda_1) \sin \dfrac{2\pi}{3}}$$

so that

$$\frac{\zeta_3}{\zeta_2} = \frac{F_1(\lambda_1) \sin \dfrac{2\pi}{3} \cos \dfrac{6\pi}{3} - \left[F_2(\lambda_1) + F_1(\lambda_1) \cos \dfrac{2\pi}{3}\right] \sin \dfrac{6\pi}{3}}{F_1(\lambda_1) \sin \dfrac{2\pi}{3} \cos \dfrac{4\pi}{3} - \left[F_2(\lambda_1) + F_1(\lambda_1) \cos \dfrac{2\pi}{3}\right] \sin \dfrac{4\pi}{3}} = \frac{F_1(\lambda_1)}{F_2(\lambda_1) - F_1(\lambda_1)}$$

and analogously

$$\frac{\zeta_2}{\zeta_1} = -\frac{F_2(\lambda_1) - F_1(\lambda_1)}{F_2(\lambda_1)}$$

$$\frac{\zeta_1}{\zeta_0} = -\frac{F_2(\lambda_1)}{F_1(\lambda_1)}$$

For the given numerical values

$$\zeta_3 : \zeta_2 : \zeta_1 : \zeta_0 \approx 1 : -0 \cdot 5 : -0 \cdot 5 : 1$$

And finally for $j = 1$

$$2F_2(\lambda_1) + F_1(\lambda_1) + 2 \cdot 4[F_2(0 \cdot 882\lambda_1) - F_1(0 \cdot 882\lambda_1)] = 0$$

$$\lambda_1 = 4 \cdot 03$$
$$\omega_{(3)} = 145 \text{ s}^{-1}$$

$$f_{(3)} = 23 \cdot 1 \text{ Hz}$$

$$\frac{\zeta_3}{\zeta_2} = \frac{F_1(\lambda_1)}{F_2(\lambda_1) + F_1(\lambda_1)}$$

$$\frac{\zeta_2}{\zeta_1} = \frac{F_2(\lambda_1) + F_1(\lambda_1)}{F_2(\lambda_1)}$$

$$\frac{\zeta_1}{\zeta_0} = -\frac{F_2(\lambda_1)}{F_1(\lambda_1)}$$

and for the given numerical values

$$\zeta_3 : \zeta_2 : \zeta_1 : \zeta_0 \approx 1 : 0 \cdot 85 : -0 \cdot 14 : -1$$

The first three natural modes are shown in Figs 5.35a, b, c.

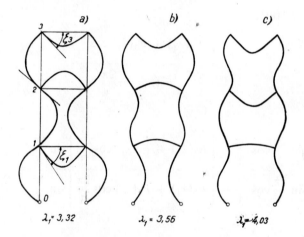

Fig. 5.35 First three modes of vibration of frame shown in Fig. 5.34

When executing symmetric vibration the multi-storey frame in Fig. 5.36a, whose top horizontal bar is of half the cross-section of the remaining horizontal bars, can be thought of as composed of two continuous frames (Fig. 5.36b); its vibration will be the same as the symmetric vibration of a frame with double the number of spans (Fig. 5.36c).

A system that has lately been gaining popularity with bridge designers is shown in Fig. 5.37. When solving it by the slope-deflection method, we choose the vertical displacements of the joints to be the unknown deformations and by applying the reciprocal theorem to the actual state of motion and the unit auxiliary state according to Fig. 5.37d we get the equations

$$bv_{k-1} + av_k + bv_{k+1} = \mathscr{Y}_k$$

where a and b denote the amplitudes of the forces necessary to make the system vibrate as shown in Fig. 5.37d. The equations are analogous to those in Table 5.7 and hold good for $k = 1$ to $n - 1$. The boundary conditions are

$$v_0 = v_n = 0$$

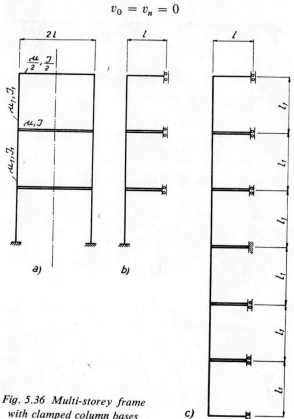

Fig. 5.36 Multi-storey frame
with clamped column bases

We get

$$a = \frac{2EJ_1}{l_1^3} \left[F_{12}(\lambda_1) - B \right] \tag{5.122}$$

$$b = \frac{EJ_1}{l_1^3} B \tag{5.123}$$

where

$$B = \frac{F_8^2(\lambda_1)}{2F_7(\lambda_1) + (Jl_1/J_1l) F_2(\lambda)} \tag{5.124}$$

The deformation is expanded in the series

$$v_k = \sum_{j=1}^{n-1} s_j \sin \frac{\pi j k}{n}$$

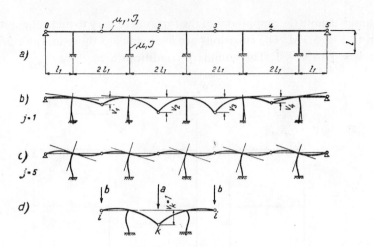

Fig. 5.37 Continuous frame with hinges mid-span

The frequency equations

$$F_{12}(\lambda_1) + B\left(\cos\frac{\pi j}{n} - 1\right) = 0 \quad \text{for} \quad j = 1 \text{ to } n - 1 \tag{5.125}$$

are arrived at in the usual way. Vibration may also occur if all the displacement amplitudes $v_k = 0$ and at the same time

$$2F_7(\lambda_1) + \frac{Jl_1}{J_1 l} F_2(\lambda) = 0 \tag{5.126}$$

or

$$F_8(\lambda_1) = \infty \tag{5.127}$$

Two of the modes of free vibration are shown in Figs 5.37b, c.

In some cases it is somewhat difficult to satisfy the boundary conditions. Consider, for example, free symmetric vibration of the frame shown in Fig. 5.38, expressed by the equation

$$a_k \zeta_k + c(\zeta_{k-1} + \zeta_{k+1}) = 0 \tag{5.128}$$

The boundary conditions are

$$\zeta_0 = 0 \tag{5.129a}$$

$$M_{n,n+1} = 0 \tag{5.129b}$$

Eq. 5.128 is satisfied by the solution

$$\zeta_k = r_j' \cos \varkappa k - r_j'' \sin \varkappa k$$

where $\varkappa$ is a hitherto undefined coefficient, generally not an integer. From the first condition — eq. 5.129a —

$$r'_j = 0$$

from the second — eq. 5.129b —

$$F_2(\lambda_1) \sin \varkappa n + F_1(\lambda_1) \sin \varkappa(n + 1) = 0 \qquad (5.130)$$

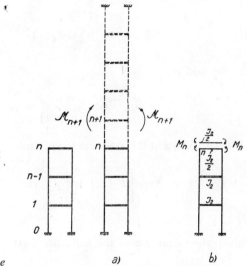

Fig. 5.38 Multi-storey frame

Furthermore it also holds by eq. 5.48 that

$$a_j = 0$$

or

$$a_j = \frac{2EJ_1}{l_1} \left[F_2(\lambda_1) + F_1(\lambda_1) \cos \varkappa \right] + \frac{EJ}{2l} \left[F_2(2\lambda) - F_1(2\lambda) \right] = 0 \qquad (5.131)$$

Eqs 5.130 and 5.131 define constants $\varkappa$ and λ.

Instead of solving the last-named equations, proceed as follows: By adding a few elements complete the system of Fig. 5.38 to a system that is easy to solve (e.g. to that shown in Fig. 5.38a). Free vibration of the original system can then be taken for forced vibration of the complemented system and the task reduced to one of finding a frequency of exciting moment $\mathscr{M}_{n+1}$ at which condition 5.129 is satisfied. Assuming that Fig. 5.38a applies we get

$$h = 4(n + 1) \qquad (5.132)$$

$$\varkappa = \frac{\pi j}{2(n + 1)} \qquad (5.133)$$

$$\zeta_n = -\sum r_j'' \sin \frac{\pi j n}{2(n + 1)} \tag{5.134}$$

$$\zeta_{n+1} = -\sum r_j'' \sin \frac{\pi j}{2} \tag{5.135}$$

$$r_j'' = \frac{R_j''}{a_j} \tag{5.136}$$

$$R_j'' = -\frac{2}{h} \sum_{k=0}^{h-1} \mathcal{M}_k \sin \frac{2\pi j k}{h} = -\frac{1}{n + 1} \mathcal{M}_{n+1} \sin \frac{\pi j}{2} \tag{5.137}$$

Further, for odd j

$$\sin \frac{\pi j}{2} \sin \frac{\pi j n}{2(n + 1)} = \cos \frac{\pi j}{2(n + 1)} \tag{5.138}$$

Eqs 5.136–5.138 yield the frequency equation

$$\sum_j \frac{1}{a_j}\left[F_1(\lambda_1) + F_2(\lambda_1) \cos \frac{\pi j}{2(n + 1)}\right] = 0 \tag{5.139}$$

where $j = 1, 3 \ldots 2n + 1$.

An illustrative example of the solution is given below.

Example 5.7

The multi-storey frame analysed in Example 3.1 is identical as to the dimensions, etc. with that shown in Fig. 5.38. The number of spans is $n = 3$. The sum in eq. 5.139 is computed for $j = 1, 3, 5, 7$. Further

$$\cos \frac{\pi}{8} = -\cos \frac{7\pi}{8} = 0.924$$

$$\cos \frac{3\pi}{8} = -\cos \frac{5\pi}{8} = 0.383$$

Eq. 5.139 is satisfied by $\lambda_1 = 3.41$ for which

$$F_1(\lambda) = 3.360$$
$$F_2(\lambda_1) = 2.306$$

and $2\lambda = \lambda_2 = 3.005$

$$F_1(\lambda_2) = 2.708$$
$$F_2(\lambda_2) = 3.094$$

Substituting the numerical values of Example 3.1 in eq. 5.131 gives

$$a_1 = 700[2\cdot306 + 3\cdot360 \times 0\cdot924] + 840[3\cdot094 - 2\cdot708] = 4106$$

$$a_3 = 2836$$

$$a_5 = 1039$$

$$a_7 = -234$$

and the above values in eq. 5.139 gives

$$\frac{5\cdot49}{4106} + \frac{4\cdot24}{2836} + \frac{2\cdot48}{1039} - \frac{1\cdot23}{234} \approx 0$$

Another way towards obtaining a solution of the system shown in Fig. 5.38 is to consider it to be composed of two parts as indicated in Fig. 5.38b. When the system as a whole executes free vibration, each part vibrates with forced vibration produced by the moment applied by the other part. The first part of the system resembles that in Fig. 5.36 whose forced vibration is found from the equations

$$h = 4n$$

$$a_j = \frac{2EJ_1}{l_1}\left[F_2(\lambda_1) + F_1(\lambda_1)\cos\frac{\pi j}{2n}\right] + \frac{EJ}{2l}[F_2(2\lambda) - F_1(2\lambda)]$$

$$R''_j = -\frac{2}{n} M_n \sin\frac{\pi j}{2}$$

$$\zeta_n = -\sum_{j=1}^{2n-1} r''_j \sin\frac{\pi j}{2} = \frac{2M_n}{n}\sum_{j=1}^{2n-1}\frac{1}{a_j} \tag{5.140}$$

for odd j

A moment of equal magnitude but of opposite sense acts on the second part, i.e. the horizontal bar of half the stiffness of the remaining ones. Its ends rotate through equal angles ζ_n. The relation that applies here is

$$M_n = \zeta_n \frac{EJ}{4l}[F_2(2\lambda) - F_1(2\lambda)] \tag{5.141}$$

For $l_2 = 2l$ the frequency equation resulting from eqs 5.140 and 5.141 is

$$\frac{1}{(EJ_2/l_2)[F_2(\lambda_2) - F_1(\lambda_2)]} = \frac{1}{n}\sum_j \frac{1}{a_j} \tag{5.142}$$

where $j = 1, 3, 5 \dots 2n - 1$ and $J_2 = J$.

Example 5.8

For the frame discussed in Examples 5.7 and 3.1, eq. 5.142 is satisfied by $\lambda_1 = 3\cdot41$. For

$$\cos\frac{\pi}{6} = \frac{\sqrt{3}}{2} = -\cos\frac{5\pi}{6}$$

$$\cos\frac{3\pi}{6} = 0$$

$a_1 = 3974$

$$a_5 = -100$$

$a_3 = 1937$

substitution in eq. 5.142 gives

$$-\frac{1}{324} \approx \frac{1}{3}\left[\frac{1}{3974} + \frac{1}{1937} - \frac{1}{100}\right]$$

Forced vibration of frames of this sort can again be solved by replacing the system according to Fig. 5.38 by that in Fig. 5.38a and finding moment $\mathcal{M}_{n+1}$ for which condition 5.129 is satisfied.

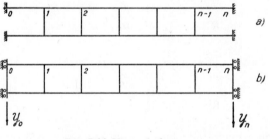

Fig. 5.39 Vierendeel truss

Lateral vibration of the Vierendeel truss shown in Fig. 5.39 can be solved as follows: consider first forced vibration of the beam in Fig. 5.39b resting on end supports permitting vertical displacement. Analogously to eq. 5.69 the deformations are described by the relations

$$v_k = \sum_j s_j'' \cos\frac{\pi jk}{n}$$

$$\zeta_k = -\sum_j r_j'' \sin\frac{\pi jk}{n}$$

$$\text{(5.143)}$$

and for steady-state forced vibration resulting from vertical forces

$$d_j s_j'' + b_j r_j'' = S_j''$$
$$b_j s_j'' + a_j r_j'' = 0$$

$$(5.144)$$

where (because $\mu l \omega^2 = (EJ/l^3)\lambda^4$)

$$a_j = \frac{EJ_1}{l_1}\left[2F_2(\lambda_1) + 2F_1(\lambda_1)\cos\frac{\pi j}{n}\right] + \frac{EJ}{l}F_7(\lambda)$$

$$b_j = -2\frac{EJ_1}{l_1^2}F_3(\lambda_1)\sin\frac{\pi j}{n}$$

$$d_j = \frac{EJ_1}{l_1^3}\left[2F_6(\lambda_1) + 2F_5(\lambda_1)\cos\frac{\pi j}{n}\right] + \frac{EJ}{l^3}\lambda^4$$

$$S_j'' = \frac{1}{n}\sum_{k=0}^{n}\mathcal{Y}_k\cos\frac{\pi jk}{n}$$

$$(5.145)$$

l denotes the half-length of the vertical bars. The formulae hold good for $j \neq 0 \neq n$. For $j = 0, j = n$, the applicable formulae are analogous to eqs 5.72 and 5.73.

From eqs 5.144

$$s_j'' = \frac{S_j''}{D_j}$$

$$(5.146)$$

where

$$D_j = d_j - \frac{b_j^2}{a_j}$$

$$(5.147)$$

Free vibration of the system in Fig. 5.39a may be regarded as forced vibration of the system in Fig. 5.39b acted upon by harmonic forces $\mathcal{Y}_0\sin\omega t$ and $\mathcal{Y}_n\sin\omega t$ with an arbitrary amplitude and hitherto unknown frequency ω. When examining the symmetric free vibration we assume that $\mathcal{Y}_n = \mathcal{Y}_0$ and consider only the fundamental modes with even j ($j = 0, 2\ldots$). For the antisymmetric vibration $\mathcal{Y}_n = -\mathcal{Y}_0$ and $j = 1, 3\ldots$. Frequency ω is found from the condition that the boundary conditions

$$v_0 = v_n = 0$$

must be satisfied. In view of eq. 5.143 it must therefore be the case that

$$\sum\frac{S_j''}{D_j} = 0$$

$$(5.148)$$

Eq. 5.148 represents the frequency equation of the system. For the symmetric free vibration we then get

$$\frac{1}{2d_0} + \frac{1}{D_2} + \frac{1}{D_4} + \ldots = 0$$

and for the antisymmetric free vibration

$$\frac{1}{D_1} + \frac{1}{D_3} + \ldots = 0$$

($j > n$ are not considered, $1/D_n$ is replaced by $1/2d_n$).

The solutions discussed in the preceding paragraphs can readily be extended to continuous multi-storey frames. Expressions 5.30 then represent a more general form of the slope-deflection equations.

5.2.6   Continuous arches with repeated elements

Continuous arches (Fig. 5.40), are amenable to the solution outlined in Section 5.1. At panel points k there arise a symmetric deformation, v_k, as well as two antisymmetric deformations ζ_k and u_k defined by relations 5.12 and 5.13. The slope-deflection equations 5.30 may be obtained through the application of the reciprocal theorem to the unit auxiliary states of the j-th mode. The equations applying to a continuous arch are

$$a_j^{vv}\, s_j(v) + b_j^{v\zeta}\, r_j(\zeta) + b_j^{vu}\, r_j(u) = S_j(v)$$
$$b_j^{v\zeta}\, s_j(v) + a_j^{\zeta\zeta}\, r_j(\zeta) + a_j^{\zeta u}\, r_j(u) = R_j(\zeta)$$
$$b_j^{vu}\, s_j(v) + a_j^{\zeta u}\, r_j(\zeta) + a_j^{uu}\, r_j(u) = R_j(u)$$

$$(5.149)$$

in which it is possible to substitute for s either s' or s'', and similarly so for r, S, R. Clearly we could also consider the deformations of the column bases; in such a case we would write two sets of eqs 5.30 for $J = I$ and $J = II$ in place of eqs 5.149.

By eqs 5.21 and 5.22 the loading terms on the right-hand sides of eqs 5.149 are

$$S_j'(v) = \frac{2}{h} \sum_{k=1}^{h-1} \mathcal{Y}_k \sin \frac{2\pi jk}{h}$$

$$S_j''(v) = \frac{2}{h} \sum_{k=0}^{h-1} \mathcal{Y}_k \cos \frac{2\pi jk}{h}$$

$$R_j'(\zeta) = \frac{2}{h} \sum_{k=0}^{h-1} \mathcal{M}_k \cos \frac{2\pi jk}{h}$$

etc.

$$(5.150)$$

A solution in the simple form of eqs 5.12 and 5.13 is not, of course, obtained except under certain boundary conditions. Whatever has been said in Section 5.2.3 applies here too. Considering that the joints yield in the vertical direction and the end supports permit vertical displacement, the stiffness of the latter must be half

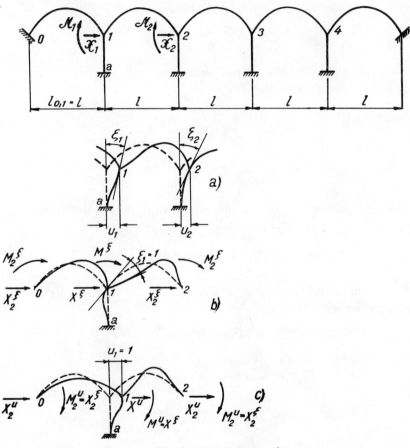

Fig. 5.40 Continuous arch

that of the intermediate supports. If the end supports are perfectly unyielding and the arches are clamped in them, the particular solution 5.13 applies only if the vertical deflection v_k is zero in all the joints. Unless these conditions are satisfied, a solution cannot be obtained except when the methods explained in the last paragraphs of the preceding section are applied.

Example 5.9

In Section 4.2.3, Example 4.3 we discussed an arch clamped at both ends, and computed the end forces and moments acting on its half gp. The continuous arch shown in Fig. 5.40 consists of arch sections computed in Example 4.3 and straight columns of constant cross-section.

Assume that the columns are clamped at the base and are incompressible, and that the end supports are unyielding, i.e.

$$v_0 = v_1 = \ldots = v_5 = 0$$
$$u_0 = u_5 = 0$$
$$\zeta_0 = \zeta_5 = 0$$

Given these assumptions the following relationships hold:

$$h = 2n$$

$$\zeta_k = -r''(\zeta) \sin \frac{\pi jk}{n}$$

$$u_k = -r''(u) \sin \frac{\pi jk}{n}$$

Eqs 5.149 describing free vibration of the system will now take on the form

$$a_j^{\zeta\zeta} r''(\zeta) + a_j^{\zeta u} r''(u) = 0$$
$$a_j^{\zeta u} r''(\zeta) + a_j^{uu} r''(u) = 0 \tag{5.151}$$

The coefficients of eq. 5.151 are computed from eq. 5.25. Referring to the notation in Fig. 5.40b, c we write

$$a_j^{\zeta\zeta} = M^\zeta + 2M_2^\zeta \cos \frac{\pi j}{n}$$

$$a_j^{\zeta u} = X^\zeta + 2X_2^\zeta \cos \frac{\pi j}{n}$$

$$a_j^{uu} = X^u + 2X_2^u \cos \frac{\pi j}{n} \tag{5.152}$$

The quantities on the right-hand sides of eqs 5.152 are obtained by first computing deformations ζ_p, u_p, v_p at the apex of bar $1-2$ (Fig. 5.41). With two spans of the system deformed according to Fig. 5.40b, c, the end moments and forces at point 1 are

$$M^\zeta = 2M_{12(\zeta_1=1)} + M_{1a(\zeta_1=1)}$$
$$X^\zeta = 2X_{12(\zeta_1=1)} + X_{1a(\zeta_1=1)}$$
$$X^u = 2X_{12(u_1=1)} + X_{1a(u_1=1)}$$

and

$$\tag{5.153}$$

$$M_2^\zeta = M_{21(\zeta_1=1)}$$

etc.

255

Table 5.8

ζ_p	u_p	v_p	Absolute term for $\zeta_1 = 1$	Absolute term for $u_1 = 1$
$2(0 \cdot 01638E - 1 \cdot 201\varrho\omega^2)$	$2(0 \cdot 01255E - 0 \cdot 776\varrho\omega^2)$		$= \ \ \ 0 \cdot 00838E - 0 \cdot 434\varrho\omega^2$	$= \ \ \ 0 \cdot 01255E + 0 \cdot 387\varrho\omega^2$
$2(0 \cdot 01255E - 0 \cdot 776\varrho\omega^2)$	$2(0 \cdot 02009E - 8 \cdot 412\varrho\omega^2)$		$= \ \ \ 0 \cdot 02155E + 1 \cdot 484\varrho\omega^2$	$= \ \ \ 0 \cdot 02009E - 0 \cdot 510\varrho\omega^2$
		$2(0 \cdot 01238E - 2 \cdot 805\varrho\omega^2)$	$= -0 \cdot 01258E + 0 \cdot 017\varrho\omega^2$	$= -0 \cdot 01493E - 0 \cdot 333\varrho\omega^2$

On substituting the numerical values of Example 4.3 we get the slope-deflection equations at the apex of the system in Fig. 5.41. They are set out in Table 5.8

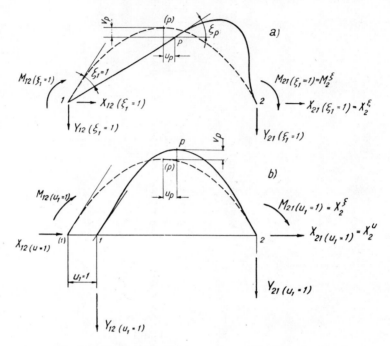

Fig. 5.41 Forces and moments acting at the ends of an arch

Writing

$$\varDelta = (0.032\ 76E - 2.402\varrho\omega^2)\ (0.040\ 18E - 16.824\varrho\omega^2) - (0.025\ 10E - 1.552\varrho\omega^2)^2$$
$$= 0.000\ 686\ 29E^2 - 0.569\ 76E\varrho\omega^2 + 38.0025\varrho^2\omega^4$$

the solution of the equations in Table 5.8 gives for the state $\zeta_1 = 1$ (Fig. 5.41a)

$$\zeta_p = [(0.008\ 38E - 0.434\varrho\omega^2)\ (0.040\ 18E - 16.824\varrho\omega^2) - (0.021\ 55E + 1.484\varrho\omega^2)$$
$$\times (0.025\ 10E - 1.552\varrho\omega^2)]/\varDelta$$
$$= [-0.000\ 204\ 20E^2 - 0.162\ 23E\varrho\omega^2 + 9.6048\varrho^2\omega^4]/\varDelta$$

and the displacement

$$u_p = [(0.021\ 55E + 1.484\varrho\omega^2)\ (0.032\ 76E - 2.402\varrho\omega^2) - (0.008\ 38E - 0.434\varrho\omega^2)$$
$$\times (0.0251E - 1.552\varrho\omega^2)]/\varDelta$$
$$= [0.000\ 495\ 64E^2 + 0.020\ 752E\varrho\omega^2 - 4.2381\varrho^2\omega^4]/\varDelta$$

$$v_p = [-0.012\ 58E + 0.017\varrho\omega^2]/[0.024\ 76E - 5.610\varrho\omega^2]$$

The end moments and forces are

$$M_{12(\zeta_1=1)} = 0{\cdot}034\,65E - 3{\cdot}469\varrho\omega^2 + (-0{\cdot}008\,38E + 0{\cdot}434\varrho\omega^2)\,\zeta_p$$
$$+ (-0{\cdot}021\,55E - 1{\cdot}484\varrho\omega^2)\,u_p + (0{\cdot}012\,58E - 0{\cdot}017\varrho\omega^2)\,v_p$$
$$= 0{\cdot}034\,65E - 3{\cdot}469\varrho\omega^2 + [-0{\cdot}000\,158\,26E^2 + 0{\cdot}000\,427\,72E\varrho\omega^2$$
$$- 0{\cdot}000\,289\varrho^2\omega^4]/[0{\cdot}024\,76E - 5{\cdot}610\varrho\omega^2]$$
$$+ \varDelta^{-1}(-0{\cdot}000\,008\,970E^3 + 0{\cdot}001\,799E^2\varrho\omega^2 - 0{\cdot}090\,358E\varrho^2\omega^4$$
$$+ 10{\cdot}458\varrho^3\omega^6)$$

$$M_{21(\zeta_1=1)} = (-0{\cdot}008\,38E + 0{\cdot}434\varrho\omega^2)\,\zeta_p + (-0{\cdot}021\,55E - 1{\cdot}484\varrho\omega^2)\,u_p$$
$$- (0{\cdot}012\,58E - 0{\cdot}017\varrho\omega^2)\,v_p$$
$$= -[-0{\cdot}000\,158\,26E^2 + 0{\cdot}000\,427\,72E\varrho\omega^2 - 0{\cdot}000\,289\varrho^2\omega^4]/[0{\cdot}024\,76E$$
$$- 5{\cdot}610\varrho\omega^2] + \varDelta^{-1}(-0{\cdot}000\,008\,97E^3 + 0{\cdot}001\,799E^2\varrho\omega^2$$
$$- 0{\cdot}090\,358E\varrho^2\omega^4 + 10{\cdot}458\varrho^3\omega^6)$$

$$X_{12(\zeta_1=1)} = 0{\cdot}021\,55E - 1{\cdot}159\varrho\omega^2 + (-0{\cdot}012\,55E - 0{\cdot}387\varrho\omega^2)\,\zeta_p$$
$$+ (-0{\cdot}020\,09E + 0{\cdot}510\varrho\omega^2)\,u_p + (0{\cdot}014\,93E + 0{\cdot}333\varrho\omega^2)\,v_p$$
$$= 0{\cdot}021\,55E - 1{\cdot}159\varrho\omega^2 + [-0{\cdot}000\,187\,819E^2 - 0{\cdot}003\,935E\varrho\omega^2$$
$$+ 0{\cdot}005\,661\varrho^2\omega^4]/[0{\cdot}024\,76E - 5{\cdot}610\varrho\omega^2] + \varDelta^{-1}(-0{\cdot}000\,007\,394E^3$$
$$+ 0{\cdot}001\,952E^2\varrho\omega^2 + 0{\cdot}038\,031E\varrho^2\omega^4 - 5{\cdot}891\varrho^3\omega^6)$$

$$X_{21(\zeta_1=1)} = (-0{\cdot}012\,55E - 0{\cdot}387\varrho\omega^2)\,\zeta_p + (-0{\cdot}020\,09E + 0{\cdot}510\varrho\omega^2)\,u_p$$
$$- (0{\cdot}014\,93E + 0{\cdot}333\varrho\omega^2)\,v_p = -[-0{\cdot}000\,187\,819E^2 - 0{\cdot}003\,935E\varrho\omega^2$$
$$+ 0{\cdot}005\,661\varrho^2\omega^4]/[0{\cdot}024\,76E - 5{\cdot}610\varrho\omega^2] + \varDelta^{-1}(-0{\cdot}000\,007\,394E^3$$
$$+ 0{\cdot}001\,952E^2\varrho\omega^2 + 0{\cdot}038\,031E\varrho^2\omega^4 - 5{\cdot}891\varrho^3\omega^6)$$

Analogously the quantities for the state $u_1 = 1$ (Fig. 5.41b) are

$$\zeta_p = [(0{\cdot}012\,55E + 0{\cdot}387\varrho\omega^2)\,(0{\cdot}040\,18E - 16{\cdot}824\varrho\omega^2) - (0{\cdot}020\,09E - 0{\cdot}510\varrho\omega^2)$$
$$\times (0{\cdot}025\,10E - 1{\cdot}552\varrho\omega^2)]/\varDelta$$

$$\bar{u}_p = [(0{\cdot}020\,09 - 0{\cdot}510\varrho\omega^2)\,(0{\cdot}032\,76E - 2{\cdot}402\varrho\omega^2) - (0{\cdot}012\,55E + 0{\cdot}387\varrho\omega^2)$$
$$\times (0{\cdot}025\,10E - 1{\cdot}552\varrho\omega^2)]/\varDelta$$

$$\bar{v}_p = [-0{\cdot}014\,93E - 0{\cdot}333\varrho\omega^2]/[0{\cdot}024\,76E - 5{\cdot}610\varrho\omega^2]$$

$$M_{12(u_1=1)} = X_{12(\zeta_1=1)}$$

$$M_{21(u_1=1)} = X_{12(\zeta_2=1)} = X_{21(\zeta_1=1)}$$

$$X_{12(u_1=1)} = 0.020\,09E - 2.880\varrho\omega^2 + \left(-0.012\,55E - 0.387\varrho\omega^2\right)\zeta_p$$
$$+ \left(-0.020\,09E + 0.510\varrho\omega^2\right)\bar{u}_p + \left(0.014\,93E + 0.333\varrho\omega^2\right)\bar{v}_p$$
$$= 0.020\,09E - 2.880\varrho\omega^2 - \left[0.000\,222\,905E^2 + 0.009\,943\,4E\varrho\omega^2\right.$$
$$+ 0.110\,89\varrho^2\omega^4]/[0.024\,76E - 5.610\varrho\omega^2] + \Delta^{-1}(-0.000\,006\,894E^3$$
$$+ 0.003\,189E^2\varrho\omega^2 + 0.085\,158E\varrho^2\omega^4 + 3.768\varrho^3\omega^6)$$

$$X_{21(u_1=1)} = -\left(0.012\,55E + 0.387\varrho\omega^2\right)\zeta_p - \left(0.020\,09E - 0.510\varrho\omega^2\right)\bar{u}_p$$
$$- \left(0.014\,93E + 0.333\varrho\omega^2\right)\bar{v}_p = \left[0.000\,222\,905E^2 + 0.009\,943\,4E\varrho\omega^2\right.$$
$$+ 0.110\,889\varrho^2\omega^4]/[0.024\,76E - 5.610\varrho\omega^2] + \Delta^{-1}(-0.000\,006\,894E^3$$
$$+ 0.003\,189E^2\varrho\omega^2 + 0.085\,158E\varrho^2\omega^4 + 3.768\varrho^2\omega^6)$$

The dimensions of the column are

$$l = 7.0 \quad \text{m}$$
$$S = 2.4 \quad \text{m}^2$$
$$J = 0.072 \, \text{m}^4$$

and by eqs 2.89 and 3.30 the end moment and horizontal force are

$$M_{1a(\zeta_1=1)} = \frac{4EJ}{l} - \frac{1}{105}\mu\omega^2 l^3 = \frac{4EJ}{l} - \frac{Sl^3}{105}\varrho\omega^2 = 0.041\,143E - 7.840\varrho\omega^2$$

$$X_{1a(\zeta_1=1)} = -\frac{6EJ}{l^2} + \frac{11}{210}\mu\omega^2 l^2 = -\frac{6EJ}{l^2} + \frac{11Sl^2}{210}\varrho\omega^2$$
$$= -0.008\,816E + 6.160\varrho\omega^2$$

$$X_{1a(u_1=1)} = \frac{12EJ}{l^3} - \frac{13}{35}\mu\omega^2 l = \frac{12EJ}{l^3} - \frac{13Sl}{35}\varrho\omega^2 = 0.002\,510E - 6.240\varrho\omega^2$$

For $E = 2.4 \times 10^6 \, \text{Mp/m}^2$, eq. 5.153 gives

$$M^\zeta = 2M_{12(\zeta_1=1)} + M_{1a(\zeta_1=1)} = 265\,063 - 14.778\varrho\omega^2$$
$$+ \frac{-18.2311 \times 10^8 + 20.530 \times 10^2\varrho\omega^4 - 5.78 \times 10^{-4}\varrho^2\omega^4}{59\,424 - 5.610\varrho\omega^2}$$
$$+ \frac{1}{\Delta}\left(-248.003 \times 10^{12} + 20.724 \times 10^9\varrho\omega^2 - 43.372 \times 10^4\varrho^2\omega^4 + 20.916\varrho^3\omega^6\right)$$

$$X^\zeta = 82\,280 + 3.842\varrho\omega^2$$
$$+ \frac{-21.6367 \times 10^8 - 188.88 \times \quad 0^2\varrho\omega^2 + 0.011\,32\varrho^2\omega^4}{59\,424 - 5.610\varrho\omega^2} + \frac{1}{\Delta}\left(-204.429 \times 10^{12}\right.$$
$$+ 22.487 \times 10^9\varrho\omega^2 + 18.255 \times 10^4\varrho^2\omega^4 - 11.782\varrho^3\omega^6)$$

$$X^u = 102\,478 - 12\varrho\omega^2$$

$$- \frac{25\cdot6787 \times 10^8 + 477\cdot282 \times 10^2\varrho\omega^2 + 0\cdot221\,78\varrho^2\omega^4}{59\,424 - 5\cdot610\varrho\omega^2} + \frac{1}{\varDelta}(-190\cdot605 \times 10^{12}$$

$$+\ 36\cdot737 \times 10^9\varrho\omega^2 + 40\cdot876 \times 10^4\varrho^2\omega^4 + 7\cdot536\varrho^3\omega^6)$$

$$M_2^\varsigma = M_{21(\varsigma_1=1)} = -\frac{-9\cdot\ 155 \times 10^8 + 10\cdot265 \times 10^2\varrho\omega^2 - 0\cdot000\,289\varrho^2\omega^4}{59\,424 - 5\cdot610\varrho\omega^2}$$

$$+\ \frac{1}{\varDelta}(-\ 24\cdot00 \ \times 10^{12} + 10\cdot362 \times \ 0^9\varrho\omega^2 - 21\cdot686 \times 10^4\varrho^2\omega^4 + 10\cdot458\varrho^3\omega^6)$$

$$X_2^\varsigma = X_{21(\varsigma_1=1)} = -\frac{-10\cdot8184 \times 10^8 - 94\cdot440 \times 10^2\varrho\omega^2 + 0\cdot005\,661\varrho^2\omega^4}{59\,424 - 5\cdot610\varrho\omega^2}$$

$$+\ \frac{1}{\varDelta}(-102\cdot215 \times 10^{12} + 11\cdot243 \times 10^9\varrho\omega^2 + 9\cdot127 \times 10^4\varrho^2\omega^4 - 5\cdot891\varrho^3\omega^6)$$

$$X_2^u = X_{21(u_1=1)} = \frac{12\cdot8393 \times 10^8 + 238\cdot641 \times 10^2\varrho\omega^2 + 0\cdot110\,889\varrho^2\omega^4}{59\,424 - 5\cdot610\varrho\omega^2}$$

$$+\ \frac{1}{\varDelta}(-95\cdot303 \times 10^{12} + 18\cdot369 \times 10^9\varrho\omega^2 + 20\cdot438 \times 10^4\varrho^2\omega^4 + 3\cdot768\varrho^3\omega^6)$$

In the above

$$\varDelta = 39\cdot5301\,.\,10^8 - 1\cdot3674\,.\,10^6\varrho\omega^2 + 38\cdot0025\varrho^2\omega^4$$

The computed numerical values are next introduced in eq. 5.152, and the natural frequencies are established from the condition that the determinant of eqs 5.151 should equal zero. That is to say

$$a_j^{\varsigma\varsigma}a_j^{uu} - (a_j^{\varsigma u})^2 = 0$$

or

$$M^\varsigma X^u - (X^\varsigma)^2 + 2(X^u M_2^\varsigma + M^\varsigma X_2^u - 2X^\varsigma X_2^\varsigma)\cos\frac{\pi j}{n}$$

$$+\ 4\left[M_2^\varsigma X_2^u - (X_2^\varsigma)^2\right]\cos^2\frac{\pi j}{n} = 0 \qquad (5.154)$$

Using abbreviated notation we write the last equation in the form

$$A(\varrho\omega^2) + 2B(\varrho\omega^2)\cos\frac{\pi j}{n} + 4C(\varrho\omega^2)\cos^2\frac{\pi j}{n} = 0$$

and solve it by evaluating its left-hand side for successively growing ω and finding the values that satisfy the equation at various j. The results are set out in Table 5.9 and Figs 5.42a, b. In the above specific mass (density) $\varrho = 2\cdot40/9\cdot81 = 0\cdot245$ Mp s^2/m^4.

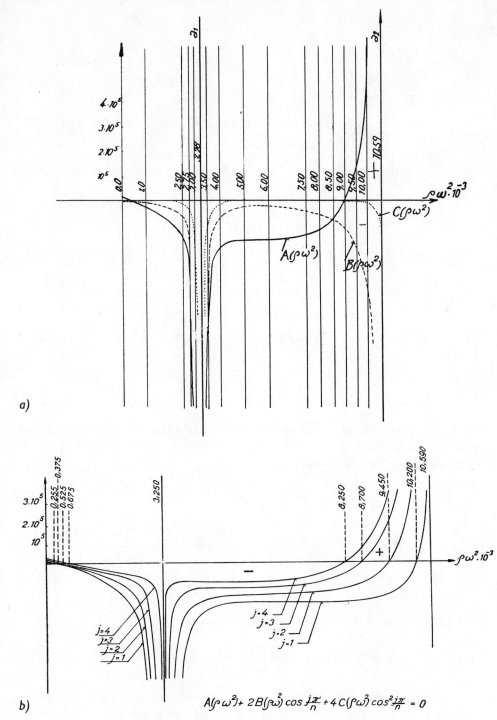

$$A(\rho\omega^2) + 2B(\rho\omega^2)\cos\frac{j\pi}{n} + 4C(\rho\omega^2)\cos^2\frac{j\pi}{n} = 0$$

Fig. 5.42 Solution of the frequency equation of a continuous arch

All the spans are apt to vibrate as though fixed at the joints. The respective frequencies were computed in Example 4.3. The lowest frequency of the antisymmetric vibration turned out as $\omega_{(1')} = 113 \text{ s}^{-1}$, and that of the symmetric vibration as $\omega_{(1)} = 208 \text{ s}^{-1}$.

Table 5.9

	$\varrho\omega^2$	ω^2	$\omega_j \ (\text{s}^{-1})$
$j = 1$	255	1040	32·2
$j = 2$	375	1532	39·1
$j = 3$	525	2150	46·3
$j = 4$	675	2760	52·5
$j = 1$	10200	41700	204
$j = 2$	9450	38600	196
$j = 3$	8700	35600	188
$j = 4$	8250	33800	184

6 SOME PROBLEMS OF DAMPED VIBRATION

Most analytic examinations of damped vibration are based on the hypothesis that the damping force is proportional to the velocity of motion. Since this hypothesis leads to results that, given certain assumptions, agree with experimental evidence, we shall discuss it in the first sections of this chapter. Section 6.1.4 will be devoted to another of the familiar hypotheses, that of Kelvin-Voigt, according to which the magnitude of the damping force depends on the rate of change in stress. The concluding parts of this chapter will present a more general conception of damping theory restricted, however, to the main problems of the linear theory.

6.1 Damped vibration of frame systems

The lateral damped motion of a length element of a uniform bar under no-load conditions is described by the equation

$$\mu dx \frac{\partial^2 v(x, t)}{\partial t^2} + b\, dx\, \frac{\partial v(x, t)}{\partial t} + EJ \frac{\partial^4 v(x, t)}{\partial x^4}\, dx = 0 \qquad (6.1)$$

where b is the damping per unit length of bar at unit velocity. In the computations that follow we shall express damping in terms of other coefficients, namely f_b and ω_b (in the form of frequency and circular frequency) defined by the formulae

$$b = 2\mu\omega_b = 4\pi\mu f_b$$

Of the possible motions likely to satisfy eq. 6.1 we shall consider harmonic vibration in the form

$$v(x, t) = {}^A v(x) \sin \omega t + {}^B v(x) \cos \omega t \qquad (6.2)$$

Substituting eq. 6.2 in 6.1 and cancelling dx gives

$$-\mu\omega^2 \left[{}^A v(x) \sin \omega t + {}^B v(x) \cos \omega t \right] + 2\mu\omega\omega_b \left[{}^A v(x) \cos \omega t - {}^B v(x) \sin \omega t \right]$$

$$+ EJ \frac{\partial^4 \left[{}^A v(x) \sin \omega t + {}^B v(x) \cos \omega t \right]}{\partial x^4} = 0 \qquad (6.3)$$

Since eq. 6.3 must be satisfied for every x, it must follow that

$$-\mu\omega^2\,^A v(x) - 2\mu\omega\omega_b\,^B v(x) + EJ\frac{\mathrm{d}^4\,^A v(x)}{\mathrm{d}x^4} = 0 \tag{6.4a}$$

$$-\mu\omega^2\,^B v(x) + 2\mu\omega\omega_b\,^A v(x) + EJ\frac{\mathrm{d}^4\,^B v(x)}{\mathrm{d}x^4} = 0 \tag{6.4b}$$

The equations could be solved by the elimination of one unknown.

A formally simpler solution is obtained by the application of the functions of the complex variable. Eq. 6.1 is satisfied by the complex expression

$$v^*(x, t) = [v(x) + i\,\bar{v}(x)]\sin \omega t + [\bar{v}(x) - i\,v(x)]\cos \omega t$$
$$= [\bar{v}(x) - i\,v(x)]\,e^{i\omega t} \tag{6.5}$$

Substituting this in eq. 6.1 we get the ordinary differential equation

$$-\mu\omega^2\left(1 - 2i\frac{\omega_b}{\omega}\right)[v(x) + i\bar{v}(x)] + EJ\frac{\mathrm{d}^4[v(x) + i\bar{v}(x)]}{\mathrm{d}x^4} = 0 \tag{6.6}$$

which differs from eq. 2.10 only by the coefficient of the first term. Its general solution is

$$v(x) + i\bar{v}(x) = A\cos\left(\varLambda + i\bar\varLambda\right)\frac{x}{l} + B\sin\left(\varLambda + i\bar\varLambda\right)\frac{x}{l} + C\cosh\left(\varLambda + i\bar\varLambda\right)\frac{x}{l}$$
$$+ D\sinh\left(\varLambda + i\bar\varLambda\right)\frac{x}{l} \tag{6.7}$$

where

$$\varLambda + i\bar\varLambda = l\left(\frac{\mu\omega^2\left(1 - 2i\dfrac{\omega_b}{\omega}\right)}{EJ}\right)^{1/4}$$

$$= l\left(\frac{\mu\omega^2}{EJ}\right)^{1/4}\left\{\left[\frac{1}{2}\left(1 + \frac{4\omega_b^2}{\omega^2}\right)^{1/2} + \frac{1}{2}\right]^{1/2} + i\left[\frac{1}{2}\left(1 + \frac{4\omega_b^2}{\omega^2}\right)^{1/2} - \frac{1}{2}\right]^{1/2}\right\}$$

$$= \lambda\left\{\frac{1}{2}\left[1 + \frac{4\omega_b^2}{\omega^2}\right]^{1/4} + \left[\frac{1}{8}\left(1 + \frac{4\omega_b^2}{\omega^2}\right)^{1/2} + \frac{1}{8}\right]^{1/2}\right\}^{1/2}$$

$$- i\lambda\left\{\frac{1}{2}\left[1 + \frac{4\omega_b^2}{\omega^2}\right]^{1/4} - \left[\frac{1}{8}\left(1 + \frac{4\omega_b^2}{\omega^2}\right)^{1/2} + \frac{1}{8}\right]^{1/2}\right\}^{1/2} \tag{6.8}$$

As formula 6.5 implies, the vibration component with amplitude $v(x)$ lags by $\varphi = \pi/2$ behind the component with amplitude $\bar{v}(x)$. Eq. 6.7 is identical with eq. 2.25 except that in the former, λ is replaced by the expression $\varLambda + i\bar\varLambda$ according to eq. 6.8. The procedure for determining the forces and moments acting at the ends of bar $\overline{gh}$ at harmonic unit end deformation, e.g. $\zeta_h(t) = 1 \sin \omega t$, is likewise the same as in the case of vibration without damping solved in Section 2.4. Substitution in eq. 6.7 gives the equations

$$A + C = 0$$
$$B + D = 0$$

$$A \cos (\varLambda + i\overline{\varLambda}) + B \sin (\varLambda + i\overline{\varLambda}) + C \cosh (\varLambda + i\overline{\varLambda}) + D \sinh (\varLambda + i\overline{\varLambda}) = 0$$

$$-A \sin (\varLambda + i\overline{\varLambda}) + B \cos (\varLambda + i\overline{\varLambda}) + C \sinh (\varLambda + i\overline{\varLambda}) + D \cosh(\varLambda + i\overline{\varLambda}) = \frac{l}{\varLambda + i\overline{\varLambda}}$$

Their solution and constants A to D are similar to those in eq. 2.40. Since the moments and shearing forces are computed from the derivatives of the deflection, formulae 2.41 − 2·46 with $\varLambda + i\overline{\varLambda}$ substituted for λ, and $M_{gh} + i\overline{M}_{gh}$ for M_{gh}, etc. will also apply. The end moment is then given by

$$M_{gh}^*(t) = (M_{gh} + i\overline{M}_{gh}) \sin \omega t + (\overline{M}_{gh} - iM_{gh}) \cos \omega t = (\overline{M}_{gh} - iM_{gh}) \, e^{i\omega t} \quad (6.9)$$

The equation applying at arbitrary end deformations, i.e.

$$M_{gh} + i\overline{M}_{gh} = \frac{EJ}{l} F_1(\varLambda + i\overline{\varLambda}) (\zeta_h + i\overline{\zeta}_h) + \frac{EJ}{l} F_2(\varLambda + i\overline{\varLambda}) (\zeta_g + i\overline{\zeta}_g)$$

$$- \frac{EJ}{l^2} F_3(\varLambda + i\overline{\varLambda}) (v_h + i\overline{v}_h) - \frac{EJ}{l^2} F_4(\varLambda + i\overline{\varLambda}) (v_g + i\overline{v}_g) \quad (6.10)$$

is analogous to eq. 2.48.

The formulae for functions $F(\varLambda + i\overline{\varLambda})$ follow directly from the formulae for $F(\lambda)$ (see Appendix) in which $\varLambda + i\overline{\varLambda}$ is substituted for λ. Thus, for example

$$F_1(\varLambda + i\overline{\varLambda}) = - \frac{(\varLambda + i\overline{\varLambda}) \left[\sinh (\varLambda + i\overline{\varLambda}) - \sin (\varLambda + i\overline{\varLambda}) \right]}{\cosh (\varLambda + i\overline{\varLambda}) \cos (\varLambda + i\overline{\varLambda}) - 1} \quad (6.11)$$

The procedure just outlined is necessary only in the case of heavy damping. For systems with light damping, $i\overline{\varLambda}$ is small, $\varLambda \approx \lambda$ and the computation simplifies accordingly. Considering ω_b an infinitely small quantity and neglecting the smalls of higher orders, we get from eq. 6.8

$$\varLambda \approx \left(1 + \frac{3\omega_b^2}{8\omega^2} \right) \lambda \approx \lambda \quad (6.12)$$

$$\overline{\varLambda} \approx - \frac{\omega_b}{2\omega} \lambda \quad (6.13)$$

$$\frac{\overline{\varLambda}}{\varLambda} \approx - \frac{\omega_b}{2\omega} = - \frac{f_b}{2f} \quad (6.14)$$

By Taylor's theorem functions $F(\varLambda + i\bar{\varLambda})$ for small $\bar{\varLambda}$ are

$$F_1(\varLambda + i\bar{\varLambda}) \approx F_1(\lambda) + \frac{i\bar{\varLambda}}{1!}\frac{dF_1(\lambda)}{d\lambda} = F_1(\lambda)$$

$$- i\bar{\varLambda}\left[\lambda\frac{(\cosh\lambda - \cos\lambda)(\cosh\lambda\cos\lambda - 1) + (\sinh\lambda - \sin\lambda)(\cosh\lambda\sin\lambda - \sinh\lambda\cos\lambda)}{(\cosh\lambda\cos\lambda - 1)^2}\right.$$

$$\left. + \frac{\sinh\lambda - \sin\lambda}{\cosh\lambda\cos\lambda - 1}\right] =$$

$$F_1(\lambda) + i\frac{\omega_b}{2\omega}\left[F_1(\lambda)F_2(\lambda) - F_3(\lambda) - F_1(\lambda)\right] = F_1(\lambda) + i\frac{2\omega_b}{\omega}\Phi_1(\lambda) \qquad (6.15)$$

$$F_2(\varLambda + i\bar{\varLambda}) \approx F_2(\lambda) + \frac{i\bar{\varLambda}}{1!}\frac{dF_2(\lambda)}{d\lambda} = F_2(\lambda) + \frac{i\omega_b}{2\omega}\left[F_1^2(\lambda) - F_2(\lambda)\right]$$

$$= F_2(\lambda) + i\frac{2\omega_b}{\omega}\Phi_2(\lambda) \qquad (6.16)$$

and generally

$$F_p(\varLambda + i\bar{\varLambda}) \approx F_p(\lambda) + \frac{i\bar{\varLambda}}{1!}\frac{dF_p(\lambda)}{d\lambda} = F_p(\lambda) + i\frac{2\omega_b}{\omega}\Phi_p(\lambda) \qquad (6.17)$$

It is evident from the formulae that

$$\frac{dF_p(\lambda)}{d\lambda} = -\frac{4}{\lambda}\Phi_p(\lambda) \qquad (6.18)$$

The proof of this equality, based on Maxwell's theorem, is presented in References 37 and 38.

6.1.1 The slope-deflection method

The slope-deflection equations for computing damped vibration of frame systems express the conditions of equilibrium at the joints and as such are given by formulae 3.7 and 3.9, Section 3.2.1. The loading terms are analogous to those for undamped vibration except that the complex expression $\varLambda + i\bar{\varLambda}$ is again substituted for λ. The external load $P\sin\omega t + \bar{P}\cos\omega t$ is replaced by load $(P + i\bar{P})\sin\omega t + (\bar{P} - iP)\cos\omega t$.

Since the coefficients of the equations as well as the unknown deformations are described by complex expressions (see eqs 6.7 and 6.10) each of eqs 3.7 will yield two equations with real expressions. The number of real slope-deflection equations for computing damped vibration is therefore double that for undamped vibration. The computational procedure will be demonstrated in the numerical example in Section 6.1.3.

Compared with the method which we shall discuss in the next section, the slope-deflection method has a special advantage in that different damping coefficients ω_b may be chosen for different bars of the system without complicating the solution in any way.

So far we have discussed only lateral vibration. Assuming that in longitudinal vibration damping obeys the same law as it does in lateral vibration, we may write the following equations for damped longitudinal motion of a length element

$$\mu \, dx \, \frac{\partial^2 u(x, t)}{\partial t^2} + b \, dx \, \frac{\partial u(x, t)}{\partial t} - ES \, \frac{\partial^2 u(x, t)}{\partial x^2} = 0 \tag{6.19}$$

and from there

$$-\mu\omega^2 \left(1 - 2i \frac{\omega_b}{\omega}\right) [u(x) + i\bar{u}(x)] - ES \frac{d^2[u(x) + i\bar{u}(x)]}{dx^2} = 0 \tag{6.20}$$

The complete solution is

$$u(x) + i\bar{u}(x) = A \cos\left(\Psi + i\overline{\Psi}\right)\frac{x}{l} + B \sin\left(\Psi + i\overline{\Psi}\right)\frac{x}{l} \tag{6.21}$$

where

$$\Psi + i\overline{\Psi} = l \left[\frac{\mu\omega^2}{ES}\left(1 - 2i\frac{\omega_b}{\omega}\right)\right]^{1/2}$$

$$= \psi \left[\frac{1}{2}\left(1 + \frac{4\omega_b^2}{\omega^2}\right)^{1/2} + \frac{1}{2}\right]^{1/2} - i\psi \left[\frac{1}{2}\left(1 + \frac{4\omega_b^2}{\omega^2}\right)^{1/2} - \frac{1}{2}\right]^{1/2} \tag{6.22}$$

Since eq. 6.20 corresponds to eq. 2.62, the formulae for the normalised forces are obtained from expressions 2.69 and 2.70 by simple substitution of complex $\Psi + i\overline{\Psi}$ for real ψ, and complex $u_g + i\bar{u}_g$, $u_h + i\bar{u}_h$ for real u_g, u_h.

6.1.2 Expansion in natural modes

The transverse damped vibration of a uniform bar subjected to a transverse continuous load $p(x, t)$ is analytically described by the equation

$$\frac{\partial^2 v(x, t)}{\partial t^2} + 2\omega_b \frac{\partial v(x, t)}{\partial t} + \frac{EJ}{\mu} \frac{\partial^4 v(x, t)}{\partial x^4} = \frac{p(x, t)}{\mu} \tag{6.23}$$

Expand $v(x, t)$ and $p(x, t)$ in a series of the natural modes of vibration $v_{(j)}(x)$ (cf. eqs 3.63 and 3.64):

$$v(x, t) = \sum_{j=1}^{\infty} q_{(j)}(t)\, v_{(j)}(x) \tag{6.24}$$

$$\frac{p(x, t)}{\mu} = \sum_{j=1}^{\infty} p_{(j)}(t)\, v_{(j)}(x) \tag{6.25}$$

and substitute in eq. 6.23. On writing in accordance with eq. 2.10

$$EJ\, \frac{d^4 v_{(j)}(x)}{dx^4} = \mu \omega_{(j)}^2\, v_{(j)}(x)$$

we get

$$\sum_{j=1}^{\infty} \frac{d^2 q_{(j)}(t)}{dt^2}\, v_{(j)}(x) + 2\omega_b \sum_{j=1}^{\infty} \frac{dq_{(j)}(t)}{dt}\, v_{(j)}(x)$$

$$+ \sum_{j=1}^{\infty} q_{(j)}(t)\, \omega_{(j)}^2\, v_{(j)}(x) = \sum_{j=1}^{\infty} p_{(j)}(t)\, v_{(j)}(x) \tag{6.26}$$

If eq. 6.26 is to be satisfied for every x it must be the case that

$$\frac{d^2 q_{(j)}(t)}{dt^2} + 2\omega_b\, \frac{dq_{(j)}(t)}{dt} + \omega_{(j)}^2\, q_{(j)}(t) = p_{(j)}(t) \tag{6.27}$$

In the case of a harmonic load $p_{(j)}(t) = p_{(j)} \sin \omega t$, the steady-state forced vibration is defined by formula 6.24 where

$$q_{(j)}(t) = \frac{p_{(j)}}{[(\omega_{(j)}^2 - \omega^2)^2 + 4\omega^2 \omega_b^2]^{1/2}} \sin(\omega t + \varphi_{(j)}) \tag{6.28}$$

and

$$\varphi_{(j)} = -\tan^{-1} \frac{2\omega \omega_b}{\omega_{(j)}^2 - \omega^2} \tag{6.29}$$

In place of eq. 6.28 we may write

$$q_{(j)}(t) = q_{(j)} \sin \omega t + \bar{q}_{(j)} \cos \omega t$$

where

$$q_{(j)} = \frac{p_{(j)} \cos \varphi_{(j)}}{[(\omega_{(j)}^2 - \omega^2)^2 + 4\omega^2 \omega_b^2]^{1/2}}$$

$$= \frac{p_{(j)}(\omega_{(j)}^2 - \omega^2)}{[(\omega_{(j)}^2 - \omega^2)^2 + 4\omega^2 \omega_b^2]} \tag{6.30}$$

and

$$\bar{q}_{(j)} = \frac{p_{(j)} \sin \varphi_{(j)}}{[(\omega_{(j)}^2 - \omega^2)^2 + 4\omega^2\omega_b^2]^{1/2}}$$

$$= \frac{p_{(j)} \cdot 2\omega\omega_b}{[(\omega_{(j)}^2 - \omega^2)^2 + 4\omega^2\omega_b^2]} \tag{6.31}$$

Coefficients $p_{(j)}$ are computed as indicated in Section 3.5.

6.1.3   Numerical examples

The slope-deflection method and the expansion in natural modes, explained in the foregoing sections, have both proved useful in practice and preference of one over the other is decided by the circumstances peculiar to the case on hand. As demonstrated earlier in connection with the calculation of undamped vibration, the slope-deflection method makes it possible to calculate the deformation of systems in forced steady-state vibration directly, without advance knowledge of the natural modes and frequencies. In calculations of damped vibration it is capable of doing even more, i.e. express the fact that the damping of the various members of the system is not of equal intensity. The end forces and moments are namely calculated separately for each member and this offers us the possibility of choosing the magnitude of the damping coefficient in the respective frequency functions at will. This possibility may be of considerable practical importance for various kinds of structures, such as frame machinery foundations where the damping of the cross beams is affected by machine parts resting on those beams while the damping of the foundation slabs depends on the subsoil, etc.

The expansion in natural modes, on the other hand, works best in cases where the natural frequencies and modes are already known or when the calculation of forced vibration is to be performed for several values of the excitation frequency. When it comes to damped vibration this method is of particular value for establishing the system response in the resonance range. As the numerical examples which follow will clearly show it is sufficient in such cases to consider no other but that mode the appertaining frequency of which is close to the excitation frequency, for the contribution of the remaining modes to the resultant vibration is insignificant. This method of calculation does not allow us to assume that the damping is different in different parts of the structure. But as evident from the way the formulae are constructed, it is possible to substitute a different coefficient of damping in each of the natural modes and thus express some specific properties of damping of certain materials and structures.

The system which we shall discuss in the numerical examples is already quite familiar from previous calculations of undamped vibration; it was chosen so on purpose to make the similarity and the differences between calculations of the two kinds of vibration as clear as possible.

6.1.3.1 Heavy damping — Example 6.1

Consider the multi-storey frame (Fig. 3.8) dealt with in Example 3.1. The first natural circular frequency of symmetric vibration is $\omega_{(1)} = 103\cdot703$ s^{-1}. The frame is loaded midspan of the topmost beam $\overline{33}'$ by a harmonic vertical force $P(t) = P \sin \omega t$, the frequency of which is equal to the natural frequency, $\omega = \omega_{(1)} = 103\cdot703$ s^{-1}. The vibration is heavily damped (damping frequency $f_b = = 10$ s^{-1}, $\omega_b = 62\cdot8318$ s^{-1}), the damping being proportional to the velocity of vibration.

Solution by the slope-deflection method

The slope-deflection equations of undamped vibration were evolved in Example 3.1. When dealing with damped vibration, complex expressions must be introduced in the coefficients as well as in the unknowns of the equations. The equations thus obtained are set out in Table 6.1.

Table 6.1

$\zeta_1 + i\bar{\zeta}_1$	$\zeta_2 + i\bar{\zeta}_2$	$\zeta_3 + i\bar{\zeta}_3$	
$a_{11} + i\bar{a}_{11}$ $a_{12} + i\bar{a}_{12}$	$a_{12} + i\bar{a}_{12}$ $a_{22} + i\bar{a}_{22}$ $a_{23} + i\bar{a}_{23}$	$a_{23} + i\bar{a}_{23}$ $a_{33} + i\bar{a}_{33}$	$= 0$ $= 0$ $= a_{03} + i\bar{a}_{03}$

The coefficients of the equations are

$$a_{11} + i\bar{a}_{11} = \frac{2EJ_1}{l_1} F_2(A + i\bar{A})_1 + \frac{EJ_2}{l_2}[F_2(A + i\bar{A})_2 - F_1(A + i\bar{A})_2]$$

$$= a_{22} + i\bar{a}_{22}$$

$$a_{12} + i\bar{a}_{12} = \frac{EJ_1}{l_1} F_1(A + i\bar{A})_1 = a_{23} + i\bar{a}_{23}$$

$$a_{33} + i\bar{a}_{33} = \frac{EJ_1}{l_1} F_2(A + i\bar{A})_1 + \frac{EJ_2}{l_2}[F_2(A + i\bar{A})_2 - F_1(A + i\bar{A})_2]$$

$$a_{03} + i\bar{a}_{03} = \frac{Pl_2}{4} \frac{F_3\left(\frac{A + i\bar{A}}{2}\right)_2}{F_6\left(\frac{A + i\bar{A}}{2}\right)_2}$$

(6.32)

In the case considered, we get for the columns

$$\lambda_1 = 3\cdot410\,612$$

and by eq. 6.8

$$A_1 = 3\cdot410\,612\left\{\frac{1}{2}\left[1+\frac{4\times62\cdot8318^2}{103\cdot703^2}\right]^{1/4}+\left[\frac{1}{8}\left(1+\frac{4\times62\cdot8318^2}{103\cdot703^2}\right)^{1/2}+\frac{1}{8}\right]^{1/2}\right\}^{1/2}$$

$$= 3\cdot410\,612(0\cdot626\,718\,4+0\cdot566\,910\,8)^{1/2} = 3\cdot410\,612(1\cdot193\,629)^{1/2} = 3\cdot726\,207$$

$$\bar{A}_1 = -3\cdot410\,612(0\cdot626\,718\,4-0\cdot566\,910\,8)^{1/2} = -3\cdot410\,612(0\cdot059\,807\,6)^{1/2}$$

$$= -0\cdot834\,085$$

Further

$$\cosh\,(3\cdot726\,207-0\cdot834\,085\,i)$$

$$= \cosh 3\cdot726\,207\cos 0\cdot834\,085 - i\sinh 3\cdot726\,207\sin 0\cdot834\,085$$

$$= 20\cdot772\,70\times0\cdot671\,86 - i\,20\cdot748\,62\times0\cdot740\,68$$

$$= 13\cdot9563 - 15\cdot3681\,i$$

$$\sinh\,(3\cdot726\,207-0\cdot834\,085\,i) = \sinh 3\cdot726\,207\cos 0\cdot834\,085 - i\cosh 3\cdot726\,207$$

$$\times\sin 0\cdot834\,085 = 13\cdot9400 - 15\cdot3859\,i$$

$$\cos\,(3\cdot726\,207-0\cdot834\,085\,i) = \cos 3\cdot726\,207\cosh 0\cdot834\,085 + i\sin 3\cdot726\,207$$

$$\times\sinh 0\cdot834\,085 = -1\cdot141\,21 - 0\cdot515\,57\,i$$

$$\sin\,(3\cdot726\,207-0\cdot834\,085\,i) = \sin 3\cdot726\,207\cosh 0\cdot834\,085 - i\cos 3\cdot726\,207$$

$$\times\sinh 0\cdot834\,085 = -0\cdot755\,23 + 0\cdot779\,06\,i$$

By eq. 6.11:

$$F_1(3\cdot726\,207-0\cdot834\,085\,i) = -\frac{A}{B} = 2\cdot450\,55 - 1\cdot897\,21\,i$$

where $A = (3\cdot726\,207 - 0\cdot834\,085\,i)\,[\sinh\,(3\cdot726\,207 - 0\cdot834\,085\,i)$

$$- \sin\,(3\cdot726\,207 - 0\cdot834\,085\,i)] = (3\cdot726\,207 - 0\cdot834\,085\,i)\,(13\cdot9400$$

$$- 15\cdot3859\,i + 0\cdot755\,23 - 0\cdot779\,06\,i)$$

$$B = \cosh\,(3\cdot726\,207 - 0\cdot834\,085\,i)\cos\,(3\cdot726\,207 - 0\cdot834\,085\,i) - 1$$

$$= (13\cdot9563 - 15\cdot3681\,i)\,(-1\cdot141\,21 - 0\cdot515\,57\,i) - 1$$

and

$$\frac{EJ_1}{l_1} F_1(A + i\bar{A})_1 = \frac{21 \times 10^6 \times 1 \times 10^{-4}}{6} (2\cdot450\,55 - 1\cdot897\,21\,i) = 857\cdot69 - 664\cdot02\,i$$

From eq. 2.46, in which $A + i\bar{A}$ is substituted for λ

$$F_2(3\cdot726\,207 - 0\cdot834\,085\,i) = -\frac{A}{B} = 3\cdot233\,41 - 2\cdot312\,98\,i$$

where $A = (3\cdot726\,207 - 0\cdot834\,085\,i)\,[\cosh(3\cdot726\,207 - 0\cdot834\,085\,i)$

$$\times \sin(3\cdot726\,207 - 0\cdot834\,085\,i) - \sinh(3\cdot726\,207 - 0\cdot834\,085\,i)$$

$$\times \cos(3\cdot726\,207 - 0\cdot834\,085\,i)]$$

$$B = \cosh(3\cdot726\,207 - 0\cdot834\,085\,i) \cos(3\cdot726\,207 - 0\cdot834\,085\,i) - 1$$

and

$$\frac{EJ_1}{l_1} F_2(A + i\bar{A})_1 = \frac{21 \times 10^6 \times 1 \times 10^{-4}}{6} (3\cdot233\,41 - 2\cdot312\,98\,i) = 1131\cdot69 + 809\cdot54\,i$$

For the horizontal bars

$$\lambda_2 = 3\cdot005\,236$$

$$A_2 = 3\cdot005\,236\,(1\cdot193\,629)^{1/2} = 3\cdot283\,320$$

$$\bar{A}_2 = -3\cdot005\,236\,(0\cdot059\,8076)^{1/2} = -0\cdot734\,948$$

$$\frac{(A + i\bar{A})_2}{2} = 1\cdot641\,660 - 0\cdot367\,474\,i$$

$$\cosh(1\cdot641\,660 - 0\cdot367\,474\,i) = \cosh 1\cdot641\,660 \cos 0\cdot367\,474 - i \sinh 1\cdot641\,660$$

$$\times \sin 0\cdot367\,474 = 2\cdot499\,84 - 0\cdot892\,78\,i$$

$$\sinh(1\cdot641\,660 - 0\cdot367\,474\,i) = 2\cdot319\,11 - 0\cdot962\,35\,i$$

$$\cos(1\cdot641\,660 - 0\cdot367\,474\,i) = -0\cdot075\,63 + 0\cdot374\,86\,i$$

$$\sin(1\cdot641\,660 - 0\cdot367\,474\,i) = 1\cdot065\,60 + 0\cdot026\,61\,i$$

Further,

$$F_2(A + i\bar{A})_2 - F_1(A + i\bar{A})_2 = \frac{A}{B} = 0\cdot890\,76 + 2\cdot203\,96\,i$$

where $A = (A + i\bar{A})_2\,2 \cosh \dfrac{(A + i\bar{A})_2}{2} \cos \dfrac{(A + i\bar{A})_2}{2}$

$$= (3\cdot283\,320 - 0\cdot734\,948\,i)\,2(2\cdot499\,84 - 0\cdot892\,78\,i)\,(-0\cdot075\,63 + 0\cdot374\,86\,i)$$

and $\quad B = \cosh \dfrac{(\varLambda + i\bar{\varLambda})_2}{2} \sin \dfrac{(\varLambda + i\bar{\varLambda})_2}{2} + \sinh \dfrac{(\varLambda + i\bar{\varLambda})_2}{2} \cos \dfrac{(\varLambda + i\bar{\varLambda})_2}{2}$

$\quad = (2{\cdot}499\,84 - 0{\cdot}892\,78\,i)(1{\cdot}065\,60 + 0{\cdot}026\,61\,i) + (2{\cdot}319\,11 - 0{\cdot}962\,35\,i)$

$\quad \times (-0{\cdot}075\,63 + 0{\cdot}374\,86\,i)$

$\dfrac{EJ_2}{l_2}\left[F_2(\varLambda + i\bar{\varLambda})_2 - F_1(\varLambda + i\bar{\varLambda})_2\right] = \dfrac{21\times10^6\times2\times10^{-4}}{5}(0{\cdot}890\,76 + 2{\cdot}203\,96\,i)$

$\quad = 748{\cdot}24 + 1851{\cdot}33\,i$

The absolute term is

$$\dfrac{Pl_2}{4}F_3\dfrac{(\varLambda + i\bar{\varLambda})_2}{2}\left[F_6\dfrac{(\varLambda + i\bar{\varLambda})_2}{2}\right]^{-1} = P\dfrac{A}{B} = P(0{\cdot}717\,861 - 0{\cdot}188\,734\,i)$$

where $A = 5\left[(2{\cdot}499\,84 - 0{\cdot}892\,78\,i) - (-0{\cdot}075\,63 + 0{\cdot}374\,86\,i)\right]$

$\quad B = 4\,(1{\cdot}641\,660 - 0{\cdot}367\,474\,i)\left[(2{\cdot}499\,84 - 0{\cdot}892\,78\,i)(1{\cdot}065\,60 + 0{\cdot}026\,61\,i)\right.$

$\quad \left. + (2{\cdot}319\,11 - 0{\cdot}962\,35\,i)(-0{\cdot}075\,63 + 0{\cdot}374\,86\,i)\right]$

Substituting the numerical values in the equations of Table 6.1 gives three equations with complex coefficients and unknowns, as shown in Table 6.2.

Table 6.2

$\zeta_1 + i\bar{\zeta}_1$	$\zeta_2 + i\bar{\zeta}_2$	$\zeta_3 + i\bar{\zeta}_3$	
$3011{\cdot}62 + 3470{\cdot}41i$	$857{\cdot}69 - 664{\cdot}02i$		$= 0$
$857{\cdot}69 - 664{\cdot}02i$	$3011{\cdot}62 + 3470{\cdot}41i$	$857{\cdot}69 - 664{\cdot}02i$	$= 0$
	$857{\cdot}69 - 664{\cdot}02i$	$1879{\cdot}93 - 2660{\cdot}87i$	$= (0{\cdot}717861 - 0{\cdot}188734i)\,P$

From the tabulated equations (with P in Mp)

$$\zeta_1 = -0{\cdot}050\,06 \times 10^{-4}P, \quad \bar{\zeta}_1 = 0{\cdot}100\,14 \times 10^{-4}P$$
$$\zeta_2 = 0{\cdot}435\,42 \times 10^{-4}P, \quad \bar{\zeta}_2 = 0{\cdot}188\,00 \times 10^{-4}P$$
$$\zeta_3 = 0{\cdot}742\,09 \times 10^{-4}P, \quad \bar{\zeta}_3 = -1{\cdot}986\,28 \times 10^{-4}P$$

and

$$\zeta_1(t) = (-0{\cdot}050\,06 \sin 103{\cdot}703t + 0{\cdot}100\,14 \cos 103{\cdot}703t) \times 10^{-4}P, \text{ etc.}$$

It is evident from the results that the phase shift $\varphi = \tan^{-1}(\bar{\zeta}/\zeta)$ is different for different joints.

The deflection at the midspan of beam $\overline{33}'$ at the point of action, p, of force $P(t)$ is computed from the slope-deflection equation

$$(v_p + i\bar{v}_p)\frac{8EJ_2}{l_2^3}F_6\frac{(A + i\bar{A})_2}{2} - (\zeta_3 + i\bar{\zeta}_3)\frac{4EJ_2}{l_3^3}F_3\frac{(A + i\bar{A})_2}{2} = \frac{P}{2}$$

From there

$$(v_p + i\bar{v}_p) = (\zeta_3 + i\bar{\zeta}_3)\frac{l_2}{2}\frac{F_3\dfrac{(A + i\bar{A})_2}{2}}{F_6\dfrac{(A + i\bar{A})_2}{2}} + \frac{Pl_2^3}{16EJ_2F_6\dfrac{(A + i\bar{A})_2}{2}}$$

Since

$$\frac{l_2}{2}\frac{F_3\dfrac{(A + i\bar{A}_2)_2}{2}}{F_6\dfrac{(A + i\bar{A})_2}{2}} = 1{\cdot}436 - 0{\cdot}377\,i$$

(the absolute term of the equations in Table 6.2, multiplied by $2/P$) and

$$\frac{Pl_2^3}{16EJ_2F_6\dfrac{(A + i\bar{A})_2}{2}} \approx -\frac{P \times 125}{16 \times 21 \times 10^6 \times 2 \times 10^{-4}}$$

$$\times \frac{\cosh(1{\cdot}642 - 0{\cdot}367\,i)\cos(1{\cdot}642 - 0{\cdot}367\,i) - 1}{(1{\cdot}642 - 0{\cdot}367\,i)^3\,[\cosh(1{\cdot}642 - 0{\cdot}367\,i)\sin(1{\cdot}642 - 0{\cdot}367\,i) + \sinh(1{\cdot}642 - 0{\cdot}367\,i)\cos(1{\cdot}642 - 0{\cdot}367\,i)]}$$

$$= (1{\cdot}75 - 0{\cdot}40\,i) \times 10^{-4}P$$

then

$$(v_p + i\bar{v}_p) = (0{\cdot}742 - 1{\cdot}986\,i) \times 10^{-4}P(1{\cdot}436 - 0{\cdot}377\,i)$$
$$+ (1{\cdot}75 - 0{\cdot}40\,i) \times 10^{-4}P \approx (2{\cdot}06 - 3{\cdot}54\,i) \times 10^{-4}P$$

(in m, if P is in Mp).

Computation by expansion in natural modes

Because ω_b is the same for all the bars of the frame system, the problem can also be solved by expansion in the natural modes of vibration. The natural modes were computed in Example 3.1, and load $p_{(j)}$ for $j = 1, 2, 3$ in Example 3.7. If including the first three terms of the infinite series is satisfactory for our purpose, we thus know all the values necessary for determining the deflections.

According to eq. 6.29, for the first natural frequency $(\omega_{(1)} \approx 103\cdot7\ s^{-1})$

$$\varphi_{(1)} = -\tan^{-1} \frac{2\omega\omega_b}{\omega_{(1)}^2 - \omega^2} = -\frac{\pi}{2}$$

for steady vibration therefore by eqs 6.30 and 6.31

$$q_{(1)} = \frac{P_{(1)}}{2\omega\omega_b} \cos \varphi_{(1)} = 0\ , \quad \bar{q}_{(1)} = \frac{P_{(1)}}{2\omega\omega_b} \sin \varphi_{(1)} = \frac{-0\cdot969P}{2 \times 103\cdot7 \times 62\cdot8}$$

$$= -0\cdot745 \times 10^{-4}P$$

Further (for P in Mp)

$$\bar{q}_{(1)}\zeta_{1(1)} \quad = -0\cdot745 \times 0\cdot584 \times 10^{-4}P = -0\cdot435 \times 10^{-4}P$$

$$\bar{q}_{(1)}\zeta_{2(1)} \quad = \quad 0\cdot745 \times 0\cdot962 \times 10^{-4}P = \quad 0\cdot717 \times 10^{-4}P$$

$$\bar{q}_{(1)}\zeta_{3(1)} \quad = -0\cdot745 \times 10^{-4}P$$

$$\bar{v}_p = \bar{q}_{(1)}v_{P(1)} = -0\cdot745 \times 1\cdot525 \times 10^{-4}P = -1\cdot136 \times 10^{-4}P\ m$$

For the second natural frequency $(\omega_{(2)} \approx 126\cdot2\ s^{-1})$

$$\varphi_{(2)} = -\tan^{-1} \frac{2 \times 103\cdot7 \times 62\cdot8}{126\cdot2^2 - 103\cdot7^2}$$

and

$$\sin \varphi_{(2)} = -\frac{2 \times 103\cdot7 \times 62\cdot8}{[(126\cdot2^2 - 103\cdot7^2)^2 + 4 \times 103\cdot7^2 \times 62\cdot8^2]^{1/2}} = -\frac{13\ 000}{14\ 000} = -0\cdot930$$

$$\cos \varphi_{(2)} = \frac{(126\cdot2^2 - 103\cdot7^2)}{14\ 000} = 0\cdot370$$

and further

$$q_{(2)} = \frac{P_{(2)} \cos \varphi_{(2)}}{[(\omega_{(2)}^2 - \omega^2)^2 + 4\omega^2\omega_b^2]^{1/2}} = \frac{1\cdot225 \times 0\cdot370P}{14\ 000} = 0\cdot324 \times 10^{-4}P$$

$$\bar{q}_{(2)} = -\frac{1\cdot225 \times 0\cdot930P}{14\ 000} = -0\cdot814 \times 10^{-4}P$$

$$q_{(2)}\zeta_{1(2)} = -0\cdot324 \times 1\cdot013 \times 10^{-4}P = -0\cdot328 \times 10^{-4}P$$

$$\bar{q}_{(2)}\zeta_{1(2)} = 0\cdot814 \times 1\cdot013 \times 10^{-4}P = 0\cdot825 \times 10^{-4}P$$

$$q_{(2)}\zeta_{2(2)} = 0\cdot324 \times 0\cdot113 \times 10^{-4}P = 0\cdot037 \times 10^{-4}P$$

$$\bar{q}_{(2)}\zeta_{2(2)} = -0\cdot814 \times 0\cdot113 \times 10^{-4}P = -0\cdot092 \times 10^{-4}P$$

$$q_{(2)}\zeta_{3(2)} = 0\cdot324 \times 10^{-4}P$$

$$\bar{q}_{(2)}\zeta_{3(2)} = -0\cdot814 \times 10^{-4}P$$

The amplitude at the midspan of the top beam (in m if P is in Mp) is

$$q_{(2)}v_{p(2)} = \quad 0\cdot324 \times 1\cdot700 \times 10^{-4}P = \quad 0\cdot551 \times 10^{-4}P$$

$$\bar{q}_{(2)}v_{p(2)} = -0\cdot814 \times 1\cdot700 \times 10^{-4}P = -1\cdot384 \times 10^{-4}P$$

And similarly so for the third natural frequency $(\omega_{(3)} \approx 155\cdot4\ \mathrm{s}^{-1})$

$$\sin \varphi_{(3)} = -\frac{2 \times 103\cdot7 \times 62\cdot8}{[(155\cdot4^2 - 103\cdot7^2)^2 + 4 \times 103\cdot7^2 \times 62\cdot8^2]^{1/2}} = -\frac{13\,000}{18\,700} = -0\cdot696$$

$$\cos \varphi_{(3)} = \frac{155\cdot4^2 - 103\cdot7^2}{18\,700} = 0\cdot718$$

$$q_{(3)} = \frac{1\cdot028 \times 0\cdot718P}{18\,700} = 0\cdot395 \times 10^{-4}P$$

$$\bar{q}_{(3)} = -\frac{1\cdot028 \times 0\cdot696P}{18\,700} = -0\cdot384 \times 10^{-4}P$$

$$q_{(3)}\zeta_{1(3)} = 0\cdot395 \times 0\cdot776 \times 10^{-4}P = 0\cdot306 \times 10^{-4}P$$

$$\bar{q}_{(3)}\zeta_{1(3)} = -0\cdot384 \times 0\cdot776 \times 10^{-4}P = -0\cdot298 \times 10^{-4}P$$

$$q_{(3)}\zeta_{2(3)} = 0\cdot395 \times 1\cdot174 \times 10^{-4}P = 0\cdot464 \times 10^{-4}P$$

$$\bar{q}_{(3)}\zeta_{2(3)} = -0\cdot384 \times 1\cdot174 \times 10^{-4}P = -0\cdot451 \times 10^{-4}P$$

$$q_{(3)}\zeta_{3(3)} = 0\cdot395 \times 10^{-4}P$$

$$\bar{q}_{(3)}\zeta_{3(3)} = -0\cdot384 \times 10^{-4}P$$

and the amplitude at the midspan of the top beam is

$$q_{(3)}v_{p(3)} = \quad 0\cdot395 \times 2\cdot067 \times 10^{-4}P = \quad 0\cdot817 \times 10^{-4}P$$

$$\bar{q}_{(3)}v_{p(3)} = -0\cdot384 \times 2\cdot067 \times 10^{-4}P = -0\cdot793 \times 10^{-4}P$$

The resultant rotations are then approximately

$$\zeta_1 = (-0\cdot328 + 0\cdot306) \times 10^{-4}P = -0\cdot022 \times 10^{-4}P$$

$$\bar{\zeta}_1 = (-0\cdot435 + 0\cdot825 - 0\cdot298) \times 10^{-4}P = 0\cdot092 \times 10^{-4}P$$

$$\zeta_2 = (0\cdot037 + 0\cdot464) \times 10^{-4}P = 0\cdot501 \times 10^{-4}P$$

$$\bar{\zeta}_2 = (0\cdot717 - 0\cdot092 - 0\cdot451) \times 10^{-4}P = 0\cdot174 \times 10^{-4}P$$

$$\zeta_3 = (0\cdot324 + 0\cdot395) \times 10^{-4}P = 0\cdot719 \times 10^{-4}P$$

$$\bar{\zeta}_3 = (-0\cdot745 - 0\cdot814 - 0\cdot384) \times 10^{-4}P = -1\cdot943 \times 10^{-4}P$$

Even with just the first three terms of the series, the results of the expansion in natural modes are in good agreement with those of the slope-deflection method, except for the amplitude at the point of force application midspan of beam $\overline{33'}$ where the difference is large, viz.

$$v_p = (\ \ 0.551 + 0.817) \times 10^{-4}P = \ \ \ \ 1.368 \times 10^{-4}P$$

$$\bar{v}_p = (-1.136 - 1.384 - 0.793) \times 10^{-4}P = -3.313 \times 10^{-4}P$$

6.1.3.2   *Light damping outside the resonance range — Example 6.2*

Consider the frame discussed in the preceding example (Fig. 3.8) subjected to vertical force $P \sin \omega t$ midspan of the horizontal bar $\overline{33'}$. The circular frequency of the force is $\omega = 62.8\ \text{s}^{-1}$ $(n = 10\ \text{s}^{-1})$, of the damping $\omega_b = 0.628\ \text{s}^{-1}$ $(f_b = = 0.1\ \text{s}^{-1})$.

Solution by the slope-deflection method

The damping being light, functions $F(\Lambda + i\bar{\Lambda})$ may be computed from eqs 6.15, 6.16 and 6.17. The equations of Table 6.1 apply again, but at $\omega_b = 0.01\ \omega$ the numerical tables in the Appendix give the following values for the coefficients:

for $\lambda_1 = 2.654$

$$F_1(\Lambda + i\bar{\Lambda})_1 = F_1(\lambda_1) + i\frac{2\omega_b}{\omega}\Phi_1(\lambda_1) = 2.397 - 0.008\ 91i \tag{6.33}$$

$$F_2(\Lambda + i\bar{\Lambda})_1 = F_2(\lambda_1) + i\frac{2\omega_b}{\omega}\Phi_2(\lambda_1) = 3.483 + 0.011\ 31i \tag{6.34}$$

for $\lambda_2 = 2.339$

$$F_1(\Lambda + i\bar{\Lambda})_2 = 2.228 - 0.004\ 90i \tag{6.35}$$

$$F_2(\Lambda + i\bar{\Lambda})_2 = 3.700 + 0.006\ 32i \tag{6.36}$$

$$F_3\frac{(\Lambda + i\bar{\Lambda})_2}{2} = F_3\left(\frac{\lambda_2}{2}\right) + i\frac{2\omega_b}{\omega}\Phi_3\left(\frac{\lambda_2}{2}\right) = 6.058 - 0.001\ 18i \tag{6.37}$$

$$F_6\frac{(\Lambda + i\bar{\Lambda})_2}{2} = F_6\left(\frac{\lambda_2}{2}\right) + i\frac{2\omega_b}{\omega}\Phi_6\left(\frac{\lambda_2}{2}\right) = 11.304 + 0.013\ 94i \tag{6.38}$$

On substituting these values in the equations of Table 6.1, we get — after multiplication — a system of six linear equations with real coefficients (Table 6.3).

Table 6.3

ζ_1	ζ_2	ζ_3	$\bar{\zeta}_1$	$\bar{\zeta}_2$	$\bar{\zeta}_3$	Absolute term
3675	840		−17·30	3·12		= 0
840	3675	840	3·12	−17·30	3·12	= 0
	840	2455		3·12	−13·34	= 0·670P
−17·30	3·12		−3675	−840		= 0
3·12	−17·30	3·12	−840	−3675	−840	= 0
	3·12	−13·34		− 840	−2455	= 0·00095P

The solution of the above equations yields the amplitudes of the unknown rotations of the joints (P is in Mp)

$$\zeta_1 = \quad 0\cdot1636 \times 10^{-4}P, \quad \bar{\zeta}_1 = -0\cdot43 \times 10^{-6}P$$

$$\zeta_2 = -0\cdot716 \quad \times 10^{-4}P, \quad \bar{\zeta}_2 = \quad 1\cdot28 \times 10^{-6}P$$

$$\zeta_3 = \quad 2\cdot97 \quad \times 10^{-4}P, \quad \bar{\zeta}_3 = -2\cdot53 \times 10^{-6}P$$

The deflection midspan of beam $\overline{33'}$, at the point of action of force P, is given by

$$v_p + i\bar{v}_p = \frac{l_2}{2} \frac{F_3 \dfrac{(\varLambda + i\bar{\varLambda})_2}{2}}{F_6 \dfrac{(\varLambda + i\bar{\varLambda})_2}{2}} (\zeta_3 + i\bar{\zeta}_3) + \frac{Pl_2^3}{16EJ_2 F_6 \dfrac{(\varLambda + i\bar{\varLambda})_2}{2}}$$

$$= \frac{5(6\cdot058 - 0\cdot00118i)}{2(11\cdot304 + 0\cdot01394i)} (2\cdot97 - 0\cdot0253i) \times 10^{-4}P$$

$$+ \frac{P \times 125}{16 \times 21 \times 10^6 \times 2 \times 10^{-4}(11\cdot304 + 0\cdot01394i)} = (5\cdot63 - 0\cdot0416i) \times 10^{-4}P$$

(in m if P is in Mp).

It is clear from the results that the values with bars, $\bar{\zeta}$, are small against those without bars, ζ. Since their coefficients in the first three equations of Table 6.3 are also small, they can be wholly omitted in the computation of the values without the bar. The real parts of values 6.33 − 6.38 being very nearly the same as in undamped vibration, the effect of light damping on forced vibration with a frequency outside the range of free vibration need not be considered at all. Accordingly, all there is to solve in the given case is the system of three equations expressing the conditions of equilibrium of undamped vibration (cf. Example 3.1, Section A).

Expansion in natural modes

The computation proceeds in the same way as in Example 6.1. For the first natural mode

$$\cos \varphi_{(1)} \approx 1; \quad \sin \varphi_{(2)} = - \frac{2 \times 62 \cdot 8 \times 0 \cdot 628}{[(103 \cdot 7^2 - 62 \cdot 8^2)^2 + 4 \times 62 \cdot 8^2 \times 0 \cdot 628^2]^{1/2}}$$

$$= \frac{78 \cdot 9}{6806} = -0 \cdot 0116$$

$$q_{(1)} = \frac{0 \cdot 969 P}{6806} = 1 \cdot 424 \times 10^{-4} P$$

$$\bar{q}_{(1)} = - \frac{0 \cdot 969 P \times 0 \cdot 0116}{6806} = -1 \cdot 65 \times 10^{-6} P$$

and

$$q_{(1)} \zeta_{1(1)} = \quad 1 \cdot 424 \times 0 \cdot 584 \times 10^{-4} P = 0 \cdot 832 \times 10^{-4} P$$

$$\bar{q}_{(1)} \zeta_{1(1)} = -1 \cdot 65 \times 0 \cdot 584 \times 10^{-6} P = -0 \cdot 966 \times 10^{-6} P$$

$$q_{(1)} \zeta_{2(1)} = -1 \cdot 424 \times 0 \cdot 962 \times 10^{-4} P = -1 \cdot 370 \times 10^{-4} P$$

$$\bar{q}_{(1)} \zeta_{2(1)} = \quad 1 \cdot 65 + 0 \cdot 962 \times 10^{-6} P = \quad 1 \cdot 59 \times 10^{-6} P$$

$$q_{(1)} \zeta_{3(1)} = \quad 1 \cdot 424 \times 10^{-4} P$$

$$\bar{q}_{(1)} \zeta_{3(1)} = -1 \cdot 65 \times 10^{-6} P$$

$$q_{(1)} v_{p(1)} = \quad 1 \cdot 424 \times 1 \cdot 525 \times 10^{-4} P = 2 \cdot 17 \times 10^{-4} P \text{ m}$$

$$\bar{q}_{(1)} v_{p(1)} = -1 \cdot 65 \times 1 \cdot 525 \times 10^{-6} P = -2 \cdot 52 \times 10^{-6} P \text{ m}$$

For the second natural mode

$$\cos \varphi_{(2)} \approx 1; \quad \sin \varphi_{(2)} = - \frac{2 \times 62 \cdot 8 \times 0 \cdot 628}{[(126 \cdot 2^2 - 62 \cdot 8^2)^2 + 4 \times 62 \cdot 8^2 + 0 \cdot 628^2]^{1/2}}$$

$$= - \frac{78 \cdot 9}{12\ 000} = -0 \cdot 00657$$

$$q_{(2)} = \frac{1 \cdot 225 P}{12\ 000} = 1 \cdot 021 \times 10^{-4} P$$

$$\bar{q}_{(2)} = - \frac{1 \cdot 225 P \times 0 \cdot 00657}{12\ 000} = -0 \cdot 671 \times 10^{-6} P$$

and

$$q_{(2)}\zeta_{1(2)} = -1{\cdot}021 \times 1{\cdot}013 \times 10^{-4}P = -1{\cdot}034 \times 10^{-4}P$$

$$\bar{q}_{(2)}\zeta_{1(2)} = 0{\cdot}671 \times 1{\cdot}013 \times 10^{-6}P = 0{\cdot}680 \times 10^{-6}P$$

$$q_{(2)}\zeta_{2(2)} = 1{\cdot}021 \times 0{\cdot}113 \times 10^{-4}P = 0{\cdot}115 \times 10^{-4}P$$

$$\bar{q}_{(2)}\zeta_{2(2)} = -0{\cdot}671 \times 0{\cdot}113 \times 10^{-6}P = -0{\cdot}076 \times 10^{-6}P$$

$$q_{(2)}\zeta_{3(2)} = 1{\cdot}021 \times 10^{-4}P$$

$$\bar{q}_{(2)}\zeta_{3(2)} = -0{\cdot}671 \times 10^{-6}P$$

$$q_{(2)}v_{p(2)} = 1{\cdot}021 \times 1{\cdot}700 \times 10^{-4}P = 1{\cdot}74 \times 10^{-4}P \text{ m}$$

$$\bar{q}_{(2)}v_{p(2)} = -0{\cdot}671 \times 1{\cdot}700 \times 10^{-6}P = -1{\cdot}14 \times 10^{-6}P \text{ m}$$

For the third natural mode

$$\cos\varphi_{(3)} \approx 1 \; ; \quad \sin\varphi_{(3)} = -\frac{2 \times 62{\cdot}8 \times 0{\cdot}628}{[(155{\cdot}4^2 - 62{\cdot}8^2)^2 + 4 \times 62{\cdot}8^2 \times 0{\cdot}628^2]^{1/2}}$$

$$= -\frac{78{\cdot}9}{20\,220} = -0{\cdot}00390$$

$$q_{(3)} = \frac{1{\cdot}028P}{20\,220} = 0{\cdot}509 \times 10^{-4}P$$

$$\bar{q}_{(3)} = -\frac{1{\cdot}028P \times 0{\cdot}00390}{20\,220} = -0{\cdot}198 \times 10^{-6}P$$

$$q_{(3)}\zeta_{1(3)} = 0{\cdot}509 \times 0{\cdot}776 \times 10^{-4}P = 0{\cdot}395 \times 10^{-4}P$$

$$\bar{q}_{(3)}\zeta_{1(3)} = -0{\cdot}198 \times 0{\cdot}776 \times 10^{-6}P = -0{\cdot}154 \times 10^{-6}P$$

$$q_{(3)}\zeta_{2(3)} = 0{\cdot}509 \times 1{\cdot}174 \times 10^{-4}P = 0{\cdot}598 \times 10^{-4}P$$

$$\bar{q}_{(3)}\zeta_{2(3)} = -0{\cdot}198 \times 1{\cdot}174 \times 10^{-6}P = -0{\cdot}233 \times 10^{-6}P$$

$$q_{(3)}\zeta_{3(3)} = 0{\cdot}509 \times 10^{-4}P$$

$$\bar{q}_{(3)}\zeta_{3(3)} = -0{\cdot}198 \times 10^{-6}P$$

$$q_{(3)}v_{p(3)} = 0{\cdot}509 \times 2{\cdot}067 \times 10^{-4}P = 1{\cdot}05 \times 10^{-4}P \text{ m}$$

$$\bar{q}_{(3)}v_{p(3)} = -0{\cdot}198 \times 2{\cdot}067 \times 10^{-6}P = -0{\cdot}409 \times 10^{-6}P \text{ m}$$

The resultant deformations are then

$$\zeta_1 \approx 0{\cdot}19 \times 10^{-4}P \qquad \bar{\zeta}_1 \approx -0{\cdot}44 \times 10^{-6}P$$

$$\zeta_2 \approx -0{\cdot}66 \times 10^{-4}P \qquad \bar{\zeta}_2 \approx 1{\cdot}28 \times 10^{-6}P$$

$$\zeta_3 \approx 2{\cdot}95 \times 10^{-4}P \qquad \bar{\zeta}_3 \approx -2{\cdot}52 \times 10^{-6}P$$

$$v_p \approx 4{\cdot}96 \times 10^{-4}P \text{ m} \qquad \bar{v}_p \approx -4{\cdot}07 \times 10^{-6}P \text{ m}$$

For systems with light damping vibrating with a frequency greatly different to a natural frequency the slope-deflection method is more expedient than the expansion in natural modes,particularly so if the natural modes of vibration are not known in advance. The converse is true of problems involving resonance.

6.1.3.3 *Light damping at resonance — Example 6.3*

Consider again the frame of Fig. 3.8 discussed in the preceding example Here again the frame is subjected to a vertical force $P \sin \omega t$ acting midspan of beam $\overline{33'}$. This time, however, $\omega = \omega_{(1)} = 103 \cdot 7 \text{ s}^{-1}$ and $\omega_b = 0 \cdot 628 \text{ s}^{-1}$.

If we applied the slope-deflection method, we would have to solve quite accurately six slope-deflection equations without omitting any of the terms with bars. In the expansion in natural modes, on the other hand, the first term of the series will be so large compared with the remaining ones that the latter may be neglected in the computation. For the case examined $\sin \varphi_{(1)} = -1$, $\cos \varphi_{(1)} = 0$ so that

$$q_{(1)} = 0, \quad \bar{q}_{(1)} = = -\frac{P_{(1)}}{2\omega\omega_b} = -\frac{0 \cdot 969 P}{2 \times 103 \cdot 7 \times 0 \cdot 628} = -0 \cdot 744 \times 10^{-2} P$$

For the second natural vibration

$$\sin \varphi_{(2)} \approx 0, \quad \cos \varphi_{(2)} \approx 1, \quad \bar{q}_{(2)} = 0$$

$$q_{(2)} = \frac{P_{(2)}}{[(\omega_{(2)}^2 - \omega^2)^2 + 4\omega^2\omega_b^2]^{1/2}}$$

$$= \frac{1 \cdot 225 P}{[(126 \cdot 2^2 - 103 \cdot 7^2)^2 + 4 \times 103 \cdot 7^2 \times 0 \cdot 628^2]^{1/2}} = 2 \cdot 36 \times 10^{-4} P$$

Since coefficient $q_{(2)}$ is smaller than $\bar{q}_{(1)}$ by a factor of about 30 and moreover is shifted in phase by $\pi/2$, its omission will result in no great error. The rotations of the joints will then be

$$\zeta_1 = -0 \cdot 744 \times 0 \cdot 584 \times 10^{-2} P = -0 \cdot 435 \times 10^{-2} P$$
$$\zeta_2 = \quad 0 \cdot 744 \times 0 \cdot 962 \times 10^{-2} P = \quad 0 \cdot 716 \times 10^{-2} P$$
$$\zeta_3 = -0 \cdot 744 \times 10^{-2} P$$

6.1.4 Damping proportional to the rate of stress change

We shall now turn to another kind of damping, namely damping caused by internal friction proportional to the rate of stress change. The pertinent equation of lateral vibration of an unloaded bar is now

$$\mu \frac{\partial^2 v(x, t)}{\partial t^2} + EJ \frac{\partial^4 v(x, t)}{\partial x^4} + \beta EJ \frac{\partial^5 v(x, t)}{\partial t \, \partial x^4} = 0 \tag{6.39}$$

The last term on the left-hand side expresses the damping force in the direction of the shear force. Substituting similarly as in Section 6.1

$$v^*(x, t) = [v(x) + i \, \bar{v}(x)] \sin \omega t + [\bar{v}(x) + i \, v(x)] \cos \omega t$$

leads to the following equation

$$-\mu\omega^2 [v(x) + i\bar{v}(x)] + EJ(1 + i\beta\omega) \frac{d^4 [v(x) + i\bar{v}(x)]}{dx^4} = 0 \tag{6.40}$$

Its complete solution is eq. 6.7 with

$$l \left[\frac{\mu\omega^2}{EJ(1 + i\beta\omega)} \right]^{1/4} = l \left[\frac{\mu\omega^2}{EJ} \frac{1 - i\beta\omega}{1 + \beta^2\omega^2} \right]^{1/4} \tag{6.41}$$

substituted for $\Lambda + i\bar{\Lambda}$.

For longitudinal vibration

$$\mu \frac{\partial^2 u(x, t)}{\partial t^2} - ES \frac{\partial^2 u(x, t)}{\partial x^2} - \beta ES \frac{\partial^3 u(x, t)}{\partial t \, \partial x^2} = 0 \tag{6.42}$$

which after the substitution

$$u(x, t) = [u(x) + i\bar{u}(x)] \sin \omega t + [\bar{u}(x) - iu(x)] \cos \omega t$$

gives

$$-\mu\omega^2 [u(x) + i\bar{u}(x)] - ES(1 + i\beta\omega) \frac{d^2 [u(x) + i\bar{u}(x)]}{dx^2} = 0 \tag{6.43}$$

The complete solution is the same as eq. 6.21 except that

$$l \left[\frac{\mu\omega^2}{ES(1 + i\beta\omega)} \right]^{1/2} = l \left[\frac{\mu\omega^2}{ES} \frac{1 - i\beta\omega}{1 + \beta^2\omega^2} \right]^{1/2} \tag{6.44}$$

is substituted for $\Psi + i\bar{\Psi}$.

It is clear that for light damping eqs 6.41 and 6.44 change respectively to eqs 6.8 and 6.22 if $\beta = 2\omega_b/\omega^2$.

6.2   A more general theory of damping

6.2.1   Introductory notes

The so-called Kelvin-Voigt hypothesis underlying the considerations of the preceding section is employed whenever damping is assumed to be caused by internal friction. But as experimental findings inform us the hypothesis is not universally

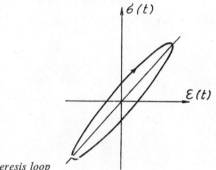

Fig. 6.1 Hysteresis loop

applicable; in certain respects it is at variance with the results of investigations of internal damping in, for example, metals, glass, wood and other materials.

All periodic vibrations are characterised by a shift of phase between stress and strain. As we know, in the stress-strain diagram damping manifests itself by the hysteresis loop (Fig. 6.1). The area L of the hysteresis loop represents the energy that is dissipated as heat during one period. The ratio of this energy to the potential energy V at maximum strain is an important constant of damping,

$$x = \frac{L}{V} \tag{6.45}$$

The forced vibration of a system with a single degree of freedom and damping proportional to the velocity of motion is described by the equation

$$m\,\ddot{v}(t) + 2m\omega_b\,\dot{v}(t) + C\,v(t) = P \sin \omega t \tag{6.46}$$

and the steady-state vibration by the formula

$$v(t) = v \sin (\omega t + \varphi) \tag{6.47}$$

In the equations C is the spring constant, v — the deflection amplitude, P — the exciting force amplitude and φ the phase shift between the exciting force and the deflection, ω — the circular frequency of the force, $\omega_0 = (C/m)^{1/2}$ the undamped natural circular frequency and ω_b the damping coefficient.

The deflection amplitude is

$$v = \frac{P\omega_0^2}{C[(\omega_0^2 - \omega^2)^2 + 4\omega^2\omega_b^2]^{1/2}} \tag{6.48}$$

and the sine of the phase shift is

$$\sin \varphi = -\frac{2\omega\omega_b}{[(\omega_0^2 - \omega^2)^2 + 4\omega^2\omega_b^2]^{1/2}} \tag{6.49}$$

The potential energy is given by

$$V = \tfrac{1}{2}Cv^2 \tag{6.50}$$

and the lost energy, i.e. energy converted to heat is

$$L = -\pi P v \sin \varphi = 2\pi C \frac{\omega\omega_b}{\omega_0^2} v^2 \tag{6.51}$$

consequently

$$\varkappa = \frac{L}{V} = 2\pi \frac{2\omega\omega_b}{\omega_0^2} \tag{6.52}$$

If $\omega = \omega_0$, i.e. at resonance

$$\varkappa = \frac{2\omega_b}{f_0} \approx 2\delta \tag{6.53}$$

where f_0 is the undamped natural frequency and δ the logarithmic decrement of damping.

What the foregoing tells us is that for damping proportional to the velocity of motion, coefficient $\varkappa$ is proportional to the frequency and in the limit case of both frequency and velocity tending to zero, becomes zero, too. The hysteresis loop then changes into a straight line. The same holds true for Voigt damping.

This, however, is at variance with experimental measurements which show that the relative loss of energy $\varkappa$ does not tend to zero as the velocity of motion decreases but remains constant over a wide spectrum of frequencies for a number of materials[66,71].

According to these measurements the shape of the hysteresis loop in steady-state harmonic vibration is independent of frequency. And for some materials the relative loss of energy $\varkappa$ is independent of frequency, as well as of amplitude over a wide range of frequencies and amplitudes. If the strain varies according to the equation

$$\varepsilon(t) = \varepsilon_0 \sin(\omega t + \varphi) \tag{6.54}$$

it holds for the stress[71] that

$$\sigma(t) = E\left[\varepsilon_0 \sin(\omega t + \varphi) + \frac{\varkappa}{2\pi}\varepsilon_0 \cos(\omega t + \varphi)\right] \tag{6.55}$$

where $\varepsilon(t)$ is the strain

$\sigma(t)$ is the stress corresponding to the strain

ε_0 is the strain amplitude

This formula holds good for harmonic vibration in which both the amplitude and the frequency remain constant. It expresses the stress-strain relation for the case of an elliptic hysteresis loop (Fig. 6.1). Its validity is extended in Reference 71 to include the case of amplitude ε_0 and frequency ω slowly varying with time, $\varepsilon_0(t)$, $\omega(t)$. In complex form eq. 6.55 may be written as

$$\sigma(t) = E\left(1 + \frac{i\varkappa}{2\pi}\right)\varepsilon(t) \tag{6.56}*$$

From relation 6.56 one can deduce the equations of motion of free and forced vibration applicable to systems with one, several or infinitely many degrees of freedom. For the sake of simplicity we shall confine our considerations to vibrating systems with a single degree of freedom whose equation of motion is

$$m\frac{d^2v^*(t)}{dt^2} + C\left(1 + \frac{\varkappa}{2\pi}i\right)v^*(t) = P^*(t) \tag{6.57}$$

where deflection $v^*(t)$ and force $P^*(t)$ are assumed to be in complex form. The equation is written on the supposition that the relation between the force (elastic plus damping) and the deflection of the system is analogous to that between stress $\sigma(t)$ and strain $\varepsilon(t)$ in eq. 6.55.

Eq. 6.57 has a unique solution provided the exciting force is harmonic and the forced vibration of the system is steady. Once the vibration stops being pure periodic, difficulties with the uniqueness of the solution set in. The difficulties are already met with in the solution of the homogeneous equation

$$m\frac{d^2v^*(t)}{dt^2} + C\left(1 + i\frac{\varkappa}{2\pi}\right)v^*(t) = 0 \tag{6.58}$$

whose general solution (with eq. 6.61) is

$$\begin{aligned}v^*(t) &= (C_1 + iC_2)\,e^{-(\varkappa/4\pi)\omega_0 t}(\cos \omega_0 t + i \sin \omega_0 t) \\ &+ (C_3 + iC_4)\,e^{+(\varkappa/4\pi)\omega_0 t}(\cos \omega_0 t - i \sin \omega_0 t)\end{aligned} \tag{6.59}$$

* Analogous equations were used by Bleich[6] in which he refers to an earlier paper by Teodorsen[73].

where C_1, C_2, C_3, C_4 are real constants. The right-hand side of the equation consists of two parts, the second of which contains a divergent term $e^{+(\varkappa/4\pi)\omega_0 t}$. It is, of course, impossible to determine four integration constants from two boundary conditions. To surmount this difficulty, the expression with the divergent term must be left out[71]. We do not approve of this modification of the equation because in more complicated cases it might give rise to ambiguities. Moreover, as we shall see later on, the divergent part, too, is apt to have a physical meaning.

Difficulties are even encountered when looking for the particular solution of the homogeneous equation 6.57. If we proceed in the customary way the expression with the term $e^{+(\varkappa/4\pi)\omega_0 t}$ cannot be neglected because otherwise the respective integral would not be the particular solution. Thus, for example, if $P^*(t) = P\,e^{i\omega t}$, eq. 6.57 may be written in the form

$$\frac{d^2 v^*(t)}{dt^2} + \omega^{*2}\, v^*(t) = \frac{P}{m}\,e^{i\omega t} \tag{6.60}$$

where at light damping

$$\omega^* = \left[\frac{C}{m}\left(1 + i\,\frac{\varkappa}{2\pi}\right)\right]^{1/2} \approx \omega_0\left(1 + i\,\frac{\varkappa}{4\pi}\right) \tag{6.61}$$

The solution of eq. 6.60 gives

$$v^*(t) = \frac{P}{2im\omega^*}\int_0^t e^{i\omega\tau}\!\left(e^{i\omega^*(t-\tau)} - e^{-i\omega^*(t-\tau)}\right)d\tau$$

$$= \frac{P}{m(\omega^{*2} - \omega^2)}\left(e^{i\omega t} - \cos\omega^* t - i\,\frac{\omega}{\omega^*}\sin\omega t\right)$$

$$= (A - iB)\left\{\cos\omega t - \frac{1}{2}\cos\omega_0 t\left[\exp\left(-\omega_0\,\frac{\varkappa}{4\pi}\,t\right) + \exp\left(\omega_0\,\frac{\varkappa}{4\pi}\,t\right)\right.\right.$$

$$\left. + \frac{\omega}{\omega_0}\exp\left(-\omega_0\,\frac{\varkappa}{4\pi}\,t\right) - \frac{\omega}{\omega_0}\exp\left(\omega_0\,\frac{\varkappa}{4\pi}\,t\right)\right]$$

$$\left. - \frac{\omega}{2\omega_0}\,\frac{\varkappa}{4\pi}\sin\omega_0 t\left[\exp\left(-\omega_0\,\frac{\varkappa}{4\pi}\,t\right) + \exp\left(\omega_0\,\frac{\varkappa}{4\pi}\,t\right)\right]\right\}$$

$$+ (B + iA)\left\{\sin\omega t - \frac{1}{2}\sin\omega_0 t\left[\exp\left(-\omega_0\,\frac{\varkappa}{4\pi}\,t\right) - \exp\left(\omega_0\,\frac{\varkappa}{4\pi}\,t\right)\right.\right.$$

$$\left. + \frac{\omega}{\omega_0}\exp\left(-\omega_0\,\frac{\varkappa}{4\pi}\,t\right) + \frac{\omega}{\omega_0}\exp\left(\omega_0\,\frac{\varkappa}{4\pi}\,t\right)\right] + \frac{\omega}{2\omega_0}\,\frac{\varkappa}{4\pi}\cos\omega_0 t$$

$$\left. \times\left[\exp\left(-\omega_0\,\frac{\varkappa}{4\pi}\,t\right) - \exp\left(\omega_0\,\frac{\varkappa}{4\pi}\,t\right)\right]\right\} \tag{6.62}$$

where

$$A = \frac{P(\omega_0^2 - \omega^2)}{m\left[(\omega_0^2 - \omega^2)^2 + \omega_0^4 \frac{\varkappa^2}{4\pi^2}\right]} \qquad (6.63)$$

$$B = \frac{P\omega_0^2 \frac{\varkappa}{2\pi}}{m\left[(\omega_0^2 - \omega^2)^2 + \omega_0^4 \frac{\varkappa^2}{4\pi^2}\right]} \qquad (6.64)$$

Eq. 6.62 contains the divergent expressions with the term $e^{(\varkappa/4\pi)\omega_0 t}$. It can readily by proved by considering the limit case with $\omega = 0$ that the equation does not correspond to damped vibration. If the divergent terms are not left out, the solution is the particular solution of eq. 6.57 but without the correct physical meaning. If they are left out, the resultant equation is not the particular integral of eq. 6.57. Other proofs concerning the insufficient clarity of the hypotheses are offered in Reference 1.

6.2.2 Linear theory of internal damping for steady-state forced vibration

In the preceding section we have pointed out some of the ambiguities of eq. 6.57. In this section we intend to formulate the equation of damped vibration in a more general but wholly unique form.

As the procedure based on the hysteresis loop is best substantiated for steady periodic vibration we shall take up this question first and discuss it at length. Let us for an instant go back to formula 6.55. The formula states the relation between strain and stress: if we know the time variation of strain, we also know the time variation of stress. In its simple form Hooke's law states that $\sigma(t) = E\,\varepsilon(t)$. Formula 6.55 is analogous to Hooke's law. So is its complex form, eq. 6.56, which associates stress to strain by means of the complex factor $C[1 + i(\varkappa/2\pi)]$. We have demonstrated, however, that this method leads to ambiguities and is not unique.

In the discussion that follows we shall associate the two quantities in a different, entirely definite way so that to every, i.e. not only harmonic, but general, strain variation there will uniquely correspond the respective stress variation. At the beginning we shall restrict ourselves to periodic motion but as we progress we shall extend the theory to non-periodic phenomena. The term 'periodic' is used here in the exact sense of the word (free damped vibration is a non-periodic motion). We shall employ the principle of superposition which in application is of greater significance than the shape of the hysteresis loop.

Let us introduce operator **A** which we shall use to associate any arbitrary periodic continuous function $f(t)$ of period T, with another periodic square integrable function $\mathbf{A} f(t)$ also of period T. Let operator $\mathbf{A}$ have the following properties:

(1) Linearity, i.e. it holds that

$$\mathbf{A}[C_1 f_1(t) + C_2 f_2(t)] = C_1 \mathbf{A} f_1(t) + C_2 \mathbf{A} f_2(t) \tag{6.65}$$

for any constants C_1 and C_2.

(2) Continuity. This means that if $\left|f(t)\right| < k$ then

$$\left\{ \frac{1}{T} \int_0^T [\mathbf{A} f(t)]^2 \, dt \right\}^{1/2} < \alpha k \tag{6.66}$$

where α is independent of both k and $f(t)$.

(3) If the relation

$$f(t) - \psi \mathbf{A} f(t) = 0 \tag{6.67}$$

holds for every t and $\psi > 0$, then $f(t) = 0$, where ψ is the damping coefficient.

(4) If $df(t)/dt \neq 0$ and is a square integrable function, then

$$\int_0^T \frac{df(t)}{dt} \mathbf{A} f(t) \, dt < 0 \tag{6.68}$$

Operator $\mathbf{A}$ which satisfies the above requirements is thus specified more definitely to suit the character of internal damping being considered.

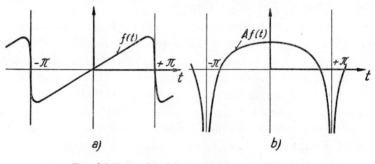

Fig. 6.2 Example of functions $f(t)$ and $\mathbf{A} f(t)$

Let stress $\sigma(t)$ depend on strain $\varepsilon(t)$ in accordance with the relation

$$\sigma(t) = E[\varepsilon(t) - \psi \mathbf{A} \, \varepsilon(t)] . \tag{6.69}$$

In place of eq. 6.69 we may also write

$$\sigma(t) = E[\mathbf{I} - \psi \mathbf{A}] \, \varepsilon(t) \tag{6.70}$$

where $\mathbf{I}$ denotes the identical operator. All the terms of the equations are real.

The properties specified above have the following meaning: The first property — linearity — enables us to use the principle of superposition. Function $f(t)$ may be expanded in a series, and a corresponding term of the series expressing function $\mathbf{A}f(t)$ may be found separately for every one of its terms. What the second property — continuity — means is that to a small change of $f(t)$ belongs a small change of $\mathbf{A}f(t)$. The continuity of operator $\mathbf{A}$, however, does not mean that function $\mathbf{A}f(t)$ has to be continuous in the ordinary sense of the word. We see in Figs 6.2a and b showing examples of associated functions that to a continuous function $f(t)$ can belong an unbounded function $\mathbf{A}f(t)$. The third property is necessary to make the solution unique. We wish the operator $\mathbf{I} - \psi\mathbf{A}$ to express a certain analogy with Hooke's law, i.e. we wish to uniquely associate a stress variation with a periodic strain variation. The third property of the operator expresses the fact that two different strains $\varepsilon(t)$ cannot produce the same stress. Of particular importance is the fourth property which tells us that physically speaking energy is lost by damping at any deformation. This energy lost during one cycle is exactly

$$\int_0^T \frac{df(t)}{dt} \, \mathbf{A}f(t) \, dt$$

If we wish to express the action of internal damping in detail we must first accept one of the hypotheses about the dependence of internal damping on strain, e.g. Sorokin's hypothesis (eq. 6.55) or that of Kelvin-Voigt. Like Sorokin we shall first assume that the magnitude of the relative loss of energy, $\varkappa$, is independent of both the strain frequency and amplitude, and denote by symbol $\mathbf{B}$ operator $\mathbf{A}$ satisfying this condition.

$\mathbf{B}$ is an operator which associates with linear trigonometric polynomials of period T, other trigonometric polynomials of the same period in such a way that

$$\mathbf{B} \cos \frac{2\pi kt}{T} = \sin \frac{2\pi kt}{T} \quad \text{for} \quad k = 0, 1, 2, \ldots$$

$$\mathbf{B} \sin \frac{2\pi kt}{T} = -\cos \frac{2\pi kt}{T} \quad \text{for} \quad k = 1, 2, 3, \ldots$$

$$(6.71)$$

Since every continuous periodic function $f(t)$ of period T may be expanded in a Fourier series, it is also possible to associate with such a function by means of operator $\mathbf{B}$ another periodic function of the same period, with operator $\mathbf{B}$ satisfying all the requirements imposed on operator $\mathbf{A}$ as well as meeting the assumptions of Sorokin's hypothesis. We shall prove it presently.

(1) Since it follows from its definition it is not necessary to prove that operator $\mathbf{B}$ is linear.

(2) Operator **B** is continuous. It is according to the assumption

$$f(t) = \frac{a_0}{2} + \sum_{k=1}^{h} a_k \cos \frac{2\pi kt}{T} + \sum_{k=1}^{h} b_k \sin \frac{2\pi kt}{T}$$

that

$$\frac{1}{T} \int_0^T f^2(t)\, dt = \frac{a_0^4}{4} + \frac{1}{2}\left(\sum_{k=1}^{h} a_k^2 + \sum_{k=1}^{h} b_k^2 \right) \le |f(t)|^2 \text{ max}$$

Further

$$\mathbf{B}\, f(t) = -\sum_{k=1}^{h} b_k \cos \frac{2\pi kt}{T} + \sum_{k=1}^{h} a_k \sin \frac{2\pi kt}{T}$$

and therefore

$$\frac{1}{T} \int_0^T [\mathbf{B}\, f(t)]^2\, dt = \frac{1}{2}\left(\sum_{k=1}^{h} a_k^2 + \sum_{k=1}^{h} b_k^2 \right) \le \frac{1}{T}\int_0^T f^2(t)\, dt \le f(t)^2 \text{ max}$$

(3) The third property is satisfied too. Assume that $f(t) = \psi \mathbf{B}\, f(t)$ (in this case $\psi = \varkappa/2\pi$). Since $f(t)$ is a continuous periodic function, it is also square integrable. It is given by

$$f(t) = \frac{a_0}{2} + \sum_{k=1}^{\infty} a_k \cos \frac{2\pi kt}{T} + \sum_{k=1}^{\infty} b_k \sin \frac{2\pi kt}{T}$$

where

$$a_k = \frac{2}{T} \int_{-T/2}^{T/2} f(t) \cos \frac{\pi kt}{T} \; ; \quad b_k = \frac{2}{T} \int_{-T/2}^{T/2} f(t) \sin \frac{\pi kt}{T}$$

Since

$$\mathbf{B}\, f(t) = \sum_{k=1}^{\infty} a_k \sin \frac{2\pi kt}{T} - \sum_{k=1}^{\infty} b_k \cos \frac{2\pi kt}{T}$$

$a_0 = 0$, $a_k = -b_k \psi$, $b_k = a_k \psi$. Therefore $a_k = b_k = 0$ and $f(t) = 0$.

(4) And finally, to prove that the fourth property of operator **A** is also satisfied by operator **B** we write similarly as in (3)

$$f(t) = \frac{a_0}{2} + \sum_{k=1}^{\infty} a_k \cos \frac{2\pi kt}{T} + \sum_{k=1}^{\infty} b_k \sin \frac{2\pi kt}{T}$$

Since it is assumed that $df(t)/dt$ is square integrable,

$$\frac{df(t)}{dt} = \left(-\sum_{k=1}^{\infty} k a_k \sin \frac{2\pi kt}{T} + \sum_{k=1}^{\infty} k b_k \cos \frac{2\pi kt}{T} \right) \frac{2\pi}{T}$$

$$\sum_{k=1}^{\infty}(kb_k)^2 + \sum_{k=1}^{\infty}(ka_k)^2 < \infty$$

Therefore

$$\int_0^T \frac{df(t)}{dt} \mathbf{B} f(t)\, dt = -\frac{2\pi}{T}\frac{T}{2}\left[\sum_{k=1}^{\infty}kb_k^2 + \sum_{k=1}^{\infty}ka_k^2\right] \leqq 0$$

However, the equality cannot hold because $df(t)/dt \neq 0$.

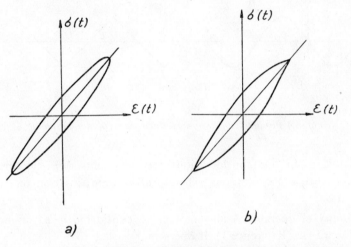

Fig. 6.3 Various shapes of hysteresis loops

We have thus proved that operator **B** satisfies all four conditions and is actually the operator of internal damping **A**. However, operator **B** is not the only one to satisfy the conditions. For example, a linear operator **A**$_1$ which satisfies the following equations:

$$\mathbf{A}_1 \cos kt = \sin\frac{2\pi kt}{T} + \tfrac{1}{3}\sin\frac{6\pi kt}{T}$$

$$\mathbf{A}_1 \sin kt = -\left(\cos\frac{2\pi kt}{T} + \tfrac{1}{3}\cos\frac{6\pi kt}{T}\right)$$

$$(6.72)$$

is likewise an operator of internal damping. In proving this we proceed in the same manner as with operator **B**.

For every operator definitely specified we can find the respective hysteresis loop. For example, the hysteresis loop appertaining to operator **B** is shown in Fig. 6.3a, and that for operator **A**$_1$ in Fig. 6.3b.

It should be noted at this point that we always use constant (i.e. time invariable) amplitudes in the expansion of functions.

If function $f(t)$ has the property that constants $1 \geqq \alpha > 0$, $C > 0$ will make $|f(t_1) - f(t_2)| \leq C|t_1 - t_2|^\alpha$, then not only $f(t)$ but $\mathbf{B} f(t)$, too, will be a continuous function. In the case where $df(t)/dt$ is a continuous function (Fig. 6.4a) this condition

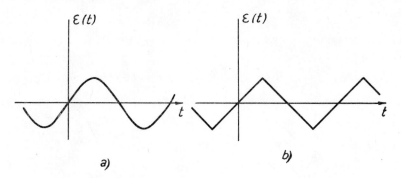

a)   b)

Fig. 6.4 Periodic vibration. (a) Harmonic, (b) non-harmonic

is always satisfied. In our considerations $f(t)$ represents $\varepsilon(t)$, the time variation of strain, and is always continuous. Therefore the time variation of stress will be continuous, too.

The continuity of operator $\mathbf{A}$ may in fact be considered in a yet more general sense of the word (see Reference 1). If operator $\mathbf{V}$ is such that

$$\mathbf{V} \cos \frac{2\pi kt}{T} = \frac{2\pi k}{T} \sin \frac{2\pi kt}{T}$$

$$\mathbf{V} \sin \frac{2\pi kt}{T} = - \frac{2\pi k}{T} \cos \frac{2\pi kt}{T}$$

$$(6.73)$$

it, too, will be an operator of internal damping.

This operator has the property that

$$\mathbf{V} f(t) = - \frac{df(t)}{dt} \tag{6.74}$$

If the solution of damped steady-state forced vibration of a system with one degree of freedom is carried out with the aid of operators of internal damping, the equation of motion is of the form

$$m \frac{d^2 v(t)}{dt^2} + C(\mathbf{I} - \psi \mathbf{A}) v(t) = P(t) \tag{6.75}$$

where $P(t)$ stands for the periodic external force with period T, and is square integrable.

It then holds that:

(1) Every continuous periodic function $v(t)$ with period T, which satisfies eq. 6.75 has the derivative $dv(t)/dt$, which is also a continuous periodic function with period T. In fact, $dv(t)/dt$ is totally continuous. This statement is obvious, for

$$\frac{d^2v(t)}{dt^2} = \frac{1}{m}[P(t) - Cl\, v(t) + C\psi A\, v(t)] \tag{6.76}$$

and the right-hand side of the equation is square integrable.

(2) There exists only one solution to eq. 6.75. Since the equation is linear, all we have to prove is that the homogeneous equation, i.e. one with $P(t) = 0$, has only a trivial solution. Let $v_0(t)$ be a periodic continuous function satisfying the equation

$$m\frac{d^2v_0(t)}{dt^2} + C[I - \psi A]\, v_0(t) = 0 \tag{6.77}$$

Then, in accord with the preceding paragraph, $[dv_0(t)]/dt$ is a continuous periodic function, and $[d^2v_0(t)]/dt^2$ a square integrable, periodic function. Multiplying eq. 6.77 by function $[dv_0(t)]/dt$ and integrating between the limits 0 to T gives

$$m\int_0^T \frac{d^2v_0(t)}{dt^2}\frac{dv_0(t)}{dt}\, dt + C\int_0^T \frac{dv_0(t)}{dt} v_0(t)\, dt = C\psi \int_0^T \frac{dv_0(t)}{dt} A\, v_0(t)\, dt$$

Since $[dv_0(t)]/dt$ and $v_0(t)$ are periodic

$$\int_0^T \frac{d^2v_0(t)}{dt^2}\frac{dv_0(t)}{dt}\, dt = \frac{1}{2}\left[\left[\frac{dv_0(t)}{dt}\right]^2\right]\Big|_0^T = 0$$

$$\int_0^T v_0(t)\frac{dv_0(t)}{dt}\, dt = \tfrac{1}{2}|v_0^2(t)|_0^T = 0$$

Therefore $\int_0^T A\, v_0(t)\, [dv_0(t)]/dt\, .\, dt = 0$. In accordance with the fourth property of operator A, $[dv_0(t)]/dt = 0$. From eq. 6.77

$$(I - \psi A)\, v_0(t) = 0\,, \quad \text{i.e.} \quad v_0(t) = \psi A\, v_0(t)$$

and consequently, in accordance with the third property of the operator of internal damping, $v_0(t) = 0$.

This establishes the truth of our statement. It is further possible to prove the existence of the solution. For information on this question the reader is referred to Reference 1. There exists in fact just one solution of eq. 6.75. The equation has a solution even in the case where the damping coefficient ψ is not small.

6.2.3 Solution of the linear problem of forced vibration

We have just demonstrated the existence of a single solution. Now we shall find this solution. The solution of the equation

$$m \frac{d^2 v(t)}{dt^2} + C(I - \psi A) v(t) = P(t) \tag{6.78}$$

is the square integrable function $v(t)$ with period T. Expand $v(t)$ in the series

$$v(t) = \frac{a_0}{2} + \sum_{k=1}^{\infty} a_k \cos \frac{2\pi kt}{T} + \sum_{k=1}^{\infty} b_k \sin \frac{2\pi kt}{T} \tag{6.79}$$

where

$$a_k = \frac{2}{T} \int_{-T/2}^{T/2} v(t) \cos \frac{2\pi kt}{T} \, dt, \quad k = 0, 1, 2, \ldots$$

$$b_k = \frac{2}{T} \int_{-T/2}^{T/2} v(t) \sin \frac{2\pi kt}{T} \, dt, \quad k = 1, 2, 3, \ldots$$

$$\tag{6.80}$$

In a like manner expand $P(t)$ in the series

$$\frac{P(t)}{m} = \frac{\alpha_0}{2} + \sum_{k=1}^{\infty} \alpha_k \cos \frac{2\pi kt}{T} + \sum_{k=1}^{\infty} \beta_k \sin \frac{2\pi kt}{T} \tag{6.81}$$

where

$$\alpha_k = \frac{2}{mT} \int_{-T/2}^{T/2} P(t) \cos \frac{2\pi kt}{T} \, dt, \quad k = 0, 1, 2, \ldots$$

$$\beta_k = \frac{2}{mT} \int_{-T/2}^{T/2} P(t) \sin \frac{2\pi kt}{T} \, dt, \quad k = 1, 2, 3, \ldots$$

$$\tag{6.82}$$

Substitute eqs 6.79 and 6.81 in eq. 6.78, multiply it by $(2/T) \cos \left[(2\pi kt)/T \right]$ (or $(2/T) \cdot \sin \left[(2\pi kt)/T \right]$) and integrate between the limits $-T/2, +T/2$. The resultant system of equations is

$$-a_k \left(\frac{2\pi k}{T} \right)^2 + \frac{C}{m} a_k - \frac{C}{m} \psi \sum_{i=0}^{\infty} K_{i,k}^{(1)} a_i - \frac{C}{m} \psi \sum_{i=0}^{\infty} K_{i,k}^{(3)} b_i = \alpha_k$$

$$-b_k \left(\frac{2\pi k}{T} \right)^2 + \frac{C}{m} b_k - \frac{C}{m} \psi \sum_{i=0}^{\infty} K_{i,k}^{(2)} a_i - \frac{C}{m} \psi \sum_{i=0}^{\infty} K_{i,k}^{(4)} b_i = \beta_k$$

$$\tag{6.83}$$

where

$$K_{i,k}^{(1)} = \frac{2}{T} \int_{-T/2}^{T/2} \cos \frac{2\pi kt}{T} A \cos \frac{2\pi it}{T} \, dt \, , \quad \begin{matrix} i = 0, 1, 2, \ldots \\ k = 0, 1, 2, \ldots \end{matrix}$$

$$K_{i,k}^{(2)} = \frac{2}{T} \int_{-T/2}^{T/2} \sin \frac{2\pi kt}{T} A \cos \frac{2\pi it}{T} \, dt \, , \quad \begin{matrix} i = 0, 1, 2, \ldots \\ k = 1, 2, 3, \ldots \end{matrix}$$

$$K_{i,k}^{(3)} = \frac{2}{T} \int_{-T/2}^{T/2} \cos \frac{2\pi kt}{T} A \sin \frac{2\pi it}{T} \, dt \, , \quad \begin{matrix} i = 1, 2, 3, \ldots \\ k = 0, 1, 2, \ldots \end{matrix}$$

$$K_{i,k}^{(4)} = \frac{2}{T} \int_{-T/2}^{T/2} \sin \frac{2\pi kt}{T} A \sin \frac{2\pi it}{T} \, dt \, , \quad \begin{matrix} i = 1, 2, 3, \ldots \\ k = 1, 2, 3, \ldots \end{matrix} \tag{6.84}$$

Expressions 6.83 represent a system of an infinite number of linear equations of coefficients a_k, b_k, which has but a single solution such that

$$\sum_{k=1}^{\infty} (a_k k)^2 + \sum_{k=1}^{\infty} (b_k k)^2 < \infty$$

In case $A = B$, system 6.83 simplifies considerably because

$$K_{i,k}^{(1)} = 0 = K_{i,k}^{(4)}$$

$$K_{i,k}^{(2)} = \begin{cases} 1 & \text{if } i = k \\ 0 & \text{if } i \neq k \end{cases}$$

$$K_{i,k}^{(3)} = \begin{cases} -1 & \text{if } i = k \\ 0 & \text{if } i \neq k \end{cases}$$

and as a result, system 6.83 reduces to a system with two unknowns

$$-a_k \left(\frac{2\pi k}{T}\right)^2 + \frac{C}{m} a_k + \frac{C}{m} \psi b_k = \alpha_k \tag{6.85}$$

$$-b_k \left(\frac{2\pi k}{T}\right)^2 + \frac{C}{m} b_k - \frac{C}{m} \psi a_k = \beta_k \tag{6.86}$$

$$\frac{C}{m} a_0 = \alpha_0 \tag{6.87}$$

By determining pairs of a_k, b_k from every two equations we find the sought-for solution. It is

$$a_k = \frac{\alpha_k(\omega_0^2 - k^2\omega^2) - \beta_k \omega_0^2 \psi}{(\omega_0^2 - k^2\omega^2)^2 + \omega_0^4 \psi^2}$$

$$b_k = \frac{\beta_k(\omega_0^2 - k^2\omega^2) + \alpha_k \omega_0^2 \psi}{(\omega_0^2 - k^2\omega^2)^2 + \omega_0^4 \psi^2} \tag{6.88}$$

where $\omega_0 = (C/m)^{1/2}$ is the undamped natural circular frequency of the system and $2\pi/T = \omega$.

The procedure is applicable to other cases, too. For example, if the operator of internal damping is $\mathbf{A}_1$ according to eq. 6.72, the problem of solving system 6.83 again reduces to that of a successive solution of systems with two unknowns.

6.2.4   The linear theory of internal damping for non-periodic vibration

6.2.4.1   *Periodic phenomena with period tending to infinity*

If a system vibrates with non-periodic motion, the solution is much more difficult. The difficulties stem from the circumstance that the theory being applied starts out from the hysteresis loop concept. When dealing with periodic vibrations the periodic forces producing them were assumed to be active for a time long enough to make the vibration perfectly steady. It is only in such a vibration that the hysteresis loop is a regular one. At the beginning of the force action, however, the vibration is not steady: the material adjusts itself to the periodic phenomenon but gradually, we may even say, runs slowly up to the hysteresis loop. This is the reason why non-periodic vibration of short duration cannot be solved without full account being taken of the loading history of the material prior to the commencement of our observations.

To overcome these difficulties we shall regard the non-periodic vibration as a periodic phenomenon with period tending to infinity. In a like manner we shall consider the external load producing this vibration to be a periodic load with period tending to infinity.

The free damped vibration of a system is a non-periodic phenomenon, too. So far we have solved it by introducing some definite initial conditions, i.e. the initial deflection and velocity, and determining from them the integration constants of the general integral of the homogeneous differential equation of free vibration of the system.

This procedure was fully justified whenever the damped vibration was computed using Voigt's hypothesis according to which the damping force is proportional to the velocity. If, however, the damping theory we plan to use is based on the concept of the hysteresis loop, for example, independent of frequency, the initial conditions, i.e. the deflection and velocity at the beginning of the process, will clearly be no longer sufficient for a unique determination of the subsequent course of the vibration. Consider the following two cases: in one the system has reached the initial state by a gradual application of force, in the other by a single impulse. The pattern of subsequent vibration will in each case be different. When it comes to attenuated vibrations, the difference will be the smaller the longer is the time since the beginning of observation. What this means in practice is that satisfactory results are more likely to be achieved by not concentrating on the phenomena that follow close behind the initial impulse.

Wholly definite results will be obtained if the non-periodic phenomenon in question is considered to be a periodic phenomenon with period tending to infinity. On the strength of this supposition free damped vibration may then be solved as though

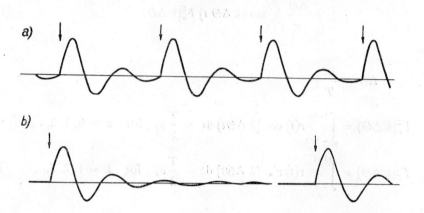

Fig. 6.5 (a) Damped vibration produced by periodically repeated impulses. (b) Vibration with growing period

the system were subjected to impulses at intervals growing beyond all limits. Imagine a system with one degree of freedom subjected to impulses acting at finite time intervals. The resulting vibration will be steady, like the one shown in Fig. 6.5a. (For reasons of simplicity the deflections were determined on the basis of Voigt's hypothesis but the results would be very much the same if we used Sorokin's hypothesis.) Were the time intervals between subsequent impulses steadily increased,

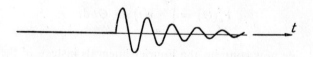

Fig. 6.6 Free damped vibration produced by single impulse

the vibration obtained in the limit case would be one produced by an impulse in time $t = 0$, that is to say free damped vibration produced by a single impulse as shown in Fig. 6.6.

6.2.4.2 Solution of the equation of motion

Rewrite expression 6.79 in the form

$$v(t) = \frac{\Delta\Theta}{2\pi} F_v^C(0) + \frac{\Delta\Theta}{\pi} \sum_{k=1}^{\infty} \cos\left(k\,\Delta\Theta t\right) F_v^C(k\,\Delta\Theta)$$

$$+ \frac{\Delta\Theta}{\pi} \sum_{k=1}^{\infty} \sin\left(k\,\Delta\Theta\,t\right) F_v^S(k\,\Delta\Theta) \tag{6.89}$$

where

$$\Delta\Theta = \frac{2\pi}{T} \tag{6.90}$$

$$F_v^C(k\,\Delta\Theta) = \int_{-T/2}^{T/2} v(t) \cos\left(k\,\Delta\Theta t\right) \mathrm{d}t = \frac{T}{2} a_k \quad \text{for} \quad k = 0, 1, 2, \ldots \tag{6.91}$$

$$F_v^S(k\,\Delta\Theta) = \int_{-T/2}^{T/2} v(t) \sin\left(k\,\Delta\Theta t\right) \mathrm{d}t = \frac{T}{2} b_k \quad \text{for} \quad k = 1, 2, 3, \ldots \tag{6.92}$$

As

$$T \to \infty, \quad \Delta\Theta \to \mathrm{d}\Theta, \quad k\,\Delta\Theta = \Theta = \frac{2\pi k}{T}$$

and eq. 6.89 changes to

$$v(t) = \frac{1}{\pi} \int_0^{\infty} \cos\Theta t\, F_v^C(\Theta)\, \mathrm{d}\Theta + \frac{1}{\pi} \int_0^{\infty} \sin\Theta t\, F_v^S(\Theta)\, \mathrm{d}\Theta \tag{6.93}$$

where

$$F_v^C(\Theta) = \int_{-\infty}^{\infty} v(t) \cos\Theta t\, \mathrm{d}t \tag{6.94}$$

$$F_v^S(\Theta) = \int_{-\infty}^{\infty} v(t) \sin\Theta t\, \mathrm{d}t \tag{6.95}$$

The right-hand side now contains the Fourier integrals instead of the Fourier series. Operator $\mathbf{B}$ then associates with function $v(t)$ the function $\mathbf{B}\,(v)t$ as follows:

$$\mathbf{B}\,v(t) = \frac{1}{\pi} \int_0^{\infty} \sin\Theta t\, F_v^C(\Theta)\, \mathrm{d}\Theta - \frac{1}{\pi} \int_0^{\infty} \cos\Theta t\, F_v^S(\Theta)\, \mathrm{d}\Theta \tag{6.96}$$

The external force is transformed in a similar manner

$$\frac{P(t)}{m} = \frac{1}{\pi} \int_0^{\infty} \cos\Theta t\, F_P^C(\Theta)\, \mathrm{d}\Theta + \frac{1}{\pi} \int_0^{\infty} \sin\Theta t\, F_P^S(\Theta)\, \mathrm{d}\Theta \tag{6.97}$$

$$F_P^C(\Theta) = \frac{1}{m} \int_{-\infty}^{\infty} P(t) \cos \Theta t \, dt \qquad (6.98)$$

$$F_P^S(\Theta) = \frac{1}{m} \int_{-\infty}^{\infty} P(t) \sin \Theta t \, dt \qquad (6.99)$$

The equation of forced vibration (periodic vibration with period tending to infinity)

$$m \frac{d^2 v(t)}{dt^2} + C\left(\mathbf{I} - \frac{\psi}{2\pi}\mathbf{B}\right) v(t) = P(t) \qquad (6.100)$$

now has but a single solution whose existence is proved in Reference 1. This solution follows directly from formulae 6.88 as their limit, after both sides have been multiplied by $T/2$, and T is increased beyond all limits. Because of eqs 6.91, 6.92, 6.98, 6.99, 6.100 and

$$\frac{2\pi k}{T} = k\omega = \Theta$$

$$F_v^C(\Theta) = \frac{F_P^C(\Theta)(\omega_0^2 - \Theta^2) - F_P^S(\Theta)\,\omega_0^2\psi}{(\omega_0^2 - \Theta^2)^2 + \omega_0^4\psi^2} \qquad (6.101)$$

$$F_v^S(\Theta) = \frac{F_P^S(\Theta)(\omega_0^2 - \Theta^2) + F_P^C(\Theta)\,\omega_0^2\psi}{(\omega_0^2 - \Theta^2)^2 + \omega_0^4\psi^2} \qquad (6.102)$$

Since integrals $F_P^C(\Theta)$, $F_P^S(\Theta)$ exist, integrals $F_v^C(\Theta)$, $F_v^S(\Theta)$ exist, also. The time variation of the deflection, $v(t)$, is described by formula 6.93.

The above deductions though illustrative enough are not wholly exact. In an exact analysis all limit transitions are a source of difficulties. In order to avoid them an exact mathematical study must adopt a procedure which — while adhering to the general idea — will work with a different apparatus. The results achieved with its aid will, of course, be the same except that they will rest on exact assumptions. The somewhat less exact procedure chosen for the purposes of our discussion has aimed at nothing more than supplying basic information about the matter. For further details the reader is referred to Reference 1. If functions $P(t)$ and $v(t)$ are assumed to be square integrable, then all the transformations and operations carried out with them have a meaning if the transformations are taken in the sense of the Fourier-Plancherer theory.

6.2.4.3 Some physical problems associated with the problem of forced vibration

Some of the questions connected with the fact that non-steady-state vibration is considered steady-state vibration with infinite period are considered at the beginning of this section. What comes next is just a supplementary note:

If we wish to solve the problem of free vibration for given initial conditions, the best way to go about it is to choose two different courses of the exciting force, $P_1(t)$ and $P_2(t)$, with both $P_1(t)$ and $P_2(t)$ zero at time $t > 0$. The initial conditions may be satisfied by a linear combination of the solutions appertaining respectively to the exciting forces $P_1(t)$ and $P_2(t)$. It goes without saying that different functions $P_1(t)$ and $P_2(t)$ will lead to different solutions. This is as it should be because the running-up to the hysteresis loop depends on the state of the material, i.e. on its previous straining history.

6.2.5 The generalised linear theory of damping

6.2.5.1 The linear theory of damping

In the preceding paragraphs dealing with the problem of linear damping we have demonstrated that this group includes both the Kelvin-Voigt hypothesis of damping being proportional to the frequency, and Sorokin's hypothesis of damping being independent of the frequency. Thus for $\varepsilon(t) = \sin \omega t$ it holds

(a) according to Voigt that

$$\sigma(t) = E(\sin \omega t + \psi \omega \cos \omega t) \tag{6.103}$$

(b) according to Sorokin that

$$\sigma(t) = E(\sin \omega t + \psi \cos \omega t) \tag{6.104}$$

It is equally possible to frame linear hypotheses such that they will make

(c)
$$\sigma(t) = E\left(\sin \omega t + \frac{\psi}{\omega} \cos \omega t\right) \tag{6.105}$$

etc. All the hypotheses including the last-named ones are in fact contained in the linear theory and can be expressed by means of the linear damping operator (see Section 6.2.2). This is true even of the hypothesis built on the linear theory of plastic flow[2]. Using the relaxation curve

$$r(t, \tau) = \tfrac{1}{2}[1 + e^{-\alpha(t-\tau)}] \tag{6.106}$$

we find that in this case

(d)
$$\sigma(t) = E\left[\frac{1}{2}\left(1 + \frac{1}{1 + \alpha^2/\omega}\right) \sin \omega t + \frac{1}{2} \frac{\alpha \omega}{\omega + \alpha^2} \cos \omega t\right] \tag{6.107}$$

The above expression (damping approximately proportional to the reciprocal of the frequency) resembles the one in (c).

It is quite obvious that any of the linear theories outlined can combine one with another and always give the linear theory discussed in general terms in Section 6.2.3. On the strength of this statement we may also include here the hypothesis that

(e) $$\sigma(t) = E\left(\sin \omega t + \psi_1 \omega \cos \omega t + \psi_2 \cos \omega t + \frac{\psi_3}{\omega} \cos \omega t\right)$$

etc.

Although the discussion has been confined to systems with one degree of freedom, the linear theory as deduced in Sections 6.2.3 and 6.2.4 can be extended to problems involving systems with many degrees of freedom. As a matter of fact, the theory is applicable even to the mathematical theory of elasticity of non-homogeneous and anisotropic media with moduli varying with time; and last but not least it can cope equally well with problems of mechanical vibrations and with some of the problems of electrical oscillations.

6.2.5.2   The linear theory of self-excited vibration

To simplify the matter somewhat assume first that the damping is proportional to the velocity (Voigt's hypothesis). The equation of motion for vibration produced by an impulse is

$$m\frac{d^2v(t)}{dt^2} + C_b\frac{dv(t)}{dt} + C\,v(t) = \delta(0) \tag{6.108}$$

where $\delta(0)$ is the Dirac function.

The impulse will cause a sudden change of velocity. The pattern of the vibration is shown in Fig. 6.6; the vibration can be considered as being the limiting case of periodic vibration with the period tending to infinity (Figs 6.5a, b).

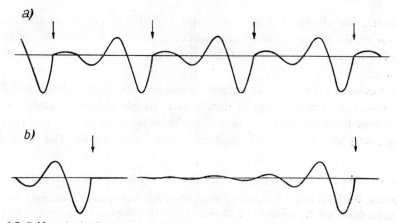

Fig. 6.7 Self-excited vibration. (a) Periodic impulses, (b) growing period of impulses

302

The vibration according to Figs 6.7a and 6.7b may be thought of as being of an opposite pattern to that shown in Figs 6.5a, b.

The pattern shown in Fig. 6.8 is for an infinitely large period and corresponds to the equation

$$m \frac{d^2v(t)}{dt^2} - C_b \frac{dv(t)}{dt} + C\,v(t) = -\delta(0)$$

where $v(t)$ is a function square integrable over the interval $(-\infty, +\infty)$ of one infinite period. Evidently, the change of sign of coefficient C_b causes the convergent pattern

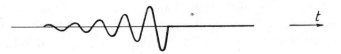

Fig. 6.8 *Self-excited vibration with period tending to infinity*

of deflection to change into a divergent one, that is to say, the equation of damped vibration changes to the equation of self-excited vibration.

As has been said earlier, the damping according to Voigt was chosen for reasons of simplicity. Assuming the damping to be independent of frequency (damping according to Sorokin) similar results would be arrived at by changing the sign of operator **B.**

Self-excited vibration is of considerable physical importance. As is well known, self-excited vibration is induced by vortices formed in the neighbourhood of an elastic body situated in a homogeneous stream of air or liquid, the magnitude and direction of the exciting force being dependent on the manner of vibration. The phenomenon is similar (but opposite) to that resulting from a damping force. It can be described by the same equations if the sign of the damping coefficient is changed from + to −.

Of course, everything that has been said here refers merely to the theoretical description of the phenomenon. The final decision about whether or not the linear theory of self-excited vibration conforms to reality, rests with the results of experimental studies* **.

In the instances mentioned so far self-excited vibration differs substantially from damped vibration. The situation in steady-state, simple harmonic vibration excited by a harmonic force is different, however, for there both the positive and the negative damping coefficients theoretically lead to one resonance curve. The result is para-

---

* Vibration induced by steady air streams are discussed in References 6, 73, 12, 14, 56 and 58.

** Self-excited vibration is not induced by wind alone but occurs quite frequently, for example the 'singing flame' (cf. Reference 61). Self-excited vibration of gates is described in Reference 69, etc.

doxical and at variance with sound judgment and seemingly also with reality. But the paradox is only imaginary. The difference between the two is that damped forced harmonic vibration is a stable phenomenon whereas the corresponding self-excited vibration is a labile, virtually non-existent case.

In concluding let us note the significance of the change of sign of operator **A** for a periodic vibration with finite period. The operator of internal damping appertaining to this case was defined in Section 6.2.2. One of its important properties was that defined in (4). The relation between strain and stress was taken to be of the form

$$\sigma(t) = E[\varepsilon(t) - \psi \mathbf{A}\, \varepsilon(t)] \tag{6.109}$$

Property 4 mentioned in connection with the negative sign expresses in eq. 6.109 the condition of damping, or in other words, the condition of the correct course of the hysteresis loop (Fig. 6.9a). If the sign in eq. 6.109 changes, the sense of the

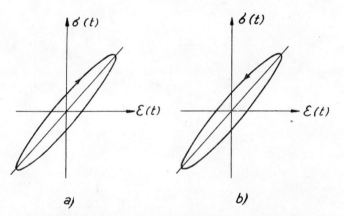

Fig. 6.9 Sense of hysteresis loop course. (a) Damped vibration, (b) self-excited vibration

loop course also changes (Fig. 6.9b), and eq. 6.109 describes self-excited vibration. The sign of the operator agrees with the sign of the phase shift of strain relative to stress.

7 MOVING LOAD

7.1 Introductory notes

The civil engineering structures in whicn the dynamic effects of a mobile load play the most important role are bridges, particularly railway bridges whose behaviour under these effects has been studied in great detail. In bridges the dynamic effects take on the most diverse forms: effects of a sudden load produced by a vehicle in rapid transit, inertia effects arising in the motion of unbalanced parts of a locomotive driving mechanism, effects of braking forces, lateral impacts, etc. In the discussion that follows we shall concentrate on the effects on a bridge of vertical periodic forces moving across the bridge with constant velocity. This problem as relating to straight girders of constant cross-section has been treated exhaustively by C. E. Inglis[28]. Although steel railway bridges are usually built of simply supported main girders, many large structures for which the dynamic effects are of particular importance have continuous beams. Continuous and arch railway bridges of the last decades, made of reinforced concrete (normal or prestressed) are nearly as sensitive to dynamic effects as steel bridges. Our studies of the dynamic effects will make use of the generalised classical methods of solution employed by Timoshenko[74] and Inglis[28], as well as of the more recent ones, especially of the slope-deflection method whose usefulness for dynamic analyses of structures has been explained in the preceding chapter of the book. In this theoretical examination we shall frequently refer to the results of extensive experimental investigations into the dynamic behaviour of bridges of the Czechoslovak Railways.

7.2 Dynamic effects on steel bridges with straight main girders of constant cross-section

7.2.1 Continuous beam subjected to a moving load

7.2.1.1 Moving force of constant magnitude

Large-span bridges — the only kind we are going to be interested in hereinafter — have a considerable mass compared to the mass of the vehicles they support. Although the natural frequency is low, the period of vibration is always but a small fraction of the time a vehicle takes to cross the bridge. If a bridge is subjected to a load moving from its end toward the centre, the deflection grows very slowly and does not differ too much from the static deflection resulting from an immobile force. Even though the bridge is set into light vibration as a load enters it, this vibration will decay even before the load reaches the centre of the bridge where the static effects are greatest. These are the reasons why we think it unnecessary to deal at length with the dynamic effects of a moving force of constant magnitude. The deflection at any point of the bridge, e.g. at the centre, plotted versus the position of the load is even at high speeds essentially identical with the deflection resulting from an infinitely slow passage. And if the weight of the load is concentrated at a single point, then according to Maxwell's theorem the deflection curve coincides with the influence line of deflection of the point at which the deflection is measured.

We shall not present here the exact solution of the problem, for the procedure leading to it is analogous with that outlined below in connection with the solution of the effects of a sinusoidally varying force.

7.2.1.2 Sinusoidally varying force with moving point of application

If the load moving along a beam is one with sinusoidally varying force, the beam is set into vibration in the vertical direction. Its motion is followed by the mass of the load thus giving rise to additional inertia forces. For the time being let

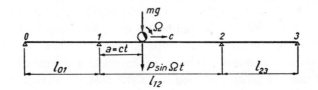

Fig. 7.1 Continuous beam subjected to a moving load — schematic representation

us neglect these forces and examine only the effects of force $P \sin \Omega t$ moving along the beam with constant velocity (see Fig. 7.1). The way of replacing the effects of concentrated forces by a substitute continuous load was shown in Section 3.5. As this method (based on expansion in natural modes) is applicable to mobile loads, too, eqs 3.66 and 3.79 can be used as they stand. Continuous main girders of large-span bridges do not have a constant cross-section. We shall assume, however, that the actual structure may be replaced by a continuous beam of constant cross-section within the spans. The equation of motion then takes on the form of eq. 6.23

$$\frac{EJ}{\mu} \frac{\partial^4 v(x, t)}{\partial x^4} + 2\omega_b \frac{\partial v(x, t)}{\partial t} + \frac{\partial^2 v(x, t)}{\partial t^2} = \frac{p(x, t)}{\mu} \tag{7.1}$$

Both the displacement and the load are expanded in the series

$$v(x, t) = \sum_j q_{(j)}(t)\, v_{(j)}(x)$$

$$\frac{p(x, t)}{\mu} = \sum_j p_{(j)}(t)\, v_{(j)}(x)$$

$$(7.2)$$

where, if the load is at point $a = ct$ (cf. 3.66)

$$p_{(j)}(t) = \frac{\sum \int p(x, t)\, v_{(j)}(x)\, dx}{\sum \int \mu\, v_{(j)}^2(x)\, dx} = \frac{P \sin \Omega t\, v_{(j)}(ct)}{\sum \int \mu\, v_{(j)}^2(x)\, dx} \qquad (7.3)$$

Expression 7.3 substituted in the right-hand side of eq. 6.27 gives

$$\ddot{q}_{(j)}(t) + 2\omega_b\, \dot{q}_{(j)}(t) + \omega_{(j)}^2\, q_{(j)}(t) = p_{(j)}(t) \qquad (7.4)$$

Since the mode of free vibration — conveniently introduced in its non-normalised form in numerical computations — is given by eq. 2.25, at the time the load moves along a span

$$p_{(j)}(t) = \frac{P}{\sum \int \mu\, v_{(j)}^2(x)\, dx} \sin \Omega t (A \cos \omega t + B \sin \omega t + C \cosh \omega t + D \sinh \omega t) \quad (7.5)$$

where A to D are constants defined by eqs 2.40, 2.47 and 2.51 for a given beam span and the j-th mode of free vibration

$$\omega = \frac{\lambda_{(j)} c}{l}$$

$$\lambda_{(j)} = l \left(\frac{\mu \omega_{(j)}^2}{EJ} \right)^{1/4}$$

and t is the time measured from the instant the load enters the span. The summation sign $\sum$ includes all the spans of the system.

Eq. 7.4 is then satisfied by the particular solution

$$\bar{q}_{(j)}(t) = K_1 \sin \Omega t \cos \omega t + K_2 \sin \Omega t \sin \omega t + K_3 \cos \Omega t \cos \omega t$$

$$+ K_4 \cos \Omega t \sin \omega t + \bar{K}_1 \sin \Omega t \cosh \omega t + \bar{K}_2 \sin \Omega t \sinh \omega t$$

$$+ \bar{K}_3 \cos \Omega t \cosh \omega t + \bar{K}_4 \cos \Omega t \sinh \omega t \qquad (7.6)$$

where K, $\overline{K}$ are constants. Substituting eqs 7.5 and 7.6 in eq. 7.4 gives

$$\left[-K_1(\omega^2 + \Omega^2) - 2K_4\omega\Omega + 2K_2\omega\omega_b - 2K_3\Omega\omega_b + K_1\omega_{(j)}^2\right] \sin \Omega t \cos \omega t$$

$$+ \left[-K_2(\omega^2 + \Omega^2) + 2K_3\omega\Omega - 2K_1\omega\omega_b - 2K_4\Omega\omega_b + K_2\omega_{(j)}^2\right] \sin \Omega t \sin \omega t$$

$$+ \left[-K_3(\omega^2 + \Omega^2) + 2K_2\omega\Omega + 2K_4\omega\omega_b + 2K_1\Omega\omega_b + K_3\omega_{(j)}^2\right] \cos \Omega t \cos \omega t$$

$$+ \left[-K_4(\omega^2 + \Omega^2) - 2K_1\omega\Omega - 2K_3\omega\omega_b + 2K_2\Omega\omega_b + K_4\omega_{(j)}^2\right] \cos \Omega t \sin \omega t$$

$$+ \left[\ \ \overline{K}_1(\omega^2 - \Omega^2) - 2\overline{K}_4\omega\Omega + 2\overline{K}_2\omega\omega_b - 2\overline{K}_3\Omega\omega_b + \overline{K}_1\omega_{(j)}^2\right] \sin \Omega t \cosh \omega t$$

$$+ \left[\ \ \overline{K}_2(\omega^2 - \Omega^2) - 2\overline{K}_3\omega\Omega + 2\overline{K}_1\omega\omega_b - 2\overline{K}_4\Omega\omega_b + \overline{K}_2\omega_{(j)}^2\right] \sin \Omega t \sinh \omega t$$

$$+ \left[\ \ \overline{K}_3(\omega^2 - \Omega^2) + 2\overline{K}_2\omega\Omega + 2\overline{K}_4\omega\omega_b + 2\overline{K}_1\Omega\omega_b + \overline{K}_3\omega_{(j)}^2\right] \cos \Omega t \cosh \omega t$$

$$+ \left[\ \ \overline{K}_4(\omega^2 - \Omega^2) + 2\overline{K}_1\omega\Omega + 2\overline{K}_3\omega\omega_b + 2\overline{K}_2\Omega\omega_b + \overline{K}_4\omega_{(j)}^2\right] \cos \Omega t \sinh \omega t$$

$$= \frac{P}{\sum \int \mu\, v_{(j)}^2(x)\, dx} \left[A \sin \Omega t \cos \omega t + B \sin \Omega t \sin \omega t + C \sin \Omega t \cosh \omega t \right.$$

$$\left. + D \sin \Omega t \sinh \omega t\right] \tag{7.7}$$

Since eq. 7.7 must be satisfied for all t, we get from it four equations defining constants K (Table 7.1) and four equations of constants $\overline{K}$ (Table 7.2). The equations in Table 7.1 can further be divided into two systems with two unknowns each (Tables 7.3 and 7.4). The other system of equations can be simplified in a like manner.

Table 7.1

K_1	K_2	K_3	K_4	
$\omega_{(j)}^2 - \Omega^2 - \omega^2$	$2\omega\omega_b$	$-2\Omega\omega_b$	$-2\omega\Omega$	$= \dfrac{PA}{\sum \int \mu v_{(j)}^2(x)\, dx}$
$-2\omega\omega_b$	$\omega_{(j)}^2 - \Omega^2 - \omega^2$	$2\omega\Omega$	$-2\Omega\omega_b$	$= \dfrac{PB}{\sum \int \mu v_{(j)}^2(x)\, dx}$
$2\Omega\omega_b$	$2\omega\Omega$	$\omega_{(j)}^2 - \Omega^2 - \omega^2$	$2\omega\omega_b$	$= 0$
$-2\omega\Omega$	$2\Omega\omega_b$	$-2\omega\omega_b$	$\omega_{(j)}^2 - \Omega^2 - \omega^2$	$= 0$

Expression 7.6 with the coefficients according to Tables 7.1 and 7.2 is only a particular solution of eq. 7.4 and fails to satisfy the initial conditions of motion. To make eq. 7.6 satisfy these conditions, we must add to it the expression of free vibration

Table 7.2

$\bar{K}_1$	$\bar{K}_2$	$\bar{K}_3$	$\bar{K}_4$	
$\omega^2_{(j)} - \Omega^2 + \omega^2$	$2\omega\omega_b$	$-2\Omega\omega_b$	$-2\omega\Omega$	$= \dfrac{PC}{\sum \int \mu v^2_{(j)}(x)\,\mathrm{d}x}$
$2\omega\omega_b$	$\omega^2_{(j)} - \Omega^2 + \omega^2$	$-2\omega\Omega$	$-2\Omega\omega_b$	$= \dfrac{PD}{\sum \int \mu v^2_{(j)}(x)\,\mathrm{d}x}$
$2\Omega\omega_b$	$2\omega\Omega$	$\omega^2_{(j)} - \Omega^2 + \omega^2$	$2\omega\omega_b$	$= 0$
$2\omega\Omega$	$2\Omega\omega_b$	$2\omega\omega_b$	$\omega^2_{(j)} - \Omega^2 + \omega^2$	$= 0$

Table 7.3

$K_1 - K_4$	$K_2 + K_3$	
$\omega^2_{(j)} - (\Omega - \omega)^2$	$-2\omega_b(\Omega - \omega)$	$= \dfrac{PA}{\sum \int \mu v^2_{(j)}(x)\,\mathrm{d}x}$
$2\omega_b(\Omega - \omega)$	$\omega^2_{(j)} - (\Omega - \omega)^2$	$= \dfrac{PB}{\sum \int \mu v^2_{(j)}(x)\,\mathrm{d}x}$

Table 7.4

$K_1 + K_4$	$K_2 - K_3$	
$\omega^2_{(j)} - (\Omega + \omega)^2$	$2\omega_b(\Omega + \omega)$	$= \dfrac{PC}{\sum \int \mu v^2_{(j)}(x)\,\mathrm{d}x}$
$-2\omega_b(\Omega + \omega)$	$\omega^2_{(j)} - (\Omega + \omega)^2$	$= \dfrac{PD}{\sum \int \mu v^2_{(j)}(x)\,\mathrm{d}x}$

of the respective natural mode j. For light damping the resultant vibration will then be described by the equation

$$q_{(j)}(t) = \bar{q}_{(j)}(t) + e^{-\omega_b t}(a_{(j)} \cos \omega_{(j)}t + b_{(j)} \sin \omega_{(j)}t) \tag{7.8}$$

The integration constants $a_{(j)}$, $b_{(j)}$ are computed from the initial conditions of motion, i.e. from the values of $q(0)$, $\dot{q}(0)$ at time $t = 0$ i.e. when the load enters the respective beam span:

$$K_3 + \bar{K}_3 + a_{(j)} = q(0)$$

$$(K_1 + \bar{K}_1)\,\Omega + (K_4 + \bar{K}_4)\,\omega - a_{(j)}\omega_b + b_{(j)}\omega_{(j)} = \dot{q}(0)$$

and from there

$$a_{(j)} = q(0) - K_3 - \bar{K}_3$$

$$b_{(j)} = \frac{1}{\omega_{(j)}}\{\dot{q}(0) - (K_1 + \bar{K}_1)\,\Omega - (K_4 + \bar{K}_4)\,\omega - [K_3 + \bar{K}_3 - q(0)]\,\omega_b\} \tag{7.9}$$

The computation is carried out for all modes of free vibration, i.e. for all j, and the computed expressions 7.8 are substituted in eq. 7.2 to obtain the resultant displacement $v(x, t)$.

Of particular importance is the case of the circular frequency of the exciting force at resonance with some (usually the first) natural circular frequency $\omega_{(j)}$. In such a case the deflections of the respective j-th mode predominate by far over the remaining modes, and the examination can consequently be confined to this mode. Because $\omega_{(j)}^2 - \Omega^2 = 0$ and ω and ω_b are usually very small compared with Ω, it follows that

$$K_1 \approx K_2 \approx \bar{K}_1 \approx \bar{K}_2 \approx 0$$

and the formulae for the other constants simplify to

$$K_3 = \varepsilon(B\omega - A\omega_b)$$
$$K_4 = \varepsilon(-A\omega - B\omega_b)$$
$$\bar{K}_3 = \bar{\varepsilon}(C\omega_b - D\omega)$$
$$\bar{K}_4 = \bar{\varepsilon}(D\omega_b - C\omega) \tag{7.10}$$

where

$$\varepsilon = \frac{P}{2\Omega(\omega^2 + \omega_b^2)\sum \int_0^l \mu\, v_{(j)}^2(x)\, \mathrm{d}x}$$

$$\bar{\varepsilon} = \frac{P}{2\Omega(\omega^2 - \omega_b^2) \sum \int_0^l \mu \, v_{(j)}^2(x) \, dx}$$

(7.11)*

Furthermore, by eq. 7.9

$$b_{(j)} \approx \frac{\dot{q}_{(j)}(0)}{\omega_{(j)}}$$

so that eq. 7.8 becomes

$$q_{(j)}(t) = \cos \Omega t [K_3 \cos \omega t + K_4 \sin \omega t + \bar{K}_3 \cosh \omega t + \bar{K}_4 \sinh \omega t]$$

$$- e^{-\omega_b t} \left\{ [K_3 + \bar{K}_3 - q_{(j)}(0)] \cos \Omega t - \frac{1}{\Omega} \dot{q}_{(j)}(0) \sin \Omega t \right\} \qquad (7.12)$$

If the load passes over several spans of a continuous beam in succession, the initial state of motion at the entrance to the next span is given by the final state of vibration at the exit from the preceding span.

The application of the equations derived for force $P \sin \Omega t$ can easily be extended to a harmonic force with arbitrarily shifted phase, $P \sin (\Omega t + \varphi)$. Eq. 7.12 then becomes

$$q_{(j)}(t) = \cos (\Omega t + \varphi) [K_3 \cos \omega t + K_4 \sin \omega t + \bar{K}_3 \cosh \omega t + \bar{K}_4 \sinh \omega t]$$

$$- e^{-\omega_b t} \left\{ [(K_3 + \bar{K}_3) \cos \varphi - q(0)] \cos \Omega t \right.$$

$$\left. - \left[\frac{\dot{q}(0)}{\Omega} + (K_3 + \bar{K}_3) \sin \varphi \right] \sin \Omega t \right\}$$

and the time variation of the deflection at point x is defined by eq. 7.2

$$v(x, t) = q_{(j)}(t) \, v_{(j)}(x)$$

7.2.1.3 Inertia forces associated with the vertical motion of the load

So far no account has been taken of the inertia forces resulting from the motion of the load vibrating in the vertical direction simultaneously with the bridge structure. Complicated enough in the case of simply supported beams, an exact solution of these problems for continuous beams becomes extremely difficult. We shall simplify the task by proceeding in a manner similar to that known from the examinations of simply supported beams[28]. Assume that the actual inertia

* If $\omega_b \to \omega$ the simplified formula for $\bar{\varepsilon}$ no longer applies, of course.

force can be replaced by a fictitious one resulting from the action of a load mass affixed in a suitable location d to the structure. This force is equal to the negative product of mass and acceleration. Because of eq. 7.2

$$-m \frac{\mathrm{d}^2 v(d, t)}{\mathrm{d}t^2} = -m \frac{\mathrm{d}^2[q_{(1)}(t)\, v_{(1)}(d) + q_{(2)}(t)\, v_{(2)}(d) + \ldots]}{\mathrm{d}t^2} \tag{7.13}$$

This concentrated force is then replaced by a continuous load in the same way as force $P \sin \Omega t$ was. Confining ourselves to the first term of series 7.3 gives the loading term

$$p_{(1)}(t) = \frac{P \sin \Omega t\, v_{(1)}(ct) - m\, q_{(1)}(t)\, v_{(1)}^2(d)}{\sum \int_0^l \mu\, v_{(1)}^2(x)\, \mathrm{d}x} \tag{7.14}$$

where the summation sign includes all the bars of the system, and its substitution in eq. 7.4 gives the equation

$$\ddot{q}_{(1)}(t) \left[1 + \frac{m\, v_{(1)}^2(d)}{\sum \int \mu\, v_{(1)}^2(x)\, \mathrm{d}x} \right] + 2\omega_b\, \dot{q}_{(1)}(t) + \omega_{(1)}^2\, q_{(1)}(t) = \frac{P \sin \Omega t\, v_{(1)}(ct)}{\sum \int \mu\, v_{(1)}^2(x)\, \mathrm{d}x}$$

or after rearrangement

$$\ddot{q}_{(1)}(t) + 2\bar{\omega}_b\, \dot{q}_{(1)}(t) + \bar{\omega}_{(1)}^2\, q_{(1)}(t) = \frac{P \sin \Omega t\, v_{(1)}(c, t)}{\sum \int \bar{\mu}\, v_{(1)}^2(x)\, \mathrm{d}x} \tag{7.15}$$

where

$$\bar{\mu} = \mu \left[1 + \frac{m\, v_{(1)}^2(d)}{\sum \int \mu\, v_{(1)}^2(x)\, \mathrm{d}x} \right] \tag{7.16}$$

$$\bar{\omega}_{(1)} = \omega_{(1)} \sqrt{\frac{\mu}{\bar{\mu}}} \tag{7.17}$$

$$\bar{\omega}_b = \omega_b \frac{\mu}{\bar{\mu}} \tag{7.18}$$

Eq. 7.15 could be obtained from eq. 7.4 by setting $j = 1$, replacing μ, $\omega_{(1)}$, ω_b by their values with bars, $\bar{\mu}$, $\bar{\omega}_{(1)}$, $\bar{\omega}_b$ and substituting the expression from eq. 7.3 for $p_{(1)}(t)$ in eq. 7.4. We are, therefore, free to use — following identical modification — all the formulae and tables arrived at in Section 7.2.1.2.

Note: By confining eq. 7.13 to the first term of the series we have introduced the assumption that the first mode of vibration undergoes no change if the vehicle mass m is placed in section d.

7.2.2　Simply supported beam

If the traversed beam is a simply supported one, the constants in eq. 7.5 are $A = C = D = 0$, and Table 7.2 gives

$$\overline{K}_1 = \overline{K}_2 = \overline{K}_3 = \overline{K}_4 = 0$$

Apart from this, for $B = 1$

$$v_{(j)}(x) = \sin\frac{j\pi x}{l}$$

and

$$\int_0^l \mu\, v_{(j)}^2(x)\, dx = \frac{\mu l}{2}$$

$$(7.19)$$

Accordingly, by eq. 7.10

$$\varepsilon = \frac{P}{\mu l\, \Omega(\omega^2 + \omega_b^2)}$$

$$K_3 = \frac{P\omega}{\mu l\, \Omega(\omega^2 + \omega_b^2)}$$

$$K_4 = -\frac{P\omega_b}{\mu l\, \Omega(\omega^2 + \omega_b^2)}$$

$$(7.20)$$

Assume that the angular velocity of the driving wheels is at resonance with the first natural frequency of the system, $\Omega = \omega_{(1)}$, and that the beam is at rest before the load enters it, i.e.

$$q(0) = 0, \quad \dot{q}(0) = 0$$

Making use of eqs 7.16–7.18, eqs 7.12 and 7.2 yield

$$v(x, t) = q_{(1)}(t)\, v_{(1)}(x)$$

$$= \frac{P\cos\overline{\omega}_{(1)}t}{\overline{\mu}l\, \overline{\omega}_{(1)}(\omega^2 + \overline{\omega}_b^2)} \left[\omega(\cos\omega t - e^{-\omega_b t}) - \overline{\omega}_b \sin\omega t\right] \sin\frac{\pi x}{l} \quad (7.21)$$

where

$$\omega = \frac{\pi c}{l}$$

When appropriate changes are made in notation, the last equation is the equation arrived at by C.E. Inglis in Reference 28.

The deflections — eq. 7.21 — resulting from the vertical component of the centrifugal force of the driving wheels ('hammer blows') are superposed upon the deflection curve resulting from the passage of the load weight. It is not too far-fetched to assume that this deflection curve will look like a sine curve whose maximum v_{st} is equal to the static deflection produced by force $G = mg$ (locomotive weight), and that for large-span bridges the total locomotive weight will act at a single point: the centre of gravity of the locomotive. Under these assumptions the deflection at the midspan $(x = l/2)$ is given by

$$v\left(\frac{l}{2}, t\right) = v_{st} \sin \omega t + \frac{P \cos \bar{\omega}_{(1)}t}{\bar{\mu} l \, \bar{\omega}_{(1)}(\omega^2 + \bar{\omega}_b^2)} \left[\omega(\cos \omega t - e^{-\bar{\omega}_b t}) - \bar{\omega}_b \sin \omega t\right] \quad (7.22)$$

7.2.3 Numerical examples

Example 7.1

The theoretical results arrived at in the preceding paragraphs have been used in the computation of the dynamic effects of a two-cylinder locomotive with

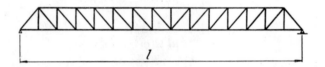

Fig. 7.2 Bridge with simply supported straight main girders

unbalanced counterweights on the driving axles, on a bridge with simply supported main girders (Fig. 7.2). The actual dynamic effects were subsequently measured on the existing structure.

Technical data of the bridge

weight	Q	$= 181$ Mp
span	l	$= 46 \cdot 86$ m
mass per unit length	μ	$= Q/gl = 0 \cdot 394$ Mpm^{-2} s^2
first natural frequency	$f_{(1)} =$	$3 \cdot 7$ Hz
damping constant	$\omega_b =$	$0 \cdot 257$ s^{-1}
static deflection midspan of the bridge resulting from load of 97 Mp	$v_{st} =$	$0 \cdot 02$ m

(The damping constant was determined experimentally. As indicated by the diagram in Fig. 7.3, after 10 vibrations the deflection of free vibration drops to half its original value; therefore $e^{-\omega_b(10/f_{(1)})} = 0.5$ and $\omega_b = 0.257 \text{ s}^{-1}$).

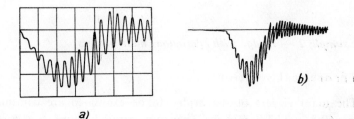

b)

a)

Fig. 7.3 Dynamic deflection of bridge shown in Fig. 7.2. (a) Theoretical curve, (b) deflectometer record

Technical data of the locomotive

weight	$G = 97 \text{ Mp}$
centrifugal force of the driving wheels	$P = 0.008\Omega^2 \text{ Mp}$
radius of the driving wheels	$r = 0.630 \text{ m}$

The substitute mass per unit length $\bar{\mu}$ (eq. 7.16) is computed for the load mass placed in section $x = d = l/3$

$$\bar{\mu} = \mu\left(1 + \frac{97 \times \frac{3}{4}}{\frac{1}{2} \times 181}\right) = 0.712 \text{ Mpm}^{-2} \text{ s}^2$$

$$\bar{\omega}_{(1)} = \omega_{(1)}\left(\frac{\mu}{\bar{\mu}}\right)^{1/2} = 17.4 \text{ s}^{-1}$$

$$\bar{\omega}_b = \omega_b \frac{\mu}{\bar{\mu}} = 0.555\omega_b = 0.143 \text{ s}^{-1}$$

The critical speed at which resonance occurs is

$$c_{cr} = \bar{\omega}_{(1)} \times r = 17.4 \times 0.63 = 10.97 \text{ m/s} \approx 40 \text{ km/h}$$

The dynamic deflection curve as a function of the instantaneous position of the locomotive, $a = ct$, is defined by eqs 7.21 and 7.22. For the given numerical values

$$v(x, t) = 0.02 \sin 0.74t + 0.0071 \cos 17.4t[0.74(\cos 0.74t - e^{-0.143t})$$

$$- 0.143 \sin 0.74t] \sin \frac{\pi x}{l}$$

The curve is plotted on the left of Fig. 7.3. On the right is the diagram of the deflection midspan of the bridge measured by a Stoppani deflectometer with the locomotive travelling at the critical speed of 40 km/h.

It can clearly be seen from the diagrams that the experimental results are in excellent agreement with the theory as well as with the data reported in Reference 28.

Example 2 — Bridge with continuous main girders

(a)   Theoretical solution

The main girders of the bridge to be examined are continuous[42], with three spans 60·8 + 77·4 + 60·8 m. They are parallel chord trusses of a simple triangular frame (Fig. 7.4). The cross-section is not constant. In the dynamic calculation the two main girders will be replaced by a single continuous beam with spans

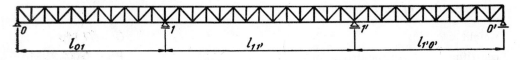

Fig. 7.4 Bridge with continuous main girders

of constant cross-section. The moments of inertia in the spans were determined on the condition that the total stiffness of each span should be the same for both the substitute beam and the actual structure. The weight per unit length μg was

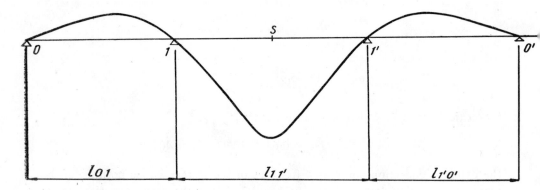

Fig. 7.5 Influence line of deflection midway of central span

assumed to be the same over whole length of the beam and equal to the average weight of the bridge per metre of length, $\mu g = 5\cdot5$ Mp/m.

The influence line of deflection at point s midway of the central span (Fig. 7.5) was taken over from the statical computation and corrected to fit the results of the loading test. Therefore it does not exactly correspond to the influence line of the substitute beam used in the dynamic computations.

Assuming the above we get the basic characteristics of the end spans

$$l_{01} = 60·8 \text{ m}, \quad J_{01} = 2·08 \text{ m}^4, \quad \mu_{01} = 5·5/9·81 = 0·560 \text{ Mpm}^{-2} \text{s}^2$$

and of the central span

$$l_{11'} = 77·4 \text{ m}, \quad J_{11'} = 3·14 \text{ m}^4, \quad \mu_{11'} = 5·5/9·81 = 0·560 \text{ Mpm}^{-2} \text{s}^2$$

Further

$$\frac{\lambda_{11'}}{\lambda_{01}} = \frac{l_{11'}}{l_{01}} \left(\frac{J_{01}}{J_{11'}}\right)^{1/4} = 1·150$$

The natural frequency of symmetric vibration is computed from the equation of equilibrium at joint 1

$$M_{10} + M_{11'} = 0 \tag{7.23}$$

substituting from Table 10 (Appendix)

$$M_{10} = \frac{EJ_0}{l_{01}} F_7(\lambda_{01}) \zeta_1$$

$$M_{11'} = \frac{EJ}{l_{11'}} \left[F_2(\lambda_{11'}) \zeta_1 + F_1(\lambda_{11'}) \zeta_{1'}\right]$$

$$\tag{7.24}$$

and

$$\zeta_1 = -\zeta_{1'} \tag{7.25}$$

leads to the frequency equation

$$\frac{J_{01}}{l_{01}} F_7(\lambda_{01}) + \frac{J_{11'}}{l_{11'}} \left[F_2(\lambda_{11'}) - F_1(\lambda_{11'})\right] = 0 \tag{7.26}$$

The lowest values of λ satisfying this equation are

$$\lambda_{01} = 2·93 \quad \text{and} \quad \lambda_{11'} = 3·37$$

The fundamental natural circular frequency is obtained from the formula

$$\omega_{(1)} = \frac{\lambda_{11'}^2}{l_{11'}^2} \left(\frac{EJ_{11'}}{\mu_{11'}}\right)^{1/2} = \frac{3·37^2}{77·4^2} \left(\frac{21 \times 10^6 \times 3·14}{0·560}\right)^{1/2} = 20·5 \text{ s}^{-1}$$

The frequency of the bridge under no-load condition is then

$$f_{(1)} = \frac{\omega_{(1)}}{2\pi} = 3·26 \text{ Hz}$$

According to Section 2.8.2 and assumption 7.25 (for $\zeta_1 = 1$)

$$\sum \int_0^l \mu\, v_{(1)}^2(x)\, dx = \frac{2E}{\omega_{(1)}^2}\left\{\frac{J_{01}}{l_{01}}\Phi_7(\lambda_{01}) + \frac{J_{11'}}{l_{11'}}[\Phi_2(\lambda_{11'}) - \Phi_1(\lambda_{11'})]\right\}$$

$$= \frac{2 \times 21 \times 10^6}{20\cdot 5^2}\left\{\frac{2\cdot 08}{60\cdot 8}\,2\cdot 78 + \frac{3\cdot 14}{77\cdot 4}[2\cdot 08 + 1\cdot 75]\right\}$$

$$= 25\,000\ \text{Mpm s}^2$$

The values of function Φ are tabulated in the Appendix.

The first natural mode is

$$v_{(1)}(x) = A \cos\frac{\lambda x}{l} + B \sin\frac{\lambda x}{l} + C \cosh\frac{\lambda x}{l} + D \sinh\frac{\lambda x}{l} \tag{7.27}$$

At $\zeta_1 = 1$ the integration constants (formula 2.51) for the first span are

$$A = C = 0$$

$$B = -\frac{l_{01}}{\lambda_{01}}\frac{\sinh\lambda_{01}}{\cosh\lambda_{01}\sin\lambda_{01} - \sinh\lambda_{01}\cos\lambda_{01}} = -\frac{60\cdot 8}{2\cdot 93}\frac{9\cdot 34}{11\cdot 10} = -17\cdot 45$$

$$D = \frac{l_{01}}{\lambda_{01}}\frac{\sin\lambda_{09}}{\cosh\lambda_{01}\sin\lambda_{01} - \sinh\lambda_{01}\cos\lambda_{01}} = \frac{60\cdot 8}{2\cdot 93}\frac{0\cdot 210}{11\cdot 10} = 0\cdot 393$$

Hence in the first span

$$v_{(1)}(x) = -17\cdot 45 \sin 2\cdot 93\frac{x}{l} + 0\cdot 393 \sinh 2\cdot 93\frac{x}{l} \tag{7.28}$$

By formula 2.47 the constants for the second span are

$$C = -A = -\frac{l_{11'}}{2\lambda_{11'}^2}[F_2(\lambda_{11'}) - F_1(\lambda_{11'})] = -\frac{77\cdot 4}{2 \times 3\cdot 37^2}(2\cdot 409 - 3\cdot 274) = 2\cdot 94$$

$$D = -\frac{l_{11'}}{2\lambda_{11'}^3}[F_4(\lambda_{11'}) + F_3(\lambda_{11'})] + \frac{l_{11'}}{2\lambda_{11'}}$$

$$= -\frac{77\cdot 4}{2 \times 3\cdot 37^3}(2\cdot 461 + 11\cdot 621) + \frac{77\cdot 4}{2 \times 3\cdot 37} = -2\cdot 87$$

$$B = -D + \frac{l_{11'}}{\lambda_{11'}} = 2\cdot 87 + \frac{77\cdot 4}{3\cdot 37} = 25\cdot 83$$

The first mode of free vibration of the second span is

$$v_{(1)}(x) = 2 \cdot 94 \cosh 3 \cdot 37 \frac{x}{l} - 2 \cdot 87 \sinh 3 \cdot 37 \frac{x}{l} - 2 \cdot 94 \cos 3 \cdot 37 \frac{x}{l}$$

$$+ 25 \cdot 83 \sin 3 \cdot 37 \frac{x}{l} \tag{7.29}$$

At midspan $(x = l_{11'}/2)$

$$v_{(1)}\left(\frac{l_{11'}}{2}\right) = 2 \cdot 94 \cosh 1 \cdot 685 - 2 \cdot 87 \sinh 1 \cdot 685 - 2 \cdot 94 \cos 1 \cdot 685$$

$$+ 25 \cdot 83 \sin 1 \cdot 685 = 26 \cdot 69$$

In the third span the integration constants (formula 2.53) are

$$C = -A = \frac{l_{01}}{2\lambda_{01}^2} F_7(\lambda_{01}) = \frac{60 \cdot 8}{2 \times 2 \cdot 93^2} 1 \cdot 035 = 3 \cdot 67$$

$$D = \frac{l_{01}}{2\lambda_{01}^3} F_9(\lambda_{01}) - \frac{l_{01}}{2\lambda_{01}} = \frac{60 \cdot 8}{2 \times 2 \cdot 93^2} 5 \cdot 536 - \frac{60 \cdot 8}{2 \times 2 \cdot 93} = -3 \cdot 68$$

$$B = -D - \frac{l_{01}}{\lambda_{01}} = 3 \cdot 68 - \frac{60 \cdot 8}{2 \cdot 93} = -17 \cdot 04$$

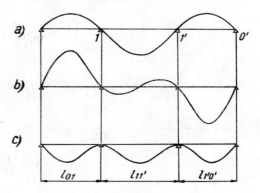

Fig. 7.6 First three modes of free
vibration of system shown in Fig. 7.4

and

$$v_{(1)}(x) = +3 \cdot 67 \cosh 2 \cdot 93 \frac{x}{l} - 3 \cdot 68 \sinh 2 \cdot 93 \frac{x}{l} - 3 \cdot 67 \cos 2 \cdot 93 \frac{x}{l}$$

$$- 17 \cdot 04 \sin 2 \cdot 93 \frac{x}{l} \tag{7.30}$$

The first mode of free vibration (eqs 7.28, 7.29 and 7.30) is shown in Fig. 7.6a.

The circular frequency of damping ω_b was estimated at $0.3\ \text{s}^{-1}$.

During the tests the bridge was traversed by a 97 t two-cylinder locomotive with five coupled axles, the driving axles producing centrifugal force $P = 0.3N^2$ $Mp\ (\approx 0.008\Omega^2 Mp)$, where N is the number of revolutions of the driving axles per second. The radius of the driving wheels was $r = 0.63$ m.

Fig. 7.5 shows the influence line of deflection at point s midway of the central span. On loading it with the locomotive axle pressures we get the deflection at point s caused by the locomotive travelling over the bridge at practically arbitrary speed,

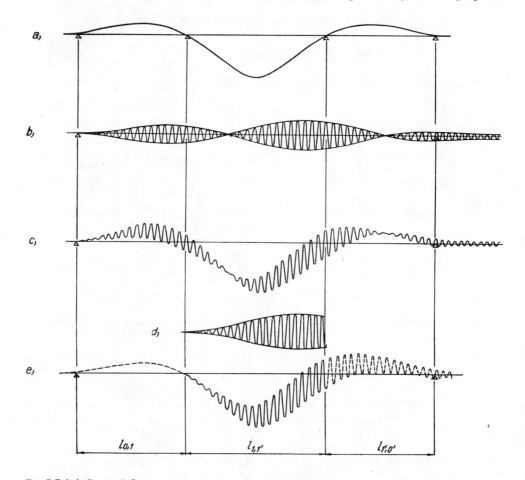

Fig. 7.7 (a) Static deflection midway of central span during passage of a 97 Mp two-cylinder steam locomotive (The deflections are magnified 2500 × compared with spans), (b) deflection midway of central span resulting from sinusoidally varying force with moving point of action; frequency of force at resonance with natural frequency of beam, (c) theoretical curve of dynamic deflection midway of central span, (d) dynamic deflection midway of central span on the assumption of the sinusoidally varying force acting only on central span, (e) resultant theoretical curve of dynamic deflection midway of central span

assuming, of course, that the centrifugal forces of the driving axles exert no action. (Fig. 7.7a.)

As to the computation of the effect of the moving periodic force of the driving axles: the only case we are going to consider here is that of $N = \bar{f}_{(1)}$ where $\bar{f}_{(1)}$ is the first natural frequency of the loaded bridge.

By eq. 7.16

$$\frac{\bar{\mu}}{\mu} = 1 + \frac{m\,v_{(1)}^2(l_{11'}/2)}{\sum \int \mu\,v_{(1)}^2(x)\,dx} = 1 + \frac{97 \times 26 \cdot 69^2}{9 \cdot 81 \times 25\,000} = 1 \cdot 28$$

so that

$$\bar{\omega}_{(1)} = \left(\frac{20 \cdot 5^2}{1 \cdot 28}\right)^{1/2} = 18 \cdot 1\ \text{s}^{-1}$$

and

$$\bar{f}_{(1)} = \frac{18 \cdot 1}{2\pi} = 2 \cdot 88\ \text{s}^{-1}$$

The critical speed

$$c = 2 \cdot 88 \times 3 \cdot 96 = 11 \cdot 4\ \text{m s}^{-1} = 41\ \text{km/h}$$

and the centrifugal force

$$P = 0 \cdot 3 \times 2 \cdot 88^2 = 2 \cdot 49\ \text{Mp}$$

Further

$$\bar{\omega}_b = \omega_b \frac{\mu}{\bar{\mu}} = \frac{0 \cdot 3}{1 \cdot 28} = 0 \cdot 235\ \text{s}^{-1}$$

and

$$\sum \int \bar{\mu}\,v_{(1)}^2(x)\,dx = 1 \cdot 28 \times 25\,000 = 32\,000\ \text{Mpm s}^2$$

In span $\overline{01}$

$$\omega = \frac{\lambda_{01}c}{l_{01}} = \frac{2 \cdot 93 \times 11 \cdot 4}{60 \cdot 8} = 0 \cdot 549\ \text{s}^{-1}$$

and by eqs 7.10 and 7.11

$$\varepsilon = \frac{P}{2\bar{\omega}_{(1)}(\omega^2 + \bar{\omega}_b^2)\sum \int \bar{\mu}\,v_{(1)}^2(x)\,dx} = \frac{2 \cdot 49}{2 \times 18 \cdot 1(0 \cdot 549^2 + 0 \cdot 235^2) \times 32\,000}$$

$$= 0 \cdot 600 \times 10^{-5}$$

$$K_3 = \varepsilon B\omega = -0 \cdot 600 \times 10^{-5} \times 17 \cdot 45 \times 0 \cdot 549 = -5 \cdot 75 \times 10^{-5}$$

$$K_4 = -\varepsilon B\bar{\omega}_b = 0 \cdot 600 \times 10^{-5} \times 17 \cdot 45 \times 0 \cdot 235 = 2 \cdot 46 \times 10^{-5}$$

$$\bar{\varepsilon} = \frac{P}{2\bar{\omega}_{(1)}(\omega^2 - \omega_b^2)\sum\int \bar{\mu}\, v_{(1)}^2(x)\,dx} = \frac{2\cdot49}{2 \times 18\cdot1(0\cdot549^2 - 0\cdot235^2) \times 32\,000}$$

$$= 0\cdot870 \times 10^{-5}$$

$$K_3 = -0\cdot870 \times 10^{-5} \times 0\cdot393 \times 0\cdot549 = -0\cdot188 \times 10^{-5}$$
$$K_4 = 0\cdot870 \times 10^{-5} \times 0\cdot393 \times 0\cdot235 = 0\cdot080 \times 10^{-5}$$

As the harmonic force passes over span $\overline{01}$ the deflection at point s midway of the central span is given by formulae 7.12 and 7.2 now written as

$$v\left(\frac{l_{11'}}{2}, t\right) = \cos \bar{\omega}_{(1)}t\, A(t) \tag{7.31a}$$

where

$$A(t) = v_{(1)}\left(\frac{l_{11'}}{2}\right)[K_3(\cos \omega t - e^{-\omega_b t}) + K_4 \sin \omega t$$

$$+ \bar{K}_3(\cosh \omega t - e^{-\omega_b t}) + \bar{K}_4 \sinh \omega t] \tag{7.31b}$$

is the envelope of the amplitudes of point s. Formula 7.31a applies whenever the centrifugal force at the instant of the locomotive entry on the beam acts downward. In the general case

$$v_{(1)}\left(\frac{l_{11'}}{2}, t\right) = \cos\left(\bar{\omega}_{(1)}t + \varphi_p\right) A(t)$$

The envelope, however, remains the same no matter what the phase shift φ_p is. Substituting in eq. 7.31b gives

$$A(t) = 26\cdot69 \times 10^{-5}[-5\cdot75(\cos 0\cdot549t - e^{-0\cdot235t}) + 2\cdot46 \sin 0\cdot549t$$

$$- 0\cdot188(\cosh 0\cdot549t - e^{-0\cdot235t}) + 0\cdot080 \sinh 0\cdot549t] \tag{7.32a}$$

Substitution of $t = a/c$, $\omega t = \lambda_{01}(a/l)$ (where a is the distance between the force and the left-hand end of the beam) and simple rearrangement yield

$$A(a) = \left[-153\left(\cos 2\cdot93\frac{a}{l} - e^{-1\cdot252a/l}\right) + 65\cdot6 \sin 2\cdot93\frac{a}{l}\right.$$

$$\left. - 5\cdot0\left(\cosh 2\cdot93\frac{a}{l} - e^{-1\cdot252a/l}\right) + 2\cdot1 \sinh 2\cdot93\frac{a}{l}\right] \times 10^{-5} \tag{7.32b}$$

For $a \doteq l_{01}$

$$A(l_{01}) \approx 180 \times 10^{-5}$$

As the harmonic force moves along the central span, we get analogously

$$\omega = \frac{3\cdot37 \times 11\cdot4}{77\cdot4} = 0\cdot496$$

$$\varepsilon = \frac{2\cdot49}{2 \times 18\cdot1(0\cdot496^2 + 0\cdot235^2) \times 32\,000} = 0\cdot714 \times 10^{-5}$$

$$K_3 = \varepsilon(B\omega - A\bar\omega_b) = 0\cdot714 \times 10^{-5}(25\cdot83 \times 0\cdot496 + 2\cdot94 \times 0\cdot235)$$
$$= 9\cdot64 \times 10^{-5}$$

$$K_4 = \varepsilon(-A\omega - B\bar\omega_b) = 0\cdot714 \times 10^{-5}(2\cdot94 \times 0\cdot496 - 25\cdot83 \times 0\cdot235)$$
$$= -3\cdot30 \times 10^{-5}$$

$$\bar\varepsilon = \frac{2\cdot49}{2 \times 18\cdot1(0\cdot496^2 - 0\cdot235^2) \times 32\,000} = 1\cdot126 \times 10^{-5}$$

$$\bar K_3 = \bar\varepsilon(C\omega_b - D\omega) = 1\cdot126 \times 10^{-5}(2\cdot94 \times 0\cdot235 + 2\cdot87 \times 0\cdot496)$$
$$= 2\cdot38 \times 10^{-5}$$

$$\bar K_4 = \bar\varepsilon(D\bar\omega_b - C\omega) = 1\cdot126 \times 10^{-5}(-2\cdot87 \times 0\cdot235 - 2\cdot94 \times 0\cdot496)$$
$$= -2\cdot40 \times 10^{-5}$$

For the amplitude of point s

$$A(t) = 26\cdot69 \times 10^{-5}[9\cdot64(\cos 0\cdot496t - e^{-0\cdot235t}) - 3\cdot30 \sin 0\cdot496t$$
$$+ 2\cdot38(\cosh 0\cdot496t - e^{-0\cdot235t}) - 2\cdot40 \sinh 0\cdot496t + A(l_{01})e^{-0\cdot235t}] \quad (7.33a)$$

$$A(a) = \left[257\left(\cos 3\cdot37\frac{a}{l} - e^{-1\cdot595a/l}\right) - 88\cdot0 \sin 3\cdot37\frac{a}{l}\right.$$

$$+ 63\cdot4\left(\cosh 3\cdot37\frac{a}{l} - e^{-1\cdot595a/l}\right) - 64\cdot0 \sinh 3\cdot37\frac{a}{l}$$

$$\left. + 180e^{-1\cdot595a/l}\right] \times 10^{-5} \quad (7.33b)$$

The last term of eqs 7.33 expresses the residual vibration caused by passage over span $\overline{01}$.

For $a = l_{11'}$

$$A(l_{11'}) \approx -300 \times 10^{-5}$$

And finally, in span $\overline{1'0'}$

$$\omega = 0.549 \text{ s}^{-1}$$

$$\varepsilon = 0.600 \times 10^{-5} \text{ (same as in span } \overline{01})$$

$$K_3 = 0.600 \times 10^{-5}(-17.04 \times 0.549 + 3.67 \times 0.235) = -5.10 \times 10^{-5}$$

$$K_4 = 0.600 \times 10^{-5}(+ 3.67 \times 0.549 + 17.04 \times 0.235) = +3.61 \times 10^{-5}$$

$$\bar{\varepsilon} = 0.870 \times 10^{-5}$$

$$\bar{K}_3 = 0.870 \times 10^{-5}(+3.67 \times 0.235 + 3.68 \times 0.549) = +2.50 \times 10^{-5}$$

$$\bar{K}_4 = 0.870 \times 10^{-5}(-3.68 \times 0.235 - 3.67 \times 0.549) = -2.50 \times 10^{-5}$$

$$A(t) = 26.69 \times 10^{-5}[-5.10(\cos 0.549t - e^{-0.235t}) + 3.61 \sin 0.549t$$
$$+ 2.50(\cosh 0.549t - e^{-0.235t}) - 2.50 \sinh 0.549t] + A(l_{11'}) e^{-0.235t} \quad (7.34a)$$

$$A(a) = \left[-136 \left(\cos 2.93 \frac{a}{l} - e^{-1.252a/l} \right) + 96.3 \sin 2.93 \frac{a}{l} \right.$$

$$+ 66.7 \left(\cosh 2.93 \frac{a}{l} - e^{-1.252a/l} \right) - 66.7 \sinh 2.93 \frac{a}{l}$$

$$\left. - 300 \times e^{-1.252a/l} \right] \times 10^{-5} \quad (7.34b)$$

The deflections according to eqs 7.32b–7.34b are shown in Fig. 7.7b. Superposing the curve of Fig. 7.7b upon that of Fig. 7.7a gives the deflection at point s for the locomotive travelling at critical speed (Fig. 7.7c).

The theoretical solution just obtained is approximate for the reason that resonance cannot last the whole time the locomotive is passing over the bridge. The natural frequency of the loaded bridge depends on the position of the locomotive mass on the bridge, and varies over fairly wide limits. To cause the resonance to last the whole of the locomotive passage over the bridge, the locomotive would have to vary its speed. The question arises whether or not one should include in the computation the residual vibration induced by passage over the preceding span.

We repeated the approximate computation of the numerical example assuming that resonance occurred only during the transit over the central span and neglected the dynamic effects brought about by passage over the end spans. The solution led to eq. 7.33 minus the last term. The result is graphically represented in Fig. 7.7d (dashed lines), and the resultant dynamic deflection is shown in Fig. 7.7e.

(b)   Results of experimental investigations

The bridge was traversed by a 97 Mp test locomotive travelling at various speeds. The dynamic effects were measured by strain gauges and deflectometers in several locations on the structure. The diagrams in Figs 7.8—7.13 show the results of deflectometer measurements taken in the middle of the central span (theoretically examined in the foregoing discussion).

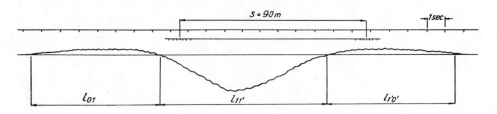

Fig. 7.8 Record taken by Geiger deflectometer midway of central span during passage of a two-cylinder locomotive travelling at 31 km/h. (s = 90 m: distance between electric contacts affixed to rail and actuated by pressure of traversing locomotive)

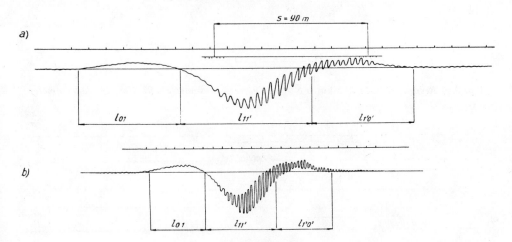

Fig. 7.9 Record taken midway of central span at locomotive speed of 37 km/h by (a) Geiger deflectometer, (b) Stoppani deflectometer

It is plain to see in the diagrams that the dynamic effects of periodic forces with moving points of application vary considerably with locomotive speed. They are hardly noticeable at 31 km/h (Fig. 7.8) but very marked at 37 km/h (Fig. 7.9). The resonance effects, still quite distinct at 44 km/h (Fig. 7.11), begin to disappear at about 49 km/h (Fig. 7.13).

The diagrams in Figs 7.9, 7.10 and 7.11 were recorded by the Geiger (top) and Stoppani (bottom) instruments. While they are in excellent agreement as regard the shape of the curves and frequency of vibration, they differ appreciably in the amplitude of vibration. Whereas the maximum amplitude of the records taken by the Geiger instrument was about 28% of the maximum static deflection, the amplitude of vibration recorded by the Stoppani instruments was much larger, nearly 40% of the maximum static deflection. The reason for this distortion of the Geiger instrument records was found to be high friction between the needle and the wax chart paper, causing a reduction of the recorded vibration amplitude.

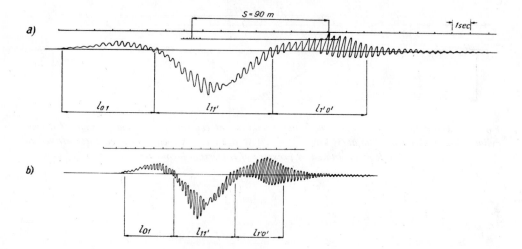

Fig. 7.10 Record taken at 42·5 km/h by (a) Geiger deflectometer, (b) Stoppani deflectometer

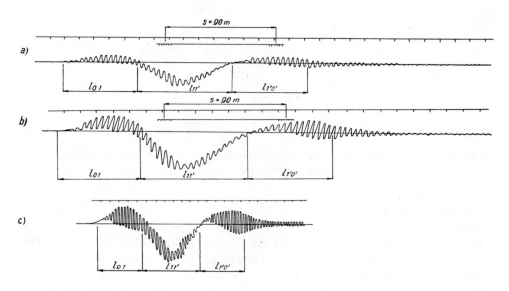

Fig. 7.11 Record taken at 44 km/h by (a) Geiger deflectometer installed on a bracket, (b) Geiger deflectometer with wire connection to a fixed point on river bed, (c) Stoppani deflectometer

The Geiger and Stoppani instruments were screwed to the bridge structure and connected to a fixed point on the bottom of the river by means of a 10 m long piece of steel wire. Such an arrangement has been regarded by many experimenters as

a)

b)

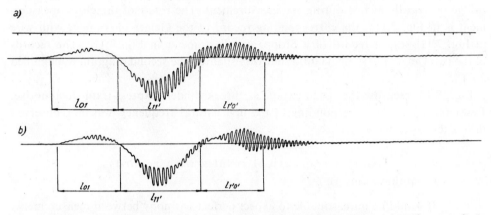

Fig. 7.12 Geiger deflectometer record taken midway of central span during passage of a two-cylinder locomotive at (a) 40 km/h, (b) 42 km/h

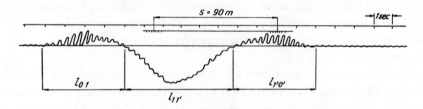

Fig. 7.13 Record taken at 49·5 km/h

a)

b)

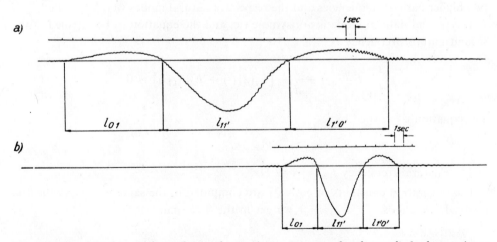

Fig. 7.14 Record taken midway of central span during passage of a three-cylinder locomotive at (a) 22·5 km/h, (b) 61 km/h

unsuitable because the wire vibration affects the results of the measurements. To find out whether or not the objections were well founded we undertook a check measurement in which the Geiger instrument was attached to a bracket of the measured bridge while the fixed point was on an opposite-lying bracket of the second track bridge practically at rest during the measurement. The result of the check measurement is in Fig. 7.11a, the corresponding record of the Geiger instrument with wire in Fig. 7.11b and a record of a Stoppani deflectometer in Fig. 7.11c. The records of the two Geiger instruments differ but very slightly (the second diagram is magnified 2×).

Fig. 7.14 records the passage of a three-cylinder express-train locomotive. Resonant vibration corresponding to the first natural frequency was not observed in this test.

(c)   Comparison of theoretical results with records of measurement

It would be unreasonable to expect perfect harmony between measurements and theory because the computation was of necessity considerably simplified and some factors were not taken into account. However, the theoretical curves in Figs 7.7c and 7.7e well illustrate the nature of vibration of a bridge with continuous beams when a two-cylinder locomotive passes over it. The differences between the computed and measured diagrams point out the dependence of the accuracy of computation on factors not accounted for.

One of them is the contribution of higher modes of vibration to the resultant deflections. The solution was carried out by expansion in natural modes of vibration, with only the first term of the series considered. It is clear that the resultant deflection is likely to be increased by the consideration of higher modes. To check this point two higher natural frequencies and the respective natural modes were computed.

The second natural mode is antisymmetric, and the equation to be satisfied at the second natural frequency is

$$\frac{J_{01}}{l_{01}} F_7(\lambda_{01}) + \frac{J_{11'}}{l_{11'}} [F_1(\lambda_{11'}) + F_2(\lambda_{11'})] = 0 \qquad (7.35)$$

The equation is satisfied by

$$\lambda_{01} = 3\cdot58, \quad \lambda_{11'} = 4\cdot12$$

and the natural frequency $f_{(2)} = 4\cdot89$ Hz.

The integration constants of eq. 7.27 are computed in the same way as for the first mode. If we choose $\zeta_1 = \zeta_{1'} = 1$ we get in the first span

$$v_{(2)}(x) = -0\cdot835 \sinh 3\cdot58 \frac{x}{l} - 35\cdot4 \sin 3\cdot58 \frac{x}{l}$$

and in the second span

$$v_{(2)}(x) = -12{\cdot}0 \cosh 4.12\,\frac{x}{l} + 12{\cdot}4 \sinh 4.12\,\frac{x}{l} + 12{\cdot}0 \cos 4.12\,\frac{x}{l} + 6{\cdot}4 \sin 4{\cdot}12\,\frac{x}{l}$$

The second mode is shown in Fig. 7.6b.

The third mode is again symmetric and the third natural frequency satisfies eq. 7.26. It is

$$\lambda_{01} = 3{\cdot}99\,, \quad \lambda_{11'} = 4{\cdot}59$$

$$f_{(3)} = 6{\cdot}08 \ \mathrm{s}^{-1}$$

In span $\overline{01}$

$$v_{(3)}(x) = 4{\cdot}82 \sinh 3{\cdot}99\,\frac{x}{l} + 17{\cdot}47 \sin 3{\cdot}99\,\frac{x}{l}$$

and in span $\overline{11'}$

$$v_{(3)}(x) = 112{\cdot}5 \cosh 4{\cdot}59\,\frac{x}{l} - 110{\cdot}2 \sinh 4{\cdot}59\,\frac{x}{l} - 112{\cdot}5 \cos 4{\cdot}59\,\frac{x}{l} + 127{\cdot}1 \sin 4{\cdot}59\,\frac{x}{l}$$

The curve is drawn in Fig. 7.6c.

It is clear that the second mode — because of its antisymmetry — will play no role in vibration at point s. The third natural frequency being nearly twice the first one, it seems in order to neglect also the effect of the third mode on the deflection at the first critical speed. However, both the second and the third critical speeds are likely to be reached in practice.

We have already pointed out the other inaccuracy caused by our neglecting the effect of the motion of the inertia mass along the beam. When the locomotive is midway over the central span, the natural frequency is lower than when it is over the end span. This is documented by the diagram in Fig. 7.9 picked up by deflectometers at a locomotive speed of 37 km/h, which shows that the most intensive vibration occurs during the passage over the central span. The diagram resembles the theoretical curve in Fig. 7.7e. At higher speeds vibration occurs during the passage over all three spans. This is plain to see in Fig. 7.10 which looks very much like the theretical curve in Fig. 7.7c. At still higher speeds the bridge is set into vibration mostly during the travel over the end spans, as suggested by the diagram in Fig. 7.11 taken at 44 km/h.

7.3   Dynamic effects on reinforced concrete arch bridges

7.3.1   Introductory notes

The incentive to carry out a theoretical study of bridges of this kind was provided by comparatively extensive experimental investigations of a newly built large-span double-track railway arch bridge stiffened with a rigid frame deck (Fig. 7.15.). Dynamic measurements preceded by detailed static measurements have not only helped to answer some of the important questions that the static analysis had failed to answer to satisfaction, but have also given an idea about the

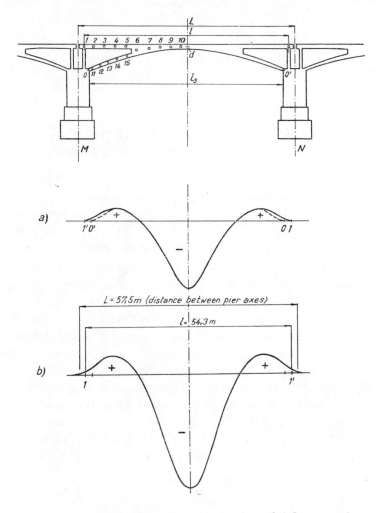

Fig. 7.15 Reinforced concrete arch bridge. (a) influence line of deflection midspan, (b) static deflection during passage of a 97 Mp test locomotive

magnitude of the effects of periodic forces on the vibration of bridges of this type and the role played by the continuity of the structure and the elasticity of the pier foundations.

The first task of the theoretical study was to determine the frequencies and modes of free vibration. It was assumed at the beginning that each span of the structure vibrates like a separate unit. The analysis of free vibration of an arch stiffened with a rigid frame deck was therefore made on the assumption of perfectly fixed springings.

However, when confronted with the results of dynamic measurements made on the structure, the assumptions of the theoretical study were found to be incorrect. The piers are not immobile but vibrate together with the supporting arch structure as do the neighbouring spans of the bridge. The theoretical computation had to be repeated, and in this second attempt the structure was considered to be continuous.

7.3.2 Arch bridge stiffened with a rigid frame deck

The bridge carries two railway tracks each on a separate supporting structure formed by a shallow arch stiffened with a rigid frame deck (Figs 7.15, 7.16). The piers are built common to both tracks. The continuous structure has fifteen spans.

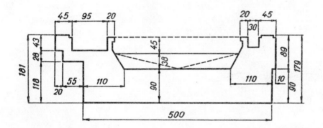

Fig. 7.16 Cross-section of bridge shown in Fig. 7.15

It was assumed in the initial analysis that the piers are perfectly rigid and immobile, and the solution was carried out for a single span supposed to be fixed in the springings.

The methods applicable to systems with curved bars were discussed in Section 4.2. The system in question was first analysed by the method of discrete mass points. As the system is symmetric, all that is needed is to solve one half of the structure. The locations of the mass points are shown in Fig. 7.15. For 15 points the square matrix $[\delta_{ik}]$ of the influence coefficients has 30 rows and 30 columns. The influence lines from which the coefficients δ were taken had been obtained by the model method of Beggs-Blažek. Some of them were checked numerically using the slope-deflection method. The agreement between the two sets of values is very good: the differences are always less than 2%.

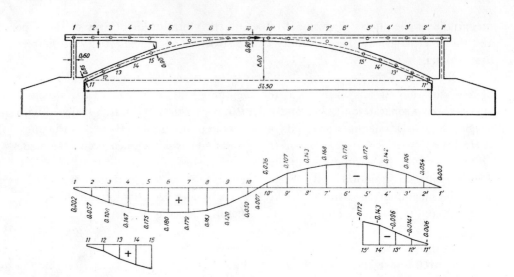

Fig. 7.17 Celluloid model of structure and one of influence lines determined by method of Beggs–Blažek (horizontal displacements of point 10 resulting from application of vertical load)

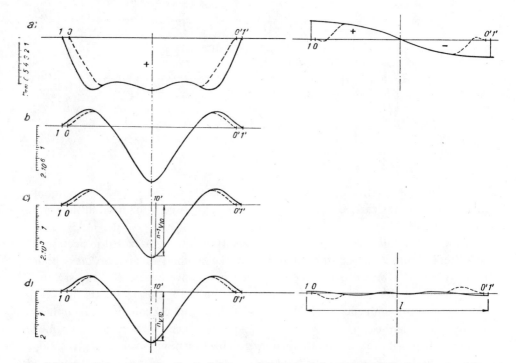

Fig. 7.18 Successive approximations of first natural mode: left — vertical displacements, right — horizontal displacements rotated 90°. Dotted lines — arches, solid lines — frames

One of the influence lines is shown in Fig. 7.17.

The equations of motion (eq. 4.86) were solved by successive approximations (Section 1.2.1), with the deflection curve resulting from self-weight chosen for the initial curve. The vertical coordinates are to the left, the horizontal ones (rotated to the vertical direction) to the right of Fig. 7.18a. The last steps of the iterative solution for the vertical direction are in Figs 7.18b, c, d, left, and for the horizontal direction in Fig. 7.18d, right. Fig. 7.18d thus represents the first symmetric mode of free vibration. The ratio of the coordinates of the last two approximations at, for example, point 10 gives the square of the natural circular frequency

i.e.

$$\omega_{(1)}^2 = \frac{2386}{2 \cdot 19} = 1090 \text{ s}^{-2}$$

$$\omega_1 = 33 \cdot 0 \text{ s}^{-1}$$

$$f_{(1)} = 5 \cdot 25 \text{ Hz}$$

It is clear to see in Fig. 7.18 that in symmetric vibration the effect of horizontal displacements is small and may be neglected.

The first mode and frequency of antisymmetric vibration are computed in an analogous manner. The computation starts out from a chosen antisymmetric deflection curve. The vertical and horizontal coordinates of the first antisymmetric mode

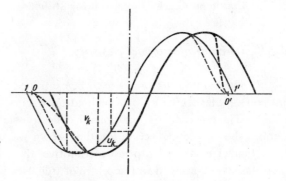

Fig. 7.19 Antisymmetric mode
of free vibration

are shown in Fig. 7.19. The first circular frequency of the antisymmetric vibration given by the ratio of the last two approximations is

$$\omega_{(1')} = 16 \cdot 6 \text{ s}^{-1}$$

Next to the natural frequencies and modes we also computed the static deflection in the vault head caused by the 97 Mp test locomotive in various positions on the bridge. This curve (Fig. 7.15b) was obtained by applying the axle thrusts of the locomotive to the influence line of deflection in Fig. 7.15a. The largest coordinate, $v_{st} = 3.0$ mm belongs to the case of the locomotive central axle coincident with the midpoint of the bridge.

The theoretical analysis was followed by measurements which showed the computed results to be at variance with the measured values. The computed fundamental frequency was higher than the one determined experimentally. The measured maximum deflection $(v_{st} = 2.0$ mm) was less than the computed one. The shape of the deflection curve picked up at low speeds of travel (Fig. 7.28a) differed from that of the curve in Fig. 7.15b with which it was supposed to coincide.

The reason for these discrepancies can be found to lie in the fact that the initial computation failed to take into account the effect of the massive parapet walls which, when acting together with the supporting structure, nearly double the stiffness of the supporting structure (cf. the cross-section in Fig. 7.16). On the other hand it neglected the continuous nature of the structure which in contrast reduces the total stiffness. In reality the piers rest on elastic foundation soil and are free to incline. To reduce the discrepancies the calculation was repeated and the actual stiffness and the continuous nature of all spans were taken into consideration.

Note: At a later date the computation was carried out using the simplified slope-deflection method outlined in sections 3.3 and 4.2.3. The results of the two methods are in good agreement. The numerical procedure was published in Reference 44.

7.3.3   Continuous arch bridge with top stiffening

7.3.3.1 Determination of pier elasticity

The continuous structure to be examined is shown in Fig. 7.20a. The determination of the degree of pier inclination begins by making use of the diagrams in Fig. 7.28a recorded by a Stoppani deflectometer at a low speed of travel. The diagram shows the deflection in the crown of arch MN. When the adjacent arch is loaded by the 97 Mp test locomotive, the crown of arch MN rises 0.25 mm. When the load is applied to the centre of arch MN its crown sags 2.0 mm. A portion of this sagging can be ascribed to the inclination of the piers founded on elastic soil. As is evident from the records, three arches are influenced significantly by a static load. Sagging produced in the crown of arch MN by the outward inclination of the piers is equally as large as the combined rise of the crowns of the two adjacent arches (Fig. 7.22). Were arch MN supported by unyielding piers, the maximum deflection of its crown would be

$$v_{st} = 2.0 - 2 \times 0.25 = 1.5 \text{ mm}$$

Comparing this deflection with the computed 3 mm deflection produced by the 97 Mp test locomotive we see that the stiffness of the reinforced concrete structure is actually double that corresponding to theoretical computation with modulus $E = 420 \, \mathrm{Mp/cm^2}$. This higher stiffness is no doubt due to the massive parapets

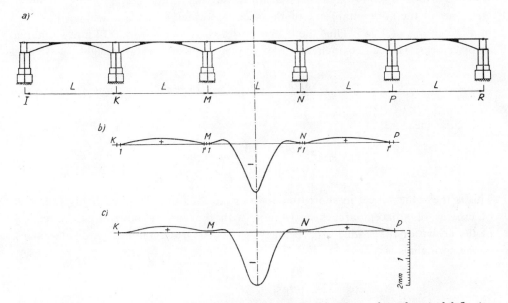

Fig. 7.20 Continuous arch structure. (a) Schematic diagram, (b) influence line of vertical deflection midway of central span, (c) vertical deflection midway of central span underneath a 97 Mp loco-motive travelling at low speed

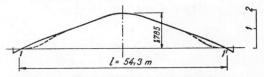

Fig. 7.21 Influence line of horizontal force

which act monolithically with the rest of the structure and whose effect has not been included in the static computation.

The deflections referred to above were measured after the load had been applied for just a few seconds. With prolonged application the deflections grow continually and do not become steady until after about 15 min. The maximum deflections recorded by a Frič deflectometer in the crowns of arches under load were 2·3 mm, i.e. about 15% larger than those produced by short-time applications. This pheno-menon can be ascribed to the plasticity of the foundation soil under the piers rather than to the imperfect elasticity of the concrete. It is noticeable even at low speeds of travel, as is evident from the record in Fig. 7.28a.

Knowing the rise of the crowns of adjacent arches — 0·25 mm in this case — we can compute the settlement of the supports and the inclination of the piers. In doing so we can omit the rotation and vertical displacement experienced by the arch springings during the inclining of the piers, and consider only the horizontal displacement which is of incomparably greater importance. As seen in Fig. 7.21 showing the influence line for the horizontal force of the arch, a springing displacement of 1 results in a crown rise of 1·785. Thus the 0·25 mm rise corresponds to a 0·14 mm springing displacement. Hence the inclination of a pier with an overall height of 18·8 m is

$$\varphi = \frac{0 \cdot 00014}{18 \cdot 8} = 0 \cdot 745 \times 10^{-5}$$

in circular measure, or

$$\varphi = \frac{0 \cdot 745 \times 10^{-5}}{0 \cdot 485 \times 10^{-5}} = 1 \cdot 54''$$

This is the value we get in short-time loading. In long-time loading the average rise of crowns of adjacent arches is 0·4 mm, with the corresponding total deflection of the crown of the loaded arch being $1 \cdot 5 + 2 \times 0 \cdot 4 = 2 \cdot 3$ mm. The displacement of the springings is then

$$\frac{0 \cdot 4}{1 \cdot 785} = 0 \cdot 224 \text{ mm}$$

and the inclination of the pier is

$$\varphi = \frac{0 \cdot 000224}{18 \cdot 8 \times 0 \cdot 485 \times 10^{-5}} \approx 2 \cdot 5''$$

This is the average inclination of the pier centreline. Since the pier is not absolutely rigid and deforms elastically, the inclination of its upper end will be somewhat larger. An approximate computation — not repeated here — yields $\varphi \approx 2 \cdot 8''$. The actual inclination of the pier was measured with levels but angles of 2″ to 3″ are on the limit of discernibility.

The conclusions arrived at in the dynamic investigation are in complete harmony with the results of detailed static measurements performed previously. According to them, the deflections of the spans differ considerably one from another depending on the height of the piers, and the deflection of the crowns of the central arches under the test locomotives is likely to be almost double the deflection of the end spans. This is due to the inclination of the central piers, which is not difficult to compute.

Information gained in the experimental investigation made it possible to devise a satisfactory computational model for the dynamic solution.

7.3.3.2 Dynamic response of continuous systems

It follows from a simple consideration that the continuity of the structure manifests itself far more markedly in vibration than in static loading. Static load applied to one span on a single track will act on the piers in the form of horizontal force imposed only on the springs of the loaded arch. But when the structure vibrates as shown in Fig. 7.22, a pier is subjected to additional forces of identical direction deriving from all four arches (two spans, two tracks) resting against it. The inclination of the piers of a vibrating structure is therefore comparatively large and observable even in the more remote piers.

(a)   Free vibration

The first thing to do in dynamic computations is to choose a suitable computational scheme. It would be possible to compute the whole structure as a two-dimensional system with repeated elements using the procedure outlined in Section 5.2. But since the structure has a large number of spans, its first natural frequency of the

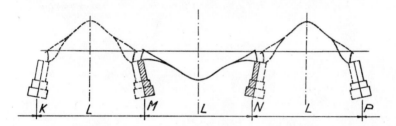

Fig. 7.22 Schematic diagram of a continuous arch structure vibrating in its first natural mode

symmetric mode will be approximately the same as that of a system with an infinite number of spans. Vibration of such a system with piers assumed to be perfectly rigid but founded on elastic soil is shown in Fig. 7.22. As the form of vibration is evidently the same for all the spans, it is possible to take out of the system a single span (marked in heavy lines in Fig. 7.22) and consider it separately. A schematic diagram of the span is in Fig. 7.23. Assume that the pier is absolutely rigid, rotates about the axis passing through point M perpendicular to the plane of the drawing, and is sprung in the base of the foundation. Consider only the portion of the mass belonging to one track. The rigid pier may be replaced by two mass points, the position of one of which may be chosen at will. Choose it at the centre of rotation I so that it is at distance ϱ_I below the centre of gravity, with $\varrho_I = 7\cdot11$ m the distance between the centre of gravity and the base of foundation. The second substitute mass point II is at a distance $\varrho_{II} = I/m\varrho_I$ above the centre of gravity (m — the total mass of the pier, I — its moment of inertia about the gravity axis parallel to the axis of rotation).

For $m = 1020 \text{ Mp m}^{-1} \text{ s}^2$, $I = 38\,800 \text{ Mp m s}^2$

$$\varrho_{\text{II}} = \frac{38\,800}{1020 \times 7 \cdot 11} = 5 \cdot 36 \text{ m}$$

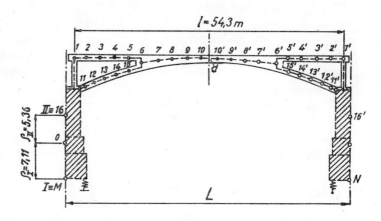

Fig. 7.23 Computational set-up for one span of continuous arch shown in Fig. 7.20

The height of point m_{II} above the base of foundation is

$$\varrho_{\text{I}} + \varrho_{\text{II}} = 7 \cdot 11 + 5 \cdot 36 = 12 \cdot 47 \text{ m}$$

and its mass is

$$m_{\text{II}} = \frac{m\varrho_{\text{I}}}{\varrho_{\text{I}} + \varrho_{\text{II}}} = \frac{1020 \times 7 \cdot 11}{12 \cdot 47} = 582 \text{ Mp m}^{-1} \text{ s}^2$$

However, point 16 of the substitute system according to Fig. 7.23 possesses but 1/4 of this mass, for it contains only a half of the pier and that for only one rail. Therefore

$$m_{16} = \frac{m_{\text{II}}}{4} = 146 \text{ Mp m}^{-1} \text{ s}^2$$

As we know from the computation of the first natural frequency of symmetric vibration of the system illustrated in Figs 7.15 and 7.18, horizontal motions are of little importance at the arches and therefore need not be considered. As to the first symmetric vibration of the system shown in Fig. 7.23: the situation is approximately the same as in the previous system, but for points 1 to 15 only. At point 16 only the horizontal motion plays a role. The matrix of the influence co-efficients based on these assumptions has 16 rows and 16 columns and is set out in Table 7.5. The influence line of deflection for the arch crown is shown in Fig. 7.24. The horizontal coordinate δ_{s16} denotes the vertical displacement of the crown, s, for a unit horizontal force acting at point 16. The influence coefficients were computed assuming Young's modulus $E = 4.2 \times 10^6$ Mp m^{-2} and a deck strengthened with parapets, as shown in Fig. 7.16.

The solution by successive approximations according to eq. 4.86 and Section 1.2.1 yields the first natural frequency as follows

$$\omega^2_{(1)} = 536 \text{ s}^{-2}$$

$$\omega_{(1)} = 23.1 \text{ s}^{-1}$$

$$f_{(1)} = 3.68 \text{ Hz}$$

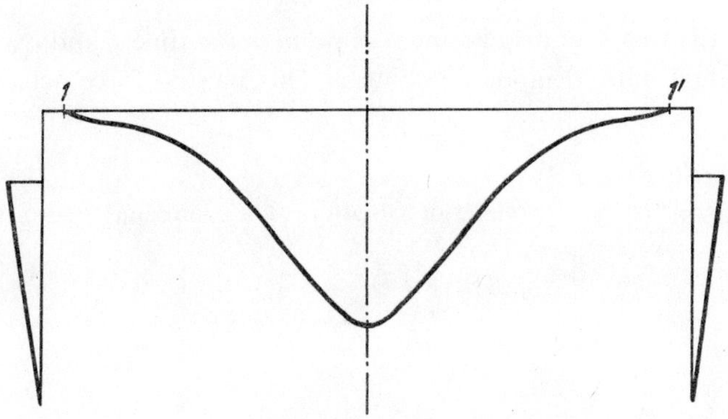

Fig. 7.24 Influence line of lateral deflection of system shown in Fig. 7.23

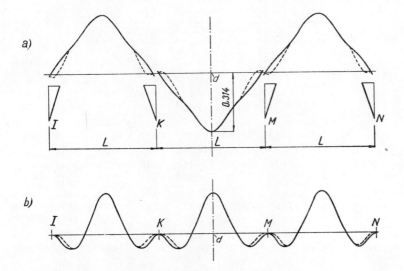

Fig. 7.25 Free vibration of continuous structure shown in Fig. 7.20. (a) Vertical coordinates of span and horizonal coordinates of pier inclination at first mode, (b) vertical coordinates at 15th mode of free vibration

The vertical coordinates of the first symmetric mode of free vibration are drawn in Fig. 7.25a (the horizontal coordinates of the vibrating pier are represented by the triangular figures).

The next task is to get information about the effect of the locomotive mass placed at some cross-section, say at d, on free vibration. To this end we shall use the modified equations 7.14—7.18. Consider the case of the system vibrating in the first natural mode, and assume approximately that this mode shape undergoes no change owing to the addition of locomotive mass m. In view of eq. 4.97 the inertia force is

$$-m \frac{d^2 r(d, t)}{dt^2} = -m \frac{d^2 q_{(1)}(t)}{dt^2} r_{(1)}(d) \qquad (7.36)$$

where $r(d, t)$ is the vector of displacement at point d and time t, and $r_{(1)}(d)$ the displacement of the first natural mode at point d. This inertial force acts on the system

Table 7.5

Computation of the first natural mode of vibration of a continuous structure by the method of

1	2	3	4	5	6	7	8	9	10	11
								Influence coefficients $\delta_{ik} = 10^6$		
Point i	Mass m_i Mpm^{-1} s^2								vertical deflections at point k	
		1	2	3	4	5	6	7	8	9
1	10·84	0	0	0	0	0	0	0	0	0
2	5·77	0	5·36	7·81	7·83	6·88	5·97	5·14	4·35	3·65
3	5·77	0	7·81	13·18	15·36	13·50	11·45	9·86	7·74	6·20
4	6·60	0	7·83	15·36	19·98	18·55	16·20	14·16	11·50	9·57
5	7·60	0	6·88	13·50	18·55	20·72	19·88	19·10	17·45	16·30
6	11·73	0	5·97	11·45	16·20	19·88	21·86	23·71	24·54	25·22
7	9·40	0	5·14	9·86	14·16	19·10	23·71	28·97	32·75	35·63
8	7·98	0	4·35	7·74	11·50	17·45	24·54	32·75	39·90	46·24
9	6·96	0	3·65	6·20	9·57	16·30	25·22	35·63	46·24	56·92
10	6·45	0	3·12	5·17	8·17	15·23	25·35	37·30	49·50	63·30
11	2·00	0	0	0	0	0	0	0	0	0
12	2·36	0	1·64	3·26	4·47	4·73	3·88	3·23	2·38	1·60
13	2·54	0	3·70	7·50	10·09	10·84	9·46	7·92	6·09	4·62
14	2·69	0	5·54	11·00	15·21	16·64	15·30	13·96	11·55	10·00
15	3·03	0	6·65	12·70	17·51	20·22	19·85	19·58	18·10	17·04
16	146	0	1·06	1·96	2·77	3·75	4·68	5·71	6·42	7·16

like a concentrated load and can be expanded in series according to eq. 4.100 in the same way as load $p(s, t)$ was. Because of the assumption we have made earlier, i.e. that the first mode shape of free vibration is not affected by the action of this force, all we have to consider in eq. 4.98 is the first term of the series, which in view of eqs 7.36 and 4.100 turns out to be

$$-m \frac{d^2 q_{(1)}(t)}{dt^2} \frac{\mu(s) \, r^2_{(1)}(d)}{\int \mu(s) \, r^2_{(1)}(s) \, ds} \, r_{(1)}(s) \tag{7.37}$$

However, this expression represents the inertia force of the continuously distributed mass

$$\Delta\mu = \mu(s) \, m \frac{r^2_{(1)}(d)}{\int \mu(s) \, r^2_{(1)}(s) \, ds} \tag{7.38}$$

successive approximations

12	13	14	15	16	17	18	19	20	21	22
mMp^{-1}							(n − 1)-th approx.		n-th approx.	
						horiz. deflect at point	$\dfrac{^{n-1}v_i}{s^{2(n-2)}}$	$\dfrac{m_i \, ^{n-1}v_i}{Mpm^{-1}\, s^{2(n-1)}}$	$\dfrac{^{n}v_i \cdot 10^2}{s^{2(n-1)}}$	19 : 21 $\omega^2_{(1)} s^{-2}$
10	11	12	13	14	15	16				
0	0	0	0	0	0	0	0	0	0	—
3·12	0	1·64	3·70	5·54	6·65	1·06	0·205	1·18	0·0389	527
5·17	0	3·26	7·50	11·00	12·70	1·96	0·377	2·18	0·0712	528
8·17	0	4·47	10·09	15·21	17·51	2·77	0·536	3·53	0·1016	528
15·23	0	4·73	10·84	16·64	20·22	3·75	0·732	5·56	0·1364	536
25·35	0	3·88	9·46	15·30	19·85	4·68	0·917	10·76	0·1702	538
37·30	0	3·23	8·92	13·96	19·58	5·71	1·139	10·70	0·2139	532
49·50	0	2·38	6·09	11·55	18·10	6·42	1·334	10·64	0·2500	533
63·30	0	1·60	4·62	10·00	17·04	7·16	1·533	10·67	0·2868	534
73·70	0	1·15	3·67	8·95	16·11	7·54	1·658	10·69	0·3084	536
0	0	0	0	0	0	0	0	0	0	—
1·15	0	4·93	6·52	6·04	4·90	0·72	0·122	0·29	0·0233	523
3·67	0	6·52	12·10	12·94	11·40	1·70	0·299	0·76	0·0568	526
8·95	0	6·04	12·94	17·56	17·45	2·85	0·521	1·40	0·0987	528
16·11	0	4·90	11·40	17·45	20·73	3·83	0·736	2·23	0·1389	528
7·54	0	0·72	1·70	2·85	3·83	1·41	0·232	33·90	0·0435	533

The system then vibrates as though the total mass

$$\bar{\mu}(s) = \mu(s) + \Delta\mu \tag{7.39}$$

were distributed over it.

According to Fig. 7.25a the deflection midway of the arch in the first natural mode is

$$r_{(1)}(d) = v_{(1)}(d) = 0.314$$

For a single span carrying one track numerical integration yields

$$\int\mu(s)\, r_{(1)}^2(s)ds = 2\sum_{i=1}^{15} m_i v_{(1)i}^2 + 2m_{16}u_{(1)}^2 = 6.15\ \text{tm}^{-1}\,\text{s}^2 \tag{7.40}$$

where $u_{(1)}$ is the horizontal displacement of point 16 in the first natural mode.

If all the 15 spans of the bridge are set into vibration, then by eqs 7.38 and 7.39

$$\frac{\bar{\mu}(s)}{\mu(s)} = 1 + \frac{97 \times 0.314^2}{9.81 \times 30 \times 6.15} = 1.005 \approx 1$$

The inertia force arising owing to the vertical motion of the locomotive was therefore neglected.

The first mode depicted in Fig. 7.25a corresponds to the first natural mode of the continuous structure. According to the theory deduced in Chapter 5 a system with n spans has n natural modes of the lowest group. The n-th mode corresponds to the mode of vibration obtained on the assumption of fixed springings. This mode is shown in Fig. 7.25b.

(b)   Forced vibration of a continuous system

The general method of solving forced vibration was described in Chapter 4, formulae 4.96–4.102. Consider a vertical harmonic force $P \sin \Omega t$ moving along the bridge at constant velocity c. $\Omega = 2\pi N$ is the circular frequency of the force, and simultaneously the angular velocity of the driving wheels, and N the revolutions per second of the locomotive axles. If time is measured from the instant the force enters the span, then at time t the force will be at distance $a = ct$ from the beginning of the span. By eq. 4.100

$$P_{(j)}(t) = \frac{P \sin \Omega t\, r_{(j)}(a)}{n \int \mu(s)\, r_{(j)}^2(s)\, ds} \tag{7.41}$$

where the integration sign includes the whole span, and n is the number of spans. Expression 7.41 may be substituted in eq. 4.102 and the latter integrated — numerically or graphically — to get $q_{(j)}(t)$. The displacements at any point in the structure could then be determined from eqs 4.102 and 4.97.

Were $\Omega = \omega_{(1)}$ and if the natural frequencies were not too close one to another, we could suppose that only the first natural mode would participate in the vibration, and compute only $p_{(1)}(t)$ and the first term of eq. 4.97. As, however, the natural frequencies of continuous structures with a large number of spans are very close one to another, it is not possible — even at resonance — to neglect the higher terms of eq. 4.97 mostly for the reason that what is involved here is not the computation of the amplitude of steady vibration but vibration with steadily growing amplitude in which the higher modes participate very intensively. Were the higher modes neglected, the deflections would be very much smaller than those of the exact solution because the denominator of formula 7.41 contains the number of spans, n. The first and foremost requirement to be satisfied in the computation of forced vibration is the determination of all n fundamental natural modes and frequencies, and the solution of eq. 4.102 for all j, from 1 to 15. As we were in no position to do this, we had to solve the problem approximately.

As in the computation of the first mode of free vibration, the system was replaced by the computational model shown in Fig. 7.23, but unlike that case the structure underneath both tracks was considered. The problem was solved only for the case of resonance $\Omega = \omega_{(1)}$, and the computation aimed at finding the deflection of the central arch crown.

It should be realised that as the load passes over it, the system of Fig. 7.23 does not start to vibrate independently but the motion is actually transmitted to the neighbouring spans, too. In our computation the energy thus lost will approximately be considered as energy lost in damping, and the damping coefficient ω_b appropriately increased as a result. (This practice is quite common in structural dynamics; for example, machine foundations are usually computed without paying attention to the fact that the whole neighbourhood of the foundation is set in vibration in the course of which a considerable mass of the subsoil is in motion).

Bearing in mind the above conditions let us compute the deflection of the centre of arch MN assuming that the force moving in this span is harmonic, $P \sin \Omega t$. Taking into account the structure of both tracks, the loading term according to eq. 7.41 is

$$P_{(1)}(t) = \frac{P \sin \Omega t \, v_{(1)}(a)}{2 \times 2 \left(\sum_i m_i v_{(1)i}^2 + m_{16} u_{(1)}^2 \right)} \tag{7.42}$$

The vertical displacement $v_{(1)}(a)$ of the bridge deck subjected to a moving load may be expanded in a Fourier series whose coefficients are computed by numerical integration as shown in Table 7.6. We get

$$v_{(1)}(a) = B_1 \sin \frac{\pi a}{l} + B_3 \sin \frac{3\pi a}{l} = 0{\cdot}272 \sin \frac{\pi a}{l} - 0{\cdot}031 \sin \frac{3\pi a}{l} \tag{7.43}$$

where $a = ct$ is measured from point 1 (Fig. 7.23) and $l = 54\cdot3$ m is double the distance between point 1 and the centre of the span. As the second term of eq. 7.43 is of little significance and may be neglected, substitution in eq. 7.42 gives

$$P_{(1)}(t) = \frac{PB_1 \sin \Omega t \sin \omega t}{2 \times 2(\sum_i m_i v_{(1)i}^2 + m_{16} u_{(1)}^2)} \tag{7.44}$$

where $\omega = \pi c/l$.

Expression 7.44 becomes identical with eq. 7.5 on substituting

$$j = 1, \quad B = B_1, \quad A = C = D = 0$$

and

$$\sum \int \mu\, v_{(j)}^2(x)\, \mathrm{d}x = 4 \sum_i m_i v_{(1)i}^2 + 4 m_{16} u_{(1)}^2$$

Consequently the equations of Tables 7.1 and 7.2 and expression 7.12 will apply. Since by Table 7.2 $\bar{K}_3 = \bar{K}_4 = 0$, eq. 7.12 will simplify to

$$q_{(1)}(t) = \cos \omega_{(1)}t \big[K_3 \cos \omega t + K_4 \sin \omega t \big] - \mathrm{e}^{-\omega bt} \Big\{ \big[K_3 - q_{(1)}(0) \big] \cos \omega_{(1)}t$$
$$- \frac{1}{\omega_{(1)}} \dot{q}_{(1)}(0) \sin \omega_{(1)}t \Big\} \tag{7.45}$$

Table 7.6

	1	2	3	4	5	6	7	8
	i	x^*	$v_{(1)}(x)$	$\sin(\pi x/l)$	Δx^{**}	$3 \times 4 \times 5$	$\sin(3\pi x/l)$	$3 \times 5 \times 7$
Segments	1	0	0	0	1·33	—	—	—
	2	2·66	0·0389	0·153	2·66	0·0158	0·446	0·0462
	3	5·32	0·0712	0·303	2·66	0·0573	0·798	0·1084
	4	8·15	0·1016	0·455	3·00	0·1385	0·988	0·301
	5	11·15	0·1364	0·601	3·00	0·246	0·934	0·382
	6	14·10	0·1702	0·728	2·90	0·360	0·642	0·137
	7	17·00	0·2139	0·832	2·90	0·516	0·191	0·1183
	8	19·90	0·2500	0·913	2·90	0·662	0·304	−0·220
	9	22·80	0·2868	0·968	2·90	0·805	−0·726	−0·605
	10	25·70	0·3096	0·996	2·90	0·893	−0·967	−0·867
	Σ					$= 3\cdot694$		$= -0\cdot419$

$$2 \sum_{i=1}^{10} v_{(1)}(x) \sin \frac{\pi x}{l}\, \Delta x \cdot \frac{P}{l/2} = 3\cdot694 \cdot \frac{4P}{54\cdot30} = 0\cdot272\,P$$

$$-0\cdot419 \cdot \frac{4P}{54\cdot30} =$$
$$= -0\cdot0309$$

* Horizontal distance between the centre of gravity of segment and point l.

** Length of segment.

where

$$K_3 = \frac{P}{2\omega_{(1)}} \frac{\omega}{\omega^2 + \omega_b^2} \frac{B_1}{4 \sum m_i v_{i(1)}^2 + 4m_{16} u_{(1)}^2} \tag{7.46}$$

and

$$K_4 = -\frac{P}{2\omega_{(1)}} \frac{\omega_b}{\omega^2 + \omega_b^2} \frac{B_1}{4 \sum m_i v_{i(1)}^2 + 4m_{16} u_{(1)}^2} \tag{7.47}$$

Assuming further that at the instant of the load entry on the bridge the system was at rest, then

$$q_{(1)}(0) = 0, \quad \dot{q}_{(1)}(0) = 0 \tag{7.48}$$

At the first mode of free vibration the midspan deflection is $v_{(1)}(l/2)$. By eq. 7.2

$$v\left(\frac{l}{2}, t\right) = q_{(1)}(t)\, v_{(1)}\left(\frac{l}{2}\right) \tag{7.49}$$

and on substituting from eqs 7.45−7.48 in eq. 7.49

$$v\left(\frac{l}{2}, t\right) = \frac{P\, v_{(1)}\left(\frac{l}{2}\right)}{2\omega_{(1)}(\omega^2 + \omega_b^2)\, 4 \sum_i m_i v_{i(1)}^2 + 4m_{16} u_{(1)}^2}$$

$$\times \left[\omega(\cos \omega t - e^{-\omega_b t}) - \omega_b \sin \omega t\right] \cos \omega_{(1)} t \tag{7.50}$$

Following addition of the term expressing the deflection produced by the locomotive weight, this equation looks like eq. 7.22 for a simply supported beam subjected to a mobile load.

After the locomotive has left the span the system continues to execute free vibration which, however, attenuates more rapidly than is indicated by the approximate theoretical analysis. The reason for this should be sought in the circumstance that as the locomotive approaches the end of the span, some of the higher modes cease to contribute positively (in Fig. 7.25b the coordinates of that natural mode are negative near the supports). Moreover, as in the case of the simply supported beam, here too the natural frequency changes somewhat. This is why the residual free vibration need not be taken into account.

The locomotive induces dynamic response not only in the span over which it passes but in the neighbouring spans as well. The amplitudes of the neighbouring spans will be smaller than those of the span under load. Referring again to Fig. 7.25 we see that in the span under load the deflections of all the modes add up while in the neighbouring span the coordinates of, say, the 15-th natural mode are of the opposite direction. An exact deflection curve corresponding to the neighbouring span being subjected to a periodic force with a moving point of action could be obtained only through an exact computation. In an approximate computation the course of vibration in the neighbouring span will be assumed to resemble that in the loaded span except for the deflections which will be smaller by factor $\varkappa$. All that is known about this reduction factor is that it is less than unity and decreases with the distance from the loaded span. In our approximate computation we shall assume it to be $\varkappa$ for the next and $\varkappa^2$ for the next-but-one span.

The approximate deflection curve thus depends to a large degree on the choice of two constants, the damping coefficient ω_b and the reduction factor $\varkappa$. In the exact computation there was just one of them, ω_b, a value easy to determine experimentally. We shall see presently that the choice of the two coefficients influences only slightly the dynamic response at the time the locomotive is over the crown of arch MN.

The locomotive used in the loading tests was a 97 Mp tank locomotive with five coupled axles, producing the total centrifugal forces

$$P = 0.3N^2 \tag{7.51}$$

where N is the rev/s of the driving wheels. At the critical velocity

$$c = \omega_{(1)}r = 23.1 \times 0.63 = 14.55 \text{ m/s} = 52 \text{ km/h}$$

(where $r = 0.63$ m is the radius of the driving wheels) the centrifugal force is

$$P = 0.3f_{(1)}^2 = 0.3 \times 3.68^2 = 4.06 \text{ Mp}$$

Further

$$\frac{\omega_{(1)}l}{c} = \frac{23.1 \times 54.3}{14.55} = 86.3$$

$$\omega = \frac{\pi c}{l} = \frac{\pi \, 14.55}{54.3} = 0.842 \text{ s}^{-1}$$

For the chosen logarithmic decrement $\delta = 0.3$

$$\omega_b \approx \delta f_{(1)} = 0.3 \times 3.68 = 1.104 \text{ s}^{-1}$$

$$\frac{\omega_b}{c} = 0.0759$$

Substituting these values in eq. 7.50 and recalling that $\omega/c = \pi/l$, and $t = a/c$ gives

$$v\left(\frac{l}{2}, t\right) = 0.000\,316 \cos 86.3 \, \frac{a}{l} \left[0.842 \left(\cos \frac{\pi a}{l} - e^{-0.0759a} \right) - 1.104 \sin \frac{\pi a}{l} \right] \tag{7.52}$$

Distance a is measured from point 1 of arch MN, and $l = 54.3$ m is double the distance between point 1 and the centre of the arch.

Expression 7.52 is graphically represented in Fig. 7.26a If the periodic force is applied to spans *KM* and *NP*, the deflection of the centre of arch *MN* will again be eq. 7.52 with the right-hand side multiplied by the reduction factor $\varkappa$. Choose $\varkappa = 0.5$ and draw the deflection as a function of a measured from points 1 of arch

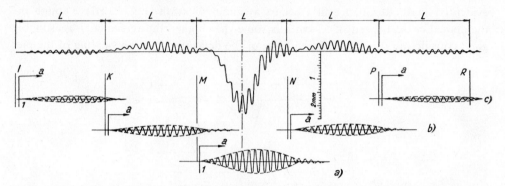

Fig. 7.26 Theoretical curve of dynamic deflection. (a) Vibration of arch crown of loaded span, (b) vibration when load applied to neighbouring spans, (c) vibration when load applied to distant spans

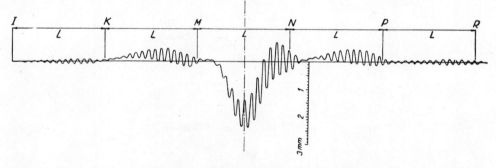

Fig. 7.27 Theoretical curve of dynamic deflection obtained for changed assumptions about damping and behaviour of spans

KM or *NP* (Fig. 7.26b). The deflection in Fig. 7.26c is that obtained analogously for the load applied to arches *IK* and *PR* and $\varkappa^2 = 0.25$. The curves from Figs 7.26a, b and c are superposed on the static deflection curve (Fig. 7.15b) to give the resultant curve (Fig. 7.26d) which is in good correspondence with the records of the Stoppani deflectometer in Figs 7.28d, e and 7.29.

To demonstrate the way the choice of coefficients ω_b and $\varkappa$ affects the solution, we repeated the computation for $\delta = 0.15$ and $\varkappa = 0.4$. Eq. 7.52 now in the form

$$v\left(\frac{l}{2}, t\right) = 0.000\,60 \cos 86.3 \frac{a}{l} \left[0.842 \left(\cos \frac{\pi a}{l} - e^{-0.038a} \right) - 0.552 \sin \frac{\pi a}{l} \right]$$

is shown in Fig. 7.27.

(c)   Results of tests

The deflections recorded midspan of arch *MN* at various speeds of a two-cylinder locomotive are shown in Figs 7.28 and 7.29. At low speeds (33 km/h — Fig. 7.28a, 38 km/h — Fig. 7.28b, 42 km/h — Fig. 7.28c) the dynamic effects are evidently small. Resonance first occurs at about 47 km/h (Fig. 7.28d). Though the theoretical critical speed of 52 km/h is somewhat higher, the agreement is reasonably good considering the strong dependence of the deflections on the yielding of the pier

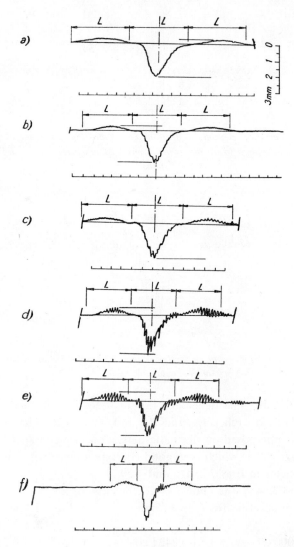

Fig. 7.28 Diagram taken by recording deflectometer in crown of central arch at various speeds of a two-cylinder locomotive. (a) 33 km/h, (b) 38 km/h, (c) 42 km/h, (d) 47 km/h, (e) 51 km/h, (f) 62 km/h

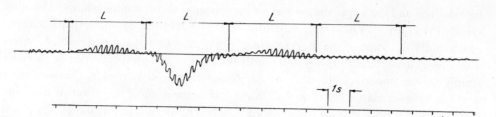

Fig. 7.29 Effect of structure continuity (deflectometer record taken at 52·5 km/h)

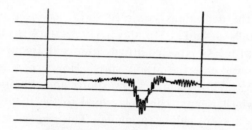

*Fig. 7.30 Record taken during passage of a two-cylinder locomotive travelling at critical speed,
by a strain gauge cemented on an arch springing*

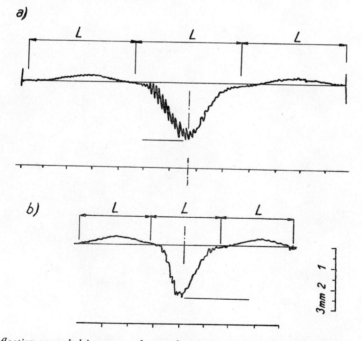

*Fig. 7.31 Deflection recorded in crown of central arch during passage of a 124 Mp three-cylinder
locomotive*

foundations and the approximate way of expressing it in our computations. At higher speeds (51 km/h — Fig. 7.28e, 62 km/h — Fig. 7.28f) the dynamic displacements again diminish. Some of the diagrams (e.g. Fig. 7.28a) hint at the presence of hysteresis. These effects, small anyway and hardly in evidence at high speeds, can be ascribed simply to the imperfect elasticity of the foundation soil. However, the system as a whole behaves elastically. Fig. 7.29 is a good illustration of the effect of dynamic loading of the more distant spans (speed — 52 km/h). While the static effects are exerted only by the adjacent spans, the dynamic effects are observable even with the load in the fourth span. Fig. 7.30 is a record of a strain-gauge cemented onto an arch springing.

Fig. 7.31 shows the deflections recorded by a deflectometer midspan of arch MN at various speeds of a 124 Mp three-cylinder tank locomotive with driving wheel radius $r = 0.81$ m. As expected no vibration with frequency equal to the rev/s of the driving axles was observed at any speed. However, at about 50 km/h i.e. at $N = 2.7 \text{ s}^{-1}$ (Fig. 7.31a) the records show appreciable vibration with a frequency of about 7.7 s^{-1} which is approximately equal to the natural frequency $f_{(15)}$ of vibration of arches with unyielding springings. This frequency is about three times the rev/s of the locomotive driving wheels. The cause of this phenomenon is not fully understood. At the maximum speed of this test locomotive, 75 km/h, the dynamic effects are negligible (Fig. 7.31b).

8   SECONDARY EFFECTS

8.1   Effect of shear deflection and rotatory inertia on lateral vibration

8.1.1   Exact solution

The deflection of a bar results not only from the longitudinal extension of fibres due to the normal stresses of bending, as hitherto assumed, but also from transverse shear displacements of cross-sections due to shear stresses. Another factor which we have neglected in the preceding chapters is the effect of rotation of the length elements of a bar in vibration. The mass of the bar is not concentrated on its gravity axis but is distributed across the bar depth so that a vertical motion of the centre of gravity of the cross-section in the vertical direction causes (in general) the cross-section to execute rotary motion about its horizontal gravity axis. (cf. Reference 74).

Both effects — the shear deflection and the rotary motion of the mass — are small except when the depth of the beam is large relative to its length, as it is, for example, in the frame structures of heavy machinery foundations.

Consider a straight bar of constant cross-section and constant mass per unit length. The bar deflection $v(x, t)$ may be thought of as composed of two parts, deflection $v_{\mathrm{I}}(x, t)$ dependent on the rotation of the planes of the cross-section, and deflection $v_{\mathrm{II}}(x, t)$ corresponding to shear displacement.

Then

$$v(x, t) = v_{\mathrm{I}}(x, t) + v_{\mathrm{II}}(x, t) \tag{8.1}$$

The bending moment depends solely on the first component and is given by

$$M(x, t) = -EJ\frac{\partial^2 v_{\mathrm{I}}(x, t)}{\partial x^2} \tag{8.2}$$

The second component depends on forces $T(x, t)$

$$\frac{\partial v_{\mathrm{II}}(x, t)}{\partial x} = \frac{T(x, t)}{kGS} \tag{8.3}$$

where k is a constant which depends on the shape of the cross-section.

Assume a harmonic motion and carry out the computation in terms of amplitudes. The shearing force is

$$\frac{\mathrm{d}T(x)}{\mathrm{d}x} = -\mu\omega^2\, v(x) \tag{8.4}$$

and the moment of the inertia forces of the rotating length element is

$$-\mu\,dx\,r^2\frac{\partial^3 v_1(x,\,t)}{\partial x\,\partial t^2} = \mu\,dx\,\omega^2 r^2\frac{\partial v_1(x)}{\partial x}\sin\omega t \tag{8.5}$$

where the amplitude is $-\mu\,dx\,\omega^2 r^2(dv_1(x)/dx)$. From the equilibrium of moments acting on the length element of the bar

$$T(x)\,dx = dM(x) - \mu\omega^2 r^2\,dv_1(x) \tag{8.6}$$

The equation resulting from expressions 8.1–8.6 is

$$\frac{d^4 v(x)}{dx^4} + \frac{\mu\omega^2(r^2 + EJ/kGS)}{EJ}\frac{d^2 v(x)}{dx^2} - \frac{\mu\omega^2(1 - \mu\omega^2 r^2/kGS)}{EJ}v(x) = 0 \tag{8.7}$$

The complete integral of this equation is

$$v(x) = A\cos\frac{ax}{l} + B\sin\frac{ax}{l} + C\cosh\frac{dx}{l} + D\sinh\frac{dx}{l} \tag{8.8}$$

where

$$a = \lambda\left\{\frac{\lambda^2}{2}\left(\frac{r^2}{l^2}+\gamma\right)+\left[\frac{\lambda^4}{4}\left(\frac{r^2}{l^2}-\gamma\right)^2+1\right]^{1/2}\right\}^{1/2} \tag{8.9}$$

$$d = \lambda\left\{-\frac{\lambda^2}{2}\left(\frac{r^2}{l^2}+\gamma\right)+\left[\frac{\lambda^4}{4}\left(\frac{r^2}{l^2}-\gamma\right)^2+1\right]^{1/2}\right\}^{1/2} \tag{8.10}$$

and

$$\gamma = \frac{EJ}{kGSl^2} = \frac{E}{kG}\frac{r^2}{l^2} \tag{8.11}$$

With any one of the factors bearing on vibration neglected, formulae 8.9 and 8.10 simplify accordingly: if it is the effect of rotatory inertia, the terms with r^2 drop out, and if it is the shear deflection, $\gamma = 0$.

In the computation of forces acting at the ends of a bar, the end conditions are given by the deflections $v(0)$, $v(l)$. Rotations of the end cross-section $v_1'(0)$, $v_1'(l)$ which are not the same as the slopes of the tangents to the deflection curve $v'(0)$, $v'(l)$, give end moments and forces. The rotation of the cross-section at point x is

$$v_1'(x) = v'(x) - v_{11}'(x) = v'(x) - \frac{T(x)}{kGS} \tag{8.12}$$

Eqs 8.1–8.4 and 8.6 give the shear force

$$T(x) = -EJ\,v_1'''(x) - \mu\omega^2 r^2\,v_1'(x)$$

$$= -EJ\,v'''(x) - \frac{\mu\omega^2 EJ}{kGS}v'(x) - \mu\omega^2 r^2\,v'(x) + \frac{\mu\omega^2 r^2\,T(x)}{kGS}$$

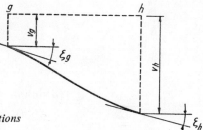

so that

$$T(x) = -\frac{EJ}{1 - \lambda^4\gamma\frac{r^2}{l^2}}\left\{v'''(x) + \frac{v'(x)}{l^2}\left[\lambda^4\left(\gamma + \frac{r^2}{l^2}\right)\right]\right\}$$

$$= -\frac{EJ}{l^3}\,\bar{a}\bar{d}\left(Ad\sin\frac{ax}{l} - Bd\cos\frac{ax}{l} + Ca\sinh\frac{dx}{l} + Da\cosh\frac{dx}{l}\right) \qquad (8.13)$$

where

$$\bar{a} = a\left(1 - d^2\,\frac{\gamma}{1 - \lambda^4\gamma\frac{r^2}{l^2}}\right) = a - \frac{\lambda^4\gamma}{a} \qquad (8.14)$$

$$\bar{d} = d\left(1 + a^2\,\frac{\gamma}{1 - \lambda^4\gamma\frac{r^2}{l^2}}\right) = d + \frac{\lambda^4\gamma}{d} \qquad (8.15)$$

Further

$$\bar{a}\bar{d} = ad\,\frac{1}{1 - \lambda^4\gamma\frac{r^2}{l^2}} \qquad (8.16)$$

Substituting eq. 8.13 in eq. 8.12 gives

$$v_I'(x) = -A\frac{\bar{a}}{l}\sin\frac{ax}{l} + B\frac{\bar{a}}{l}\cos\frac{ax}{l} + C\frac{\bar{d}}{l}\sinh\frac{dx}{l} + D\frac{\bar{d}}{l}\cosh\frac{dx}{l} \qquad (8.17)$$

Fig. 8.1 Schematic diagram of bar deformations

By 8.2 and 8.15 the bending moment is

$$M(x) = -EJ\,v_I''(x) = -\frac{EJ}{l^2}\left(-Aa\bar{a}\cos\frac{ax}{l} - Ba\bar{a}\sin\frac{ax}{l} + Cd\bar{d}\cosh\frac{dx}{l}\right.$$

$$\left. + Dd\bar{d}\sinh\frac{dx}{l}\right) \qquad (8.18)$$

Eqs 8.8, 8.13, 8.17 and 8.18 yield constants A, B, C, D for the computation of end moments and forces (cf. Chapter 2).

For a bar clamped at both ends performing harmonic motion according to Fig. 8.1 the end conditions are

$$v(0) = v_g$$
$$v(l) = v_h$$
$$v_1'(0) = \zeta_g$$
$$v_1'(l) = \zeta_h$$

(8.19)

Substitution in eqs 8.8 and 8.17 gives

$$A + C = v_g$$
$$B\bar{a} + D\bar{d} = l\zeta_g$$
$$A \cos a + B \sin a + C \cosh d + D \sinh d = v_h$$
$$-A\bar{a} \sin a + B\bar{a} \cos a + C\bar{d} \sinh d + D\bar{d} \cosh d = l\zeta_h$$

(8.20)

and from there the integration constants are

$$A = \frac{l}{a^2 + d^2}\left[F_1(a, d)\, \zeta_h + F_2(a, d)\, \zeta_g - F_3(a, d)\frac{v_h}{l} - F_4(a, d)\frac{v_g}{l}\right] + \frac{d\bar{d}}{a^2 + d^2}\, v_g$$

$$B = \frac{l}{\bar{a}(a^2 + d^2)}\left[-F_3(a, d)\, \zeta_h + F_4(a, d)\, \zeta_g - F_5(a, d)\frac{v_h}{l} - F_6(a, d)\frac{v_g}{l}\right]$$
$$+ \frac{al}{a^2 + d^2}\, \zeta_g$$

$$C = -A + v_g$$

$$D = \frac{l}{\bar{d}}\, \zeta_g - B\frac{\bar{a}}{\bar{d}}$$

(8.21)

where

$$F_1(a, d) = -\frac{(a^2 + d^2)\,(\bar{a} \sinh d - \bar{d} \sin a)}{f_1(a, d)}$$

$$F_2(a, d) = -\frac{(a^2 + d^2)\,(\bar{d} \cosh d \sin a - \bar{a} \sinh d \cos a)}{f_1(a, d)}$$

$$F_3(a, d) = -\frac{\bar{a}\bar{d}(a^2 + d^2)\,(\cosh d - \cos a)}{f_1(a, d)}$$

$$F_4(a, d) = \frac{\bar{a}\bar{d}(d\bar{d} - a\bar{a})\,(\cosh d \cos a - 1) + \bar{a}\bar{d}(\bar{a}d + a\bar{d}) \sinh d \sin a}{f_1(a, d)}$$

$$F_5(a, d) = \frac{\bar{a}\bar{d}(a^2 + d^2)(\bar{d} \sinh d + \bar{a} \sin a)}{f_1(a, d)}$$

$$F_6(a, d) = -\frac{\bar{a}\bar{d}(a^2 + d^2)(\bar{a} \cosh d \sin a + \bar{d} \sinh d \cos a)}{f_1(a, d)}$$

(8.22)

$$f_1(a, d) = 2\bar{a}\bar{d}(\cosh d \cos a - 1) + (\bar{a}^2 - \bar{d}^2) \sinh d \sin a$$ (8.23)

The amplitudes of end moments and forces are computed using eqs 8.18 and 8.13. The pertinent formulae analogous to eqs 2.48 and 2.49 are

$$M_{gh} = \frac{EJ}{l}\left[F_1(a, d)\,\zeta_h + F_2(a, d)\,\zeta_g - F_3(a, d)\frac{v_h}{l} - F_4(a, d)\frac{v_g}{l}\right]$$ (8.24)

$$Y_{gh} = \frac{EJ}{l^2}\left[F_3(a, d)\,\zeta_h - F_4(a, d)\,\zeta_g + F_5(a, d)\frac{v_h}{l} + F_6(a, d)\frac{v_g}{l}\right]$$ (8.25)

The values of end moments and forces are summarised in Tables 10.1 and 10.2 in the Appendix. In contrast to functions 2.46, functions 8.22 depend on two parameters, a, d; their numerical values, therefore, are not easy to tabulate.

For a bar hinged at its right-hand end the end conditions are

$$v(0) = v_g$$
$$v(l) = v_h$$
$$v_1'(0) = \zeta_g$$
$$v_1''(l) = 0$$

(8.26)

While the first three of eqs 8.20 apply again, the fourth is replaced by (see eq. 8.18)

$$-Aa\bar{a} \cos a - Ba\bar{a} \sin a + Cd\bar{d} \cosh d + Dd\bar{d} \sinh d = 0$$

The integration constants are

$$A = \frac{l}{a^2 + d^2}\left[F_7(a, b)\,\zeta_g - F_8(a, b)\frac{v_h}{l} - F_9(a, b)\frac{v_g}{l}\right] + \frac{d\bar{d}}{a^2 + d^2}v_g$$

$$B = \frac{l}{\bar{a}(a^2 + d^2)}\left[F_9(a, b)\,\zeta_g - F_{10}(a, b)\frac{v_h}{l} - F_{11}(a, b)\frac{v_g}{l}\right] + \frac{la}{a^2 + d^2}\zeta_g$$

$$C = v_g - A$$

$$D = \frac{l}{d}\zeta_g - B\frac{\bar{a}}{\bar{d}}$$

(8.27)

where the frequency functions $F(a, d)$ are given by formulae 10.35 in the Appendix. The formulae for the end moments and forces are summarised in Tables 10.3 – 10.5 in the Appendix.

For a bar hinged at both ends the integration constants are analogously

$$A = \frac{d\bar{d}}{a^2 + d^2} v_g$$

$$B = \frac{d\bar{d}}{a^2 + d^2} (v_h \cosec a - v_g \cot a)$$

$$C = \frac{a\bar{a}}{a^2 + d^2} v_g$$

$$D = \frac{a\bar{a}}{a^2 + d^2} (v_h \cosech d - v_g \coth d)$$

$$(8.28)$$

The end forces are described by formulae 10.36 and Table 10.6 in the Appendix. And finally, for a cantilever bar with the end conditions

$$v(0) = v_g$$
$$v_1'(0) = \zeta_g$$
$$v_1''(l) = 0$$
$$T(l) = 0$$

$$(8.29)$$

the equations turn out to be

$$A + C = v_g$$
$$B\bar{a} + D\bar{d} = l\zeta_g$$
$$-Aa\bar{a} \cos a - Ba\bar{a} \sin a + Cd\bar{d} \cosh d + Dd\bar{d} \sinh d = 0$$
$$Ad \sin a - Bd \cos a + Ca \sinh d + Da \cosh d = 0$$

and the integration constants are

$$A = \frac{l}{a^2 + d^2}\left[F_{15}(a, d)\, \zeta_g - F_{16}(a, d)\frac{v_g}{l}\right] + \frac{d\bar{d}}{a^2 + d^2} v_g$$

$$B = \frac{l}{\bar{a}(a^2 + d^2)}\left[F_{16}(a, d)\, \zeta_g - F_{17}(a, d)\frac{v_g}{l}\right] + \frac{al}{a^2 + d^2} \zeta_g$$

$$C = v_g - A$$

$$D = \frac{l}{d}\zeta_g - B\frac{\bar{a}}{d}$$

$$(8.30)$$

The frequency functions $F_{15}(a, b)$ to $F_{17}(a, b)$ are given by formulae 10.37 in the Appendix.

If the integration constants A to D are known, formulae 8.8, 8.13, 8.17 and 8.18 may be used for computing the displacements, the slope of the tangent, the bending moment and the shear force in any cross-section of the bar. The formulae for the end forces and moments are set out in Tables 10.7 and 10.8 in the Appendix.

If the velocity of vibration is zero, the limit of the first of eqs 8.22 is

$$F_1(0, 0) = 2\frac{1 - 6\gamma}{1 + 12\gamma} \qquad (8.31)$$

where γ is defined by eq. 8.11. The remaining functions $F(a, d)$ will simplify in a like manner. The respective formulae are reviewed in Tables 10.1–10.8 in the Appendix; they can be used for computing the end forces and moments under static load in the case where shear displacements of cross-sections are to be taken into consideration.

Example 8.1

We shall examine the symmetric forced and free vibrations of the simple frame with fixed columns shown in Fig. 8.2 (Reference 37). In our examination

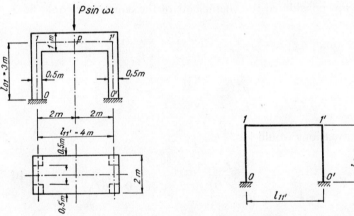

Fig. 8.2 Frame foundation

Fig. 8.3 Computational set-up of system shown in Fig. 8.2

the frame will be replaced by the two-dimensional system in Fig. 8.3, with dimensions as indicated and

$$S_{01} = 2 \times 0.5 \times 0.5 = 0.5 \text{ m}^2 ; \quad J_{01} = \frac{1.0}{12}0.5^3 = 0.01042 \text{ m}^4$$

$$\mu_{01} = 0.125 \text{ Mpm}^{-2} \text{ s}^2$$

$$S_{11'} = 1.0 \times 2.0 = 2.0 \text{ m}^2 ; \quad J_{11'} = \frac{2.0}{12}1.0^3 = 0.1667 \text{ m}^4 ;$$

$$\mu_{11'} = 0.500 \text{ Mpm}^{-2} \text{ s}^2$$

$E = 2.1 \times 10^6 \text{ Mpm}^{-2}$. Because of the large dimensions of the horizontal plate the effect of shear cannot be neglected.

(a)   Static deformation

Before examining the dynamic effects we shall compute the static deformations of the frame induced by a vertical constant force acting at the midspan of the horizontal bar. Using the slope-deflection method we get the equations set out in Table 8.1

Table 8.1

ζ_1	v_1	
a_{11}^*		$= Pl_{11'}/8$
	a_{22}^*	$= P/2$

Were shear not considered the coefficients of the equations would be

$$a_{11}^* = \frac{4EJ_{01}}{l_{01}} + \frac{2EJ_{11'}}{l_{11'}}$$

$$a_{22}^* = \frac{ES_{01}}{l_{01}}$$

$$(8.32)$$

and the deformations would be

$$\zeta_1 = \frac{P \times 4}{8 \times 2 \cdot 1 \times 10^6 \left(\dfrac{4 \times 0 \cdot 01042}{3} + \dfrac{2 \times 0 \cdot 1667}{4} \right)} = 2 \cdot 45 \times 10^{-6} P$$

$$v_1 = \frac{0 \cdot 5 \times P \times 3}{2 \cdot 1 \times 10^6 \times 0 \cdot 5} = 1 \cdot 429 \times 10^{-6} P \text{ (in m, if } P \text{ is in Mp)}$$

with the deflection midspan of the horizontal bar given by

$$v_p = v_1 + \zeta_1 \frac{l_{11'}}{4} + \frac{Pl_{11'}^3}{192EJ_{11'}}$$

$$= (1 \cdot 429 + 2 \cdot 45 + 0 \cdot 952) \times 10^{-6} P = 4 \cdot 83 \times 10^{-6} P \qquad (8.33)$$

Next we shall determine the effects of shear stresses and rotatory inertia. We shall approximately assume that $E/kG = 2 \cdot 4$. Then by eq. 8.11 we get for the columns

$$\gamma_{01} = \frac{2 \cdot 4 \times 0 \cdot 010\,42}{3^2 \times 0 \cdot 5} = 0 \cdot 005\,56$$

and for the horizontal bars

$$\gamma_{11'} = \frac{2 \cdot 4 \times 0 \cdot 1667}{4^2 \times 2} = 0 \cdot 012\,50$$

$$\gamma_{1p} = 4\gamma_{11'} = 0 \cdot 0500$$

The displacement is like the one above, $v_1 = 1 \cdot 429 \times 10^{-6}\,P$; the rotation

$$\zeta_1 = \frac{Pl_{11'}}{8\left[\dfrac{EJ_{01}}{l_{01}}\dfrac{4 + 12\gamma_{01}}{1 + 12\gamma_{01}} + \dfrac{EJ_{11'}}{l_{11'}}\dfrac{2 + 24\gamma_{11'}}{1 + 12\gamma_{11'}}\right]}$$

$$= \frac{P \times 4}{8 \times 2 \cdot 1 \times 10^6 \dfrac{0 \cdot 01042}{3}\left[\dfrac{4 + 12 \times 0 \cdot 00556}{1 + 12 \times 0 \cdot 00556} + \dfrac{2 \times 0 \cdot 1667}{4}\right]} = 2 \cdot 46 \times 10^{-6}P$$

$$(8.34)$$

The deflection midspan of the horizontal bar is

$$\bar{v}_p = v_1 + \zeta_1 \frac{l_{11'}}{4} + \frac{Pl_{11'}^3}{192EJ_{11'}\dfrac{1}{1 + 12\gamma_{1p}}}$$

$$= (1 \cdot 429 + 2 \cdot 46 + 0 \cdot 952 \times 1 \cdot 6) \times 10^{-6} = 5 \cdot 41 \times 10^{-6}P \text{ (in m, if } P \text{ is in Mp)}$$

$$(8.35)$$

It can be seen that the effect of shear displacements has caused the static deflection midspan of the horizontal bar to increase by about 12%.

(b)   Forced vibration

Working on the two assumptions used in the computation of static deformations we shall now determine the amplitudes of steady-state forced vibration.

Table 8.2

ξ_1	v_1	
a_{11}^*	a_{21}^*	$= a_{01}^*$
a_{12}^*	a_{22}^*	$= a_{02}^*$

The shear deflections will not be considered at first. Consider a vertical harmonic force $P \sin \omega t$ acting at the midspan of the horizontal bar, circular frequency $\omega = 209 \text{ s}^{-1}$ and corresponding frequency $n = 2\,000 \text{ min}^{-1}$. The motion of the joints is governed by the equations set out in Table 8.2

The coefficients a^* are expressed by formulae 8.37 with $F(\lambda)$ substituted for $F(a, d)$. On substituting the numerical values we get

$$\lambda_{01} = 3 \left(\frac{0 \cdot 125 \times 209^2}{2 \cdot 1 \times 10^6 \times 0 \cdot 010\,42} \right)^{1/4} = 2 \cdot 12$$

$$\psi_{01} = 3 \left(\frac{0 \cdot 125 \times 209^2}{2 \cdot 1 \times 10^6 \times 0 \cdot 5} \right)^{1/2} = 0 \cdot 216$$

$$\cot \psi_{01} = 4 \cdot 56$$

$F_2(\lambda_{01}) = 3 \cdot 801$ (from the numerical tables in the Appendix)

$$\lambda_{11'} = 4 \left(\frac{0 \cdot 500 \times 209^2}{2 \cdot 1 \times 10^6 \times 0 \cdot 1667} \right)^{1/4} \doteq 2 \cdot 0$$

The numerical values of functions $F_1(\lambda_{11'})$ to $F_6(\lambda_{11'})$ are taken from the tables in the Appendix. With the computed values substituted in, Table 8.2 becomes Table 8.3.

Table 8.3

$\zeta_1 \times 10^4$	$v_1 \times 10^4$	
17·87	−3·005	= 0·518P
−3·005	30·0	= 0·522P

From there the amplitudes are

$$v_1 = 2 \cdot 06 \times 10^{-6} P$$
$$\zeta_1 = 3 \cdot 24 \times 10^{-6} P$$

(again in m and Mp).

The amplitude midspan of the horizontal bar is given by

$$v_p = -v_1 \frac{F_5 \left(\dfrac{\lambda_{11'}}{2} \right)}{F_6 \left(\dfrac{\lambda_{11'}}{2} \right)} + \zeta_1 \frac{l_{11'}}{2} \frac{F_3 \left(\dfrac{\lambda_{11'}}{2} \right)}{F_6 \left(\dfrac{\lambda_{11'}}{2} \right)} + \frac{P l_{11'}^3}{16 E J_{11'} F_6 \left(\dfrac{\lambda_{11'}}{2} \right)}$$

$$= 2.06 \times 10^{-6} P \frac{12 \cdot 129}{11 \cdot 628} + 3 \cdot 24 \times 10^{-6} P \times 2 \frac{6 \cdot 031}{11 \cdot 628}$$

$$+ \frac{P \times 4^3}{16 \times 11 \cdot 628 \times 2 \cdot 1 \times 10^6 \times 0 \cdot 1667} = 6 \cdot 5 \times 10^{-6} P \qquad (8.36)$$

The determinant of the coefficients on the left-hand side of the equations is of value $5 \cdot 27 \times 10^{10}$ Mp².

If shear deflection and rotatory inertia are considered, the equations of Table 8.2 will again apply but the coefficients a will be

$$a_{11}^* = \frac{EJ_{01}}{l_{01}} F_2(a, d)_{01} + \frac{EJ_{11'}}{l_{11'}} [F_2(a, d)_{11'} - F_1(a, d)_{11'}]$$

$$a_{12}^* = -\frac{EJ_{11'}}{l_{11'}^2} [F_4(a, d)_{11'} + F_3(a, d)_{11'}]$$

$$a_{22}^* = \frac{ES_{01}}{l_{01}} \psi_{01} \cot \psi_{01} + \frac{EJ_{11'}}{l_{11'}^3} [F_6(a, d)_{11'} + F_5(a, d)_{11'}]$$

$$a_{01}^* = \frac{Pl_{11'}}{4} \frac{F_3(\tfrac{1}{2}a, \tfrac{1}{2}d)_{11'}}{F_6(\tfrac{1}{2}a, \tfrac{1}{2}d)_{11'}}$$

$$a_{02}^* = -\frac{P}{2} \frac{F_5(\tfrac{1}{2}a, \tfrac{1}{2}d)_{11'}}{F_6(\tfrac{1}{2}a, \tfrac{1}{2}d)_{11'}}$$

$$(8.37)$$

The r^2/l^2 ratio is; for the column

$$\frac{r_{01}^2}{l_{01}^2} = \frac{J_{01}}{S_{01} l_{01}^2} = \frac{0 \cdot 010\,42}{0 \cdot 5 \times 3^2} = 0 \cdot 002\,32$$

and for the horizontal bar

$$\frac{r_{11'}^2}{l_{11'}^2} = \frac{J_{11'}}{S_{11'} l_{11'}^2} = \frac{0 \cdot 1667}{2 \times 4^2} = 0 \cdot 005\,21$$

For $\omega = 209$ s^{-1} we get for the column:

by eqs 8.9 and 8.10

$$a_{01} = 2 \cdot 12 \left[\frac{2 \cdot 12^2}{2} (0 \cdot 002\,32 + 0 \cdot 005\,56) + \left(\frac{2 \cdot 12^4}{4} (0 \cdot 002\,32 - 0 \cdot 005\,56)^2 + 1 \right)^{1/2} \right]^{1/2}$$

$$= 2 \cdot 12 (0 \cdot 017\,75 + 1)^{1/2} = 2 \cdot 14$$

$$d_{01} = 2 \cdot 12 (-0 \cdot 017\,75 + 1)^{1/2} = 2 \cdot 10$$

and by eqs 8.14 und 8.15

$$\bar{a}_{01} = 2 \cdot 14 - \frac{2 \cdot 12^4 \times 0 \cdot 005\ 56}{2 \cdot 14} = 2 \cdot 09$$

$$\bar{d}_{01} = 2 \cdot 10 + \frac{2 \cdot 12^4 \times 0 \cdot 005\ 56}{2 \cdot 10} = 2 \cdot 15$$

By eq. 8.22

$$F_2(a, d)_{01} = -\frac{(2 \cdot 14^2 + 2 \cdot 10^2)(2 \cdot 15\cosh 2 \cdot 10 \sin 2 \cdot 14 - 2 \cdot 09 \sinh 2 \cdot 10 \cos 2 \cdot 14)}{\begin{array}{c} 2 \times 2 \cdot 09 \times 2 \cdot 15(\cosh 2 \cdot 10 \cos 2 \cdot 14 - 1) + \\ + (2 \cdot 09^2 - 2 \cdot 15^2)\sinh 2 \cdot 10 \sin 2 \cdot 14 \end{array}}$$

$$= 3 \cdot 61$$

Similarly

$$a_{11'} = 2 \cdot 035, \qquad \bar{a}_{11'} = 1 \cdot 938$$

$$d_{11'} = 1 \cdot 963, \qquad \bar{d}_{11'} = 2 \cdot 065$$

$$F_1(a, d)_{11'} = 1 \cdot 74, \quad F_4(a, d)_{11'} = -4 \cdot 38$$

$$F_2(a, d)_{11'} = 3 \cdot 44, \quad F_5(a, d)_{11'} = -12 \cdot 64$$

$$F_3(a, d)_{11'} = 5 \cdot 77, \quad F_6(a, d)_{11'} = 4 \cdot 34$$

$$F_3(\tfrac{1}{2}a, \tfrac{1}{2}d)_{11'} = 3 \cdot 82$$

$$F_5(\tfrac{1}{2}a, \tfrac{1}{2}d)_{11'} = -7 \cdot 71$$

$$F_6(\tfrac{1}{2}a, \tfrac{1}{2}d)_{11'} = 7 \cdot 20$$

Substituting the computed values in eq. 8.37 and those expressions in turn in Table 8.2 results in the equations set out in Table 8.4.

Table 8.4

$\bar{\zeta}_1 \times 10^4$	$\bar{v}_1 \times 10^4$	
17·51	−3·05	= 0·531P
−3·05	30·0	= 0·535P

From there

$$\bar{v}_1 = 2 \cdot 13 \times 10^{-6}P\ \text{(m, Mp)}$$

$$\bar{\zeta}_1 = 3 \cdot 40 \times 10^{-6}P$$

and midspan of the beam

$$\bar{v}_p = -\bar{v}_1 \frac{F_5(\tfrac{1}{2}a, \tfrac{1}{2}d)_{11'}}{F_6(\tfrac{1}{2}a, \tfrac{1}{2}d)_{11'}} - \zeta_1 \frac{l_{11'}}{2} \frac{F_3(\tfrac{1}{2}a, \tfrac{1}{2}d)_{11'}}{F_6(\tfrac{1}{2}a, \tfrac{1}{2}d)_{11'}} + \frac{Pl_{11'}^3}{16EJ_{11}.F_6(\tfrac{1}{2}a, \tfrac{1}{2}d)_{11'}}$$

$$= 2 \cdot 13 \times 10^{-6}P \frac{7 \cdot 71}{7 \cdot 20} + 3 \cdot 40 \times 10^{-6}P \frac{4}{2} \times \frac{3 \cdot 82}{7 \cdot 20}$$

$$+ \frac{P \times 4^3}{16 \times 2 \cdot 1 \times 10^6 \times 0 \cdot 1667 \times 7 \cdot 20} = 7 \cdot 48 \times 10^{-6}P \qquad (8.38)$$

The determinant of the coefficients on the left-hand side of the equations is

$$\bar{\varDelta} = 5 \cdot 17 \times 10^{10} \text{ Mp}^2$$

(c)   Free vibration

The method outlined in the foregoing paragraphs may be used for computing the coefficients on the left-hand side of the slope-deflection equations and their determinants for any ω. Determinants $\varDelta$ and $\bar{\varDelta}$ (without and with consideration given to the secondary effects) may then be plotted relative to the circular frequency ω (Fig. 8.4). At the points of intersection of curves $\varDelta$ and $\bar{\varDelta}$ with the zero axis lie the

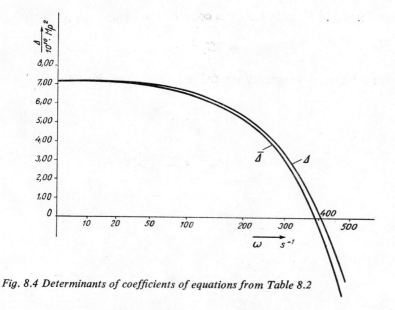

Fig. 8.4 Determinants of coefficients of equations from Table 8.2

natural circular frequencies belonging to the cases without and with consideration given to the effects of shear deflection and rotatory inertia ($\omega_{(1)} \approx 410 \text{ s}^{-1}$, $\bar{\omega}_{(1)} = 390 \text{ s}^{-1}$).

8.1.2 Approximate solution

As in Section 3.3 the formulae for the frequency functions $F(a, d)$ may be simplified also in the case where consideration is given to the secondary effects. When computing the end forces and moments it will again be approximately assumed that the dynamic deflection curve does not substantially differ from the static curve determined without regard to inertia forces and shear stresses. An example of such a curve is $v_c(x)$ in Fig. 3.16c or $v_d(x)$ in Fig. 3.16d. Let us start, for example, with the end moment M_{gh} of bar gh vibrating as shown in Fig. 3.16a. In the approximate solution curve $v(x)$ will be replaced by curve $v_d(x)$ according to Fig. 3.16d. The work of inertia forces will manifest itself not only as the work done on displacements $v_c(x)$ but also as the work of their moments on rotations $dv_c(x)/dx$. The moments of the inertia forces are described by formula 8.5. In harmonic motion the end moment will have the amplitude

$$M_{gh} \approx M_{ghst} - \omega^2 \left[\int \mu \, v_d(x) \, v_c(x) \, dx + \int \mu r^2 \, v_d'(x) \, v_{(c)}'(x) \, dx \right] \tag{8.39}$$

where by eq. 8.31

$$M_{ghst} = \frac{EJ}{l} \left(2 - \frac{36\gamma}{1 + 12\gamma} \right) \tag{8.40}$$

The first integral in formula 8.39 was computed in Section 3.4; further,

$$v'(x) = 3C_1 x^2 + 2C_2 x + C_3$$

where the integration constants are defined by eq. 3.44. Hence

$$v_d'(x) = \frac{3x^2}{l^2} - \frac{2x}{l} \tag{8.41}$$

$$v_c'(x) = \frac{3x^2}{l^2} - \frac{4x}{l} + 1 \tag{8.42}$$

and

$$\mu\omega^2 r^2 \int_0^l \left(\frac{3x^2}{l^2} - \frac{2x}{l} \right) \left(\frac{3x^2}{l^2} - \frac{4x}{l} + 1 \right) dx =$$

$$\mu\omega^2 r^2 \left| \frac{9}{5} \frac{x^5}{l^4} - \frac{9}{2} \frac{x^4}{l^3} + \frac{11}{3} \frac{x^3}{l^3} - x \right|_0^l = -\frac{\mu\omega^2 r^2 l}{30} = -\frac{l}{EJ} \frac{\lambda^4 r^2}{30 \, l^2}$$

Therefore, for

$$M_{gh} = \frac{EJ}{l} F_1(a, d)$$

the approximate value of the frequency function is

$$F_1(a, d) \approx 2 - \frac{36\gamma}{1 + 12\gamma} + \lambda^4 \left(\frac{1}{140} + \frac{1}{30} \frac{r^2}{l^2} \right) \tag{8.43}$$

and analogously

$$F_2(a, d) \approx \quad 4 - \frac{36\gamma}{1 + 12\gamma} - \lambda^4 \left(\frac{1}{105} + \frac{2}{15} \frac{r^2}{l^2} \right)$$

$$F_3(a, d) \approx \quad 6 - \frac{72\gamma}{1 + 12\gamma} + \lambda^4 \left(\frac{13}{420} - \frac{1}{10} \frac{r^2}{l^2} \right)$$

$$F_4(a, d) \approx -\ 6 + \frac{72\gamma}{1 + 12\gamma} + \lambda^4 \left(\frac{11}{210} + \frac{1}{10} \frac{r^2}{l^2} \right)$$

$$F_5(a, d) \approx -12 + \frac{144\gamma}{1 + 12\gamma} - \lambda^4 \left(\frac{9}{70} - \frac{12}{10} \frac{r^2}{l^2} \right)$$

$$F_6(a, d) \approx \quad 12 - \frac{144\gamma}{1 + 12\gamma} - \lambda^4 \left(\frac{13}{35} + \frac{12}{10} \frac{r^2}{l^2} \right)$$

$$\tag{8.44}$$

Example 8.2

Let us apply this simplified method to the computation of forced vibration of the frame discussed in Example 8.1. For

$$\gamma_{01} = 0.005\,56, \quad \gamma_{11'} = 0.012\,50, \quad \gamma_{1p} = 0.0500$$

$$\lambda_{01} = 2.12, \quad \lambda_{11'} = 2.0, \quad \lambda_{1p} = 1.0$$

$$\frac{r_{01}^2}{l_{01}^2} = 0.002\,32, \quad \frac{r_{11'}^2}{l_{11'}^2} = 0.005\,21, \quad \frac{r_{1p}^2}{l_{1p}^2} = 0.020\,84$$

we get

$$F_2(a, d)_{01} \quad \approx \quad 4 - \frac{36 \times 0.005\,56}{1 + 12 \times 0.005\,56} - 2.12^4 \left(\frac{1}{105} + \frac{2 \times 0.002\,32}{15} \right) = 3.63$$

$$F_1(a, d)_{11'} \quad \approx \quad 2 - 0.391 + 0.117 = \quad 1.73$$

$$F_2(a, d)_{11'} \quad \approx \quad 4 - 0.391 - 0.163 = \quad 3.45$$

$$F_3(a, d)_{11'} \quad \approx \quad 6 - 0.782 + 0.504 = \quad 5.72$$

$$F_4(a, d)_{11'} \quad \approx -\ 6 + 0.782 + 0.847 = -\ 4.37$$

$$F_5(a, d)_{11'} \quad \approx -12 + 1.564 - 1.956 = -12.39$$

$$F_6(a, d)_{11'} \quad \approx \quad 12 - 1.564 - 6.05 \ = \quad 4.39$$

$$F_3(\tfrac{1}{2}a, \tfrac{1}{2}d)_{11'} \approx \quad 6 - 2\cdot25 \quad + 0\cdot029 = \quad 3\cdot78$$

$$F_5(\tfrac{1}{2}a, \tfrac{1}{2}d)_{11'} \approx -12 + 4\cdot50 \quad - 0\cdot104 = - 7\cdot60$$

$$F_6(\tfrac{1}{2}a, \tfrac{1}{2}d)_{11'} \approx \quad 12 - 4\cdot50 \quad - 0\cdot397 = \quad 7\cdot10$$

As the above agree with the values obtained in Example 8.1 by the exact method to about 1%, the results of the simplified procedure are satisfactory.

8.1.3 Non-prismatic bars

Computations relating to the vibration of non-prismatic bars can advantageously be carried out with the aid of the simplified method described in the preceding section. In numerical computations one may replace formula 8.39 by the relation

$$M_{gh} \approx M_{ghst} - \omega^2 \sum m_i v_{di} v_{ci} - \omega^2 \sum m_i r_i^2 v'_{di} v'_{ci} \tag{8.45}$$

When determining the static moment, M_{ghst}, consideration should be given to shear deflections, and rotations v'_{di}, v'_{ci} at the centre of gravity of each segment established simultaneously with the static displacements, v_{di}, v_{ci}.

The other end moments and forces are computed in an analogous manner (cf. Section 4.1.3).

8.2 Effect of static axial force on lateral vibration

8.2.1 Equation of motion and its solution

Whenever a bar is subjected to a static (time invariable) axial force N_{st}, the dynamic forces acting on its ends differ from those of a bar not sustaining such a force. The axial force produces yet another effect, namely a change of the frequency of free vibration: when it is compressive, the frequency diminishes, when it is tensile, it grows. The positive sign applies to tensile axial forces. An analysis of the effects of axial forces follows the same pattern as the examination of the effects of shear[34,36,37].

Consider a length element of a bar vibrating under the action of axial force (Fig. 8.5). The conditions of equilibrium of forces acting on the element give

$$T(x, t)\,dx - \frac{\partial M(x, t)}{\partial x}\,dx - N_{st}\frac{\partial v(x, t)}{\partial x}\,dx = 0 \tag{8.46}$$

$$\mu\frac{\partial^2 v(x, t)}{\partial t^2} - N_{st}\frac{\partial^2 v(x, t)}{\partial x^2} + EJ\frac{\partial^4 v(x, t)}{\partial x^4} = 0 \tag{8.47}$$

Assuming harmonic motion

$$v(x, t) = v(x) \sin \omega t$$

the amplitudes are defined by the equation

$$v''''(x) - \frac{N_{st}}{EJ} v''(x) - \frac{\mu\omega^2}{EJ} v(x) = 0 \tag{8.48}$$

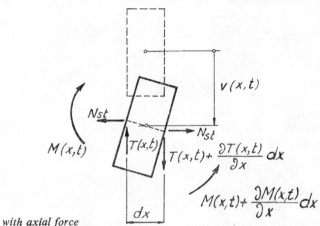

Fig. 8.5 Element of bar loaded with axial force

Since eq. 8.48 is analogous to eq. 8.7 its solution is likewise given by eq. 8.8 where however,

$$a = \left[\frac{\alpha^2}{2} + \left(\frac{\alpha^4}{4} + \lambda^4 \right)^{1/2} \right]^{1/2} \tag{8.49}$$

$$d = \left[-\frac{\alpha^2}{2} + \left(\frac{\alpha^4}{4} + \lambda^4 \right)^{1/2} \right]^{1/2} \tag{8.50}$$

and

$$\alpha = l \left(-\frac{N_{st}}{EJ} \right)^{1/2} \tag{8.51}$$

The effect of static axial force comes most into play in slender bars in which, fortunately, one can neglect the effects of shear deflection and rotatory inertia. If the bar is clamped at both ends, the angle of rotation of the tangent at the end points coincides with the angle of rotation of the respective joints so that

$$
\begin{aligned}
v(0) &= v_g \\
v(l) &= v_h \\
v'(0) &= \zeta_g \\
v'(l) &= \zeta_h
\end{aligned}
\tag{8.52}
$$

The amplitude of the bending moments is given by

$$M(x) = -EJ\, v''(x)$$

$$= \frac{EJ}{l^2}\left[Aa^2 \cos \frac{ax}{l} + Ba^2 \sin \frac{ax}{l} - Cd^2 \cosh \frac{dx}{l} - Dd^2 \sinh \frac{dx}{l} \right]$$

The amplitude of the shear force follows from condition 8.46 which gives

$$T(x) = \frac{dM(x)}{dx} + N_{st}\frac{dv(x)}{dx}$$

$$= -\frac{EJ}{l^3}\,ad\left[Ad \sin \frac{ax}{l} - Bd \cos \frac{ax}{l} + Ca \sinh \frac{dx}{l} + Da \cosh \frac{dx}{l} \right] \quad (8.53)$$

As the above relations imply, formulae 8.21 – 8.25 continue to apply but with the following substitutions made

$$\bar{a} = a\ ;\quad \bar{d} = d\ ;\quad v_1'(0) = v'(0)\ ;\quad v_1'(l) = v'(l) \quad (8.54)$$

where a, d are defined by eqs 8.49 and 8.50. Formulae 8.26 – 8.30 analogously modified apply to bars hinged at one and both ends, and to a cantilever bar, respectively.

8.2.2   Buckling of frame systems

The above formulae are applicable generally to any N_{st} and ω, and will simplify in the case where one or both of those quantities equal zero. If it is the axial force that equals zero, then $a = d = \lambda$ and the end moments and forces are described by the formulae shown in column 5, Tables 10.1 to 10.8, Appendix. If both N_{st} and ω equal zero, the applicable formulae are those arranged in column 1.

If only the vibration velocity tends to zero, i.e. $\omega = 0$, expressions 8.49 and 8.50 turn out to be

$$d = 0$$
$$a = \alpha$$

$$(8.55)$$

for negative N_{st} (compression), and

$$d = l\left(\frac{N_{st}}{EJ}\right)^{1/2} = \delta$$

$$a = 0$$

$$(8.56)$$

for positive N_{st} (tension). Substituting 8.55 in eq. 8.22 or in the formulae shown in column 6, Tables 10.1 – 10.8 in the Appendix, and computing the limits of the indefinite expressions gives

$$F_1(\alpha, 0) = \alpha \frac{\sin \alpha - \alpha}{\alpha \sin \alpha + 2(\cos \alpha - 1)} = \Gamma_1(\alpha) \tag{8.57}$$

as well as the other formulae for computing end moments and forces of bars subjected to the action of axial force (column 3, Tables 10.1 – 10.8 in the Appendix). The same results may be arrived at by solving directly the differential equation

$$N_{st} \, v''(x) - EJ \, v''''(x) = 0 \tag{8.58}$$

whose complete integral for a negative axial force is

$$v(x) = A \cos \frac{\alpha x}{l} + B \sin \frac{\alpha x}{l} + C + Dx \tag{8.59}$$

(cf. Reference 8).

The angle of rotation of the tangent is

$$v'(x) = -A \frac{\alpha}{l} \sin \frac{\alpha x}{l} + B \frac{\alpha}{l} \cos \frac{\alpha x}{l} + D \tag{8.60}$$

the bending moment is given by

$$M(x) = -EJ \, v''(x) = \frac{EJ\alpha^2}{l^2} \left(A \cos \frac{\alpha x}{l} + B \sin \frac{\alpha x}{l} \right) \tag{8.61}$$

and the shear force by

$$T(x) = -EJ \, v'''(x) + N_{st} \, v'(x) = -EJ \frac{\alpha^2}{l^2} D \tag{8.62}$$

The deflection of a bar of constant cross-section is defined by eq. 8.59. If such a bar is clamped at both ends,

$$A + C = v_g$$

$$A \cos \alpha + B \sin \alpha + C + Dl = v_h$$

$$B \frac{\alpha}{l} + D = \zeta_g$$

$$-A \frac{\alpha}{l} \sin \alpha + B \frac{\alpha}{l} \cos \alpha + D = \zeta_h$$

$$\tag{8.63}$$

and the integration constants are

$$A = \frac{\zeta_h\, l(\sin \alpha - \alpha) + \zeta_g\, l(\alpha \cos \alpha - \sin \alpha) + (v_g - v_h)\, \alpha(\cos \alpha - 1)}{2\alpha(\cos \alpha - 1) + \alpha^2 \sin \alpha}$$

$$B = \frac{-\zeta_h\, l(\cos \alpha - 1) + \zeta_g\, l(\cos \alpha - 1 + \alpha \sin \alpha) + (v_g - v_h)\, \alpha \sin \alpha}{2\alpha(\cos \alpha - 1) + \alpha^2 \sin \alpha}$$

$$C = v_g - A$$

$$D = \zeta_g - B\frac{\alpha}{l} \tag{8.64}$$

When examining the effect of axial force one may sometimes come across the case of zero shear force in some of the bars of the system. In such a case it is convenient to use the rotations of ends g and h of bar gh and the zero shear forces in them as the end conditions and write

$$T(0) = T(l) = -EJ\frac{\alpha^2}{l^2}\, D = 0$$

$$v'(0) = B\frac{\alpha}{l} = \zeta_g$$

$$v'(l) = -A\frac{\alpha}{l}\sin \alpha + B\frac{\alpha}{l}\cos \alpha = \zeta_h$$

$$\tag{8.65}$$

and the integration constants are

$$D = 0$$

$$A = \frac{l}{\alpha}(\zeta_g \cot \alpha - \zeta_h\, \mathrm{cosec}\, \alpha)$$

$$B = \frac{l}{\alpha}\zeta_g$$

$$\tag{8.66}$$

In this case eq. 8.61 gives

$$M_{gh} = \frac{EJ\alpha}{l}(\zeta_g \cot \alpha - \zeta_h \cos \mathrm{ec}\, \alpha)$$

$$M_{hg} = -\frac{EJ\alpha}{l}(\zeta_g \cos \mathrm{ec}\, \alpha - \zeta_h \cot \alpha)$$

$$\tag{8.67}$$

By eqs 8.59 and 8.60 the difference between the end displacements is

$$v_h - v_g = -A + A \cos \alpha + B \sin \alpha = l(\zeta_g + \zeta_h) \frac{\operatorname{cosec} \alpha - \cot \alpha}{\alpha}$$

(8.68)

If bar gh is hinged at end g, the integration constants are

$$C = v_g$$

$$D = -\frac{v_h - v_g}{l} \frac{\alpha \cos \alpha}{\sin \alpha - \alpha \cos \alpha} + \zeta_h \frac{\sin \alpha}{\sin \alpha - \alpha \cos \alpha}$$

$$A = 0$$

$$B = \frac{v_h - v_g - \zeta_h l}{\sin \alpha - \alpha \cos \alpha}$$

(8.69)

so that the deflection (setting $(v_h - v_g)/l = \varepsilon$) is

$$v(x) = v_g + \varepsilon x + (\zeta_h - \varepsilon) \frac{\sin \alpha}{\sin \alpha - \alpha \cos \alpha} \left[x - \frac{l}{\sin \alpha} \sin \left(\frac{\alpha x}{l} \right) \right]$$

(8.70)

Differentiation of eq. 8.70 and substitution give, for the rotation at point g,

$$\zeta_g - \varepsilon = (\zeta_h - \varepsilon) \frac{\sin \alpha - \alpha}{\sin \alpha - \alpha \cos \alpha}$$

(8.71)

Note: Consider a frame system subjected to static axial forces. The natural frequencies of such a system depend on the magnitude of the axial forces. If the axial forces are negative (compression) and their magnitude is gradually increased, i.e. N_{st} is multiplied by a steadily growing coefficient $\varkappa$, the frequency of free vibration of the system decreases and at $N_{cr} = \varkappa_{(k)} N_{st}$ unstable equilibrium and buckling will take place at zero angular frequency of vibration ω. The ratio of the critical axial forces to the actual ones is called the safety factor $\varkappa_{(k)}$ of the system. Coefficient $\varkappa_{(k)}$ is a parameter whose significance is similar to that of $\omega_{(j)}$: to a given $\varkappa$ we can find circular frequencies $\bar{\omega}_{(j)}$ at which the system vibrates in free vibration, and vice versa. To $\omega = 0$ we can find $\varkappa_{(k)}$ at which the system will buckle. Only the lowest value, $\varkappa_{(1)}$, is of practical importance.

8.2.3   Numerical examples

Example 8.3

Consider the multi-storey frame shown in Fig. 3.8, which we have analysed in Example 3.1, Section 3.2.2.

(a)   Symmetric vibration

In the absence of axial forces, the applicable equations are those arranged in Table 3.4 with their right-hand sides set to zero. The computed lowest circular frequency of symmetric vibration is $\omega_{(1)} = 103\cdot7\ \mathrm{s}^{-1}$, the first natural mode being given by the ratios

$$\frac{\zeta_1}{\zeta_3} = 0\cdot584 \quad \text{and} \quad \frac{\zeta_2}{\zeta_3} = -0\cdot962$$

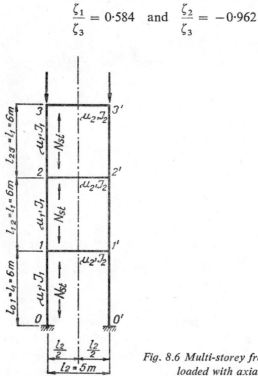

Fig. 8.6 *Multi-storey frame with bars loaded with axial forces*

If the system is subjected to constant forces

$$Y_{3\mathrm{st}} = Y_{3'\mathrm{st}} = 200\ \mathrm{Mp}$$

as shown in Fig. 8.6, static axial forces will arise in the frame columns. Neglecting the relatively low self weight gives

$$N_{01\mathrm{st}} = N_{12\mathrm{st}} = N_{23\mathrm{st}} = N_{\mathrm{st}} = -200\ \mathrm{Mp}$$

In the horizontal bars the axial forces are zero, i.e.

The task is to examine symmetric free vibration of the system under the action of these forces. The equations arranged in Table 3.4 continue to apply. The terms on their right-hand sides are equal to zero, i.e.

$$a_{10}^* = a_{20}^* = a_{30}^* = 0$$

and in the coefficients on the left-hand sides functions $F(\lambda)$ of the vertical bars are replaced by functions $F(a, d)$:

$$a_{11}^* = \frac{2EJ_1}{l_1} F_2(a, d)_1 + \frac{EJ_2}{l_2} [F_2(\lambda_2) - F_1(\lambda_2)] = a_{22}^*$$

$$a_{12}^* = \frac{EJ_1}{l_1} F_1(a, d)_1 = a_{23}^*$$

$$a_{13}^* = 0$$

$$a_{33}^* = \frac{EJ_1}{l_1} F_2(a, d)_1 + \frac{EJ_2}{l_2} [F_2(\lambda_2) - F_1(\lambda_2)]$$

$$(8.72)$$

A solution different from zero is obtained from the condition that the determinant of the coefficients be equal to zero. The lowest frequency at which the condition is satisfied, is

$$\bar{\omega}_{(1)} = 95 \cdot 6 \text{ s}^{-1}$$

For this value eqs 8.49−8.51 give

$$\alpha_1^2 = \frac{6^2 \times 200}{21 \times 10^6 \times 1 \cdot 10^{-4}} = 3 \cdot 43$$

$$\lambda_1^4 = 6^4 \frac{0 \cdot 2 \times 95 \cdot 6^2}{9 \cdot 81 \times 21 \times 10^6 \times 1 \cdot 10^{-4}} = 115$$

$$a_1 = \left[\frac{3 \cdot 43}{2} + \left(\frac{3 \cdot 43^2}{4} + 115 \right)^{1/2} \right]^{1/2} = 3 \cdot 546$$

$$d_1 = \left[-\frac{3 \cdot 43}{2} + \left(\frac{3 \cdot 43^2}{4} + 115 \right)^{1/2} \right]^{1/2} = 3 \cdot 024$$

Further, by eq. 8.23

$$f_1(a, d) = 2 \times 3 \cdot 546 \times 3 \cdot 024 (-10 \cdot 483) + (3 \cdot 546^2 - 3 \cdot 024^2)(-4 \cdot 038) = -238 \cdot 7$$

and by eq. 8.22

$$F_1(a, d)_1 = \frac{(3 \cdot 546^2 + 3 \cdot 024^2)(3 \cdot 546 \times 10 \cdot 264 + 3 \cdot 024 \times 0 \cdot 3934)}{238 \cdot 7} = 3 \cdot 420$$

$$\frac{1}{l_1} F_1(a, d)_1 = 0 \cdot 5700$$

$F_2(a, d)_1$

$$= \frac{(3\cdot546^2 + 3\cdot024^2)(-3\cdot024 \times 10\cdot315 \times 0\cdot3934 + 3\cdot546 \times 10\cdot264 \times 0\cdot9194)}{238\cdot7}$$

$$= 1\cdot928$$

$$\frac{1}{l_1} F_2(a, d)_1 = 0\cdot3213$$

$$\lambda_2 = l_2 \left(\frac{\mu_2 \omega^2}{EJ_2}\right)^{1/4} = 5 \left(\frac{0\cdot5 \times 95\cdot6^2}{9\cdot81 \times 21 \times 10^6 \times 2 \times 10^{-4}}\right)^{1/4} = 2\cdot885$$

$$\frac{J_2}{l_2 J_1} [F_2(\lambda_2) - F_1(\lambda_2)] = 0\cdot2666$$

After all its coefficients have been divided by EJ_1 and the numerical values substituted in, Table 3.4 becomes Table 8.5.

Table 8.5

ζ_1	ζ_2	ζ_3	
0·3213	0·5700		= 0
0·3213			
0·2666			
0·9092			
0·5700	0·9092	0·5700	= 0
	0·5700	0·3213	= 0
		0·2666	
		0·5879	

The determinant of the coefficients (in Mp, m) is

$$\Delta = 0\cdot9092^2 \times 0\cdot5879 - 0\cdot5700^2(0\cdot9092 + 0\cdot5879) = 0\cdot4860 - 0\cdot4864 \approx 0$$

and the complements are

$$A_{11} = \quad 0\cdot9092 \times 0\cdot5879 - 0\cdot5700^2 = 0\cdot2096$$
$$A_{12} = -0\cdot5700 \times 0\cdot5879 = -0\cdot3351$$
$$A_{13} = \quad 0\cdot5700^2 = 0\cdot3249$$
$$A_{22} = \quad 0\cdot9092 \times 0\cdot5879 = 0\cdot5345$$
$$A_{23} = -0\cdot5700 \times 0\cdot9092 = -0\cdot5182$$
$$A_{33} = \quad 0\cdot9092^2 - 0\cdot5700^2 = 0\cdot5017$$

The ratios of the rotations of the joints are

$$\frac{\zeta_1}{\zeta_3} = \frac{0 \cdot 2096}{0 \cdot 3249} = \frac{-0 \cdot 3351}{-0 \cdot 5182} = \frac{0 \cdot 3249}{0 \cdot 5017} = 0 \cdot 646$$

$$\frac{\zeta_2}{\zeta_3} = \frac{-0 \cdot 3351}{0 \cdot 3249} = \frac{0 \cdot 5345}{-0 \cdot 5182} = \frac{-0 \cdot 5182}{0 \cdot 5017} = -1 \cdot 030$$

(b) Buckling and mode shape of symmetric lateral deflection

The system of equations for solving the lateral deflection of frames, induced by buckling is formally the same as that used in the computation of vibration. The conditions of equilibrium of the moments in the various joints are expressed by the equations in Table 8.6, with the coefficients taken from column 3, Table 10.1 in the Appendix. The condition that the determinant of the coefficients should equal zero, is in this case used for computing the parameter $\varkappa_{(1)}$ i.e. the factor of safety against buckling:

$$\alpha_1 = l_1 \left(\frac{-N_{cr}}{EJ_1} \right)^{1/2} = l_1 \left(\frac{-\varkappa_{(1)} N_{st}}{EJ_1} \right)^{1/2} = 4 \cdot 45$$

Table 8.6

ζ_1	ζ_2	ζ_3	
$\dfrac{2EJ_1}{l_1} \Gamma_2(\alpha_1) + \dfrac{2EJ_2}{l_2}$	$\dfrac{EJ_1}{l_1} \Gamma_1(\alpha_1)$		$= 0$
$\dfrac{EJ_1}{l_1} \Gamma_1(\alpha_1)$	$\dfrac{2EJ_1}{l_1} \Gamma_2(\alpha_1) + \dfrac{2EJ_2}{l_2}$	$\dfrac{EJ_1}{l_1} \Gamma_1(\alpha_1)$	$= 0$
	$\dfrac{EJ_1}{l_1} \Gamma_1(\alpha_1)$	$\dfrac{EJ_1}{l_1} \Gamma_2(\alpha_1) + \dfrac{2EJ_2}{l_2}$	$= 0$

and from there

$$\varkappa_{(1)} = \frac{\alpha_1^2}{l_1^2} \frac{EJ}{(-N_{st})} = \frac{4 \cdot 45^2}{6^2} \frac{21 \times 10^6 \times 1 \times 10^{-4}}{200} = 5 \cdot 78$$

and the ratios of the rotations of the joints are

$$\frac{\zeta_1}{\zeta_3} = 0 \cdot 98, \quad \frac{\zeta_2}{\zeta_3} = -1 \cdot 40$$

As shown in Section 5.2.5, Example 5.7, the solution may be simplified by the fact that the system is one with repeated elements. The formulae derived there can be used even in the case where the bars of the system are subjected to static axial forces. All that is necessary is to replace the frequency functions $F(\lambda)$ by functions $F(a, d)$. The frequency equations 5.139 and 5.142 continue to apply

(c)   Antisymmetric vibration

The task is to find the first frequency and mode of free vibration of the system under the action of axial forces. In this case we shall assume that the vertical forces are half the value they had in Example 8.3a, i.e.

$$Y_{3st} = Y_{3'st} = 100 \text{ Mp}$$

and therefore also

$$N_{01st} = N_{12st} = N_{23st} = N_{st} = -100 \text{ Mp}$$

(The example was of course chosen without a view to the practical feasibility of such a load at the given mass distribution.) The slope-deflection method applied analogously as in Example 3.1 again leads to the system of equations shown in Table 3.7, this time with the coefficients

$$a_{11}^* = \frac{2EJ_1}{l_1} F_2(a, d)_1 + \frac{EJ_2}{l_2} [F_1(\lambda_2) + F_2(\lambda_2)] = a_{22}^*$$

$$a_{12}^* = \frac{EJ_1}{l_1} F_1(a, d)_1 = a_{23}^*$$

$$a_{15}^* = -\frac{EJ_1}{l_1^2} F_3(a, d)_1 = -a_{24}^* = a_{26}^* = -a_{35}^*$$

$$a_{33}^* = \frac{EJ_1}{l_1} F_2(a, d)_1 + \frac{EJ_2}{l_2} [F_1(\lambda_2) + F_2(\lambda_2)]$$

$$a_{36}^* = \frac{EJ_1}{l_1^2} F_4(a, d)_1$$

$$a_{44}^* = \frac{2EJ_1}{l_1^3} F_6(a, d)_1 - \frac{EJ_2}{l_2^3} \frac{\lambda_2^4}{2} = a_{55}^*$$

$$a_{45}^* = \frac{EJ_1}{l_1^3} F_5(a, d)_1 = a_{56}^*$$

$$a_{66}^* = \frac{EJ_1}{l_1^3} F_6(a, d)_1 - \frac{EJ_2}{l_2^3} \frac{\lambda_2^4}{2}$$

$$P = 0$$

To determine the natural frequency of the system, we find by trial such values of $\omega = \bar{\omega}_{(1')}$ for which the determinant Δ of the coefficients will be zero. The first circular frequency turns out to be $\omega_{(1')} = 7 \cdot 31 \text{ s}^{-1}$ ($a_1 = 1 \cdot 429$, $d_1 = 0 \cdot 573$), and the ratio of the respective complements gives the ratio of the deformations of the joints in the first mode of free vibration as follows

$$\frac{\zeta_1}{u_3} = 0 \cdot 0326 \text{ m}^{-1}, \quad \frac{\zeta_2}{u_3} = 0 \cdot 0239 \text{ m}^{-1}, \quad \frac{\zeta_3}{u_3} = 0 \cdot 0083 \text{ m}^{-1}$$

$$\frac{u_1}{u_3} = 0 \cdot 376, \quad \frac{u_2}{u_3} = 0 \cdot 784$$

(d)   Buckling and shape of antisymmetric lateral deflection

One way of obtaining the slope-deflection equations for lateral deflection is to use Table 3.7, with functions $\Gamma(\alpha_1)$ (column 3, Tables 10.1 and 10.2 in the Appendix) substituted for $F(\lambda_1)$, and $F_2(\lambda_2) - F_1(\lambda_2) = 2$, $\omega = 0$.

A quicker way to attain this end is to recall that because of the symmetry of the system, the shear force in the frame columns is zero, and use expressions 8.67. The conditions of equilibrium of the moments at the joints result in the system of equations shown in Table 8.7.

In the next step we find the lowest value $\alpha_1 = l_1(-\varkappa_{(1')}N_{st}/EJ_1)^{1/2}$ for which determinant Δ is zero, and compute the respective value of $\varkappa_{(1')}$ ($\alpha_1 = 2 \cdot 582$, $\varkappa_{(1')} = 3 \cdot 89$). The ratios of the rotations of the joints are obtained from the ratio of the respective complements, and the displacements from eq. 8.68

$$\frac{\zeta_1}{u_3} = 0 \cdot 0241 \text{ m}^{-1}, \quad \frac{\zeta_2}{u_3} = 0 \cdot 0305 \text{ m}^{-1}, \quad \frac{\zeta_3}{u_3} = 0 \cdot 0144 \text{ m}^{-1}$$

$$\frac{u_1}{u_3} = 0 \cdot 195, \quad \frac{u_2}{u_3} = 0 \cdot 637$$

Table 8.7

ζ_1	ζ_2	ζ_3	
$\dfrac{2EJ_1}{l_1} \alpha_1 \cot \alpha_1 + \dfrac{6EJ_2}{l_2}$	$-\dfrac{EJ_1}{l_1} \alpha_1 \operatorname{cosec} \alpha_1$		$= 0$
$-\dfrac{EJ_1}{l_1} \alpha_1 \operatorname{cosec} \alpha_1$	$\dfrac{2EJ_1}{l_1} \alpha_1 \cot \alpha_1 + \dfrac{6EJ_2}{l_2}$	$-\dfrac{EJ_1}{l_1} \alpha_1 \operatorname{cosec} \alpha_1$	$= 0$
	$-\dfrac{EJ_1}{l_1} \alpha_1 \operatorname{cosec} \alpha_1$	$\dfrac{EJ_1}{l_1} \alpha_1 \cot \alpha_1 + \dfrac{6EJ_2}{l_2}$	$= 0$

378

In Fig. 8.7, the solid line *a* is the first mode of free vibration when axial forces are present, computed in paragraph *c* of this example; the dashed line *b* is the mode of free vibration in the absence of axial forces, computed in paragraph *c*, Example 3.1;

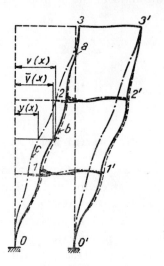

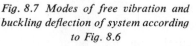

Fig. 8.7 *Modes of free vibration and buckling deflection of system according to Fig. 8.6*

and the dash-and-dot line *c* is the shape of lateral deflection owing to buckling, computed in this paragraph.

Example 8.4

Fig. 8.8 shows a two-span continuous beam on elastic supports (a guyed antenna mast — cf. Reference 35). The characteristics of this tubular mast are

$$E = 21 \cdot 5 \times 10^6 \text{ Mpm}^{-2}$$

$l_{01} = 19 \cdot 0 \text{ m},$ $\qquad\qquad$ $l_{12} = 16 \cdot 0 \text{ m}$

$J_{01} = 0 \cdot 5899 \times 10^{-4} \text{ m}^4,$ $\qquad$ $J_{12} = 0 \cdot 2814 \times 10^{-4} \text{ m}^4$

$\mu_{01} = 0 \cdot 049\,90 \text{ Mpm}^{-1}/9 \cdot 81 \text{ m s}^{-2},$ $\quad$ $\mu_{12} = 0 \cdot 035\,44 \text{ Mpm}^{-1}/9 \cdot 81 \text{ m s}^{-2}$

The elasticity of the supports is defined by the values

$$C_0 = 0, \quad C_1 = 11 \cdot 35 \text{ Mpm}^{-1}, \quad C_2 = 5 \cdot 71 \text{ Mpm}^{-1}$$

where C is the spring constant of the elastic supports, i.e. the force necessary to produce unit displacement of the point of the support and the subscript is the serial number of the support. The vibration will be computed assuming very small deformations. At large amplitudes the vibration is no longer linear.

First we computed the mast vibration without considering the axial forces. On substituting in the numerical values we get

$$\lambda_{01} = l_{01}\left(\frac{\mu_{01}\omega^2}{EJ_{01}}\right)^{1/4} = 0{\cdot}850\omega^{1/2}$$

$$\lambda_{12} = l_{12}\left(\frac{\mu_{12}\omega^2}{EJ_{12}}\right)^{1/4} = 0{\cdot}791\omega^{1/2}$$

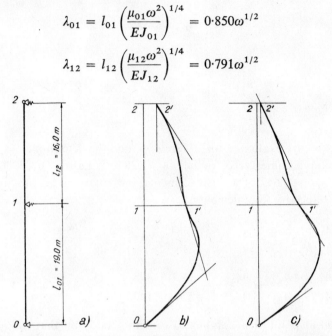

Fig. 8.8 Continuous beam on elastic supports (antenna mast). (a) schematic diagram, (b) first natural mode with static axial forces neglected, (c) first natural mode with static axial forces considered

The coefficients of the slope-deflection equations set out in Table 8.8 are as follows

$$a_{11}^* = \frac{EJ_{01}}{l_{01}}F_7(\lambda_{01}) + \frac{EJ_{12}}{l_{12}}F_7(\lambda_{12})$$

$$a_{12}^* = \frac{EJ_{01}}{l_{01}^2}F_9(\lambda_{01}) - \frac{EJ_{12}}{l_{12}^2}F_9(\lambda_{12})$$

$$a_{13}^* = -\frac{EJ_{12}}{l_{12}^2}F_8(\lambda_{12})$$

$$a_{22}^* = \frac{EJ_{01}}{l_{01}^3}F_{11}(\lambda_{01}) + \frac{EJ_{12}}{l_{12}^3}F_{11}(\lambda_{12}) + C_1$$

$$a_{23}^* = \frac{EJ_{12}}{l_{12}^3}F_{10}(\lambda_{12})$$

$$a_{33}^* = \frac{EJ_{12}}{l_{12}^3}F_{12}(\lambda_{12}) + C_2$$

$$(8.73)$$

Table 8.8

ζ_1	u_1	u_2	
a_{11}^*	a_{12}^*	a_{13}^*	$= 0$
a_{12}^*	a_{22}^*	a_{23}^*	$= 0$
a_{13}^*	a_{23}^*	a_{33}^*	$= 0$

Coefficients a^* are evaluated for different ω, and the determinant Δ is computed. The determinant becomes zero for $\omega_{(1)} = 11 \cdot 28 \text{ s}^{-1}$, i.e. for $f_{(1)} = 1 \cdot 80 \text{ s}^{-1}$, with $\lambda_{01} = 2 \cdot 856$, $\lambda_{12} = 2 \cdot 657$. The displacements and rotations are obtained from the ratios of the respective complements:

$$\frac{\zeta_1}{u_2} = \frac{A_{11}}{A_{13}} = \frac{A_{21}}{A_{23}} = \frac{A_{31}}{A_{33}} = -\frac{0 \cdot 81}{5 \cdot 8} = -\frac{20 \cdot 0}{143} = -\frac{5 \cdot 8}{42} = -0 \cdot 140 \text{ m}^{-1}$$

$$\frac{u_1}{u_2} = \frac{A_{12}}{A_{13}} = \frac{A_{22}}{A_{23}} = \frac{A_{32}}{A_{33}} = \frac{-20 \cdot 0}{-5 \cdot 8} = \frac{478}{143} = \frac{143}{42} = 3 \cdot 35$$

The rotation at point O is computed from the condition that the moment there should be zero. By eq. 2.48

$$M_{01} = \frac{EJ_{01}}{l_{01}} \left[F_2(\lambda_{01}) \zeta_0 + F_1(\lambda_{01}) \zeta_1 - F_3(\lambda_{01}) \frac{u_1}{l_{01}} \right] = 0$$

and from there

$$\frac{\zeta_0}{u_2} = \frac{F_1(2 \cdot 856)}{F_2(2 \cdot 856)} 0 \cdot 140 + \frac{F_3(2 \cdot 856)}{19 \cdot 0 F_2(2 \cdot 856)} 3 \cdot 35 = 0 \cdot 562 \text{ m}^{-1}$$

and analogously,

$$\frac{\zeta_2}{u_2} = \frac{F_1(2 \cdot 657)}{F_2(2 \cdot 657)} 0 \cdot 140 - \frac{F_3(2 \cdot 657)}{16 \cdot 0 F_2(2 \cdot 657)} 3 \cdot 35 - \frac{F_4(2 \cdot 657)}{16 \cdot 0 F_2(2 \cdot 657)} = -0 \cdot 313 \text{ m}^{-1}$$

The first mode of vibration is shown in Fig. 8.8b.

Next we shall consider the effects of static axial forces which are considerable in this case since they are the result of the vertical component of tension in the anchoring cables and the self weight of the mast. We shall take their mean values

$$N_{01st} = -6 \cdot 4 \text{ Mp}$$

$$N_{12st} = -4 \cdot 0 \text{ Mp}$$

The problem is to find the first frequency and mode of natural vibration under this load. The equations of Table 8.8 continue to apply except that in the coefficients a^* according to eq. 8.73 functions $F(\lambda)$ must be replaced by $F(a, d)$. In span $\overline{01}$ we have

$$\frac{\alpha_{01}^2}{2} = \frac{l_{01}^2(-N_{01st})}{2EJ_{01}} = \frac{19^2 \times 6\cdot4}{2 \times 21\cdot5 \times 10^6 \times 0\cdot5899 \times 10^{-4}} = 0\cdot910$$

$$\lambda_{01}^4 = 0\cdot521\omega^2$$

$$a_{01} = [0\cdot910 + (0\cdot910^2 + 0\cdot521\omega^2)^{1/2}]^{1/2}$$

$$d_{01} = [-0\cdot910 + (0\cdot910^2 + 0\cdot521\omega^2)^{1/2}]^{1/2}$$

in span $\overline{12}$

$$\frac{\alpha_{12}^2}{2} = \frac{16^2 \times 4\cdot0}{2 \times 21\cdot5 \times 10^6 \times 0\cdot2814 \times 10^{-4}} = 0\cdot846$$

$$\lambda_{12}^4 = 0\cdot391\omega^2$$

$$a_{12} = [0\cdot846 + (0\cdot846 + 0\cdot391\omega^2)^{1/2}]^{1/2}$$

$$d_{12} = [-0\cdot846 + (0\cdot846 + 0\cdot391\omega^2)^{1/2}]^{1/2}$$

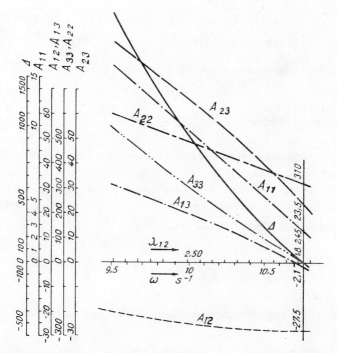

Fig. 8.9 Determinants and complements of coefficients of equations set out in Table 8.8

The natural frequencies are obtained from the condition that the determinant Δ should be zero. The curves of Δ as a function of ω, and of complements A are drawn in Fig. 8.9. The curve of Δ intersects the zero axis at

$$\omega = \bar{\omega}_{(1)} = 10 \cdot 8 \ s^{-1}$$

The ratios of the deformations are

$$\frac{\zeta_1}{u_2} = -\frac{2 \cdot 45}{2 \cdot 1} = -\frac{27 \cdot 5}{23 \cdot 5} = -\frac{2 \cdot 1}{1 \cdot 8} = -1 \cdot 17 \ m^{-1}$$

$$\frac{u_1}{u_2} = \frac{-27 \cdot 5}{-2 \cdot 1} = \frac{310}{23 \cdot 5} = \frac{23 \cdot 5}{1 \cdot 8} = 13 \cdot 1$$

Rotation ζ_0 is computed from the formula

$$\frac{\zeta_0}{u_2} = -\frac{F_1(a, d)_{01}}{F_2(a, d)_{01}} \frac{\zeta_1}{u_2} + \frac{F_3(a, d)_{01}}{l_{01} F_2(a, d)_{01}} \frac{u_1}{u_2}$$

$$= 0 \cdot 847 \times 1 \cdot 17 + 0 \cdot 140 \times 13 \cdot 1 = 2 \cdot 83 \ m^{-1}$$

and analogously

$$\frac{\zeta_2}{u_2} = -\frac{F_1(a, d)_{12}}{F_2(a, d)_{12}} \frac{\zeta_1}{u_2} - \frac{F_3(a, d)_{12}}{l_{12} F_2(a, d)_{12}} \frac{u_1}{u_2} - \frac{F_4(a, d)_{12}}{l_{12} F_2(a, d)_{12}}$$

$$= 0 \cdot 748 \times 1 \cdot 17 - 0 \cdot 143 \times 13 \cdot 1 + 0 \cdot 0555 = -0 \cdot 943 \ m^{-1}$$

The mode of free vibration is shown in Fig. 8.8c.

8.2.4 Simplified solution

The simplified approximate solution outlined in Section 8.1.2 for the case of secondary effects can also be applied to bars subjected to static axial force. The end moments and forces (see Fig. 8.10) are computed using the reciprocal theorem. For example, the amplitude of end moment M_{gh} is obtained by applying the theorem to the states shown in Fig. 8.10b and Fig. 8.10c

$$M_{gh} = M_{hg}^c - \omega^2 \int \mu \, v(x) \, v_c(x) \, dx + N_{st} \int v'(x) \, v_c'(x) \, dx \tag{8.74}$$

where $M_{hg}^c = M_{ghst}$.

The curve $v(x)$ is approximately replaced by the static curve $v_d(x)$, so that

$$M_{gh} \approx M_{ghst} - \omega^2 \int \mu \, v_d(x) \, v_c(x) \, dx + N_{st} \int v_d'(x) \, v_c'(x) \, dx \tag{8.75}$$

In addition,

$$M_{gh} = \frac{EJ}{l} F_1(a, d) \tag{8.76}$$

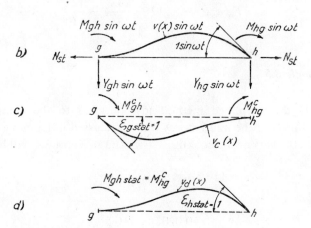

Fig. 8.10 Simplified slope-deflection method. Determination of end moments, static axial force considered

The first integral of eq. 8.75 was computed in Section 3.4; in view of eqs 8.41 and 8.42 the second integral is

$$N_{st} \int v'_d(x) v'_c(x) \, dx = -\frac{l}{30} N_{st} \tag{8.77}$$

and from eqs 8.51, 8.75, 8.76 and 8.77 the approximate value of function $F_1(a, d)$ is

$$F_1(a, d) \approx 2 + \frac{\lambda^4}{140} + \frac{\alpha^2}{30} \tag{8.78}$$

Similarly so the remaining approximate formulae are

$$F_2(a, d) \approx 4 - \frac{\lambda^4}{105} - \frac{2\alpha^2}{15}$$

$$F_3(a, d) \approx 6 + \frac{13\lambda^4}{420} - \frac{\alpha^2}{10}$$

$$\tag{8.79}$$

$$F_4(a, d) \approx -6 + \frac{11\lambda^4}{210} + \frac{\alpha^2}{10}$$

$$F_5(a, d) \approx -12 - \frac{9\lambda^4}{70} + \frac{12\alpha^2}{10}$$

$$F_6(a, d) \approx 12 - \frac{13\lambda^4}{35} - \frac{12\alpha^2}{10}$$

$$(8.79)$$

Example 8.5

Apply the simplified method just outlined to the computation of free symmetric vibration of the multi-storey frame discussed in Example 8.3. The determinant of the coefficients of the equations arranged in Table 3.4 will be zero for

$$\bar{\omega}_{(1)} = 110\cdot5 \text{ s}^{-1}$$

for which

$$\lambda_1^4 = 6^4 \frac{0\cdot0204 \times 110\cdot5^2}{21 \times 10^6 \times 1 \times 10^{-4}} = 154$$

$$\lambda_2^4 = 5^4 \frac{0\cdot0510 \times 110\cdot5^2}{21 \times 10^6 \times 2 \times 10^{-4}} = 92\cdot8$$

$$\alpha^2 = 3\cdot43$$

$$F_1(a, d)_1 \approx 2 + \frac{154}{140} + \frac{3\cdot43}{30} = 3.214$$

$$\frac{1}{l_1} F_1(a, d)_1 = 0\cdot536$$

$$F_2(a, d)_1 \approx 4 - \frac{154}{105} - \frac{2 \times 3\cdot43}{15} = 2\cdot078$$

$$\frac{1}{l_1} F_2(a, d)_1 = 0\cdot346$$

$$\frac{J_2}{l_2 J_1} [F_2(\lambda_2) - F_1(\lambda_2)] = \frac{2}{5} \left[4 - 2 - \frac{\lambda_2^4}{60} \right] = 0\cdot178$$

Table 8.5 now becomes Table 8.9

Table 8.9

ζ_1	ζ_2	ζ_3	
0·870	0·536		= 0
0·536	0·524 0·346 ⎯⎯⎯ 0·870	0·536	= 0
	0·536	0·346 0·178 0·524	= 0

and the determinant of the coefficients becomes

$$\Delta = 0{\cdot}870^2 \times 0{\cdot}524 - (0{\cdot}870 + 0{\cdot}524) \times 0{\cdot}536^2 \approx 0$$

Referring to the correction Table 3.9 we see that the error in $\bar{\omega}_{(1)}$ should be about 11%,

i.e. $\quad \bar{\omega}_{(1)} = 0{\cdot}89 \times 110{\cdot}5 = 98{\cdot}3 \text{ s}^{-1}$

as compared with the exact value of $95{\cdot}6 \text{ s}^{-1}$ (Example 8.3). Evidently, the simplified method works out less accurately for systems subjected to axial forces than it does for systems without them.

8.2.5 Approximate formulae

The first circular frequency of free vibration $\bar{\omega}_{(1)}$ of a simply supported bar of constant cross-section subjected to a static axial force is given by the formula

$$\bar{\omega}_{(1)} = \omega_{(1)} \left(1 - \frac{1}{\varkappa_{(1)}}\right)^{1/2} \tag{8.80}$$

where $\omega_{(1)}$ is the first circular frequency of free vibration of the bar not loaded by an axial force, and $\varkappa_{(1)}$ is the safety factor in the first mode of lateral deflection induced by buckling. The equation holds good in the range of elastic deformations. The higher modes of free vibration and lateral deflection induced by buckling are defined by analogous relations. Conversely, the safety factor $\varkappa_{(1)}$ is given by

$$\varkappa_{(1)} = \frac{1}{1 - \dfrac{\bar{\omega}_{(1)}^2}{\omega_{(1)}^2}} \tag{8.81}$$

Consequently $\varkappa_{(k)}$ of a simply supported beam can be computed from two known natural frequencies $\omega_{(k)}$, $\bar{\omega}_{(k)}$ established, say experimentally, for an unloaded beam and for a beam acted upon by an axial force.

The exact computation of natural circular frequencies in the presence of static axial forces, presented in Sections 8.2.2, 8.2.3 is quite time consuming. The natural circular frequencies of systems whose bars are not subjected to axial forces, and the safety factors against buckling are computed far more readily because in the determination of the coefficients of the slope-deflection equations we can make use of tables to a large extent. The solution with axial force would therefore become less tedious, even for more complicated systems, were it possible to compute $\bar{\omega}_{(1)}$ from $\omega_{(1)}$ and $\varkappa_{(1)}$. By way of a few examples we shall demonstrate the magnitude of errors stemming from the use of formulae 8.80 and 8.81 in the solution of frames and continuous beams.

In the absence of axial forces free vibration of a frame is governed by the equation

$$\sum \int_0^l \mu \omega_{(1)}^2 \, v_{(1)}^2(x) \, dx = \sum \int_0^l EJ \, v_{(1)}''^2(x) \, dx \qquad (8.82)$$

which states that the work of inertia forces arising in free vibration in the first mode, done on displacements of the same mode is equal to the work done by corresponding internal forces. If the frame is subjected to axial forces then

$$\sum \int_0^l \mu \bar{\omega}_{(1)}^2 \, \bar{v}_{(1)}^2(x) \, dx = \sum \int_0^l EJ \, \bar{v}_{(1)}''^2(x) \, dx + \sum \int_0^l N_{st} \, \bar{v}_{(1)}'^2(x) \, dx \qquad (8.83)$$

And finally, with regard to the deformation on the buckling limit, the equality between work done by external and internal forces as well as eq. 8.83 for $\bar{\omega}_{(1)} = 0$ imply that

$$\sum \int_0^l -\varkappa_{(1)} N_{st}^` \, y_{(1)}'^2(x) \, dx = \sum \int_0^l EJ \, y_{(1)}''(x) \, dx \qquad (8.84)$$

where $y_{(1)}$ is the first mode of lateral deflection induced by buckling. In all cases the summation sign includes all the bars of the system.

Were the curves $v_{(1)}(x)$, $\bar{v}_{(1)}(x)$, $y_{(1)}(x)$ of the same shape as is the case with a simply supported beam where all three are sine curves, eqs 8.82–8.84 would lead to expressions 8.80 and 8.81. Actually, however, the shapes of those curves are not identical and formulae 8.80 and 8.81 do not apply. If, on the other hand, the three curves display some degree of similarity, the two formulae apply approximately at least.

Example 8.6

Let us compare the symmetric deformations of the multi-storey frame shown in Fig. 3.8, computed in Example 3.1b with those arrived at in Example 8.3a and b. In all three cases the ratios ζ_1/ζ_3, ζ_2/ζ_3 agree as to the sign and are of comparable magnitude. Substituting the numerical values in eq. 8.80 gives

$$\bar{\omega}_{(1)} \approx \omega_1 \left(1 - \frac{1}{\varkappa_{(1)}}\right)^{1/2} = 103\cdot7 \left(1 - \frac{1}{5\cdot78}\right)^{1/2} = 94\cdot3 \text{ s}^{-1}$$

Compared with the exact value of $\bar{\omega}_{(1)} = 95\cdot6 \text{ s}^{-1}$, the error is about $1\cdot5\%$. If conversely $\varkappa_{(1)}$ is computed from the given $\bar{\omega}_{(1)}$ and $\omega_{(1)}$ values, the error is much larger. By eq. 8.81

$$\varkappa_{(1)} \approx \frac{1}{1 - \dfrac{\bar{\omega}_{(1)}^2}{\omega_{(1)}^2}} = \frac{1}{1 - \dfrac{95\cdot6^2}{103\cdot7^2}} = 6\cdot66$$

instead of $\varkappa_{(1)} = 5\cdot78$. The error here is 15%.

As to antisymmetric vibration: using the values obtained in Example 3.1d and Example 8.3c and d, eq. 8.80 gives

$$\bar{\omega}_{(1')} \approx 8\cdot41 \left(1 - \frac{1}{3\cdot89}\right)^{1/2} = 7\cdot25 \text{ s}^{-1}$$

which differs from the exact $\bar{\omega}_{(1')} = 7\cdot31$ by about 1%. By eq. 8.81

$$\varkappa_{(1')} = \frac{1}{1 - \dfrac{7\cdot31^2}{8\cdot41^2}} = 4\cdot09$$

Compared with the exact value, $\varkappa_{(1')} = 3\cdot89$, the above is correct to about 5%. When the modes of free lateral vibration are of completely different shapes, the results are less accurate still.

9 OTHER APPLICATIONS OF FREQUENCY FUNCTIONS $F(a, d)$

9.1 Transverse impact

Consider a beam with uniformly distributed mass and support conditions not specified for the present, on which falls from height h a mass point with mass m (Fig. 9.1).

The problem of transverse impact will simplify considerably if the beam cross-section has a vertical plane of symmetry and the falling mass point falls in this plane, and if, moreover, the whole system including the striking body can be replaced by the elastic system shown in Fig. 9.1b. Following the impact the system is set into

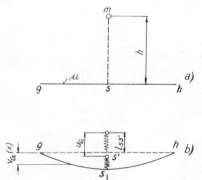

Fig. 9.1 Impact of mass falling on a beam

free vibration whose initial conditions are known. No exact solution of the contact problem will be attempted here. It is assumed that the forces between the beam and the striking mass are transmitted by an elastic medium which can be replaced by a spring of length $l_{ss'}$ and spring constant C. In this method of solution the energy losses arising during the impact may be considered to be a consequence of viscous damping.

Section 1.2.1 has shown the general solution of free vibration of systems with a finite number of degrees of freedom. The motion is described by eqs 1.51 or 1.52 whose integration constants are computed from the initial conditions of motion. The solution can readily be extended to systems with continuously distributed mass.

Considering only the lateral vibration we may write for the beam

$$v(x, t) = \sum_{j=1}^{\infty} q_{(j)} V_{(j)}(x) \sin \left(\omega_{(j)} t + \varphi_{(j)} \right) \tag{9.1}$$

or

$$v(x, t) = \sum_{j=1}^{\infty} V_{(j)}(x) \left[A_{(j)} \cos \omega_{(j)} t + B_{(j)} \sin \omega_{(j)} t \right] \tag{9.2}$$

where

$$A_{(j)}(x) = q_{(j)} \sin \varphi_{(j)}$$
$$B_{(j)}(x) = q_{(j)} \cos \varphi_{(j)} \tag{9.3}$$

and $V_{(j)}(x)$ are the normalised modes of free vibration of the system shown in Fig. 9.1b. All displacements are measured from the equilibrium position in which the system is loaded with its own weight μg and the weight of the fallen body mg.

The initial state of motion is defined by displacement $v(x, 0)$ and velocity $\dot{v}(x, 0)$,

$$v(x, 0) = \sum_{j} A_{(j)} V_{(j)}(x)$$
$$\dot{v}(x, 0) = \sum_{j} \omega_{(j)} B_{(j)} V_{(j)}(x) \tag{9.4}$$

Denoting by $U_{(j)}$ the normalised displacement of the mass point in the j-th mode of free vibration we may write

$$u(t) = \sum_{j=1}^{\infty} q_{(j)} U_{(j)} \sin \left(\omega_{(j)} t + \varphi_{(j)} \right) \tag{9.5}$$

or

$$u(t) = \sum_{j=1}^{\infty} U_{(j)} \left[A_{(j)} \cos \omega_{(j)} t + B_{(j)} \sin \omega_{(j)} t \right] \tag{9.6}$$

where $u(t)$ is the displacement of the mass point measured from the equilibrium position.

The displacement and velocity at the beginning of the beam motion are

$$u(0) = \sum_{j=1}^{\infty} A_{(j)} U_{(j)}$$
$$\dot{u}(0) = \sum_{j=1}^{\infty} \omega_{(j)} B_{(j)} U_{(j)} \tag{9.7}$$

Coefficients $A_{(j)}$, $B_{(j)}$ are computed as follows: both sides of eq. 9.4 are multiplied by $\mu\,V_{(j)}(x)\,\mathrm{d}x$ and integrated over the whole system. To the ensuing equation are added eqs 9.7 multiplied by $m\,U_{(j)}$. Due to orthogonality of the natural modes

$$A_{(j)} = m\,u(0)\,U_{(j)} + \int \mu\,v(x, 0)\,V_{(j)}(x)\,\mathrm{d}x \qquad (9.8)$$

$$B_{(j)} = \frac{1}{\omega_{(j)}}\left[m\,\dot{u}(0)\,U_{(j)} + \int \mu\,\dot{v}(x, 0)\,V_{(j)}(x)\,\mathrm{d}x\right] \qquad (9.9)$$

Assume that the beam is at rest, and at instant $t = 0$ is struck by mass point m falling from height h with velocity $\dot{u}(0) = (2gh)^{1/2}$. The impact will take place at

$$u(0) = -u_G \qquad (9.10)$$

where u_G is the static displacement of the system at the point of incidence of load induced by force mg. If $v_G(x)$ is the static displacement at point x of the system, produced by the same load eqs 9.8 and 9.9 will give

$$A_{(j)} = -mu_G U_{(j)} - \int \mu\,v_G(x)\,V_{(j)}(x)\,\mathrm{d}x \qquad (9.11)$$

$$B_{(j)} = \frac{1}{\omega_{(j)}}\,m(2gh)^{1/2}\,U_{(j)} \qquad (9.12)$$

The foregoing equations enable us to compute the motion of the system from the instant the mass point contacts the system to the instant it separates from it. As only pressures can act between the mass point and the beam, the former will separate from the latter as soon as the fictitious spring becomes as long as it was in its initial unstretched state. After the separation of the mass point in time $t = t_1$, the system — now with $m = 0$ — vibrates in free vibration until the next impact. Denoting by $\bar{v}(x, t)$ the displacements of this unloaded system and recalling that its equilibrium position differs from that of the original system, we write

$$\bar{v}(x, t) = v(x, t) - v_G(x)$$

Once separated from the system, the mass point m moving with velocity $\dot{u}(t_1)$ will ascend to height $h' = \dot{u}^2(t_1)/2g$ and strike the beam again at instant t_2, for which

$$u(t_2) - v_s(t_2) = l_{ss}.$$

The meaning of the symbols is evident from Fig. 9.1b, $v_s(t)$ is the deflection at the midspan of the beam in time t.

The solution just outlined has not considered plastic deformations. This can to some extent be remedied by a simple modification taking care of damping. The computation can also be carried out so as to take account of the effects of shear deflection and rotatory inertia.

9.1.1 Load falling midspan of a simply supported beam

The natural modes of the system shown in Fig. 9.1b may be computed by the slope-deflection method described in Chapter 2. The sought-for displacements are u and v_s, and the end forces are

$$Y_{sg} = Y_{sh} = \frac{8EJ}{l^3} F_{11}\left(\frac{\lambda}{2}\right) v_s \tag{9.13}$$

$$Y_{ss'} = C(v_s - u) \tag{9.14}$$

$$Y_{s's} = m\omega^2 u - C(v_s - u) \tag{9.15}$$

where C is the spring constant of the fictitious spring. The slope-deflection equations are then of the form

$$2Y_{sg} + Y_{ss'} = 0$$
$$Y_{s's} = 0 \tag{9.16}$$

and from there

$$u = \frac{C}{C + m\omega^2} v_s$$

and

$$\left[\frac{16EJ}{l^3} F_{11}\left(\frac{\lambda}{2}\right) + \frac{Cm\omega^2}{C + m\omega^2}\right] v_s = 0 \tag{9.17}$$

Since

$$\omega^2 = \frac{\lambda^4}{l^4} \frac{EJ}{\mu} \tag{9.18}$$

it is possible to find $\lambda = \lambda_{(j)}$ corresponding to the j-th natural mode from the equation

$$\frac{16EJ}{l^3} F_{11}\left(\frac{\lambda_{(j)}}{2}\right) + \frac{CEJm\lambda_{(j)}^4}{C\mu l^4 + EJm\lambda_{(j)}^4} = 0 \tag{9.19}$$

The natural mode is defined by eq. 2.25 with the integration constants according to eqs 2.51 and 2.53. It is normalised with the aid of function $\Phi(\lambda)$ and eq. 2.96. Since the system is symmetric and the origin of the coordinates can be placed at point s, the equations of the natural mode need be written for one half of the system only. We may then write

$$V_{(j)}(x) = \frac{1}{\left(16\mu l \Phi_{11}\left(\frac{\lambda}{2}\right) + m \dfrac{C\lambda^4}{C + \dfrac{m\lambda^4}{l^4} \dfrac{EJ}{\mu}}\right)^{1/2}} \left[2F_9\left(\frac{\lambda}{2}\right)\left(\cosh\frac{\lambda x}{l} - \cos\frac{\lambda x}{l}\right)\right]$$

$$+ \frac{4}{\lambda} F_{11} \left(\frac{\lambda}{2} \right) \left(\sinh \frac{\lambda x}{l} - \sin \frac{\lambda x}{l} \right) \Bigg]$$ (9.20)

$$U_{(j)} = \frac{C}{C + \frac{m\lambda^4 EJ}{\mu l^4}} V_{s(j)}$$ (9.21)

where the respective $\lambda_{(j)}$ is substituted for λ. The variation of the displacement is then computed from eqs 9.2, 9.6, 9.8, 9.9, 9.11 and 9.12. It is necessary to establish the time variations of displacements $v(s, t)$ and $u(t)$ and determine instant t_1 at which the mass point will separate from the beam. To do so we write

$$v(s, t_1) - u(t_1) = v(s, 0) - u(0)$$ (9.22)

Thereafter the system vibrates in accordance with the equation

$$\bar{v}(x, \bar{t}) = \sum_j \left(\frac{2}{\mu l} \right)^{1/2} \cos \frac{j\pi x}{l} (\bar{A}_{(j)} \cos \bar{\omega}_{(j)} \bar{t} + \bar{B}_{(j)} \sin \bar{\omega}_{(j)} \bar{t})$$ (9.23)

where $\bar{t} = t - t_1$ and $\bar{\omega}_{(j)} = \frac{j^2 \pi^2}{l^2} \left(\frac{EJ}{\mu} \right)^{1/2}$. Constants $\bar{A}_{(j)}$, $\bar{B}_{(j)}$ are computed from the conditions

$$\bar{v}(x, 0) = v(x, t_1) + v_G$$
$$\dot{\bar{v}}(x, 0) = \dot{v}(x, t_1)$$
(9.24)

consequently

$$\bar{A}_{(j)} = 2 \left(\frac{2\mu}{l} \right)^{1/2} \int_0^{l/2} \bar{v}(x, 0) \cos \frac{j\pi x}{l} \, dx$$

$$\bar{B}_{(j)} = \frac{2}{\omega_{(j)}} \left(\frac{2\mu}{l} \right)^{1/2} \int_0^{l/2} \dot{\bar{v}}(x, 0) \cos \frac{j\pi x}{l} \, dx$$

(9.25)

9.1.2 Load falling midspan of a beam fixed at both ends

The procedure of solving this case remains essentially the same as that outlined in the last paragraph, except that $F_6(\lambda/2)$ is substituted for $F_{11}(\lambda/2)$ in eqs 9.13, 9.17 and 9.19, and eq. 9.20 is replaced by

$$V_{(j)}(x) = \cfrac{1}{\left(16\mu l\Phi_6\left(\cfrac{\lambda}{2}\right) + m\cfrac{C\lambda^4}{C + \cfrac{m\lambda^4}{l^4}\cfrac{EJ}{\mu}}\right)^{1/2}}\left[2F_4\left(\frac{\lambda}{2}\right)\left(\cosh\frac{\lambda x}{l} - \cos\frac{\lambda x}{l}\right)\right.$$

$$\left. + \frac{4F_6(\lambda)}{\lambda}\left(\sinh\frac{\lambda x}{l} - \sin\frac{\lambda x}{l}\right)\right] \tag{9.26}$$

The equation governing vibration after time t_1 is

$$\bar{v}(x, \bar{t}) = \sum \bar{V}_{(j)}(x)\left(\bar{A}_{(j)}\cos\bar{\omega}_{(j)}\bar{t} + \bar{B}_{(j)}\sin\bar{\omega}_{(j)}\bar{t}\right) \tag{9.27}$$

where the natural mode $\bar{V}_{(j)}(x)$ and frequency $\bar{\omega}_{(j)}$ are computed without regard to mass m.

Coefficients $\bar{A}_{(j)}$, $\bar{B}_{(j)}$ are

$$\bar{A}_{(j)} = 2\mu \int_0^{l/2} \bar{v}(x, 0)\,\bar{V}_{(j)}(x)\,\mathrm{d}x \tag{9.28}$$

$$\bar{B}_{(j)} = \frac{2\mu}{\bar{\omega}_{(j)}} \int_0^{l/2} \dot{\bar{v}}(x, 0)\,\bar{V}_{(j)}(x)\,\mathrm{d}x \tag{9.29}$$

9.1.3   Load falling on a system composed of prismatic bars

By way of an example consider a mass point falling on point s midway of the central span $\overline{12}$ of a uniform beam continuous over three spans $(\overline{01}, \overline{12}, \overline{23})$. In this case the slope-deflection equations are as shown in Table 9.1.

The equations are used in the determination of the frequencies and deformations of the joints in free vibration. The modes of the various bars are again computed from eqs 2.25 with the integration constants defined by eqs 2.47, 2.51 and 2.53 and normalised with the aid of function $\Phi(\lambda)$ (Section 2.8.2) in the manner indicated in the preceding examples. Eqs 9.2–9.12 and 9.27–9.29 continue to apply.

In case of violent impacts due account should also be taken of the effects of shear and rotatory inertia. The procedure is the same as that just outlined except that functions $F(a, d)$ (Chapter 8) are employed in place of functions $F(\lambda)$. The effect of shear on the mode of free vibration must also be considered when normalising the frequency functions.

As to the number of terms in the series in eq. 9.2: it is necessary to consider as many as will make the period of the highest forms shorter by far than the duration of the first stage of impact, i.e. than the time interval 0 to t_1.

Table 9.1

	ζ_1	ζ_s	v_s	ζ_2	u	
	$\dfrac{1}{l_{01}}F_7(\lambda_{01}) + \dfrac{2}{l_{12}}F_2\left(\dfrac{\lambda_{12}}{2}\right)$	$\dfrac{2}{l_{12}}F_1\left(\dfrac{\lambda_{12}}{2}\right)$	$-\dfrac{4}{l_{12}^2}F_3\left(\dfrac{\lambda_{12}}{2}\right)$			$= 0$
	$\dfrac{2}{l_{12}}F_1\left(\dfrac{\lambda_{12}}{2}\right)$	$\dfrac{4}{l_{12}}F_2\left(\dfrac{\lambda_{12}}{2}\right)$		$\dfrac{2}{l_{12}}F_1\left(\dfrac{\lambda_{12}}{2}\right)$		$= 0$
	$-\dfrac{4}{l_{12}^2}F_3\left(\dfrac{\lambda_{12}}{2}\right)$		$\dfrac{16}{l_{12}^3}F_6\left(\dfrac{\lambda_{12}}{2}\right) + \dfrac{C}{EJ}$	$\dfrac{4}{l_{12}^2}F_3\left(\dfrac{\lambda_{12}}{2}\right)$	$-\dfrac{C}{EJ}$	$= 0$
		$\dfrac{2}{l_{12}}F_1\left(\dfrac{\lambda_{12}}{2}\right)$	$\dfrac{4}{l_{12}^2}F_3\left(\dfrac{\lambda_{12}}{2}\right)$	$\dfrac{2}{l_{12}}F_2\left(\dfrac{\lambda_{12}}{2}\right) + \dfrac{1}{l_{23}}F_7(\lambda_{23})$		$= 0$
			$-\dfrac{C}{EJ}$		$\dfrac{C - m\omega^2}{EJ}$	$= 0$

9.1.4   Effect of damping

If damping is proportional to velocity (as in Section 6.1), eq. 9.2 changes to

$$v(x, t) = \sum_j e^{-\omega_b t} V_{(j)}(x) \left[A_{(j)} \cos \omega'_{(j)} t + B_{(j)} \sin \omega'_{(j)} t \right] \tag{9.30}$$

where

$$\omega'_{(j)} = (\omega_{(j)}^2 - \omega_b^2)^{1/2}$$

The natural modes $V_{(j)}(x)$ are not changed by damping. Since the derivative of eq. 9.30 is

$$\dot{v}(x, t) = \sum_j V_{(j)}(x) e^{-\omega_b t} \left[-\omega_b (A_{(j)} \cos \omega'_{(j)} t + B_{(j)} \sin \omega'_{(j)} t) \right.$$
$$\left. + \omega'_{(j)} (-A_{(j)} \sin \omega'_{(j)} t + B_{(j)} \cos \omega'_{(j)} t) \right] \tag{9.31}$$

we get in place of eq. 9.4

$$v(x, 0) = \sum_j A_{(j)} V_{(j)}(x)$$
$$\dot{v}(x, 0) = \sum_j (-A_{(j)} \omega_b + B_{(j)} \omega'_{(j)}) V_{(j)}(x) \tag{9.32}$$

and in place of eq. 9·7

$$u(0) = \sum_j A_{(j)} U_{(j)}$$
$$\dot{u}(0) = \sum_j (-A_{(j)} \omega_b + B_{(j)} \omega'_{(j)}) U_{(j)} \tag{9.33}$$

From there eq. 9.8 and

$$(-A_{(j)} \omega_b + B_{(j)} \omega'_{(j)}) = m \, \dot{u}(0) \, U_{(j)} + \int \mu \, \dot{v}(x, 0) \, V_{(j)}(x) \, dx \tag{9.34}$$

are used for computing the values of $A_{(j)}$ and $B_{(j)}$.

9.2   Lateral vibration of plates

Accepting the usual hypotheses we may write the equation of motion of a plate performing free vibration as

$$H \, \nabla^4 w(x, y, t) + \mu \, \frac{\partial^2 w(x, y, t)}{\partial t^2} = 0 \tag{9.35}$$

where

$$\nabla^4 = \frac{\partial^4}{\partial x^4} + 2\frac{\partial^4}{\partial x^2 \, \partial y^2} + \frac{\partial^4}{\partial y_4} \quad \text{is the Laplace operator of the fourth order,}$$

$H = Eh^3/12(1 - v^2)$ the bending or cylindrical stiffness of the plate

$w(x, y, t)$ the vertical displacement (in the direction of axis Z) at position x, y and time t

h the thickness of the plate

μ the mass per unit area

v Poisson's ratio

Assuming harmonic motion

$$w(x, y, t) = w_0(x, y) \sin \omega_0 t \tag{9.36}$$

eq. 9.35 gives for the amplitude of motion

$$\nabla^4 w_0(x, y) - \frac{\mu \omega_0^2}{H} w_0(x, y) = 0 \tag{9.37}$$

The way of solving eq. 9.37 depends on the edge conditions which can be as follows: displacements $w_0(x, y)$ on plate edges; rotations $\partial w_0(x, y)/\partial x$, $\partial w_0(x, y)/\partial y$ in the directions of the coordinate axes, bending moments and shear forces in directions x or y.

The bending moments (per unit length of edge) are

$$M_x(x, y) = -H \left(\frac{\partial^2 w_0(x, y)}{\partial x^2} + v \frac{\partial^2 w_0(x, y)}{\partial y^2} \right) \tag{9.38}$$

$$M_y(x, y) = -H \left(\frac{\partial^2 w_0(x, y)}{\partial y^2} + v \frac{\partial^2 w_0(x, y)}{\partial x^2} \right) \tag{9.39}$$

The torque is

$$M_{xy}(x, y) = -H(1 - v) \frac{\partial^2 w_0(x, y)}{\partial x \, \partial y} \tag{9.40}$$

and the shear forces are

$$T_x(x, y) = -H \left(\frac{\partial^3 w_0(x, y)}{\partial x^3} + \frac{\partial^3 w_0(x, y)}{\partial x \, \partial y^2} \right) \tag{9.41}$$

$$T_y(x, y) = -H \left(\frac{\partial^3 w_0(x, y)}{\partial y^3} + \frac{\partial^3 w_0(x, y)}{\partial x^2 \, \partial y} \right) \tag{9.42}$$

On free edges we introduce the concept of reduced shear force

$$\bar{T}_x(x, y) = T_x(x, y) + \frac{\partial M_{xy}}{\partial y} = -H\left[\frac{\partial^3 w_0(x, y)}{\partial x^3} + (2 - v)\frac{\partial^3 w_0(x, y)}{\partial x \, \partial y^2}\right] \quad (9.43)$$

$$\bar{T}_y(x, y) = T_y(x, y) + \frac{\partial M_{xy}}{\partial x} = -H\left[\frac{\partial^3 w_0(x, y)}{\partial y^3} + (2 - v)\frac{\partial^3 w_0(x, y)}{\partial x^2 \, \partial y}\right] \quad (9.44)$$

9.2.1 Rectangular plates

In the case of rectangular plates a direct solution is possible whenever two opposite edges (e.g. edges parallel to axis X) are simply supported. Then for $y = 0$ and $y = l_y$

$$w(x, y) = 0$$
$$M_x(x, y) = 0$$

The conditions on the remaining edges may be of any kind whatsoever. The particular solution of eq. 9.37 is then of the form

$$w_0(x, y) = K_j \sin\frac{j\pi y}{l_y} X_i(x) \quad (9.45)$$

where i, j are integers, K_j a coefficient of arbitrary magnitude, l_y the span in the direction of axis Y (similarly, l_x is the span in the direction of axis X), and $X_i(x)$ is a function dependent on x (to be specified later on). Substituting eq. 9.45 in 9.37 gives

$$\frac{d^4 X_i(x)}{dx^4} - \frac{2j^2\pi^2}{l_y^2}\frac{d^2 X_i}{dx^2} + \left(\frac{j^4\pi^4}{l_y^2} - \frac{\mu\omega^2}{H}\right)X_i(x) = 0 \quad (9.46)$$

The form of this equation resembles that of eq. 8.7, Section 8.1; consequently, its solution is analogous to eq. 8.8:

$$X_i(x) = A\cos\frac{ax}{l_x} + B\sin\frac{ax}{l_x} + C\cosh\frac{dx}{l_x} + D\sinh\frac{dx}{l_x} \quad (9.47)$$

where

$$a = l_x\left[-\frac{j^2\pi^2}{l_y^2} + \left(\frac{\mu\omega^2}{H}\right)^{1/2}\right]^{1/2} \quad (9.48)$$

and

$$d = l_x\left[\frac{j^2\pi^2}{l_y^2} + \left(\frac{\mu\omega^2}{H}\right)^{1/2}\right]^{1/2} \quad (9.49)$$

A, B, C, D are integration constants which depend on the conditions on edges $x = 0$ and $x = l_x$.

If a rectangular plate is simply supported on all edges, then on edges $x = 0$, $x = l_x$

$$w_0(x, y) = 0$$

$$\frac{\partial^2 w_0(x, y)}{\partial y^2} = 0$$

$$M_x(x, y) = 0$$

It follows from eq. 9.38 that (for $x = 0$, $x = l_x$)

$$\frac{\partial^2 w_0(x, y)}{\partial x^2} = 0$$

and from eq. 9.45 that

$$X_i(x) = 0$$

$$\frac{\partial^2 X_i(x)}{\partial x^2} = 0$$

Substituting eq. 9.47 in the above leads to the following four equations:

$$A + C = 0$$
$$-Aa^2 + Cd^2 = 0$$
$$B \sin a + D \sinh d = 0$$
$$Ba^2 \sin a - Dd^2 \sinh d = 0$$

From there

$$A = C = 0$$
$$D = 0$$
$$\sin a = 0$$
$$a = 0, \pi, 2\pi \dots i\pi \dots$$

B is a constant of arbitrary magnitude.

With

$$K_j B \equiv q_{ij}$$

eq. 9.45 will give the familiar equation of the mode of free vibration of a simply supported plate

$$w_0(x, y) = w_{ij}(x, y) = q_{ij} \sin \frac{j\pi y}{l_y} \sin \frac{i\pi x}{l_x}$$

The natural frequencies are obtained from eqs 9.48 and 9.49

$$\omega_0 = \omega_{ij} = \pi^2 \left(\frac{i^2}{l_x^2} + \frac{j^2}{l_y^2} \right) \left(\frac{H}{\mu} \right)^{1/2}$$

If edges $x = 0$, $x = l_x$ are clamped (Fig. 9.2), the edge conditions there are

$$X(x) = 0$$

$$\frac{dX(x)}{dx} = 0$$

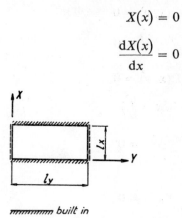

built in

simply supported   Fig. 9.2 Rectangular plate

and the equations of the integration constants are

$$A + C = 0$$

$$aB + dD = 0$$

$$A(\cos a - \cosh d) + B\left(\sin a - \frac{a}{d}\sinh d\right) = 0$$

$$-A(a \sin a + d \sinh d) + Ba(\cos a - \cosh d) = 0$$

$$(9.50)$$

From the condition that for a non-trivial solution the determinant of the coefficients of the last two equations should equal zero we get the frequency equation

$$f_1(a, d) = 0$$

where

$$f_1(a, d) = 2ad(\cosh d \cos a - 1) + (a^2 - d^2)\sinh d \sin a \qquad (9.51)$$

from which the natural frequencies may be computed by successive substitution. Eqs 9.50 then give the ratios of constants A, B, C, D and eqs 9.45 and 9.47 the natural modes of vibration.

If edges $x = 0$, $x = l_x$ are free, the edge conditions are

$$M_x(x, y) = 0$$

$$T_x(x, y) = 0 \quad \text{for} \quad x = 0, x = l_x$$

Eqs 9.38 and 9.44 then give

$$\frac{d^2X(x)}{dx^2} - \frac{vj^2\pi^2}{l_y^2} X(x) = 0$$

$$\frac{d^3X(x)}{dx^3} - (2-v)\frac{j^2\pi^2}{l_y^2}\frac{dX(x)}{dx} = 0 \qquad (9.52)$$

for $x = 0$, $x = l_x$.

Therefore

$$-A\bar{a} + C\bar{d} = 0$$

$$-Ba\bar{d} + Dd\bar{a} = 0$$

$$-A\bar{a}\cos a - B\bar{a}\sin a + C\bar{d}\cosh d + D\bar{d}\sinh d = 0$$

$$Aa\bar{d}\sin a - Ba\bar{d}\cos a + Cd\bar{a}\sinh d + Dd\bar{a}\cosh d = 0$$

$$(9.53)$$

where

$$\bar{a} = a^2 + \frac{vj^2\pi^2 l_x^2}{l_y^2} \qquad (9.54)$$

and

$$\bar{d} = d^2 - \frac{vj^2\pi^2 l_x^2}{l_y^2} \qquad (9.55)$$

so that

$$C = \frac{\bar{a}}{\bar{d}} A$$

$$D = \frac{a\bar{d}}{d\bar{a}} B$$

$$B = -Ad\bar{a}^2 \frac{\cosh d - \cos a}{a\bar{d}^2 \sinh d - d\bar{a}^2 \sin a} \qquad (9.56)$$

and the frequency equation is

$$2ad\bar{a}^2\bar{d}^2(\cosh d \cos a - 1) + (a^2\bar{d}^4 - d^2\bar{a}^4)\sinh d \sin a = 0 \qquad (9.57)$$

Plates with edge conditions of other kinds should be treated in an analogous manner.

9.2.2 Forced vibration induced by edge rotation

Consider a plate simply supported on edges $y = 0$, $y = l_y$ and clamped on edge $x = 0$. The plate vibrates in harmonic motion

$$w(x, y, t) = w(x, y) \sin \omega t \qquad (9.58)$$

with the amplitude

$$w(x, y) = \sin\frac{j\pi y}{l_y} X(x) \qquad (9.59)$$

Let rotation on edge $x = l_x$ be such that

$$\frac{dX(x)}{dx} = 1$$

From eq. 9.47 and its derivatives we get

$$A + \quad C = 0$$

$$aB + dD = 0$$

$$Ad(\cos a - \cosh d) + B(d \sin a - a \sinh d) = 0$$

$$-A(a \sin a + d \sinh d) + Ba(\cos a - \cosh d) = l_x \qquad (9.60)$$

and in turn the integration constants are

$$A = -l_x \frac{a \sinh d - d \sin a}{f_1(a, d)} = -C \qquad (9.61)$$

$$B = l_x d \frac{\cosh d - \cos a}{f_1(a, d)} = -\frac{d}{a} D \qquad (9.62)$$

In view of eqs 9.38 and 9.58 and because $X(0) = 0$ the bending moment on edge $x = 0$ is given by

$$M_x(0, y) = -H \sin\frac{j\pi y}{l_y} X''(0) \qquad (9.63)$$

where

$$X''(x) = \frac{d^2 X}{dx^2}$$

In view of eqs 9.61 and 9.62

$$X''(0) = \frac{1}{l_x} \frac{1}{f_1(a, d)} (a \sinh d - d \sin a)(a^2 + d^2) = -\frac{1}{l_x} F_1(a, d)$$

where

$$F_1(a, d) = -\frac{(a^2 + d^2)(a \sinh d - d \sin a)}{f_1(a, d)} \qquad (9.64)$$

Formula 9.64 is analogous to the first of formulae 8.22. Further

$$M_x(l_x,y) = -H \sin \frac{j\pi y}{l_y} X''(l_x)$$

where

$$X''(l_x) = \frac{1}{l_x} F_2(a, d) \tag{9.65}$$

$$F_2(a, d) = -\frac{(a^2 + d^2)(d \cosh d \sin a - a \sinh d \cos a)}{f_1(a, d)} \tag{9.66}$$

For a plate simply supported on edge $x = 0$ and clamped on edge $x = l_x$ the integration constants are computed from the equations

$$\left.\begin{array}{l} A \quad + C \quad = 0 \\ Aa^2 - Cd^2 = 0 \end{array}\right\} \rightarrow A = C = 0$$

$$B \sin a + D \sinh d = 0$$

$$Ba \cos a + Dd \cosh d = l_x$$

$$B = -l_x \frac{\sinh d}{d \cosh d \sin a - a \sinh d \cos a} \tag{9.67}$$

$$D = l_x \frac{\sin a}{d \cosh d \sin a - a \sinh d \cos a} \tag{9.68}$$

Then

$$X''(l_x) = \frac{1}{l_x} F_7(a, d) \tag{9.69}$$

$$F_7(a, d) = \frac{(a^2 + d^2) \sinh d \sin a}{d \cosh d \sin a - a \sinh d \cos a} \tag{9.70}$$

If edge $x = 0$ is free the equations of the integration constants will be

$$-A\bar{a} + C\bar{d} = 0$$

$$-Ba\bar{d} + Dd\bar{a} = 0$$

$$A \cos a + B \sin a + C \cosh d + D \sinh d = 0$$

$$-Aa \sin a + Ba \cos a + Cd \sinh d + Dd \cosh d = l_x$$

$$\tag{9.71}$$

Then

$$A = l_x \frac{-\bar{d}(a\bar{d} \sinh d + d\bar{a} \sin a)}{f_2(a, d)} \tag{9.72}$$

$$B = l_x \frac{d\bar{a}(\bar{a} \cosh d + \bar{d} \cos a)}{f_2(a, d)} \tag{9.73}$$

$$X''(l_x) = \frac{1}{l_x} F_{15}^*(a, d) \tag{9.74}$$

where

$$F_{15}^*(a, d) = \frac{(a^2 + d^2)(a\bar{d}^2 \sinh d \cos a - d\bar{a}^2 \cosh d \sin a)}{f_2(a, d)} \tag{9.75}$$

$$f_2(a,d) = 2ad\bar{a}\bar{d} + ad(\bar{a}^2 + \bar{d}^2) \cosh d \cos a + (a^2\bar{d}^2 - d^2\bar{a}^2) \sinh d \sin a \tag{9.76}$$

9.2.3 Vibration of continuous plates

In a manner similar to that employed with continuous beams the slope-deflection method can be applied to continuous plates. The pertinent equations are derived from the conditions of equilibrium on plate edges. With notation $X'(l_{01}) = = \zeta_1$, $X'(l_{12}) = \zeta_2$, etc. the slope-deflection equations will formally be the same

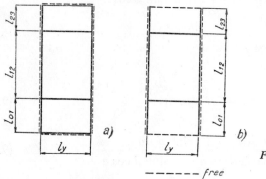

Fig. 9.3 Plates. (a) Continuous, (b) with overhung edges

$----$ free

as those of a continuous beam. Thus, for example, for a plate in free vibration according to Fig. 9.3a

$$\left[\frac{1}{l_{01}} F_7(a, d)_{01} + \frac{1}{l_{12}} F_2(a, d)_{12}\right] \zeta_1 + \frac{1}{l_{12}} F_1(a, d)_{12} \zeta_2 = 0$$

$$\frac{1}{l_{12}} F_1(a, d)_{12} \zeta_1 + \left[\frac{1}{l_{12}} F_2(a, d)_{12} + \frac{1}{l_{23}} F_7(a, d)_{23}\right] \zeta_2 = 0$$

$$\tag{9.77}$$

For the plate shown in Fig. 9.3b the equations will be the same except that they will contain $F_{15}^*(a, d)$ in place of $F_7(a, d)$.

When analysing plates continuous over several equal spans, one can take advantage of their cyclosymmetric properties and analogously apply formulae 5.48a. This gives

$$F_2(a, d) + F_1(a, d) \cos \frac{\pi j}{n} = 0$$

where n is the number of spans.

10 APPENDIX

10.1 Comprehensive tables of end moments and forces

10.1.1 Bar clamped at both ends (folder)

10.1.2 Bar hinged at one end

10.1.3 Bar hinged at both ends (folder)

10.1.4 Cantilever beam

10.1.5 Frequency functions formulae

10.1.6 Beam under uniform load

10.1.7 Longitudinal vibration. Torsional vibration

10.2 Numerical tables of frequency functions

10.2.1 Functions $F(\lambda)$

10.2.2 Functions $\Phi(\lambda)$

10.1.2 Bar hinged at one end

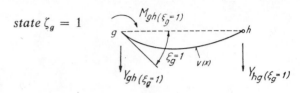

state $\zeta_g = 1$

		Static state				Vibration
		$N_{st} = 0$		$N_{st} \neq 0$		
		shear not con- sidered	shear considered	$N_{st} < 0$ (compression)	$N_{st} > 0$ (tension)	shear, N_{st} not considered
End moments and forces	M_{gh}	$\dfrac{3EJ}{l}$	$\dfrac{3EJ}{l}\dfrac{1}{1+3\gamma}$	$\dfrac{EJ}{l}\Gamma_5(\alpha)$	$\dfrac{EJ}{l}\bar{\Gamma}_5(\delta)$	$\dfrac{EJ}{l}F_7(\lambda)$
	Y_{gh}	$\dfrac{3EJ}{l^2}$	$\dfrac{3EJ}{l^2}\dfrac{1}{1+3\gamma}$	$\dfrac{EJ}{l^2}\Gamma_5(\alpha)$	$\dfrac{EJ}{l^2}\bar{\Gamma}_5(\delta)$	$-\dfrac{EJ}{l^2}F_9(\lambda)$
	Y_{hg}	$-\dfrac{3EJ}{l^2}$	$-\dfrac{3EJ}{l^2}\dfrac{1}{1+3\gamma}$	$-\dfrac{EJ}{l^2}\Gamma_5(\alpha)$	$-\dfrac{EJ}{l^2}\bar{\Gamma}_5(\delta)$	$-\dfrac{EJ}{l^2}F_8(\lambda)$
Integration constants	A			$\dfrac{l\,\Gamma_5(\alpha)}{\alpha^2}$	$-\dfrac{l\,\bar{\Gamma}_5(\delta)}{\delta^2}$	$\dfrac{l\,F_7(\lambda)}{2\lambda^2}$
	B			$l\,\dfrac{-\Gamma_5(\alpha)+\alpha^2}{\alpha^3}$	$l\,\dfrac{\bar{\Gamma}_5(\delta)+\delta^2}{\delta^3}$	$l\,\dfrac{F_9(\lambda)+\lambda^2}{2\lambda^3}$
	C			$-A$	$-A$	$-A$
	D			$1 - B\dfrac{\alpha}{l}$	$1 - B\dfrac{\delta}{l}$	$\dfrac{l}{\lambda} - B$

state $\bar{\zeta}_h = 1$

$\bar{v}(x) = -v(l-x)$

$M_{gh}(\bar{\xi}_g = 1)$

$Y_{gh}(\bar{\xi}_h = 1) = -Y_{hg}(\xi_g = 1)$ $Y_{hg}(\bar{\xi}_h = 1) = -Y_{gh}(\xi_g = 1)$

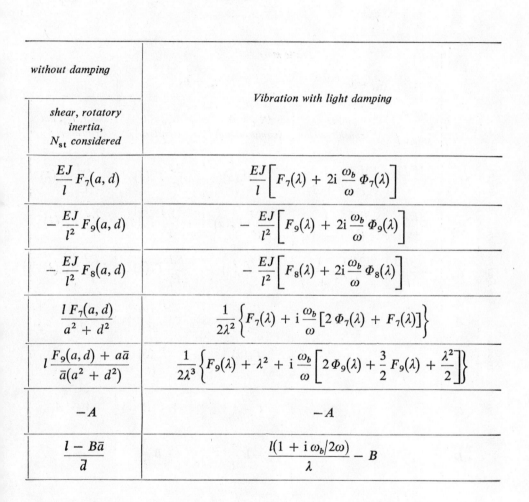

without damping *shear, rotatory inertia, N_{st} considered*	*Vibration with light damping*
$\dfrac{EJ}{l} F_7(a, d)$	$\dfrac{EJ}{l}\left[F_7(\lambda) + 2i\,\dfrac{\omega_b}{\omega}\,\Phi_7(\lambda) \right]$
$-\dfrac{EJ}{l^2} F_9(a, d)$	$-\dfrac{EJ}{l^2}\left[F_9(\lambda) + 2i\,\dfrac{\omega_b}{\omega}\,\Phi_9(\lambda) \right]$
$-\dfrac{EJ}{l^2} F_8(a, d)$	$-\dfrac{EJ}{l^2}\left[F_8(\lambda) + 2i\,\dfrac{\omega_b}{\omega}\,\Phi_8(\lambda) \right]$
$\dfrac{l\,F_7(a, d)}{a^2 + d^2}$	$\dfrac{1}{2\lambda^2}\left\{ F_7(\lambda) + i\,\dfrac{\omega_b}{\omega}\left[2\,\Phi_7(\lambda) + F_7(\lambda) \right] \right\}$
$l\,\dfrac{F_9(a, d) + a\bar{a}}{\bar{a}(a^2 + d^2)}$	$\dfrac{1}{2\lambda^3}\left\{ F_9(\lambda) + \lambda^2 + i\,\dfrac{\omega_b}{\omega}\left[2\,\Phi_9(\lambda) + \dfrac{3}{2}\,F_9(\lambda) + \dfrac{\lambda^2}{2} \right] \right\}$
$-A$	$-A$
$\dfrac{l - B\bar{a}}{\bar{d}}$	$\dfrac{l(1 + i\,\omega_b/2\omega)}{\lambda} - B$

As evident from the figures, the formulae of the above table are equally applicable to state $\zeta_h = 1$

10.1.2 (contd.)

state $v_g = 1$

		Static state				Vibration
		$N_{st} = 0$		$N_{st} \neq 0$		shear, N_{st} not considered
		shear not considered	shear considered	$N_{st} < 0$ (compression)	$N_{st} > 0$ (tension)	
End moments and forces	M_{gh}	$\dfrac{3EJ}{l^2}$	$\dfrac{3EJ}{l^2}\dfrac{1}{1+3\gamma}$	$\dfrac{EJ}{l^2}\Gamma_5(\alpha)$	$\dfrac{EJ}{l^2}\bar\Gamma_5(\delta)$	$-\dfrac{EJ}{l^2}F_9(\lambda)$
	Y_{gh}	$\dfrac{3EJ}{l^3}$	$\dfrac{3EJ}{l^3}\dfrac{1}{1+3\gamma}$	$-\dfrac{EJ}{l^3}\Gamma_6(\alpha)$	$-\dfrac{EJ}{l^3}\bar\Gamma_6(\delta)$	$\dfrac{EJ}{l^3}F_{11}(\lambda)$
	Y_{hg}	$-\dfrac{3EJ}{l^3}$	$-\dfrac{3EJ}{l^3}\dfrac{1}{1+3\gamma}$	$\dfrac{EJ}{l^2}\Gamma_6(\alpha)$	$\dfrac{EJ}{l^3}\bar\Gamma_6(\delta)$	$\dfrac{EJ}{l^3}F_{10}(\lambda)$
Integration constants	A			$\dfrac{\Gamma_5(\alpha)}{\alpha^2}$	$-\dfrac{\bar\Gamma_5(\delta)}{\delta^2}$	$-\dfrac{F_9(\lambda)}{2\lambda^2}+\dfrac{1}{2}$
	B			$\dfrac{\Gamma_6(\alpha)}{\alpha^3}$	$-\dfrac{\bar\Gamma_6(\delta)}{\delta^3}$	$-\dfrac{F_{11}(\lambda)}{2\lambda^3}$
	C			$1-A$	$1-A$	$1-A$
	D			$-B\dfrac{\alpha}{l}$	$-B\dfrac{\delta}{l}$	$-B$

state $\bar v_h = 1$

without damping	*Vibration with light damping*
shear, rotatory inertia, N_{st} considered	
$-\dfrac{EJ}{l^2} F_9(a, d)$	$-\dfrac{EJ}{l^2}\left[F_9(\lambda) + 2\mathrm{i}\dfrac{\omega_b}{\omega}\Phi_9(\lambda)\right]$
$\dfrac{EJ}{l^3} F_{11}(a, d)$	$\dfrac{EJ}{l^3}\left[F_{11}(\lambda) + 2\mathrm{i}\dfrac{\omega_b}{\omega}\Phi_{11}(\lambda)\right]$
$\dfrac{EJ}{l^3} F_{10}(a, d)$	$\dfrac{EJ}{l^3}\left[F_{10}(\lambda) + 2\mathrm{i}\dfrac{\omega_b}{\omega}\Phi_{10}(\lambda)\right]$
$\dfrac{-F_9(a, d) + d\bar{d}}{a^2 + d^2}$	$-\dfrac{1}{2\lambda^2}\left\{F_9(\lambda) + \mathrm{i}\dfrac{\omega_b}{\omega}\left[2\Phi_9(\lambda) + F_9(\lambda)\right]\right\} + \dfrac{1}{2}$
$-\dfrac{F_{11}(a, d)}{\bar{a}(a^2 + d^2)}$	$-\dfrac{1}{2\lambda^3}\left\{F_{11}(\lambda) + \mathrm{i}\dfrac{\omega_b}{\omega}\left[2\Phi_{11}(\lambda) + \dfrac{3}{2}F_{11}(\lambda)\right]\right\}$
$1 - A$	$1 - A$
$-B\dfrac{\bar{a}}{d}$	$-B$

As evident from the figures, the formulae of the above table are equally applicable to state $\bar{v}_h = 1$

10.1.2 (contd.)

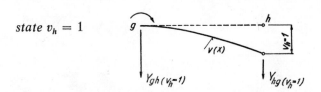

state $v_h = 1$

		Static state				Vibration
		$N_{st} = 0$		$N_{st} \neq 0$		
		shear not considered	shear considered	$N_{st} < 0$ (compression)	$N_{st} > 0$ (tension)	shear, N_{st} not considered
End moments and forces	M_{gh}	$-\dfrac{3EJ}{l^2}$	$-\dfrac{3EJ}{l^2}\dfrac{1}{1+3\gamma}$	$-\dfrac{EJ}{l^2}\Gamma_5(\alpha)$	$-\dfrac{EJ}{l^2}\bar{\Gamma}_5(\delta)$	$-\dfrac{EJ}{l^2}F_8(\lambda)$
	Y_{gh}	$-\dfrac{3EJ}{l^3}$	$-\dfrac{3EJ}{l^3}\dfrac{1}{1+3\gamma}$	$\dfrac{EJ}{l^3}\Gamma_6(\alpha)$	$\dfrac{EJ}{l^3}\bar{\Gamma}_6(\delta)$	$\dfrac{EJ}{l^3}F_{10}(\lambda)$
	Y_{hg}	$\dfrac{3EJ}{l^3}$	$\dfrac{3EJ}{l^3}\dfrac{1}{1+3\gamma}$	$-\dfrac{EJ}{l^3}\Gamma_6(\alpha)$	$-\dfrac{EJ}{l^3}\bar{\Gamma}_6(\delta)$	$\dfrac{EJ}{l^3}F_{12}(\lambda)$
Integration constants	A			$-\dfrac{\Gamma_5(\alpha)}{\alpha^2}$	$\dfrac{\bar{\Gamma}_5(\delta)}{\delta^2}$	$-\dfrac{F_8(\lambda)}{2\lambda^2}$
	B			$-\dfrac{\Gamma_6(\alpha)}{\alpha^3}$	$\dfrac{\bar{\Gamma}_6(\delta)}{\delta^3}$	$-\dfrac{F_{10}(\lambda)}{2\lambda^3}$
	C			$-A$	$-A$	$-A$
	D			$-B\dfrac{\alpha}{l}$	$-B\dfrac{\delta}{l}$	$-B$

state $\bar{v}_g = 1$

without damping	*Vibration with light damping*
shear, rotatory inertia, N_{st} considered	
$-\dfrac{EJ}{l^2}\,F_8(a,d)$	$-\dfrac{EJ}{l^2}\left[F_8(\lambda)+2\mathrm{i}\,\dfrac{\omega_b}{\omega}\,\Phi_8(\lambda)\right]$
$\dfrac{EJ}{l^3}\,F_{10}(a,d)$	$\dfrac{EJ}{l^3}\left[F_{10}(\lambda)+2\mathrm{i}\,\dfrac{\omega_b}{\omega}\,\Phi_{10}(\lambda)\right]$
$\dfrac{EJ}{l^3}\,F_{12}(a,d)$	$\dfrac{EJ}{l^3}\left[F_{12}(\lambda)+2\mathrm{i}\,\dfrac{\omega_b}{\omega}\,\Phi_{12}(\lambda)\right]$
$-\dfrac{F_8(a,d)}{a^2+d^2}$	$-\dfrac{1}{2\lambda^2}\left\{F_8(\lambda)+\mathrm{i}\,\dfrac{\omega_b}{\omega}\left[2\,\Phi_8(\lambda)+F_8(\lambda)\right]\right\}$
$-\dfrac{F_{10}(a,d)}{\bar{a}(a^2+d^2)}$	$-\dfrac{1}{2\lambda^3}\left\{F_{10}(\lambda)+\mathrm{i}\,\dfrac{\omega_b}{\omega}\left[2\,\Phi_{10}(\lambda)+\dfrac{3}{2}\,F_{10}(\lambda)\right]\right\}$
$-A$	$-A$
$-B\,\dfrac{\bar{a}}{d}$	$-B$

As evident from the figures, the formulae of the above table are equally applicable to state $\bar{v}_g = 1$

10.1.4 Cantilever beam

state $\zeta_g = 1$

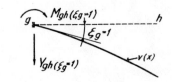

		Static state				Vibration
		$N_{st} = 0$		$N_{st} \neq 0$		
		shear not con- sidered	shear considered	$N_{st} < 0$ (compression)	$N_{st} > 0$ (tension)	shear, N_{st} not considered
End moments and forces	M_{gh}	0	0	$-\dfrac{EJ}{l}\alpha\tan\alpha$	$\dfrac{EJ}{l}\delta\tanh\delta$	$\dfrac{EJ}{l}F_{15}(\lambda)$
	Y_{gh}	0	0	0	0	$-\dfrac{EJ}{l^2}F_{16}(\lambda)$
Integration constants	A			$-\dfrac{l}{\alpha}\tan\alpha$	$-\dfrac{l}{\delta}\tanh\delta$	$\dfrac{l\,F_{15}(\lambda)}{2\lambda^2}$
	B			$\dfrac{l}{\alpha}$	$\dfrac{l}{\delta}$	$l\,\dfrac{F_{16}(\lambda)+\lambda^2}{2\lambda^3}$
	C			$-A$	$-A$	$-A$
	D			0	0	$\dfrac{l}{\lambda}-B$

state $\zeta_h = 1$

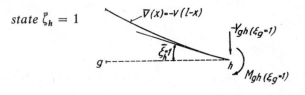

without damping	*Vibration with light damping*
shear, rotatory inertia, N_{st} considered	
$\dfrac{EJ}{l} F_{15}(a, d)$	$\dfrac{EJ}{l}\left[F_{15}(\lambda) + 2\mathrm{i}\dfrac{\omega_b}{\omega}\Phi_{15}(\lambda)\right]$
$-\dfrac{EJ}{l^2} F_{16}(a, d)$	$-\dfrac{EJ}{l^2}\left[F_{16}(\lambda) + 2\mathrm{i}\dfrac{\omega_b}{\omega}\Phi_{16}(\lambda)\right]$
$\dfrac{l\, F_{15}(a, d)}{a^2 + d^2}$	$\dfrac{1}{2\lambda^2}\left\{F_{15}(\lambda) + \mathrm{i}\dfrac{\omega_b}{\omega}\left[2\Phi_{15}(\lambda) + F_{15}(\lambda)\right]\right\}$
$l\,\dfrac{F_{16}(a, d) + a\bar{a}}{\bar{a}(a^2 + d^2)}$	$\dfrac{1}{2\lambda^3}\left\{F_{16}(\lambda) + \lambda^2 + \mathrm{i}\dfrac{\omega_b}{\omega}\left[2\Phi_{16}(\lambda) + \dfrac{3}{2}F_{16}(\lambda) + \dfrac{\lambda^2}{2}\right]\right\}$
$-A$	$-A$
$\dfrac{l - B\bar{a}}{d}$	$\dfrac{l}{\lambda}\left(l + \mathrm{i}\dfrac{\omega_b}{2\omega}\right) - B$

As evident from the figures, the formulae of the above table are equally applicable to state $\bar{\zeta}_h = 1$

10.1.4 (contd.)

state $v_g = 1$

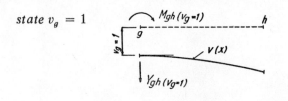

		Static state				Vibration
		$N_{st} = 0$		$N_{st} \neq 0$		
		shear not con- sidered	shear considered	$N_{st} < 0$ (compression)	$N_{st} > 0$ (tension)	shear not considered $N_{st} = 0$
End moments and forces	M_{gh}	0	0	0 for $\cos \alpha \neq 0$; indeterminate for $\cos \alpha = 0$	0	$-\dfrac{EJ}{l^2} F_{16}(\lambda)$
	Y_{gh}	0	0	0	0	$\dfrac{EJ}{l^3} F_{17}(\lambda)$
Integration constants	A			0 for $\cos \alpha \neq 0$; arbitrary for $\cos \alpha = 0$	0	$-\dfrac{F_{16}(\lambda)}{2\lambda^2} + \dfrac{1}{2}$
	B			0	0	$-\dfrac{F_{17}(\lambda)}{2\lambda^3}$
	C			$1 - A$	1	$1 - A$
	D			0	0	$-B$

state $\bar{v}_h = 1$

$\bar{v}(x) = v(l-x)$

$-M_{gh}(v_g=1)$

$Y_{gh}(v_g=1)$

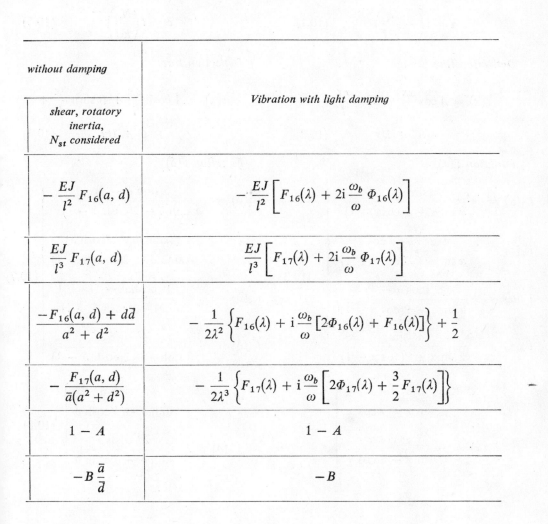

without damping	Vibration with light damping
shear, rotatory inertia, N_{st} considered	
$-\dfrac{EJ}{l^2} F_{16}(a, d)$	$-\dfrac{EJ}{l^2}\left[F_{16}(\lambda) + 2i\dfrac{\omega_b}{\omega}\, \Phi_{16}(\lambda) \right]$
$\dfrac{EJ}{l^3} F_{17}(a, d)$	$\dfrac{EJ}{l^3}\left[F_{17}(\lambda) + 2i\dfrac{\omega_b}{\omega}\, \Phi_{17}(\lambda) \right]$
$\dfrac{-F_{16}(a, d) + d\bar{d}}{a^2 + d^2}$	$-\dfrac{1}{2\lambda^2}\left\{ F_{16}(\lambda) + i\dfrac{\omega_b}{\omega}\left[2\Phi_{16}(\lambda) + F_{16}(\lambda) \right] \right\} + \dfrac{1}{2}$
$-\dfrac{F_{17}(a, d)}{\bar{a}(a^2 + d^2)}$	$-\dfrac{1}{2\lambda^3}\left\{ F_{17}(\lambda) + i\dfrac{\omega_b}{\omega}\left[2\Phi_{17}(\lambda) + \dfrac{3}{2}F_{17}(\lambda) \right] \right\}$
$1 - A$	$1 - A$
$-B\dfrac{\bar{a}}{d}$	$-B$

As evident from the figures, the formulae of the above table are equally
applicable to state $\bar{v}_h = 1$

10.1.5 Frequency functions formulae

Static state under load by axial force N_{st}

compression $N_{st} < 0$	*tension $N_{st} > 0$*

$$\alpha = l\left(-\frac{N_{st}}{EJ}\right)^{1/2} \qquad (10.1)$$

$$\delta = l\left(\frac{N_{st}}{EJ}\right)^{1/2} \qquad (10.5)$$

Deflection line

$$v(x) = A\cos\frac{\alpha x}{l} + B\sin\frac{\alpha x}{l} +$$

$$+ C + Dx \qquad (10.2)$$

Deflection line

$$v(x) = A\cosh\frac{\delta x}{l} + B\sinh\frac{\delta x}{l} +$$

$$+ C + Dx \qquad (10.6)$$

Function $\Gamma(\alpha)$

$$\Gamma_1(\alpha) = \frac{\alpha(\sin\alpha - \alpha)}{\alpha\sin\alpha + 2(\cos\alpha - 1)}$$

$$\Gamma_2(\alpha) = -\frac{\alpha(\sin\alpha - \alpha\cos\alpha)}{\alpha\sin\alpha + 2(\cos\alpha - 1)}$$

$$\Gamma_3(\alpha) = \frac{\alpha^2(\cos\alpha - 1)}{\alpha\sin\alpha + 2(\cos\alpha - 1)}$$

$$\Gamma_4(\alpha) = \frac{\alpha^3\sin\alpha}{\alpha\sin\alpha + 2(\cos\alpha - 1)}$$

$$\left.\right\} \quad (10.3)$$

Function $\bar{\Gamma}(\delta)$

$$\bar{\Gamma}_1(\delta) = \frac{\delta(\sinh\delta - \delta)}{\delta\sinh\delta - 2(\cosh\delta - 1)}$$

$$\bar{\Gamma}_2(\delta) = -\frac{\delta(\sinh\delta - \delta\cosh\delta)}{\delta\sinh\delta - 2(\cosh\delta - 1)}$$

$$\bar{\Gamma}_3(\delta) = \frac{\delta^2(\cosh\delta - 1)}{\delta\sinh\delta - 2(\cosh\delta - 1)}$$

$$\bar{\Gamma}_4(\delta) = -\frac{\delta^3\sinh\delta}{\delta\sinh\delta - 2(\cosh\delta - 1)}$$

$$\left.\right\} \quad (10.7)$$

$$\Gamma_5(\alpha) = \frac{\alpha^2\sin\alpha}{\sin\alpha - \alpha\cos\alpha}$$

$$\Gamma_6(\alpha) = -\frac{\alpha^3\cos\alpha}{\sin\alpha - \alpha\cos\alpha}$$

$$\left.\right\} \quad (10.4)$$

$$\bar{\Gamma}_5(\delta) = -\frac{\delta^2\sinh\delta}{\sinh\delta - \delta\cosh\delta}$$

$$\bar{\Gamma}_6(\delta) = \frac{\delta^3\cosh\delta}{\sinh\delta - \delta\cosh\delta}$$

$$\left.\right\} \quad (10.8)$$

Vibration without damping; shear and rotatory inertia not considered,

$$N_{st} = 0$$

$$\lambda = l \left(\frac{\mu \omega^2}{EJ} \right)^{1/4} \tag{10.9}$$

Deflection amplitudes

$$v(x) = A \cos \frac{\lambda x}{l} + B \sin \frac{\lambda x}{l} + C \cosh \frac{\lambda x}{l} + D \sinh \frac{\lambda x}{l} \tag{10.10}$$

Frequency functions $F(\lambda)$

$$F_1(\lambda) = - \lambda \frac{\sinh \lambda - \sin \lambda}{\cosh \lambda \cos \lambda - 1}$$

$$F_2(\lambda) = - \lambda \frac{\cosh \lambda \sin \lambda - \sinh \lambda \cos \lambda}{\cosh \lambda \cos \lambda - 1}$$

$$F_3(\lambda) = - \lambda^2 \frac{\cosh \lambda - \cos \lambda}{\cosh \lambda \cos \lambda - 1}$$

$$F_4(\lambda) = \lambda^2 \frac{\sinh \lambda \sin \lambda}{\cosh \lambda \cos \lambda - 1} \tag{10.11}$$

$$F_5(\lambda) = \lambda^3 \frac{\sinh \lambda + \sin \lambda}{\cosh \lambda \cos \lambda - 1}$$

$$F_6(\lambda) = - \lambda^3 \frac{\cosh \lambda \sin \lambda + \sinh \lambda \cos \lambda}{\cosh \lambda \cos \lambda - 1}$$

$$F_7(\lambda) = \lambda \frac{2 \sinh \lambda \sin \lambda}{\cosh \lambda \sin \lambda - \sinh \lambda \cos \lambda}$$

$$F_8(\lambda) = \lambda^2 \frac{\sinh \lambda + \sin \lambda}{\cosh \lambda \sin \lambda - \sinh \lambda \cos \lambda}$$

$$F_9(\lambda) = - \lambda^2 \frac{\cosh \lambda \sin \lambda + \sinh \lambda \cos \lambda}{\cosh \lambda \sin \lambda - \sinh \lambda \cos \lambda}$$

$$F_{10}(\lambda) = - \lambda^3 \frac{\cosh \lambda + \cos \lambda}{\cosh \lambda \sin \lambda - \sinh \lambda \cos \lambda} \tag{10.12}$$

$$F_{11}(\lambda) = \lambda^3 \frac{2 \cosh \lambda \cos \lambda}{\cosh \lambda \sin \lambda - \sinh \lambda \cos \lambda}$$

$$F_{12}(\lambda) = \lambda^3 \frac{\cosh \lambda \cos \lambda + 1}{\cosh \lambda \sin \lambda - \sinh \lambda \cos \lambda}$$

$$F_{13}(\lambda) = -\lambda^3 \frac{\sinh \lambda - \sin \lambda}{2 \sinh \lambda \sin \lambda}$$

$$F_{14}(\lambda) = -\lambda^3 \frac{\cosh \lambda \sin \lambda - \sinh \lambda \cos \lambda}{2 \sinh \lambda \sin \lambda}$$

$$\left. \right\} \quad (10.13)$$

$$F_{15}(\lambda) = -\lambda \frac{\cosh \lambda \sin \lambda - \sinh \lambda \cos \lambda}{\cosh \lambda \cos \lambda + 1}$$

$$F_{16}(\lambda) = \lambda^2 \frac{\sinh \lambda \sin \lambda}{\cosh \lambda \cos \lambda + 1}$$

$$F_{17}(\lambda) = -\lambda^3 \frac{\cosh \lambda \sin \lambda + \sinh \lambda \cos \lambda}{\cosh \lambda \cos \lambda + 1}$$

$$\left. \right\} \quad (10.14)$$

Frequency functions F(λ) expanded in series

$$F_1(\lambda) = 2 + 0.007\,142\,857\lambda^4 + 0.000\,015\,704\lambda^8 + 0.000\,000\,032\lambda^{12} + \ldots$$

$$F_2(\lambda) = 4 - 0.009\,523\,810\lambda^4 - 0.000\,016\,262\lambda^8 - 0.000\,000\,032\lambda^{12} - \ldots$$

$$F_3(\lambda) = 6 + 0.030\,952\,381\lambda^4 + 0.000\,072\,193\lambda^8 + 0.000\,000\,148\lambda^{12} - \ldots$$

$$F_4(\lambda) = -6 + 0.052\,380\,952\lambda^4 + 0.000\,076\,617\lambda^8 + 0.000\,000\,149\lambda^{12} + \ldots$$

$$F_5(\lambda) = -12 - 0.128\,571\,429\lambda^4 - 0.000\,329\,571\lambda^8 - 0.000\,000\,684\lambda^{12} - \ldots$$

$$F_6(\lambda) = 12 - 0.371\,428\,571\lambda^4 - 0.000\,364\,873\lambda^8 - 0.000\,000\,693\lambda^{12} - \ldots$$

$$\left. \right\} \quad (10.15)$$

$$F_7(\lambda) = 3 - 0.019\,047\,619\lambda^4 - 0.000\,071\,463\lambda^8 - 0.000\,000\,298\lambda^{12} - \ldots$$

$$F_8(\lambda) = 3 + 0.039\,285\,714\lambda^4 + 0.000\,188\,127\lambda^8 + 0.000\,000\,803\lambda^{12} + \ldots$$

$$F_9(\lambda) = -3 + 0.085\,714\,286\lambda^4 + 0.000\,283\,103\lambda^8 + 0.000\,001\,172\lambda^{12} + \ldots$$

$$F_{10}(\lambda) = -3 - 0.139\,285\,714\lambda^4 - 0.000\,730\,455\lambda^8 - 0.000\,003\,150\lambda^{12} - \ldots$$

$$F_{11}(\lambda) = 3 - 0.485\,714\,285\lambda^4 - 0.001\,129\,664\lambda^8 - 0.000\,004\,611\lambda^{12} - \ldots$$

$$F_{12}(\lambda) = 3 - 0.235\,714\,286\lambda^4 - 0.000\,534\,426\lambda^8 - 0.000\,002\,178\lambda^{12} - \ldots$$

$$\left. \right\} \quad (10.16)$$

$$F_{13}(\lambda) = -0.166\,666\,667\lambda^4 - 0.002\,050\,265\lambda^8 - 0.000\,021\,336\lambda^{12} - \ldots$$

$$F_{14}(\lambda) = -0.333\,333\,333\lambda^4 - 0.002\,116\,402\lambda^8 - 0.000\,021\,378\lambda^{12} - \ldots$$

$$\left. \right\} \quad (10.17)$$

Approximate values of the frequency functions for $\lambda > 6$

$$F_1(\lambda) \approx -\frac{\lambda}{\cos \lambda}$$

$$F_2(\lambda) \approx \lambda(1 - \tan \lambda)$$

$$F_3(\lambda) \approx -\frac{\lambda^2}{\cos \lambda}$$

$$F_4(\lambda) \approx \lambda^2 \tan \lambda \qquad\qquad (10.18)$$

$$F_5(\lambda) \approx \frac{\lambda^3}{\cos \lambda}$$

$$F_6(\lambda) \approx -\lambda^3(1 + \tan \lambda)$$

$$F_7(\lambda) \approx \frac{2\lambda \sin \lambda}{\sin \lambda - \cos \lambda}$$

$$F_8(\lambda) \approx \frac{\lambda^2}{\sin \lambda - \cos \lambda}$$

$$F_9(\lambda) \approx -\lambda^2 \frac{\sin \lambda + \cos \lambda}{\sin \lambda - \cos \lambda}$$

$$F_{10}(\lambda) \approx -\frac{\lambda^3}{\sin \lambda - \cos \lambda} \qquad\qquad (10.19)$$

$$F_{11}(\lambda) \approx \frac{2\lambda^3 \cos \lambda}{\sin \lambda - \cos \lambda}$$

$$F_{12}(\lambda) \approx \frac{\lambda^3 \cos \lambda}{\sin \lambda - \cos \lambda}$$

$$F_{13}(\lambda) \approx -\frac{\lambda^3}{2 \sin \lambda}$$

$$F_{14}(\lambda) \approx -\lambda^3 \frac{\sin \lambda - \cos \lambda}{2 \sin \lambda} \qquad\qquad (10.20)$$

$$F_{15}(\lambda) \approx F_2(\lambda)$$

$$F_{16}(\lambda) \approx F_4(\lambda) \qquad\qquad (10.21)$$

$$F_{17}(\lambda) \approx F_6(\lambda)$$

422

Vibration without damping: shear deflection, rotatory inertia (case 1), possibly also static axial force N_{st} (case 2) considered

$$\gamma = \frac{Er^2}{kGl^2} \tag{10.22}$$

$$r = \sqrt{\frac{J}{S}} \text{ (gyration radius)} \tag{10.23}$$

$$v(x) = A \cos \frac{ax}{l} + B \sin \frac{ax}{l} + C \cosh \frac{dx}{l} + D \sinh \frac{dx}{l} \tag{10.24}$$

Case 1:
$$r \neq 0, \quad \gamma \neq 0, \quad N_{st} = 0$$

$$a = \lambda \left\{ \frac{\lambda^2}{2} \left(\frac{r^2}{l^2} + \gamma \right) + \left[\frac{\lambda^4}{4} \left(\frac{r^2}{l^2} - \gamma \right)^2 + 1 \right]^{1/2} \right\}^{1/2} \tag{10.25}$$

$$d = \lambda \left\{ -\frac{\lambda^2}{2} \left(\frac{r^2}{l^2} + \gamma \right) + \left[\frac{\lambda^4}{4} \left(\frac{r^2}{l^2} - \gamma \right)^2 + 1 \right]^{1/2} \right\}^{1/2} \tag{10.26}$$

$$\bar{a} = a - \frac{\lambda^4 \gamma}{a} \tag{10.27}$$

$$\bar{d} = d + \frac{\lambda^4 \gamma}{d} \tag{10.28}$$

Case 2:
$$N_{st} \neq 0, \quad r \to 0, \quad \gamma = 0$$

$$\alpha = l \left(-\frac{N_{st}}{EJ} \right)^{1/2} \tag{10.29}$$

$$a = \left[\frac{\alpha^2}{2} + \left(\frac{\alpha^4}{4} + \lambda^4 \right)^{1/2} \right]^{1/2} \tag{10.30}$$

$$d = \left[-\frac{\alpha^2}{2} + \left(\frac{\alpha^4}{4} + \lambda^4 \right)^{1/2} \right]^{1/2} \tag{10.31}$$

$$\left. \begin{array}{l} \bar{a} = a \\ \bar{d} = d \end{array} \right\} \tag{10.32}$$

Frequency functions F(a,d)

$$F_1(a, d) = -\frac{(a^2 + d^2)(\bar{a} \sinh d - \bar{d} \sin a)}{f_1(a, d)}$$

$$F_2(a, d) = -\frac{(a^2 + d^2)(\bar{d} \cosh d \sin a - \bar{a} \sinh d \cos a)}{f_1(a, d)}$$

$$F_3(a, d) = -\frac{\bar{a}\bar{d}(a^2 + d^2)(\cosh d - \cos a)}{f_1(a, d)}$$

$$F_4(a, d) = \frac{\bar{a}\bar{d}(d\bar{d} - a\bar{a})(\cosh d \cos a - 1) + \bar{a}\bar{d}(\bar{a}d + a\bar{d})\sinh d \sin a}{f_1(a, d)}$$

$$F_5(a, d) = \frac{\bar{a}\bar{d}(a^2 + d^2)(\bar{d} \sinh d + \bar{a} \sin a)}{f_1(a, d)}$$

$$F_6(a, d) = -\frac{\bar{a}\bar{d}(a^2 + d^2)(\bar{a} \cosh d \sin a + \bar{d} \sinh d \cos a)}{f_1(a, d)}$$

$$\left.\right\} \quad (10.33)$$

$$f_1(a, d) = 2\bar{a}\bar{d}(\cosh d \cos a - 1) + (\bar{a}^2 - \bar{d}^2)\sinh d \sin a \qquad (10.34)$$

$$F_7(a, d) = \frac{(a^2 + d^2)\sinh d \sin a}{\bar{d} \cosh d \sin a - \bar{a} \sinh d \cos a}$$

$$F_8(a, d) = \frac{\bar{a}\bar{d}(d \sinh d + a \sin a)}{\bar{d} \cosh d \sin a - \bar{a} \sinh d \cos a}$$

$$F_9(a, d) = -\frac{\bar{a}\bar{d}(a \cosh d \sin a + d \sinh d \cos a)}{\bar{d} \cosh d \sin a - \bar{a} \sinh d \cos a}$$

$$F_{10}(a, d) = -\frac{\bar{a}\bar{d}(d\bar{d} \cosh \bar{d} + a\bar{a} \cos a)}{\bar{d} \cosh d \sin a - \bar{a} \sinh d \cos a}$$

$$F_{11}(a, d) = \frac{\bar{a}\bar{d}(a^2 + d^2)\cosh d \cos a}{\bar{d} \cosh d \sin a - \bar{a} \sinh d \cos a}$$

$$F_{12}(a, d) = \frac{\bar{a}\bar{d}\left[2ad\bar{a}\bar{d} + (a^2\bar{a}^2 + d^2\bar{d}^2)\cosh d \cos a - \bar{a}\bar{d}(a^2 - d^2)\sinh d \sin a\right]}{(a^2 + d^2)(\bar{d} \cosh d \sin a - \bar{a} \sinh d \cos a)}$$

$$\left.\right\} \quad (10.35)$$

$$F_{13}(a, d) = -\frac{\bar{a}\bar{d}(d^2\bar{d} \sinh d - a^2\bar{a} \sin a)}{(a^2 + d^2)\sinh d \sin a}$$

$$F_{14}(a, d) = -\frac{\bar{a}\bar{d}(a^2\bar{a} \cosh d \sin a - d^2\bar{d} \sinh d \cos a)}{(a^2 + d^2)\sinh d \sin a}$$

$$\left.\right\} \quad (10.36)$$

$$F_{15}(a, d) = -\frac{(a^2 + d^2)(a^2\bar{a}\cosh d \sin a - d^2\bar{d}\sinh d \cos a)}{f_2(a,d)}$$

$$F_{16}(a, d) = \frac{\bar{a}\bar{d}[ad(a\bar{a} - d\bar{d})(\cosh d \cos a - 1) + (a^3\bar{a} + d^3\bar{d})\sinh d \sin a]}{f_2(a, d)}$$

$$F_{17}(a, d) = -\frac{\bar{a}\bar{d}(a^2 + d^2)(d^2\bar{d}\cosh d \sin a + a^2\bar{a}\sinh d \cos a)}{f_2(a, d)}$$

(10.37)

$$f_2(a, d) = 2\,ad\bar{a}\bar{d} + (a^2\bar{a}^2 + d^2\bar{d}^2)\cosh d \cos a + \bar{a}\bar{d}(d^2 - a^2)\sinh d \sin a \quad (10.38)$$

Approximate formulae for frequency functions $F(a,d)$

Case 1 $(r \neq 0, \quad \gamma \neq 0, \quad N_{st} = 0)$

$$F_1(a, d) \approx \quad 2 - \frac{36\gamma}{1 + 12\gamma} + \lambda^4\left(\frac{1}{140} + \frac{1}{30}\frac{r^2}{l^2}\right)$$

$$F_2(a, d) \approx \quad 4 - \frac{36\gamma}{1 + 12\gamma} - \lambda^4\left(\frac{1}{105} + \frac{2}{15}\frac{r^2}{l^2}\right)$$

$$F_3(a, d) \approx \quad 6 - \frac{72\gamma}{1 + 12\gamma} + \lambda^4\left(\frac{13}{420} - \frac{1}{10}\frac{r^2}{l^2}\right)$$

$$F_4(a, d) \approx \quad -6 + \frac{72\gamma}{1 + 12\gamma} + \lambda^4\left(\frac{11}{210} + \frac{1}{10}\frac{r^2}{l^2}\right)$$

$$F_5(a, d) \approx -12 + \frac{144\gamma}{1 + 12\gamma} - \lambda^4\left(\frac{9}{70} - \frac{12}{10}\frac{r^2}{l^2}\right)$$

$$F_6(a, d) \approx \quad 12 - \frac{144\gamma}{1 + 12\gamma} - \lambda^4\left(\frac{13}{35} + \frac{12}{10}\frac{r^2}{l^2}\right)$$

(10.39)

Case 2 $(N_{st} \neq 0, \quad r = 0, \quad \gamma = 0)$

$$F_1(a, d) \approx \quad 2 + \frac{\lambda^4}{140} + \frac{\alpha^2}{30}$$

$$F_2(a, d) \approx \quad 4 - \frac{\lambda^4}{105} - \frac{2\alpha^2}{15}$$

$$F_3(a, d) \approx \quad 6 + \frac{13\lambda^4}{420} - \frac{\alpha^2}{10}$$

$$F_4(a, d) \approx \quad -6 + \frac{11\lambda^4}{210} + \frac{\alpha^2}{10}$$

$$F_5(a, d) \approx -12 - \frac{9\lambda^4}{70} + \frac{12\alpha^2}{10}$$

$$F_6(a, d) \approx \quad 12 - \frac{13\lambda^4}{35} - \frac{12\alpha^2}{10}$$

(10.40)

Damped vibration

Vibration with any degree of damping is described by the equations applying to vibration without damping, in which $\Lambda + i\bar{\Lambda}$ is substituted for λ (cf. eq. 6.8, Section 6.1).

For light damping

$$\Lambda \approx \lambda \tag{10.41}$$

$$\bar{\Lambda} \approx - i \frac{\omega_b}{2\omega} \lambda \tag{10.42}$$

$$F_i(\Lambda + i\bar{\Lambda}) = F_i(\lambda) + 2i \frac{\omega_b}{\omega} \Phi_i(\lambda), \quad i = 1 \ldots 17 . \tag{10.43}$$

Functions $\Phi(\lambda)$

$$
\left.
\begin{aligned}
\Phi_1(\lambda) &= 1/4 \left[F_1(\lambda) F_2(\lambda) - F_3(\lambda) - F_1(\lambda) \right] \\
\Phi_2(\lambda) &= 1/4 \left[F_1^2(\lambda) - F_2(\lambda) \right] \\
\Phi_3(\lambda) &= -1/4 \left[F_1(\lambda) F_4(\lambda) + 2 F_3(\lambda) \right] \\
\Phi_4(\lambda) &= -1/4 \left[F_1(\lambda) F_3(\lambda) + 2 F_4(\lambda) \right] \\
\Phi_5(\lambda) &= 1/4 \left[F_3(\lambda) F_4(\lambda) - 3 F_5(\lambda) \right] \\
\Phi_6(\lambda) &= 1/4 \left[F_3^2(\lambda) - 3 F_6(\lambda) \right]
\end{aligned}
\right\} \tag{10.44}
$$

$$
\left.
\begin{aligned}
\Phi_7(\lambda) &= 1/4 \left[-F_7(\lambda) + F_7^2(\lambda) + 2 F_9(\lambda) \right] \\
\Phi_8(\lambda) &= 1/4 \left[-2 F_8(\lambda) + F_7(\lambda) F_8(\lambda) + F_{10}(\lambda) \right] \\
\Phi_9(\lambda) &= 1/4 \left[-2 F_9(\lambda) + F_7(\lambda) F_9(\lambda) + F_{11}(\lambda) \right] \\
\Phi_{10}(\lambda) &= 1/4 \left[-3 F_{10}(\lambda) + F_8(\lambda) F_9(\lambda) \right] \\
\Phi_{11}(\lambda) &= 1/4 \left[-3 F_{11}(\lambda) + 2 F_9^2(\lambda) - F_7(\lambda) F_{11}(\lambda) \right] \\
\Phi_{12}(\lambda) &= 1/4 \left[-3 F_{12}(\lambda) + F_8^2(\lambda) \right]
\end{aligned}
\right\} \tag{10.45}
$$

$$
\left.
\begin{aligned}
\Phi_{13}(\lambda) &= -1/4 \left[3 F_{13}(\lambda) + \frac{F_{10}(\lambda) F_{14}(\lambda)}{F_4(\lambda)} \right] \\
\Phi_{14}(\lambda) &= -1/4 \left[3 F_{14}(\lambda) + \frac{2 F_8(\lambda) F_{13}(\lambda)}{F_7(\lambda)} \right]
\end{aligned}
\right\} \tag{10.46}
$$

$$
\left.
\begin{aligned}
\Phi_{15}(\lambda) &= 1/4 \left[-F_{15}(\lambda) + F_{15}^2(\lambda) + 2 F_{16}(\lambda) \right] \\
\Phi_{16}(\lambda) &= 1/4 \left[-2 F_{16}(\lambda) + F_{15}(\lambda) F_{16}(\lambda) + F_{17}(\lambda) \right] \\
\Phi_{17}(\lambda) &= 1/4 \left[-3 F_{17}(\lambda) + F_{15}(\lambda) F_{17}(\lambda) - F_{15}(\lambda) F_{11}(\lambda) \right]
\end{aligned}
\right\} \tag{10.47}
$$

Frequency functions $\Phi(\lambda)$ expanded in series

$$\left.\begin{aligned}
\Phi_1(\lambda) &= -0{\cdot}007\ 142\ 857\lambda^4 - 0{\cdot}000\ 031\ 408\lambda^8 - 0{\cdot}000\ 000\ 095\lambda^{12} - \ldots \\
\Phi_2(\lambda) &= +0{\cdot}009\ 523\ 810\lambda^4 + 0{\cdot}000\ 032\ 525\lambda^8 + 0{\cdot}000\ 000\ 096\lambda^{12} + \ldots \\
\Phi_3(\lambda) &= -0{\cdot}030\ 952\ 381\lambda^4 - 0{\cdot}000\ 144\ 386\lambda^8 - 0{\cdot}000\ 000\ 443\lambda^{12} - \ldots \\
\Phi_4(\lambda) &= -0{\cdot}052\ 380\ 952\lambda^4 - 0{\cdot}000\ 153\ 234\lambda^8 - 0{\cdot}000\ 000\ 447\lambda^{12} - \ldots \\
\Phi_5(\lambda) &= +0{\cdot}128\ 571\ 429\lambda^4 + 0{\cdot}000\ 659\ 142\lambda^8 + 0{\cdot}000\ 002\ 053\lambda^{12} + \ldots \\
\Phi_6(\lambda) &= +0{\cdot}371\ 428\ 571\lambda^4 + 0{\cdot}000\ 729\ 746\lambda^8 + 0{\cdot}000\ 002\ 080\lambda^{12} + \ldots
\end{aligned}\right\} \quad (10.48)$$

$$\left.\begin{aligned}
\Phi_7(\lambda) &= 0{\cdot}019\ 047\ 619\lambda^4 + 0{\cdot}000\ 142\ 926\lambda^8 + 0{\cdot}000\ 000\ 894\lambda^{12} + \ldots \\
\Phi_8(\lambda) &= -0{\cdot}039\ 285\ 714\lambda^4 - 0{\cdot}000\ 376\ 254\lambda^8 - 0{\cdot}000\ 002\ 408\lambda^{12} - \ldots \\
\Phi_9(\lambda) &= -0{\cdot}085\ 714\ 286\lambda^4 - 0{\cdot}000\ 566\ 206\lambda^8 - 0{\cdot}000\ 003\ 516\lambda^{12} - \ldots \\
\Phi_{10}(\lambda) &= +0{\cdot}139\ 285\ 714\lambda^4 + 0{\cdot}001\ 460\ 910\lambda^8 + 0{\cdot}000\ 009\ 450\lambda^{12} + \ldots \\
\Phi_{11}(\lambda) &= 0{\cdot}485\ 714\ 285\lambda^4 + 0{\cdot}002\ 259\ 328\lambda^8 + 0{\cdot}000\ 013\ 833\lambda^{12} + \ldots \\
\Phi_{12}(\lambda) &= 0{\cdot}235\ 714\ 286\lambda^4 + 0{\cdot}001\ 068\ 852\lambda^8 + 0{\cdot}000\ 006\ 533\lambda^{12} + \ldots
\end{aligned}\right\} \quad (10.49)$$

$$\left.\begin{aligned}
\Phi_{13}(\lambda) &= 0{\cdot}166\ 666\ 667\lambda^4 + 0{\cdot}004\ 100\ 529\lambda^8 + 0{\cdot}000\ 064\ 008\lambda^{12} + \ldots \\
\Phi_{14}(\lambda) &= 0{\cdot}333\ 333\ 333\lambda^4 + 0{\cdot}004\ 232\ 804\lambda^8 + 0{\cdot}000\ 064\ 133\lambda^{12} + \ldots
\end{aligned}\right\} \quad (10.50)$$

10.1.6 Beam under uniform load

Table 10.9

	Supports	Constants	End moments and forces
1	(Clamped at both ends)	$A = \dfrac{pl^4}{2EJ\lambda^6}[\lambda^2 - F_3(\lambda) - F_4(\lambda)]$ $B = -D =$ $= -\dfrac{pl^4}{2EJ\lambda^7}[F_5(\lambda) + F_6(\lambda)]$ $C = \dfrac{pl^4}{2EJ\lambda^6}[\lambda^2 + F_3(\lambda) + F_4(\lambda)]$	$M_{gh} =$ $= -\dfrac{l^2 p}{\lambda^4}[F_3(\lambda) + F_4(\lambda)] =$ $= -M_{hg}$ $Y_{gh} =$ $= \dfrac{lp}{\lambda^4}[F_5(\lambda) + F_6(\lambda)] = Y_{hg}$
2	(Hinged at one end)	$A = C = \dfrac{p}{2\mu\omega^2}$ $B = \dfrac{p}{2\mu\omega^2}\left[-\dfrac{\cosh\lambda\cos\lambda + 1}{f(\lambda)} \right.$ $\left. -\dfrac{\sinh\lambda\sin\lambda - 2\cosh\lambda}{f(\lambda)} \right]$ $D = \dfrac{p}{2\mu\omega^2}\left[\dfrac{\cosh\lambda\cos\lambda + 1}{f(\lambda)} \right.$ $\left. -\dfrac{\sinh\lambda\sin\lambda + 2\cos\lambda}{f(\lambda)} \right]_*$	$M_{hg} = \dfrac{l^2 p}{\lambda^4}[F_8(\lambda) + F_9(\lambda)]$ $Y_{gh} = \dfrac{lp}{\lambda^4}[F_{10}(\lambda) + F_{12}(\lambda)]$ $Y_{hg} = \dfrac{lp}{\lambda^4}[F_{10}(\lambda) + F_{11}(\lambda)]$
3	(Hinged at both ends)	$A = C = \dfrac{p}{2\mu\omega^2}$ $B = \dfrac{p}{2\mu\omega^2}\dfrac{1 - \cos\lambda}{\sin\lambda}$ $D = \dfrac{p}{2\mu\omega^2}\dfrac{1 - \cosh\lambda}{\sinh\lambda}$	$Y_{gh} = \dfrac{lp}{\lambda^4}[F_{13}(\lambda) + F_{14}(\lambda)] =$ $= Y_{hg}$
4	(Cantilever beam)	$A = C =$ $= \dfrac{p}{2\mu\omega^2}\dfrac{\cosh\lambda + \cos\lambda}{\cosh\lambda\cos\lambda + 1}$ $B = D =$ $= -\dfrac{p}{2\mu\omega^2}\dfrac{\sinh\lambda - \sin\lambda}{\cosh\lambda\cos\lambda + 1}$	$M_{hg} = \dfrac{l^2 p}{\lambda^4}F_{16}(\lambda)$ $Y_{hg} = \dfrac{lp}{\lambda^4}F_{17}(\lambda)$

* $f(\lambda) = \cosh\lambda\sin\lambda - \sinh\lambda\cos\lambda$

10.1.7 Longitudinal vibration. Torsional vibration

Table 10.10

	1	2	3	4	5	6	7
	Kind of vibration	Deformation	Boundary conditions	End force, moment	Beam at rest	Beam vibrating without damping	Beam vibrating with light damping
1	*Longitudinal*		$u(0) = 0$ $u(l) = 1$	$X_{gh} =$ $= -N(0)$	$-\dfrac{ES}{l}$	$-\dfrac{ES}{l} \cdot$ $\cdot \psi \operatorname{cosec} \psi$	$-\dfrac{ES}{l}\left[\psi \operatorname{cosec} \psi \right.$ $-i\dfrac{\omega_b}{2\omega}(\psi \operatorname{cosec} \psi$ $\left. - \psi^2 \operatorname{cosec} \psi \cot \psi) \right]$
2				$X_{hg} =$ $= N(l)$	$\dfrac{ES}{l}$	$\dfrac{ES}{l} \psi \cot \psi$	$\dfrac{ES}{l}\left[\psi \cot \psi \right.$ $-i\dfrac{\omega_b}{2\omega}(\psi \cot \psi$ $\left. - \psi^2 \operatorname{cosec}^2 \psi) \right]$
3	*Torsional*		$\xi(0) = 0$ $\xi(l) = 1$	$K_{gh} =$ $= -K(0)$	$-\dfrac{GJ}{l}$	$-\dfrac{GJ}{l} \cdot$ $\cdot \vartheta \operatorname{cosec} \vartheta$	$-\dfrac{GJ}{l}\left[\vartheta \operatorname{cosec} \vartheta \right.$ $-i\dfrac{\omega_b}{2\omega}(\vartheta \operatorname{cosec} \vartheta$ $\left. - \vartheta^2 \operatorname{cosec} \vartheta \cot \vartheta) \right]$
4				$K_{hg} = K(l)$	$\dfrac{GJ}{l}$	$\dfrac{GJ}{l} \vartheta \cot \vartheta$	$\dfrac{GJ}{l}\left[\vartheta \cot \vartheta \right.$ $-i\dfrac{\omega_b}{2\omega}(\vartheta \cot \vartheta$ $\left. - \vartheta^2 \operatorname{cosec}^2 \vartheta) \right]$

$$\psi = l \left(\frac{\mu\omega^2}{ES} \right)^{1/2} \tag{10.51}$$

$$\vartheta = l \left(\frac{\chi\omega^2}{GJ} \right)^{1/2} \tag{10.52}$$

10.2 Numerical tables of frequency functions

10.2.1 Functions $F(\lambda)$

Tables of functions

$$F_1(\lambda),\ F_2(\lambda),\ F_3(\lambda),\ F_4(\lambda),\ F_5(\lambda),\ F_6(\lambda)$$

for calculating the end moments and forces of a vibrating beam clamped at both ends $(\lambda = 0{\cdot}00 - 6{\cdot}00)$

λ	Δ₂	Δ₁	$F_3(\lambda)$	Δ₂	Δ₁	$F_2(\lambda)$	Δ₂	Δ₁	$F_1(\lambda)$	λ
0·00	—	0·00000	6·00000	—	−0·00000	4·00000	—	0·00000	2·00000	0·00
·02	—	0	6·00000	—	0	4·00000	—	0	2·00000	·02
·04	—	0	6·00000	—	0	4·00000	—	0	2·00000	·04
·06	—	0	6·00000	—	0	4·00000	—	0	2·00000	·06
·08	—	0	6·00000	—	0	4·00000	—	0	2·00000	·08
0·10	—	1	6·00000	—	0	4·00000	—	0	2·00000	0·10
·12	—	0	6·00001	—	0	4·00000	—	0	2·00000	·12
·14	—	1	6·00001	—	1	4·00000	—	0	2·00000	·14
·16	0·00001	1	6·00002	—	0	3·99999	—	1	2·00000	·16
·18	0	2	6·00003	—	1	3·99999	—	0	2·00001	·18
0·20	1	2	6·00005	—	0	3·99998	—	1	2·00001	0·20
·22	0	3	6·00007	—	1	3·99998	—	0	2·00002	·22
·24	1	4	6·00010	−0·00001	1	3·99997	—	1	2·00002	·24
·26	1	5	6·00014	0	2	3·99996	—	1	2·00003	·26
·28	1	6	6·00019	1	2	3·99994	—	2	2·00004	·28
0·30	1	7	6·00025	0	2	3·99992	—	1	2·00006	0·30
·32	2	9	6·00032	1	3	3·99990	—	3	2·00007	·32
·34	2	11	6·00041	0	3	3·99987	—	2	2·00010	·34
·36	2	13	6·00052	1	4	3·99984	0·00001	3	2·00012	·36
·38	1	14	6·00065	0	4	3·99980	0	3	2·00015	·38
0·40	3	17	6·00079	2	6	3·99976	1	4	2·00018	0·40
·42	3	20	6·00096	0	6	3·99970	1	5	2·00022	·42
·44	3	23	6·00116	1	7	3·99964	0	5	2·00027	·44
·46	2	25	6·00139	1	8	3·99957	1	6	2·00032	·46
·48	4	29	6·00164	1	9	3·99949	1	7	2·00038	·48

arg	Δ²	Δ	value	Δ²		Δ		value	Δ²	Δ	value	arg
0.50	4	33	6·00193	1	—	10	—	3·99940	0	7	2·00045	0.50
·52	4	37	6·00226	1	—	11	—	3·99930	2	9	2·00052	·52
·54	4	41	6·00263	2	—	13	—	3·99919	0	9	2·00061	·54
·56	5	46	6·00304	1	—	14	—	3·99906	2	11	2·00070	·56
·58	5	51	6·00350	1	—	15	—	3·99892	1	12	2·00081	·58
0.60	6	57	6·00401	3	—	18	—	3·99877	1	13	2·00093	0.60
·62	4	61	6·00458	1	—	19	—	3·99859	1	14	2·00106	·62
·64	8	69	6·00519	2	—	21	—	3·99840	2	16	2·00120	·64
·66	5	74	6·00588	2	—	23	—	3·99819	1	17	2·00136	·66
·68	8	82	6·00662	2	—	25	—	3·99796	2	19	2·00153	·68
0.70	6	88	6·00744	2	—	27	—	3·99771	1	20	2·00172	0.70
·72	9	97	6·00832	3	—	30	—	3·99744	2	22	2·00192	·72
·74	7	104	6·00929	2	—	32	—	3·99714	2	24	2·00214	·74
·76	10	114	6·01033	3	—	35	—	3·99682	3	27	2·00238	·76
·78	8	122	6·01147	2	—	37	—	3·99647	1	28	2·00265	·78
0.80	10	132	6·01269	4	—	41	—	3·99610	2	30	2·00293	0.80
·82	10	142	6·01401	3	—	44	—	3·99569	3	33	2·00323	·82
·84	10	152	6·01543	5	—	46	—	3·99525	2	35	2·00356	·84
·86	12	164	6·01695	5	—	51	—	3·99479	3	38	2·00391	·86
·88	11	175	6·01859	3	—	54	—	3·99428	2	40	2·00429	·88
0.90	12	187	6·02034	5	—	57	—	3·99374	4	44	2·00469	0.90
·92	13	200	6·02221	5	—	62	—	3·99317	2	46	2·00513	·92
·94	13	213	6·02421	3	—	65	—	3·99255	3	49	2·00559	·94
·96	14	227	6·02634	5	—	70	—	3·99190	3	52	2·00608	·96
·98	14	241	6·02861	4	—	74	—	3·99120	4	56	2·00660	·98
1·00	16		6·03102	5	—		—	3·99046	3		2·00716	1·00

λ	$F_4(\lambda)$	Δ_1	Δ_2	$F_5(\lambda)$	Δ_1	Δ_2	$F_6(\lambda)$	Δ_1	Δ_2	λ
0·00	−6·00000	0·00000	—	−12·00000	−0·00000	—	12·00000	−0·00000	—	0·00
·02	−6·00000	0	—	−12·00000	0	—	12·00000	0	−0·00000	·02
·04	−6·00000	0	—	−12·00000	0	—	12·00000	0	0	·04
·06	−6·00000	0	—	−12·00000	−1	—	12·00000	−2	−2	·06
·08	−6·00000	1	—	−12·00001	0	—	11·99998	−2	0	·08
0·10	−5·99999	0	—	−12·00001	−2	−0·00002	11·99996	−4	−2	0·10
·12	−5·99999	1	0·00001	−12·00003	−2	0	11·99992	−6	−2	·12
·14	−5·99998	1	0	−12·00005	−3	−1	11·99986	−10	−4	·14
·16	−5·99997	2	1	−12·00008	−5	−2	11·99976	−15	−5	·16
·18	−5·99995	3	1	−12·00013	−8	−3	11·99961	−20	−5	·18
0·20	−5·99992	4	1	−12·00021	−9	−1	11·99941	−28	−8	0·20
·22	−5·99988	5	2	−12·00030	−13	−4	11·99913	−36	−8	·22
·24	−5·99983	7	1	−12·00043	−16	−3	11·99877	−47	−11	·24
·26	−5·99976	8	2	−12·00059	−20	−4	11·99830	−58	−11	·26
·28	−5·99968	10	3	−12·00079	−25	−5	11·99772	−73	−15	·28
0·30	−5·99958	13	2	−12·00104	−31	−6	11·99699	−88	−15	0·30
·32	−5·99945	15	3	−12·00135	−37	−6	11·99611	−107	−19	·32
·34	−5·99930	18	3	−12·00172	−44	−7	11·99504	−128	−21	·34
·36	−5·99912	21	4	−12·00216	−52	−8	11·99376	−150	−22	·36
·38	−5·99891	25	4	−12·00268	−61	−9	11·99226	−177	−27	·38
0·40	−5·99866	29	4	−12·00329	−71	−10	11·99049	−205	−28	0·40
·42	−5·99837	33	6	−12·00400	−82	−11	11·98844	−236	−31	·42
·44	−5·99804	39	4	−12·00482	−94	−12	11·98608	−271	−35	·44
·46	−5·99765	43	6	−12·00576	−107	−13	11·98337	−309	−38	·46
·48	−5·99722	49	—	−12·00683	−121	−14	11·98028	−350	−41	·48

0·50	44	394	11·97678	15	136	—12·00804	7	56	—5·99673	0·50
·52	49	443	11·97284	17	153	—12·00940	6	62	—5·99617	·52
·54	51	494	11·96841	19	172	—12·01093	8	70	—5·99555	·54
·56	57	551	11·96347	18	190	—12·01265	8	78	—5·99485	·56
·58	59	610	11·95796	22	212	—12·01455	8	86	—5·99407	·58
0·60	65	675	11·95186	22	234	—12·01667	9	95	—5·99321	0·60
·62	69	744	11·94511	23	257	—12·01901	10	105	—5·99226	·62
·64	72	816	11·93767	26	283	—12·02158	10	115	—5·99121	·64
·66	78	894	11·92951	27	310	—12·02441	11	126	—5·99006	·66
·68	83	977	11·92057	28	338	—12·02751	12	138	—5·98880	·68
0·70	87	1064	11·91080	31	369	—12·03089	12	150	—5·98742	0·70
·72	93	1157	11·90016	31	400	—12·03458	13	163	—5·98592	·72
·74	98	1255	11·88859	35	435	—12·03858	14	177	—5·98429	·74
·76	102	1357	11·87604	36	471	—12·04293	15	192	—5·98252	·76
·78	110	1467	11·86247	37	508	—12·04764	15	207	—5·98060	·78
0·80	114	1581	11·84780	40	548	—12·05272	16	223	—5·97853	0·80
·82	119	1700	11·83199	41	589	—12·05820	17	240	—5·97630	·82
·84	127	1827	11·81499	45	634	—12·06409	18	258	—5·97390	·84
·86	133	1960	11·79672	45	679	—12·07043	18	276	—5·97132	·86
·88	137	2097	11·77712	49	728	—12·07722	20	296	—5·96856	·88
0·90	146	2243	11·75615	50	778	—12·08450	20	316	—5·96560	0·90
·92	151	2394	11·73372	52	830	—12·09228	22	338	—5·96244	·92
·94	158	2552	11·70978	56	886	—12·10058	22	360	—5·95906	·94
·96	164	2716	11·68426	57	943	—12·10944	24	384	—5·95546	·96
·98	173	2889	11·65710	60	1003	—12·11887	24	408	—5·95162	·98
1·00	179		11·62821	63		—12·12890	25		—5·94754	1·00

λ	$F_1(\lambda)$	Δ_1	Δ_2	$F_2(\lambda)$	Δ_1	Δ_2	$F_3(\lambda)$	Δ_1	Δ_2	λ
1·00	2·00716	0·00059	0·00003	3·99046	−0·00079	−0·00005	6·03102	0·00257	0·00016	1·00
·02	2·00775	63	4	3·98967	83	4	6·03359	272	15	·02
·04	2·00838	66	3	3·98884	89	6	6·03631	288	16	·04
·06	2·00904	71	5	3·98795	94	5	6·03919	305	17	·06
·08	2·00975	74	3	3·98701	99	5	6·04224	323	18	·08
1·10	2·01049	79	5	3·98602	105	6	6·04547	341	18	1·10
·12	2·01128	83	4	3·98497	110	5	6·04888	360	19	·12
·14	2·01211	87	4	3·98387	117	7	6·05248	380	20	·14
·16	2·01298	93	6	3·98270	123	6	6·05628	400	20	·16
·18	2·01391	97	4	3·98147	129	6	6·06028	421	21	·18
1·20	2·01488	102	5	3·98018	136	7	6·06449	444	23	1·20
·22	2·01590	108	6	3·97882	143	7	6·06893	465	21	·22
·24	2·01698	112	4	3·97739	150	7	6·07358	490	25	·24
·26	2·01810	119	7	3·97589	157	7	6·07848	513	23	·26
·28	2·01929	124	5	3·97432	165	8	6·08361	539	26	·28
1·30	2·02053	130	6	3·97267	173	8	6·08900	564	25	1·30
·32	2·02183	136	6	3·97094	182	9	6·09464	591	27	·32
·34	2·02319	143	7	3·96912	189	7	6·10055	619	28	·34
·36	2·02462	149	6	3·96723	199	10	6·10674	647	28	·36
·38	2·02611	156	7	3·96524	207	8	6·11321	677	30	·38
1·40	2·02767	163	7	3·96317	216	9	6·11998	707	30	1·40
·42	2·02930	171	8	3·96101	226	10	6·12705	739	32	·42
·44	2·03101	177	6	3·95875	236	10	6·13444	770	31	·44
·46	2·03278	186	9	3·95639	246	10	6·14214	804	34	·46
·48	2·03464	193	7	3·95393	257	11	6·15018	839	35	·48

1·50	34	873	6·15857	10	—	267	—	3·95136	8	201	2·03657	1·50
·52	37	910	6·16730	11	—	278	—	3·94869	9	210	2·03858	·52
·54	38	948	6·17640	11	—	289	—	3·94591	8	218	2·04068	·54
·56	38	986	6,18588	12	—	301	—	3·94302	9	227	2·04286	·56
·58	39	1025	6·19574	12	—	313	—	3·94001	9	236	2·04513	·58
1·60	42	1067	6·20599	13	—	326	—	3·93688	10	246	2·04749	1·60
·62	41	1108	6·21666	12	—	338	—	3·93362	10	256	2·04995	·62
·64	44	1152	6·22774	13	—	351	—	3·93024	9	265	2·05251	·64
·66	44	1196	6·23926	13	—	364	—	4·92673	10	275	2·05516	·66
·68	46	1242	6·25122	15	—	379	—	3·92309	11	286	2·05791	·68
1·70	47	1289	6·26364	13	—	392	—	3·91930	11	297	2·06077	1·70
·72	48	1337	6·27653	15	—	407	—	3·91538	11	308	2·06374	·72
·74	50	1387	6·28990	15	—	422	—	3·91131	11	319	2·06682	·74
·76	51	1438	6·30377	15	—	437	—	3·90709	12	331	2·07001	·76
·78	53	1491	6 31815	16	—	453	—	3 90272	12	343	2·07332	·78
1·80	53	1544	6·33306	16	—	469	—	3·89819	12	355	2·07675	1·80
·82	56	1600	6·34850	16	—	485	—	4·89350	14	369	2·08030	·82
·84	57	1657	6·36450	17	—	502	—	3·88865	12	381	2·08399	·84
·86	58	1715	6·38107	18	—	520	—	3·88363	13	394	2·08780	·86
·88	60	1775	6·39822	18	—	538	—	3·87843	15	409	2·09174	·88
1·90	62	1837	6·41597	18	—	556	—	3·87305	13	422	2·09583	1·90
·92	64	1901	6·43434	19	—	575	—	3·86749	15	437	2·10005	·92
·94	64	1965	6·45335	19	—	594	—	3·86174	15	452	2·10442	·94
·96	68	2033	6·47300	20	—	614	—	3·85580	15	467	2·10894	·96
·98	68	2101	6·49333	20	—	634	—	3·84966	16	483	2·11361	·98
2·00	71		6·51434	21				3·84332	16		2·11844	2·00

λ	Δ_2	Δ_1	$F_6(\lambda)$	Δ_2	Δ_1	$F_5(\lambda)$	Δ_2	Δ_1	$F_4(\lambda)$	λ
1·00	—0·00179	—0·03068	11·62821	—0·00063	—0·01066	—12·12890	0·00025	0·00433	—5·94754	1·00
·02	—187	—3255	11·59753	—64	—1130	—12·13956	26	459	—5·94321	·02
·04	—193	—3448	11·56498	—69	—1199	—12·15086	28	487	—5·93862	·04
·06	—202	—3650	11·53050	—69	—1268	—12·16285	29	516	—5·93375	·06
·08	—209	—3859	11·49400	—74	—1342	—12·17553	29	545	—5·92859	·08
1·10	—218	—4077	11·45541	—76	—1418	—12·18895	30	575	—5·92314	1·10
·12	—224	—4301	11·41464	—79	—1497	—12·20313	33	608	—5·91739	·12
·14	—234	—4535	11·37163	—81	—1578	—12·21810	33	641	—5·91131	·14
·16	—242	—4777	11·32628	—86	—1664	—12·23388	33	674	—5·90490	·16
·18	—251	—5028	11·27851	—87	—1751	—12·25052	37	711	—5·89816	·18
1·20	—259	—5287	11·22823	—91	—1842	—12·26803	36	747	—5·89105	1·20
·22	—268	—5555	11·17536	—95	—1937	—12·28645	38	785	—5·88358	·22
·24	—277	—5832	11·11981	—98	—2035	—12·30582	39	824	—5·87573	·24
·26	—286	—6118	11·06149	—100	—2135	—12·32617	41	865	—5·86749	·26
·28	—296	—6414	11·00031	—105	—2240	—12·34752	42	907	—5·85884	·28
1·30	—305	—6719	10·93617	—107	—2347	—12·36992	44	951	—5·84977	1·30
·32	—316	—7035	10·86898	—113	—2460	—12·39339	44	995	—5·84026	·32
·34	—324	—7359	10·79863	—114	—2574	—12·41799	46	1041	—5·83031	·34
·36	—336	—7695	10·72504	—119	—2693	—12·44373	48	1089	—5·81990	·36
·38	—344	—8039	10·64809	—123	—2816	—12·47066	49	1138	—5·80901	·38
1·40	—357	—8396	10·56770	—127	—2943	—12·49882	50	1188	—5·79763	1·40
·42	—366	—8762	10·48374	—130	—3073	—12·52825	53	1241	—5·78575	·42
·44	—377	—9139	10·39612	—135	—3208	—12·55898	53	1294	—5·77334	·44
·46	—388	—9527	10·30473	—139	—3347	—12·59106	56	1350	—5·76040	·46
·48	—399	—9926	10·20946	—143	—3490	—12·62453	56	1406	—5·74690	·48

x										x
1·50	411	10337	10·11020	147	3637	—12·65943	59	1465	—5·73284	1·50
·52	422	10759	10·00683	153	3790	—12·69580	61	1526	—5·71819	·52
·54	435	11194	9·89924	156	3946	—12·73370	61	1587	—5·70293	·54
·56	445	11639	9·78730	161	4107	—12·77316	64	1651	—5·68706	·56
·58	458	12097	9·67091	166	4273	—12·81423	66	1717	—5·67055	·58
1·60	471	12568	9·54994	170	4443	—12·85696	67	1784	—5·65338	1·60
·62	484	13052	9·42426	177	4620	—12·90139	69	1853	—5·63554	·62
·64	495	13547	9·29374	180	4800	—12·94759	71	1924	—5·61701	·64
·66	509	14056	9·15827	187	4987	—12·99559	73	1997	—5·59777	·66
·68	522	14578	9·01771	191	5178	—13·04546	75	2072	—5·57780	·68
1·70	536	15114	8·87193	197	5375	—13·09724	78	2150	—5·55708	1·70
·72	550	15664	8·72079	202	5577	—13·15099	78	2228	—5·53558	·72
·74	563	16227	8·56415	209	5786	—13·20676	81	2309	—5·51330	·74
·76	578	16805	8·40188	214	6000	—13·26462	83	2392	—5·49021	·76
·78	592	17397	8·23383	220	6220	—13·32462	86	2478	—5·46629	·78
1·80	606	18003	8·05986	226	6446	—13·38682	88	2566	—5·44151	1·80
·82	623	18626	7·87983	234	6680	—13·45128	89	2655	—5·41585	·82
·84	636	19262	7·69357	238	6918	—13·51808	93	2748	—5·38930	·84
·86	653	19915	7·50095	247	7165	—13·58726	94	2842	—5·36182	·86
·88	669	20584	7·30180	252	7417	—13·65891	96	2938	—5·33340	·88
1·90	684	21268	7·09596	260	7677	—13·73308	101	3039	—5·30402	1·90
·92	700	21968	6·88328	268	7945	—13·80985	101	3140	—5·27363	·92
·94	719	22687	6·66360	274	8219	—13·88930	104	3244	—5·24223	·94
·96	733	23420	6·43673	282	8501	—13·97149	108	3352	—5·20979	·96
·98	753	24173	6·20253	290	8791	—14·05650	109	3461	—5·17627	·98
2·00	769		5·96080	298		—14·14441	112		—5·14166	2·00

λ	$F_1(\lambda)$	Δ_1	Δ_2	$F_2(\lambda)$	Δ_1	Δ_2	$F_3(\lambda)$	Δ_1	Δ_2	λ
2·00	2·11844	0·00499	0·00016	3·84332	−0·00655	−0·00021	6·51434	0·02172	0·00071	2·00
·02	2·12343	516	17	3·83677	676	21	6·53606	2245	73	·02
·04	2·12859	533	17	3·83001	698	22	6·55851	2319	74	·04
·06	2·13392	550	17	3·82303	721	23	6·58170	2396	77	·06
·08	2·13942	568	18	3·81582	744	23	6·60566	2474	78	·08
2·10	2·14510	586	18	3·80838	767	23	6·63040	2555	81	2·10
·12	2·15096	606	20	3·80071	791	24	6·65595	2638	83	·12
·14	2·15702	625	19	3·79280	816	25	6·68233	2724	86	·14
·16	2·16327	645	20	3·78464	842	26	6·70957	2811	87	·16
·18	2·16972	665	20	2·77622	868	26	6·73768	2901	90	·18
2·20	2·17637	687	22	3·76754	894	26	6·76669	2993	92	2·20
·22	2·18324	708	21	3·75860	922	28	6·79662	3089	96	·22
·24	2·19032	730	22	3·74938	950	28	6·82751	3186	97	·24
·26	2·19762	754	24	3·73988	979	29	6·85937	3287	101	·26
·28	2·20516	776	22	3·73009	1008	29	5·89224	3390	103	·28
2·30	2·21292	801	25	3·72001	1039	31	6·92614	3496	106	2·30
·32	2·22093	826	25	3·70962	1070	31	6·96110	3605	109	·32
·34	2·22919	851	25	3·69892	1102	32	6·99715	3718	113	·34
·36	2·23770	877	26	3·68790	1135	33	6·03433	3832	114	·36
·38	2·24647	904	27	3·67655	1168	33	7·07265	3952	120	·38
2·40	2·25551	932	28	3·66487	1203	35	7·11217	4074	122	2·40
·42	2·26483	961	29	3·65284	1238	35	7·15291	4199	125	·42
·44	2·27444	989	28	3·64046	1274	36	7·19490	4328	129	·44
·46	2·28433	1020	31	3·62772	1312	38	7·23818	4462	134	·46
·48	2·29453	1051	31	3·61460	1350	38	7·28280	4598	136	·48

2·50	2·30504	1082	31	3·60110	1390	40	7·32878	4740	142	2·50
·52	2·31586	1116	34	3·58720	1430	40	7·37618	4885	145	·52
·54	2·32702	1149	33	3·57290	1471	41	7·42503	5034	149	·54
·56	2·33851	1184	35	3·55819	1514	43	7·47537	5188	154	·56
·58	2·35035	1220	36	3·54305	1558	44	7·52725	5347	159	·58
2·60	2·36255	1256	36	3·52747	1603	45	7·58072	5510	163	2·60
·62	2·37511	1295	39	3·51144	1649	46	7·63582	5679	169	·62
·64	2·38806	1334	39	3·49495	1697	48	7·69261	5853	174	·64
·66	2·40140	1374	40	3·47798	1745	48	7·75114	6032	179	·66
·68	2·41514	1416	42	3·46053	1796	51	7·81146	6217	185	·68
2·70	2·42930	1458	42	3·44257	1848	52	7·87363	6408	191	2·70
·72	2·44388	1503	45	3·42409	1900	52	7·93771	6605	197	·72
·74	2·45891	1549	46	3·40509	1956	56	8·00376	6808	203	·74
·76	2·47440	1595	46	3·38553	2012	56	8·07184	7017	209	·76
·78	2·49035	1645	50	3·36541	2070	58	8·14201	7235	218	·78
2·80	2·50680	1694	49	3·34471	2129	59	8·21436	7458	223	2·80
·82	2·52374	1746	52	3·32342	2192	63	8·28894	7690	232	·82
·84	2·54120	1800	54	3·30150	2254	62	8·36584	7929	239	·84
·86	2·55920	1855	55	3·27896	2320	66	8·44513	8176	247	·86
·88	2·57775	1912	57	3·25576	2388	68	8·52689	8432	256	·88
2·90	2·59687	1972	60	3·23188	2457	69	8·61121	8697	265	2·90
·92	2·61659	2033	61	3·20731	2529	72	8·69818	8971	274	·92
·94	2·63692	2096	63	3·18202	2603	74	8·78789	9255	284	·94
·96	2·65788	2161	65	3·15599	2679	76	8·88044	9549	294	·96
·98	2·67949	2230	69	3·12920	2759	80	8·97593	9854	305	·98
3·00	2·70179		70	3·10161		81	9·07447		315	3·00

λ	$F_4(\lambda)$	Δ_1	Δ_2	$F_5(\lambda)$	Δ_1	Δ_2	$F_6(\lambda)$	Δ_1	Δ_2	λ
2·00	−5·14166	0·03573	0·00112	−14·14441	−0·09089	−0·00298	5·96080	−0·24942	−0·00769	2·00
·02	−5·10593	3689	116	−14·23530	9395	306	5·71138	25729	787	·02
·04	−5·06904	3807	118	−14,32925	9711	316	5·45409	26535	806	·04
·06	−5·03097	3928	121	−14·42636	10034	323	5·18874	27360	825	·06
·08	−4·99169	4052	124	−14·52670	10366	332	4·91514	28203	843	·08
2·10	−4·95117	4179	127	−14·63036	10709	343	4·63311	29066	863	2·10
·12	−4·90938	4308	129	−14·73745	11060	351	4·34245	29949	883	·12
·14	−4·86630	4443	135	−14·84805	11422	362	4·04296	30852	903	·14
·16	−4·82187	4579	136	−14·96227	11793	371	3·73444	31775	923	·16
·18	−4·77608	4719	140	−15·08020	12175	382	3·41669	32720	945	·18
2·20	−4·72889	4862	143	−15·20195	12568	393	3·08949	33687	967	2·20
·22	−4·68027	5009	147	−15·32763	12973	405	2·75262	34674	987	·22
·24	−4·63018	5160	151	−15·45736	13387	414	2·40588	35685	1011	·24
·26	−4·57858	5314	154	−15·59123	13816	429	2·04903	36719	1034	·26
·28	−4·52544	5473	159	−15·72939	14255	439	1·68184	37776	1057	·28
2·30	−4·47071	5634	161	−15·87194	14707	452	1·30408	38857	1081	2·30
·32	−4·41437	5800	166	−16·01901	15173	466	0·91551	39962	1105	·32
·34	−4·35637	5971	171	−16·17074	15653	480	0·51589	41093	1131	·34
·36	−4·29666	6145	174	−16·32727	16145	492	+0·10496	42250	1157	·36
·38	−4·23521	6324	179	−16·48872	16653	508	−0·31754	43432	1182	·38
2·40	−4·17197	6507	183	−16·65525	17175	522	−0·75186	44641	1209	2·40
·42	−4·10690	6695	188	−16·82700	17714	539	−1·19827	45878	1237	·42
·44	−4·03995	6888	193	−17·00414	18268	554	−1·65705	47144	1266	·44
·46	−3·97107	7085	197	−17·18682	18839	571	−2·12849	48438	1294	·46
·48	−3·90022	7288	203	−17·37521	19426	587	−2·61287	49762	1324	·48

x		Δ	δ^2		Δ	δ^2		Δ	δ^2	x
2·50	—3·82734	7495	207	—17·56947	—20035	509	—3·11049	—51116	1354	2·50
·52	—3·75239	7709	214	—17·76982	—20656	621	—3·62165	—52503	1387	·52
·54	—3·67530	7927	218	—17·97638	—21302	646	—4·14668	—53919	1416	·54
·56	—3·59603	8151	224	—18·18940	—21965	663	—4·68587	—55371	1452	·56
·58	—3·51452	8382	231	—18·40905	—22650	685	—5·23958	—56855	1484	·58
2·60	—3·43070	8618	236	—18·63555	—23357	707	—5·80813	—58375	1520	2·60
·62	—3·34452	8860	242	—18·86912	—24085	728	—6·39188	—59929	1554	·62
·64	—3·25592	9110	250	—19·10997	—24837	752	—6·99117	—61522	1593	·64
·66	—3·16482	9365	255	—19·35834	—25613	776	—7·60639	—63152	1630	·66
·68	—3·07117	9627	262	—19·61447	—26415	802	—8·23791	—64821	1669	·68
2·70	—2·97490	9898	271	—19·87862	—27242	827	—8·88612	—66530	1709	2·70
·72	—2·87592	10174	276	—20·15104	—28098	856	—9·55142	—68280	1750	·72
·74	—2·77418	10459	285	—20·43202	—28981	883	—10·23422	—70074	1794	·74
·76	—2·66959	10752	293	—20·72183	—29894	913	—10·93496	—71911	1837	·76
·78	—2·56207	11053	301	—21·02077	—30839	945	—11·65407	—73794	1883	·78
2·80	—2·45154	11362	309	—21·32916	—31815	976	—12·39201	—75725	1931	2·80
·82	—2·33792	11681	319	—21·64731	—32825	1010	—13·14926	—77702	1977	·82
·84	—2·22111	12008	327	—21·97556	—33870	1045	—13·92628	—79732	2030	·84
·86	—2·10103	12346	338	—22·31426	—34952	1082	—14·72360	—81812	2080	·86
·88	—1·97757	12693	347	—22·66378	—36073	1121	—15·54172	—83946	2134	·88
2·90	—1·85064	13050	357	—23·02451	—37234	1161	—16·38118	—86137	2191	2·90
·92	—1·72014	13418	368	—23·39685	—38436	1202	—17·24255	—88384	2247	·92
·94	—1·58596	13798	380	—23·78121	—39683	1247	—18·12639	—90692	2308	·94
·96	—1·44798	14189	391	—24·17804	—40975	1292	—19·03331	—93062	2370	·96
·98	—1·30609	14592	403	—24·58779	—42316	1341	—19·96393	—95496	2434	·98
3·00	—1·16017		416	—25·01095		1392	—20·91889		2501	3·00

λ	$F_1(\lambda)$	$\varDelta_1$	$\varDelta_2$	$F_2(\lambda)$	$\varDelta_1$	$\varDelta_2$	$F_3(\lambda)$	$\varDelta_1$	$\varDelta_2$	λ
3·00	2·70179	0·02300	0·00070	3·10161	−0·02840	−0·00081	9·07447	0·10169	0·00315	3·00
·02	2·72479	2372	72	3·07321	2924	84	9·17616	10497	328	·02
·04	2·74851	2449	77	3·04397	3011	87	9·28113	10837	340	·04
·06	2·77300	2526	77	3·01386	3102	91	9·38950	11189	352	·06
·08	2·79826	2608	82	2·98284	3195	93	9·50139	11556	367	·08
3·10	2,82434	2693	85	2·95089	3291	96	9·61695	11935	379	3·10
·12	2·85127	2780	87	2·91798	3392	101	9·73630	12331	396	·12
·14	2·87907	2871	91	2·88406	3495	103	9·85961	12742	411	·14
·16	2·90778	2966	95	2·84911	3602	107	9·98703	13169	427	·16
·18	2·93744	3064	98	2·81309	3714	112	10·11872	13614	445	·18
3·20	2·96808	3167	103	2·77595	3830	116	10·25486	14076	462	3·20
·22	2·99975	3273	106	2·73765	3949	119	10·39562	14560	484	·22
·24	3·03248	3384	111	2·69816	4074	125	10·54122	15061	501	·24
·26	3·06632	3500	116	2·65742	4204	130	10·69183	15585	524	·26
·28	3·10132	3621	121	2·61538	4338	134	10·84768	16134	549	·28
3·30	3·13753	3747	126	2·57200	4478	140	11·00902	16704	570	3·30
·32	3·17500	3878	131	2·52722	4624	146	11·17606	17300	596	·32
·34	3·21378	4016	138	2·48098	4776	152	11·34906	17924	624	·34
·36	3·25394	4159	143	2·43322	4935	159	11·52830	18575	651	·36
·38	3·29553	4309	150	2·38387	5099	164	11·71405	19257	682	·38
3·40	3·33862	4466	157	2·33288	5272	173	11·90662	19972	715	3·40
·42	3·38328	4630	164	2·28016	5452	180	12·10634	20720	748	·42
·44	3·42958	4803	173	2·22564	5640	188	12·31354	21504	784	·44
·46	3·47761	4983	180	2·16924	5836	196	12·52858	22328	824	·46
·48	3·52744	5173	190	2·11088	6043	207	12·75186	23193	865	·48

x												x
3·50	908	24101	12·98379	215	—	6258	—	2·05045	199	5372	3·57917	3·50
·52	956	25057	13·22480	226	—	6484	—	1·98787	209	5581	3·63289	·52
·54	1006	26063	13·47537	237	—	6721	—	1·92303	221	5802	3·68870	·54
·56	1069	27122	13·73600	250	—	6971	—	1·85582	231	6033	3·74672	·56
·58	1117	28239	14·00722	261	—	7232	—	1·78611	245	6278	3·80705	·58
3·60	1180	29419	14·28961	276	—	7508	—	1·71379	257	6535	3·86983	3·60
·62	1244	30663	14·58380	290	—	7798	—	1·63871	272	6807	3·93518	·62
·64	1317	31980	14·89043	305	—	8103	—	1·56073	288	7095	4·00325	·64
·66	1394	33374	15·21023	323	—	8426	—	1·47970	303	7398	4·07420	·66
·68	1475	34849	15·54397	341	—	8767	—	1·39544	323	7721	4·14818	·68
3·70	1567	36416	15·89246	361	—	9128	—	1·30777	340	8061	4·22539	3·70
·72	1661	38077	16·25662	381	—	9509	—	1·21649	363	8424	4·30600	·72
·74	1768	39845	16·63739	404	—	9913	—	1·12140	384	8808	4·39024	·74
·76	1880	41725	17·03584	430	—	10343	—	1·02227	409	9217	4·47832	·76
·78	2005	43730	17·45309	457	—	10800	—	0·91884	437	9654	4·57049	·78
3·80	2138	45868	17·89039	485	—	11285	—	0·81084	464	10118	4·66703	3·80
·82	2285	48153	18·34907	517	—	11802	—	0·69799	496	10614	4·76821	·82
·84	2444	50597	18·83060	553	—	12355	—	0·57997	532	11146	4·87435	·84
·86	2621	53218	19·33657	590	—	12945	—	0·45642	568	11714	4·98581	·86
·88	2812	56030	19·86875	633	—	13578	—	0·32697	610	12324	5·10295	·88
3·90	3023	59053	20·42905	678	—	14256	—	0·19119	656	12980	5·22619	3·90
·92	3257	62310	21·01958	728	—	14984	—	+0·04863	706	13686	5·35599	·92
·94	3512	65822	21·64268	785	—	15769	—	−0,10121	761	14447	5·49285	·94
·96	3797	69619	22·30090	845	—	16614	—	−0·25890	822	15269	5·63732	·96
·98	4112	73731	22·99709	916	—	17530	—	−0·42504	891	16160	5·79001	·98
4·00	4465		23·73440	989	—		—	−0·60034	965		5·95161	4·00

λ	$F_4(\lambda)$	Δ_1	Δ_2	$F_5(\lambda)$	Δ_1	Δ_2	$F_6(\lambda)$	Δ_1	Δ_2	λ
3·00	−1·16017	0·15008	0·00416	−25·01095	−0·43708	−0·01392	−20·91889	−0·97997	−0·02501	3·00
·02	−1·01009	15437	429	−25·44803	45152	1444	−21·89886	−1·00568	2571	·02
·04	−0·85572	15881	444	−25·89955	46652	1500	−22·90454	−1·03210	2642	·04
·06	−0·69691	16338	457	−26·36607	48210	1558	−23·93664	−1·05930	2720	·06
·08	−0·53353	16810	472	−26·84817	49831	1621	−24·99594	−1·08726	2796	·08
3·10	−0·36543	17299	489	−27·34648	51516	1685	−26·08320	−1·11605	2879	3·10
·12	−0·19244	17804	505	−27·86162	53267	1751	−27·19925	−1·14569	2964	·12
·14	−0·01440	18326	522	−28·39429	55090	1823	−28·34494	−1·17623	3054	·14
·16	+0·16886	18867	541	−28·94519	56989	1899	−29·52117	−1·20769	3146	·16
·18	0·35753	19427	560	−29·51508	58966	1977	−30·72886	−1·24013	3244	·18
3·20	0·55180	20006	579	−30·10474	61028	2062	−31·96899	−1·27358	3345	3·20
·22	0·75186	20607	601	−30·71502	63178	2150	−33·24257	−1·30811	3453	·22
·24	0·95793	21230	623	−31·34680	65420	2242	−34·55068	−1·34374	3563	·24
·26	1·17023	21876	646	−32·00100	67757	2337	−35·89442	−1·38056	3682	·26
·28	1·38899	22548	672	−32·67857	70209	2452	−37·27498	−1·41861	3805	·28
3·30	1·61447	23243	695	−33·38066	72760	2551	−38·69359	−1·45795	3934	3·30
·32	1·84690	23968	725	−34·10826	75432	2672	−40·15154	−1·49865	4070	·32
·34	2·08658	24721	753	−34·86258	78226	2794	−41·65019	−1·54079	4214	·34
·36	2·33379	25504	783	−35·64484	81151	2925	−43·19098	−1·58443	4364	·36
·38	2·58883	26319	815	−36·45635	84215	3064	−44·77541	−1·62966	4523	38
3 40	2·85202	27170	851	−37·29850	87428	3213	−46·40507	−1·67659	4693	3·40
·42	3·12372	28055	885	−38·17278	90796	3368	−48·08166	−1·72529	4870	·42
·44	3·40427	28980	925	−39·08074	94332	3536	−49·80695	−1·77587	5058	·44
·46	3·69407	29945	965	−40·02406	98046	3714	−51·58282	−1·82844	5257	·46
·48	3·99352	30953	1008	−41·00452	−1·01948	3902	−53·41126	−1·88313	5469	·48

3·50	5693	—	—1·94006	55·29439	4107	—	—1·06055	—42·02400	1055	32008	4·30305	3·50
·52	5932	—	—1·99938	57·23445	4321	—	—1·10376	—43·08455	1104	33112	4·62313	·52
·54	6186	—	—2·06124	59·23383	4554	—	—1·14930	—44·18831	1156	34268	4·95425	·54
·56	6455	—	—2·12579	61·29507	4800	—	—1·19730	—45·33761	1212	35480	5·29693	·56
·58	6745	—	—2·19324	63·42086	5067	—	—1·24797	—46·53491	1272	36752	5·65173	·58
3·60	7053	—	—2·26377	65·61410	5351	—	—1·30148	—47·78288	1337	38089	6·01925	3·60
·62	7383	—	—2·33760	67·87787	5657	—	—1·35805	—49·08436	1404	39493	6·40014	·62
·64	7736	—	—2·41496	70·21547	5986	—	—1·41791	—50·44241	1479	40972	6·79507	·64
·66	8114	—	—2·49610	72·63043	6341	—	—1·48132	—51·86032	1558	42530	7·20479	·66
·68	8523	—	—2·58133	75·12653	6723	—	—1·54855	—53·34164	1643	44173	7·63009	·68
3·70	8960	—	—2·67093	77·70786	7137	—	—1·61992	—54·89019	1735	45908	8·07182	3·70
·72	9433	—	—2·76526	80·37879	7584	—	—1·69576	—56·51011	1834	47742	8·53090	·72
·74	9943	—	—2·86469	83·14405	8067	—	—1·77643	—58·20587	1942	49684	9·00832	·74
·76	10494	—	—2·96963	86·00874	8594	—	—1·86237	—59·98230	2058	51742	9·50516	·76
·78	11091	—	—3·08054	88·97837	9163	—	—1·95400	—61·84467	2184	53926	10·02258	·78
3·80	11741	—	—3·19795	92·05891	9786	—	—2·05186	—63·79867	2321	56247	10·56184	3·80
·82	12445	—	—3·32240	95·25686	10463	—	—2·15649	—65·85053	2469	58716	11·12431	·82
·84	13214	—	—3·45454	98·57926	11205	—	—2·26854	—68·00702	2635	61351	11·71147	·84
·86	14052	—	—3·59506	102·03380	12015	—	—2·38869	—70·27556	2810	64161	12·32498	·86
·88	14972	—	—3·74478	105·62886	12905	—	—2·51774	—72·66425	3006	67167	12·96659	·88
3·90	15977	—	—3·90455	109·37364	13884	—	—2·65658	—75,18199	3220	70387	13·63826	3·90
·92	17086	—	—4·07541	113·27819	14961	—	—2·80619	—77·83857	3456	73843	14·34213	·92
·94	18306	—	—4·25847	117·35360	16151	—	—2·96770	—80·64476	3716	77559	15·08056	·94
·96	19654	—	—4·45501	121·61207	17471	—	—3·14241	—83·61246	4002	81561	15·85615	·96
·98	21148	—	—4·66649	126·06708	18933	—	—3·33174	—86·75487	4323	85884	16·67176	·98
4·00	22808	—		130·73357	20563	—		—90·08661	4676		17·53060	4·00

λ	$F_1(\lambda)$	Δ_1	Δ_2	$F_2(\lambda)$	Δ_1	Δ_2	$F_3(\lambda)$	Δ_1	Δ_2	λ
4·00	5·95161	0·17125	0·00965	0·60034	0·18519	0·00989	23·73440	0·78196	0·04465	4·00
·02	6·12286	18177	1052	0·78553	19595	1076	24·51636	83052	4856	·02
·04	6·30463	19321	1144	0·98148	20765	1170	25·34688	88347	5295	·04
·06	6·49784	20573	1252	1·18913	22042	1277	26·23035	94135	5788	·06
·08	6·70357	21945	1372	1·40955	23439	1397	27·17170	1·00480	6345	·08
4·10	6·92302	23450	1505	1·64394	24972	1533	28·17650	1·07452	6972	4·10
·12	7·15752	25112	1662	1·89366	26660	1688	29·25102	1·15141	7689	·12
·14	7·40864	26947	1835	2·16026	28522	1862	30·40243	1·23642	8501	·14
·16	7·67811	28984	2037	2·44548	30587	2065	31·63885	1·33079	9437	·16
·18	7·96795	31252	2268	2·75135	32883	2296	32·96964	1·43589	10510	·18
4·20	8·28047	33788	2536	3·08018	35447	2564	34·40553	1·55342	11753	4·20
·22	8·61835	36635	2847	3·43465	38324	2877	35·95895	1·68543	13201	·22
·24	8·98470	39847	3212	3·81789	41565	3241	37·64438	1·83437	14894	·24
·26	9·38317	43487	3640	4·23354	45236	3671	39·47875	2·00323	16886	·26
·28	9·81804	47638	4151	4·68590	49416	4180	41·48198	2·19576	19253	·28
4·30	10·29442	52396	4758	5·18006	54206	4790	43·67774	2·41655	22079	4·30
·32	10·81838	57888	5492	5·72212	59729	5523	46·09429	2·67142	25487	·32
·34	11·39726	64271	6383	6·31941	66145	6416	48·76571	2·96772	29630	·34
·36	12·03997	71751	7480	6·98086	73656	7511	51·73343	3·31495	34723	·36
·38	12·75748	80591	8840	7·71742	82530	8874	55·04838	3·72543	41048	·38
4·40	13·56339	91145	10554	8·54272	93118	10588	58·77381	4·21550	49007	4·40
·42	14·47484	1·03883	12738	9·47390	1·05890	12772	62·98931	4·80711	59161	·42
·44	15·51367	1·19455	15572	10·53280	1·21497	15607	67·79642	5·53037	72326	·44
·46	16·70822	1·38765	19310	11·74777	1·40843	19346	73·32679	6·42737	89700	·46
·48	18·09587	1·63114	24349	13·15620	1·65228	24385	79·75416	7·55856	1·13119	·48

x										x
4·50	19·72701	1·94423	31309	14·80848	1·96573	31345	87·31272	9·01314	1·45458	4·50
·52	21·67124	2·35621	41198	16·77421	2·37808	41235	96·32586	10·92726	1·91412	·52
·54	24·02745	2·91358	55737	19·15229	2·93584	55776	107·25312	13·51712	2·58986	·54
·56	26·94103	3·69390	78032	22·08813	3·71655	78071	120·77024	17·14303	3·62591	56
·58	30·63493	4·83431	1·14041	25·80468	4·85735	1·14080	137·91327	22·44239	5·29936	·58
4·60	35·46924	6·59660	1·76229	30·66203	6·62005	1·76270	160·35566	30·63179	8·18940	4·60
·62	42·06584	9·53332	2·93672	37·28208	9·55717	2·93712	190·98745	44·27905	13·64726	·62
·64	51·59916	14·98588	5·45256	46·83925	15·01014	5·45297	235·26650	69·61809	25·33904	·64
·66	66·58504	26·97633	11·99045	61·84939	27·00102	11·99088	304·88459	125·34058	55·72249	·66
·68	93·56137	62·91549	35·93916	88·85041	62·94061	35·93959	430·22517	292·35921	167·01863	·68
4·70	156·47686	313·61541	+250·69992	−151·79102	−313·64097	250·70036	722·58438	1457·4303.	+1165·0711.	4·70
·72	+470·09227	—	—	−465·43199	—	—	+2180·0147.	—	—	·72
·74	−475·94681	317·04137	−253·85243	+480·58107	−317·06783	+253·85196	−2216·5169.	1473·3506.	−1179·7227.	·74
·76	−158·90544	63·18894	36·13806	163·51324	63·21587	36·13759	743·16628	293·62791	167·94405	·76
·78	95·71650	27·05088	12·03522	100·29737	27·07828	12·03474	449·53837	125·68386	55·93158	·78
4·80	68·66562	15·01566	5·46823	73·21909	15·04354	5·46773	323·85451	69·75228	25·41292	4·80
·82	53·64996	9·54743	2·94372	58·17555	9·57581	2·94322	254·10223	44·33936	13·68090	·82
·84	44·10253	6·60371	1·76600	48·59974	6·63259	1·76549	209·76287	30·65846	8·20765	·84
·86	37·49882	4·83771	1·14262	41·96715	4·86710	1·14210	179·10441	22·45081	5·31070	·86
·88	32·66111	3·69509	0·78178	37·10005	3·72500	0·78125	156·65360	17·14011	3·63380	·88
4·90	28·96602	2·91331	55843	33·37505	2·94375	55789	139·51349	13·50631	2·59584	4·90
·92	26·05271	2·35488	41279	30·43130	2·38586	41224	126·00718	10·91047	1·91900	·92
·94	23·69783	1·94209	31376	28·04544	1·97362	31319	115·09671	8·99147	1·45879	·94
·96	21·75574	1·62833	24407	26·07182	1·66043	24351	106·10524	7·53268	1·13498	·96
·98	20·12741	1·38426	19363	24·41139	1·41692	19302	98·57256	6·39770	0·90051	·98
5·00	18·74315	—	—	22·99447	—	—	92·17486	—	—	5·00

λ	$F_4(\lambda)$	Δ_1	Δ_2	$F_5(\lambda)$	Δ_1	Δ_2	$F_6(\lambda)$	Δ_1	Δ_2	λ
4·00	17·53060	0·90560	0·04676	90·08661	3·53737	0·20563	130·73357	4·89457	0·22808	4·00
·02	18·43620	0·95632	5072	93·62398	3·76118	22381	135·62814	5·14116	24659	·02
·04	19·39252	1·01146	5514	97·38516	4·00537	24419	140·76930	5·40843	26727	·04
·06	20·40398	1·07156	6010	101·39053	4·27242	26705	146·17773	5·69892	29049	·06
·08	21·47554	1·13726	6570	105·66295	4·56529	29287	151·87665	6·01553	31661	·08
4·10	22·61280	1·20929	7203	110·22824	4·88735	32206	157·89218	6·36168	34615	4·10
·12	23·82209	1·28850	7921	115·11559	5·24256	35521	164·25386	6·74132	37964	·12
·14	25·11059	1·37588	8738	120·35815	5·63561	39305	170·99518	7·15915	41783	·14
·16	26·48647	1·47265	9677	125·99376	6·07200	43639	178·15433	7·62067	46152	·16
·18	27·95912	1·58019	10754	132·06576	6·55828	48628	185·77500	8·13243	51176	·18
4·20	29·53931	1·70020	12001	138·62404	7·10230	54402	193·90743	8·70232	56989	5·20
·22	31·23951	1·83473	13453	145·72634	7·71350	61120	202·60975	9·33974	63742	·22
·24	33·07424	1·98622	15149	153·43984	8·40333	68983	211·94949	10·05617	71643	·24
·26	35·06046	2·15769	17147	161·84317	9·18579	78246	222·00566	10·86562	80945	·26
·28	37·21815	2·35286	19517	171·02896	10·07811	89232	232·87128	11·78532	91970	·28
4·30	39·57101	2·57633	22347	181·10707	11·10177	1·02366	244·65660	12·83675	1·05143	4·30
·32	42·14734	2·83393	25760	192·20884	12·28372	1·18195	257·49335	14·04688	1·21013	·32
·34	44·98127	3·13300	29907	204·49256	13·65820	1·37448	271·54023	15·44996	1·40308	·34
·36	48·11427	3·48304	35004	218·15076	15·26926	1·61106	286·99019	17·09003	1·64007	·36
·38	51·59731	3·89639	41335	233·42002	17·17425	1·90499	304·08022	19·02445	1·93442	·38
4·40	55·49370	4·38937	49298	250·59427	19·44905	2·27480	323·10467	21·32914	2·30469	4·40
·42	59·88307	4·98394	59457	270·04332	22·19571	2·74666	344·43381	24·10610	2·77696	·42
·44	64·86701	5·71020	72626	292·23903	25·55404	3·35833	368·53991	27·49521	3·38911	·44
·46	70·57721	6·61026	90006	317·79307	29·71981	4·16577	396·03512	31·69220	4·19699	·46
·48	77·18747	7·74455	1·13429	347·51288	34·97385	5·25404	427·72732	36·97793	5·28573	·48

4·50									4·50
84·93202	9·20230	1·45775	382·48673	41·73070	6·75685	464·70525	43·76697	6·78904	4·50
94·13432	11·11963	1·91733	424·21743	50·62309	8·89239	508·47222	52·69200	8·92503	·52
105·25395	13·71275	2·59312	474·84052	62·65578	12·03269	561·16422	64·75786	12·06586	·54
118·96670	17·34200	3·62925	537·49630	79·50324	16·84746	625·92208	81·63897	16·88111	·56
136·30870	22·64473	5·30273	616·99954	104·12755	24·62431	707·56105	106·29746	24·65849	·58
158·95343	30·83757	8·19284	721·12709	142·18268	38·05513	813·85851	144·38730	38·08984	4·60
189·79100	44·48833	13·65076	863·30977	205·6018	63·4191	958·24581	207·8417	63·4544	·62
234·27933	69·83093	25·34260	1068·9116	323·3559	117·7541	1166·0875	325·6316	117·7899	·64
304·11026	125·55703	55·72610	1392·2675	582·3101	258·9542	1491·7191	584·6220	258·9904	·66
429·66729	292·57935	167·02232	1974·5776	1358·4881	776·1780	2076·3411	1360·8371	776·2151	·68
722·24664	1457·6543	+1165·0749	3333·0657	-6772·8909	-5414·4028	3437·1782	-6775·2773	-5414·4402	4·70
+2179·9009	—	—	-10105·9566	—	—	-10212·4555	—	—	·72
-2216·4031	-1473·5823	-1179·7188	+10326·1303	-6846·8694	+5482·5005	+10217·2069	-6849·3326	+5482·4611	·74
742·82081	293·86347	167·94004	3479·2609	1364·3689	780·4864	3367·8743	1366·8715	780·4465	·76
448·95734	125·92343	55·92746	2114·8920	583·8825	259·9334	2001·0028	586·4250	259·8928	·78
323·03391	69·99597	25·40876	1531·0095	323·9491	118·1052	1414·5778	326·5322	118·0639	4·80
253·03794	44·58721	13·67663	1207·0604	205·8439	63·5832	1088·0456	208·4683	63·5412	·82
208·45073	30·91058	8·20332	1001·2165	142·2607	38·1476	879·57728	144·92714	38·10501	·84
177·54015	22·70726	5·30631	858·95575	104·11311	24·68481	734·65014	106·82213	24·64150	·86
154·83289	17·40095	3·62927	754·84264	79·42830	16·89184	627·82801	82·18063	16·84762	·88
137·43194	13·77168	2·59127	675·41434	62·53646	12·06833	545·64738	65·33301	12·02355	4·90
123·66026	11·18041	1·91433	612·87788	50·46813	8·92294	480·31437	53·30946	8·87736	·92
112·47985	9·26608	1·45404	562·40975	41·54519	6·78437	427·00491	44·43210	6·73797	·94
103·21377	7·81204	1·13012	520·86456	34·76082	5·27961	382·57281	37·69413	5·23245	·96
95·40173	6·68192	0·89557	486·10374	29·48121	4·19014	344·87868	32·46168	4·14212	·98
88·71981	—	—	456·62253	—	—	312·41700	—	—	5·00

λ	$F_1(\lambda)$	Δ_1	Δ_2	$F_2(\lambda)$	Δ_1	Δ_2	$F_3(\lambda)$	Δ_1	Δ_2	λ
5·00	−18·74315	1·19063	−0·19363	22·99447	−1·22390	19302	−92·17486	5·49719	−0·90051	5·00
·02	−17·55252	1·03445	— 15618	21·77057	−1·06830	15560	−86·67767	4·77061	— 72658	·02
·04	−16·51807	0·90661	— 12784	20·70227	−0·94108	12722	−81·90706	4·17578	— 59483	·04
·06	−15·61146	80065	— 10596	19·76119	83574	10534	−77·73128	3·68258	— 49320	·06
·08	−14·81081	71183	— 8882	18·92545	74755	8819	−74·04870	3·26901	— 41357	·08
5·10	−14·09898	63662	— 7521	18·17790	67300	7455	−70·77969	2·91874	— 35027	5·10
·12	−13·46236	57239	— 6423	17·50490	60942	6358	−67·86095	2·61939	— 29935	·12
·14	−12·88997	51707	— 5532	16·89548	55479	5463	−65·24156	2·36148	— 25791	·14
·16	−12·37290	46909	— 4798	16·34069	50750	4729	−62·88008	2·13764	— 22384	·16
·18	−11·90381	42719	— 4190	15·83319	46630	4120	−60·74244	1·94205	— 19559	·18
5·20	−11·47662	39038	— 3681	15·36689	43022	3608	−58·80039	1·77009	— 17196	5·20
·22	−11·08624	35787	— 3251	14·93667	39844	3178	−57·03030	1·61805	— 15204	·22
·24	−10·72837	32898	— 2889	14·53823	37032	2812	−55·41225	1·48289	— 13516	·24
·26	−10·39939	30322	— 2576	14·16791	34532	2500	−53·92936	1·36218	— 12071	·26
·28	−10·09617	28012	— 2310	13·82259	32301	2231	−52·56718	1·25384	— 10834	·28
5·30	− 9·81605	25934	— 2078	13·49958	30303	1998	−51·31334	1·15621	— 9763	5·30
·32	− 9·55671	24055	— 1879	13·19655	28507	1796	−50·15713	1·06788	— 8833	·32
·34	− 9·31616	22351	— 1704	12·91148	26889	1618	−49·08925	0·98763	— 8025	·34
·36	− 9·09265	20801	— 1550	12·64259	25424	1465	−48·10162	91448	— 7315	·36
·38	− 8·88464	19386	— 1415	12·38835	24097	1327	−47·18714	84755	— 6693	·38
5·40	− 8·69078	18089	— 1297	12·14738	22891	1206	−46·33959	78614	— 6141	5·40
·42	− 8·50989	16897	— 1192	11·91847	21792	1099	−45·55345	72958	— 5656	·42
·44	− 8·34092	15800	— 1097	11·70055	20790	1002	−44·82387	67735	— 5223	·44
·46	− 8·18292	14786	— 1014	11·49265	19873	917	−44·14652	62896	— 4839	·46
·48	− 8·03506	13847	— 939	11·29392	19033	840	−43·51756	58403	— 4493	·48

5·50	4187	—	54216	−42·93353	770	18263	—	11·10359	872	—	12975	—7·89659	5·50
·52	3911	—	50305	−42·39137	707	17556	—	10·92096	813	—	12162	—7·76684	·52
·54	3661	—	46644	−41·88832	651	16905	—	10·74540	757	—	11405	—7·64522	·54
·56	3438	—	43206	−41·42188	598	16307	—	10·57635	709	—	10696	—7,53117	·56
·58	3236	—	39970	−40·98982	552	15755	—	10·41328	664	—	10032	—7·42421	·58
5·60	3055	—	36915	−40·59012	508	15247	—	10·25573	624	—	9408	—7·32389	5·60
·62	2888	—	34027	−40·22097	469	14778	—	10·10326	588	—	8820	—7·22981	·62
·64	2739	—	31288	−39·88070	433	14345	—	9·95548	554	—	8266	—7·14161	·64
·66	2605	—	28683	−39·56782	399	13946	—	9·81203	525	—	7741	—7·05895	·66
·68	2480	—	26203	−39·28099	368	13578	—	9·67257	497	—	7244	—6·98154	·68
5·70	2371	—	23832	−39·01896	340	13238	—	9·53679	471	—	6773	—6·90910	5·70
·72	2268	—	21564	−38·78064	314	12924	—	9·40441	450	—	6323	—6·84137	·72
·74	2177	—	19387	−38·56500	288	12636	—	9·27517	427	—	5896	—6·77814	·74
·76	2094	—	17293	−38·37113	266	12370	—	9·14881	409	—	5487	—6·71918	·76
·78	2019	—	15274	−38·19820	243	12127	—	9·02511	391	—	5096	—6·66431	·78
5·80	1950	—	13324	−38·04546	225	11902	—	8·90384	376	—	4720	—6·61335	5·80
·82	1891	—	11433	−37·91222	204	11698	—	8·78482	361	—	4359	—6·56615	·82
·84	1834	—	9599	−37·79789	187	11511	—	8·66784	347	—	4012	—6·52256	·84
·86	1786	—	7813	−37·70190	170	11341	—	8·55273	335	—	3677	—6·48244	·86
·88	1741	—	6072	−37·62377	155	11186	—	8·43932	325	—	3352	—6·44567	·88
5·90	1702	—	4370	−37·56305	138	11048	—	8·32746	313	—	3039	—6·41215	5·90
·92	1667	—	2703	−37·51935	125	10923	—	8·21698	306	—	2733	—6·38176	·92
·94	1639	—	+1064	−37·49232	110	10813	—	8·10775	297	—	2436	—6·35443	·94
·96	1612	—	− 548	−37·48168	98	10715	—	7·99962	288	—	2148	—6·33007	·96
·98	1590	—	−2138	−37·48716	84	10631	—	7·89247	284	—	1864	—6·30859	·98
6·00				−37·50854				7·78616				—6·28995	6·00

452

λ	Δ₂	Δ₁	F₆(λ)	Δ₂	Δ₁	F₅(λ)	Δ₂	Δ₁	F₄(λ)	λ
5·00	4·14212	—28·31956	312·41700	4·19014	—25·29107	456·62253	—0·89557	5·78635	—88·71981	5·00
·02	3·33309	—24·98647	284·09744	3·38192	—21·90915	431·33146	72154	5·06481	—82·93346	·02
·04	2·72007	—22·26640	259·11097	2·76976	—19·13939	409·42231	58968	4·47513	—77·86865	·04
·06	2·24704	—20·01936	236·84457	2·29762	—16·84177	390·28292	48795	3·98718	—73·39352	·06
0·8	1·87617	—18·14319	216·82521	1·92767	—14·91410	373·44115	40822	3·57896	—69·40634	·08
5·10	1·58128	—16·56191	198·68202	1·63367	—13·28043	358·52705	34482	3·23414	—65·82738	5·10
·12	1·34379	—15·21812	182·12011	1·39717	—11·88326	345·24662	29378	2·94036	—62·59324	·12
·14	1·15037	—14·06775	166·90199	1·20471	—10·67855	333·36336	25222	2·68814	—59·65288	·14
·16	0·99123	—13·07652	152·83424	1·04657	—9·63198	322·68481	21804	2·47010	—56·96474	·16
·18	85905	—12·21747	139·75772	0·91543	—8·71655	313·05283	18967	2·28043	—54·49464	·18
5·20	74835	—11·46912	127·54025	80577	—7·91078	304·33628	16591	2·11452	—52·21421	5·20
·22	65489	—10·81423	116·07113	71342	—7·19736	296·42550	14588	1·96864	—50·09969	·22
·24	57543	—10·23880	105·25690	63506	—6·56230	289·22814	12884	1·83980	—48·13105	·24
·26	50741	—9·73139	95·01810	56819	—5·99411	282·66584	11427	1·72553	—46·29125	·26
·28	44883	—9·28256	85·28671	51080	—5·48331	276·67173	10176	1·62377	—44·56572	·28
5·30	39807	—8·88449	76·00415	46125	—5·02206	271·18842	9090	1·53287	—42·94195	5·30
·32	35388	—8·53061	67·11966	41829	—4·60377	266·16636	8146	1·45141	—41·40908	·32
·34	31516	—8·21545	58·58905	38088	—4·22289	261·56259	7322	1·37819	—39·95767	·34
·36	28111	—7·93434	50·37360	34814	—3·87475	257·33970	6596	1·31223	—38·57948	·36
·38	25100	—7·68334	42·43926	31941	—3·55534	253·46495	5958	1·25265	—37·26725	·38
5·40	22428	—7·45906	34·75592	29407	—3·26127	249·90961	5391	1·19874	—36·01460	5·40
·42	20046	—7·25860	27·29686	27170	—2·98957	246·64834	4886	1·14988	—34·81586	·42
·44	17912	—7·07948	20·03826	25186	—2·73771	243·65877	4437	1·10551	—33·66598	·44
·46	15994	—6·91954	12·95878	23422	—2·50349	240·92106	4032	1·06519	—32·56047	·46
·48	14265	—6·77689	+ 6·03924	21850	—2·28499	238·41757	3671	1·02848	—31·49528	·48

x										x
5·50	12698	−6·64991	−0·73765	20447	−2·08052	236·13258	−3342	99506	−30·46680	5·50
·52	11274	−6·53717	−7·38756	19193	−1·88859	234·05206	−3046	96460	−29·47174	·52
·54	9977	−6·43740	−13·92473	18067	−1·70792	232·16347	−2776	93684	−28·50714	·54
·56	8786	−6·34954	−20·36213	17059	−1·53733	230·45555	−2531	91153	−27·57030	·56
·58	7695	−6·27259	−26·71167	16152	−1·37581	228·91822	−2306	88847	−26·65877	·58
5·60	6688	−6·20571	−32·98426	15339	−1·22242	227·54241	−2102	86745	−25·77030	5·60
·62	5757	−6·14814	−39·18997	14604	−1·07638	226·31999	−1911	84834	−24·90285	·62
·64	4893	−6·09921	−45·33811	13947	−93691	225·24361	−1737	83097	−24·05451	·64
·66	4088	−6·05833	−51·43732	13354	−80337	224·30670	−1577	81520	−23·22354	·66
·68	3337	−6·02496	−57·49565	12822	−67515	223·50333	−1425	80095	−22·40834	·68
5·70	2630	−5·99866	−63·52061	12344	−55171	222·82818	−1288	78807	−21·60739	5·70
·72	1968	−5·97898	−69·51927	11916	−43255	222·27647	−1157	77650	−20·81932	·72
·74	1340	−5·96558	−75·49825	11533	−31722	221·84392	−1036	76614	−20·04282	·74
·76	746	−5·95812	−81·46383	11192	−20530	221·52670	−921	75693	−19·27668	·76
·78	+182	−5·95630	−87·42195	10891	−9639	221·32140	−814	74879	−18·51975	·78
5·80	−358	−5·95988	−93·37825	10623	−984	221·22501	−714	74165	−17·77096	5·80
·82	−874	−5·96862	−99·33813	10389	+11373	221·23485	−616	73549	−17·02931	·82
·84	−1371	−5·98233	−105·30675	10187	21560	221·34858	−526	73023	−16·29382	·84
·86	−1849	−6·00082	−111·28908	10012	31572	221·56418	−439	72584	−15·56359	·86
·88	−2313	−6·02395	−117·28990	9864	41436	221·87990	−355	72229	−14·83775	·88
5·90	−2764	−6·05159	−123·31385	9744	51180	222·29426	−277	71952	−14·11546	5·90
·92	−3204	−6·08363	−129·36544	9646	60826	222·80606	−198	71754	−13·39594	·92
·94	−3633	−6·11996	−135·44907	9573	70399	223·41432	−126	71628	−12·67840	·94
·96	−4056	−6·16052	−141·56903	9521	79920	224·11831	−53	71575	−11·96212	·96
·98	−4472	−6·20524	−147·72955	9490	89410	224·91751	+16	71591	−11·24637	·98
6·00			−153·93479			225·81161			−10·53046	6·00

Tables of functions

$$F_7(\lambda),\ F_8(\lambda),\ F_9(\lambda),\ F_{10}(\lambda),\ F_{11}(\lambda),\ F_{12}(\lambda)$$

for calculating the end moments and forces of a vibrating beam clamped at one and hinged at other end $(\lambda = 0{\cdot}00 - 6{\cdot}00)$

λ	Δ_2	Δ_1	$F_9(\lambda)$	Δ_2	Δ_1	$F_8(\lambda)$	Δ_2	Δ_1	$F_7(\lambda)$	λ
0·00	—	0·00000	−3·00000	—	0·00000	3·00000	—	−0·00000	3·00000	0·00
·02	0·0000	0	−3·00000	0·00000	0	3·00000	—	0	3·00000	·02
·04	0	0	−3·00000	0	0	3·00000	—	0	3·00000	·04
·06	0	0	−3·00000	0	0	3·00000	—	0	3·00000	·06
·08	1	1	−3·00000	0	0	3·00000	—	0	3·00000	·08
0·10	0	1	−2·99999	1	1	3·00000	—	0	3·00000	0·10
·12	0	1	−2·99998	0	1	3·00001	—	−1	3·00000	·12
·14	2	3	−2·99997	0	1	3·00002	—	0	2·99999	·14
·16	0	3	−2·99994	0	1	3·00003	—	−1	2·99999	·16
·18	2	5	−2·99991	1	2	3·00004	−0·00001	−1	2·99998	·18
0·20	1	6	−2·99986	1	3	3·00006	0	−1	2·99997	0·20
·22	2	8	−2·99980	1	4	3·00009	0	−2	2·99996	·22
·24	3	11	−2·99972	1	5	3·00013	1	−3	2·99994	·24
·26	3	14	−2·99961	1	6	3·00018	1	−3	2·99991	·26
·28	2	16	−2·99947	2	8	3·00024	0	−3	2·99988	·28
0·30	5	21	−2·99931	1	9	3·00032	0	−5	2·99985	0·30
·32	4	25	−2·99910	3	12	3·00041	2	−5	2·99980	·32
·34	4	29	−2·99885	1	13	3·00053	0	−7	2·99975	·34
·36	6	35	−2·99856	3	16	3·00066	2	−8	2·99968	·36
·38	5	40	−2·99821	3	19	3·00082	1	−9	2·99960	·38
0·40	8	48	−2·99781	2	21	3·00101	1	−10	2·99951	0·40
·42	6	54	−2·99733	4	25	3·00122	2	−12	2·99941	·42
·44	9	63	−2·99679	4	29	3·00147	2	−14	2·99929	·44
·46	8	71	−2·99616	4	33	3·00176	2	−16	2·99915	·46
·48	10	81	−2·99545	4	37	3·00209	2	−18	2·99899	·48

x												x
0·50	10	91	−2·99464	4	41	3·00246	2	—	20	—	2·99881	0·50
·52	11	102	−2·99373	6	47	3·00287	3	—	23	—	2·99861	·52
·54	12	114	−2·99271	6	53	3·00334	2	—	25	—	2·99838	·54
·56	13	127	−2·99157	5	58	3·00387	4	—	29	—	2·99813	·56
·58	14	141	−2·99030	6	64	3·00445	2	—	31	—	2·99784	·58
0·60	15	156	−2·98889	8	72	3·00509	4	—	35	—	2·99753	0·60
·62	16	172	−2·98733	7	79	3·00581	3	—	38	—	2·99718	·62
·64	16	188	−2·98561	7	86	3·00660	4	—	42	—	2·99680	·64
·66	19	207	−2·98373	9	95	3·00746	4	—	46	—	2·99638	·66
·68	19	226	−2·98166	8	103	3·00841	4	—	50	—	2·99592	·68
0·70	20	246	−2·97940	10	113	3·00944	4	—	54	—	2·99542	0·70
·72	21	267	−2·97694	10	123	3·01057	6	—	60	—	2·99488	·72
·74	23	290	−2·97427	10	133	3·01180	4	—	64	—	2·99428	·74
·76	24	314	−2·97137	11	144	3·01313	6	—	70	—	2·99364	·76
·78	25	339	−2·96823	11	155	3·01457	5	—	75	—	2·99294	·78
0·80	26	365	−2·96484	13	168	3·01612	7	—	82	—	2·99219	0·80
·82	29	394	−2·96119	13	181	3·01780	5	—	87	—	2·99137	·82
·84	28	422	−2·95725	13	194	3·01961	7	—	94	—	2·99050	·84
·86	31	453	−2·95303	14	208	3·02155	7	—	101	—	2·98956	·86
·88	33	486	−2·94850	15	223	3·02363	7	—	108	—	2·98855	·88
0·90	33	519	−2·94364	15	238	3·02586	7	—	115	—	2·98747	0·90
·92	35	554	−2·93845	17	255	3·02824	9	—	124	—	2·98632	·92
·94	38	592	−2·93291	16	271	3·03079	7	—	131	—	2·98508	·94
·96	37	629	−2·92699	19	290	3·03350	9	—	140	—	2·98377	·96
·98	41	670	−2·92070	17	307	3·03640	9	—	149	—	2·98237	·98
1·00	41		−2·91400	21		3·03947	9	—		—	2·98088	1·00

458

λ	Δ_2	Δ_1	$F_{12}(\lambda)$	Δ_2	Δ_1	$F_{11}(\lambda)$	Δ_2	Δ_1	$F_{10}(\lambda)$	λ
0·00	—	-0·00000	3·00000	—	-0·00000	3·00000	—	-0·00000	-3·00000	0·00
·02	-0·00000	0	3·00000	-0·00000	0	3·00000	—	0	-3·00000	·02
·04	0	0	3·00000	-1	-1	3·00000	—	0	-3·00000	·04
·06	-1	-1	3·00000	0	-1	2·99999	—	-1	-3·00000	·06
·08	0	-1	2·99999	-2	-3	2·99998	—	0	-3·00001	·08
0·10	-2	-3	2·99998	-2	-5	2·99995	-0·00002	-2	-3·00001	0·10
·12	-1	-4	2·99995	-4	-9	2·99990	0	-2	-3·00003	·12
·14	-2	-6	2·99991	-4	-13	2·99981	-2	-4	-3·00005	·14
·16	-4	-10	2·99985	-6	-19	2·99968	-2	-6	-3·00009	·16
·18	-3	-13	2·99975	-8	-27	2·99949	-1	-7	-3·00015	·18
0·20	-4	-17	2·99962	-9	-36	2·99922	-4	-11	-3·00022	0·20
·22	-6	-23	2·99945	-11	-47	2·99886	-2	-13	-3·00033	·22
·24	-7	-30	2·99922	-14	-61	2·99839	-5	-18	-3·00046	·24
·26	-7	-37	2·99892	-16	-77	2·99778	-4	-22	-3·00064	·26
·28	-9	-46	2·99855	-17	-94	2·99701	-5	-27	-3·00086	·28
0·30	-10	-56	2·99809	-22	-116	2·99607	-6	-33	-3·00113	0·30
·32	-12	-68	2·99753	-24	-140	2·99491	-7	-40	-3·00146	·32
·34	-13	-81	2·99685	-27	-167	2·99351	-8	-48	-3·00186	·34
·36	-15	-96	2·99604	-30	-197	2·99184	-8	-56	-3·00234	·36
·38	-15	-111	2·99508	-34	-231	2·98987	-11	-67	-3·00290	·38
0·40	-20	-131	2·99397	-37	-268	2·98756	-9	-76	-3·00357	0·40
·42	-19	-150	2·99266	-41	-309	2·98488	-13	-89	-3·00433	·42
·44	-22	-172	2·99116	-45	-354	2·98179	-13	-102	-3·00522	·44
·46	-23	-195	2·98944	-50	-404	2·97825	-14	-116	-3·00624	·46
·48	-27	-222	2·98749	-53	-457	2·97421	-15	-131	-3·00740	·48

0·50	29	—	251	—	2·98527	59	—	516	—	2·96964	17	—	148	—	—3·00871	0·50
·52	30	—	281	—	2·98276	63	—	579	—	2·96448	18	—	166	—	3·01019	·52
·54	33	—	314	—	2·97995	68	—	647	—	2·95869	20	—	186	—	3·01185	·54
·56	35	—	349	—	2·97681	73	—	720	—	2·95222	20	—	206	—	3·01371	·56
·58	39	—	388	—	2·97332	79	—	799	—	2·94502	23	—	229	—	3·01577	·58
0·60	40	—	428	—	2·96944	84	—	883	—	2·93703	25	—	254	—	3·01806	0·60
·62	44	—	472	—	2·96516	89	—	972	—	2·92820	25	—	279	—	3·02060	·62
·64	47	—	519	—	2·96044	96	—	1068	—	2·91848	28	—	307	—	3·02339	·64
·66	48	—	567	—	2·95525	102	—	1170	—	2·90780	28	—	335	—	3·02646	·66
·68	54	—	621	—	2·94958	109	—	1279	—	2·89610	32	—	367	—	3·02981	·68
0·70	54	—	675	—	2·94337	113	—	1392	—	2·88331	33	—	400	—	3·03348	0·70
·72	60	—	735	—	2·93662	122	—	1514	—	2·86939	35	—	435	—	3·03748	·72
·74	62	—	797	—	2·92927	128	—	1642	—	2·85425	37	—	472	—	3·04183	·74
·76	65	—	862	—	2·92130	135	—	1777	—	2·83783	39	—	511	—	3·04655	·76
·78	70	—	932	—	2·91268	143	—	1920	—	2·82006	40	—	551	—	3·05166	·78
0·80	72	—	1004	—	2·90336	149	—	2069	—	2·80086	44	—	595	—	3·05717	0·80
·82	77	—	1081	—	2·89332	158	—	2227	—	2·78017	46	—	641	—	3·06312	·82
·84	80	—	1161	—	2·88251	166	—	2393	—	2·75790	47	—	688	—	3·06953	·84
·86	84	—	1245	—	2·87090	173	—	2566	—	2·73397	50	—	738	—	3·07641	·86
·88	88	—	1333	—	2·85845	181	—	2747	—	2·70831	53	—	791	—	3·08379	·88
0·90	93	—	1426	—	2·84512	191	—	2938	—	2·68084	55	—	846	—	3·09170	0·90
·92	96	—	1522	—	2·83086	199	—	3137	—	2·65146	57	—	903	—	3·10016	·92
·94	101	—	1623	—	2·81564	208	—	3345	—	2·62009	61	—	964	—	3·10919	·94
·96	105	—	1728	—	2·79941	216	—	3561	—	2·58664	63	—	1027	—	3·11883	·96
·98	110	—	1838	—	2·78213	227	—	3788	—	2·55103	65	—	1092	—	3·12910	·98
1·00	114	—		—	2·76375	235	—		—	2·51315	69	—		—	3·14002	1·00

λ	$F_7(\lambda)$	Δ_1	Δ_2	$F_8(\lambda)$	Δ_1	Δ_2	$F_9(\lambda)$	Δ_1	Δ_2	λ
1·00	2·98088	−0·00158	−0·00009	3·03947	0·00328	0·00021	−2·91400	0·00711	0·00041	1·00
·02	2·97930	168	10	3·04275	347	19	−2·90689	755	44	·02
·04	2·97762	178	10	3·04622	368	21	−2·89934	801	46	·04
·06	2·97584	189	11	3·04990	390	22	−2·89133	847	46	·06
·08	2·97395	199	10	3·05380	412	22	−2·88286	896	49	·08
1·10	2·97196	211	12	3·05792	437	25	−2·87390	948	52	1·10
·12	2·96985	223	12	3·06229	460	23	−2·86442	1000	52	·12
·14	2·96762	234	11	3·06689	486	26	−2·85442	1055	55	·14
·16	2·96528	248	14	3·07175	513	27	−2·84387	1112	57	·16
·18	2·96280	261	13	3·07688	540	27	−2·83275	1171	59	·18
1·20	2·96019	274	13	3·08228	568	28	−2·82104	1233	62	1·20
·22	2·95745	289	15	3·08796	598	30	−2·80871	1295	62	·22
·24	2·95456	303	14	3·09394	629	31	−2·79576	1362	67	·24
·26	2·95153	318	15	3·10023	660	31	−2·78214	1429	67	·26
·28	2·94835	334	16	3·10683	693	33	−2·76785	1500	71	·28
1·30	2·94501	350	16	3·11376	727	34	−2·75285	1572	72	1·30
·32	2·94151	368	18	3·12103	762	35	−2·73713	1647	75	·32
·34	2·93783	384	16	3·12865	798	36	−2·72066	1725	78	·34
·36	2·93399	403	19	3·13663	836	38	−2·70341	1805	80	·36
·38	2·92996	421	18	3·14499	875	39	−2·68536	1889	84	·38
1·40	2·92575	440	19	3·15374	916	41	−2·66647	1973	84	1·40
·42	2·92135	460	20	3·16290	956	40	−2·64674	2062	89	·42
·44	2·91675	480	20	3·17246	1000	44	−2·62612	2154	92	·44
·46	2·91195	502	22	3·18246	1045	45	−2·60458	2247	93	·46
·48	2·90693	523	21	3·19291	1090	45	−2·58211	2345	98	·48

1·50	100	2445	−2·55866	48	1138	3·20381	23	—	546	—	2·90170	1·50
·52	103	2548	−2·53421	49	1187	3·21519	23	—	569	—	2·89624	·52
·54	106	2654	−2·50873	51	1238	3·22706	24	—	593	—	2·89055	·54
·56	111	2765	−2·48219	52	1290	3·23944	24	—	617	—	2·88462	·56
·58	113	2878	−2·45454	54	1344	3·25234	27	—	644	—	2·87845	·58
1·60	116	2994	−2·42576	55	1399	3·26578	25	—	669	—	2·87201	1·60
·62	121	3115	−2·39582	59	1458	3·27977	28	—	697	—	2·86532	·62
·64	123	3238	−2·36467	59	1517	3·29435	27	—	724	—	2·85835	·64
·66	128	3366	−2·33229	62	1579	3·30952	29	—	753	—	2·85111	·66
·68	132	3498	−2·29863	63	1642	3·32531	30	—	783	—	2·84358	·68
1·70	135	3633	−2·26365	65	1707	3·34173	31	—	814	—	2·83575	1·70
·72	140	3773	−2·22732	69	1776	3·35880	31	—	845	—	2·82761	·72
·74	144	3917	−2·18959	69	1845	3·37656	33	—	878	—	2·81916	·74
·76	148	4065	−2·15042	73	1918	3·39501	33	—	911	—	2·81038	·76
·78	153	4218	−2·10977	74	1992	3·41419	35	—	946	—	2·80127	·78
1·80	157	4375	−2·06759	77	2069	3·43411	36	—	982	—	2·79181	1·80
·82	162	4537	−2·02384	80	2149	3·45480	36	—	1018	—	2·78199	·82
·84	167	4704	−1·97847	82	2231	3·47629	38	—	1056	—	2·77181	·84
·86	172	4876	−1·93143	86	2317	3·49860	40	—	1096	—	2·76125	·86
·88	177	5053	−1·88267	87	2404	3·52177	40	—	1136	—	2·75029	·88
1·90	183	5236	−1·83214	90	2494	3·54581	41	—	1177	—	2·73893	1·90
·92	188	5424	−1·77978	95	2589	3·57075	43	—	1220	—	2·72716	·92
·94	194	5618	−1·72554	97	2686	3·59664	45	—	1265	—	2·71496	·94
·96	199	5817	−1·66936	100	2786	3·62350	45	—	1310	—	2·70231	·96
·98	207	6024	−1·61119	104	2890	3·65136	48	—	1358	—	2·68921	·98
2·00	212		−1·55095	107		3·68026	48				2·67563	2·00

λ	$F_{10}(\lambda)$	Δ_1	Δ_2	$F_{11}(\lambda)$	Δ_1	Δ_2	$F_{12}(\lambda)$	Δ_1	Δ_2	λ
1·00	−3·14002	−0·01161	−0·00069	2·51315	−0·04023	−0·00235	2·76375	−0·01952	−0·00114	1·00
·02	−3·15163	−1232	−71	2·47292	−4269	−246	2·74423	−2072	−120	·02
·04	−3·16395	−1307	−75	2·43023	−4524	−255	2·72351	−2195	−123	·04
·06	−3·17702	−1384	−77	2·38499	−4790	−266	2·70156	−2324	−129	·06
·08	−3·19086	−1464	−80	2·33709	−5066	−276	2·67832	−2458	−134	·08
1·10	−3·20550	−1549	−85	2·28643	−5353	−287	2·65374	−2597	−139	1·10
·12	−3·22099	−1636	−87	2·23290	−5650	−297	2·62777	−2742	−145	·12
·14	−3·23735	−1726	−90	2·17640	−5958	−308	2·60035	−2891	−149	·14
·16	−3·25461	−1820	−94	2·11682	−6279	−321	2·57144	−3046	−155	·16
·18	−3·27281	−1918	−98	2·05403	−6611	−332	2·54098	−3207	−161	·18
1·20	−3·29199	−2019	−101	1·98792	−6953	−342	2·50891	−3374	−167	1·20
·22	−3·31218	−2125	−106	1·91839	−7310	−357	2·47517	−3547	−173	·22
·24	−3·33343	−2233	−108	1·84529	−7677	−367	2·43970	−3724	−177	·24
·26	−3·35576	−2346	−113	1·76852	−8058	−381	2·40246	−3909	−185	·26
·28	−3·37922	−2463	−117	1·68794	−8451	−393	2·36337	−4100	−191	·28
1·30	−3·40385	−2584	−121	1·60343	−8858	−407	2·32237	−4298	−198	1·30
·32	−3·42969	−2709	−125	1·51485	−9278	−420	2·27939	−4500	−202	·32
·34	−3·45678	−2840	−131	1·42207	−9712	−434	2·23439	−4712	−212	·34
·36	−3·48518	−2973	−133	1·32495	−10159	−447	2·18727	−4928	−216	·36
·38	−3·51491	−3113	−140	1·22336	−10622	−463	2·13799	−5152	−224	·38
1·40	−3·54604	−3257	−144	1·11714	−11098	−476	2·08647	−5384	−232	1·40
·42	−3·57861	−3405	−148	1·00616	−11590	−492	2·03263	−5622	−238	·42
·44	−3·61266	−3560	−155	0·89026	−12097	−507	1·97641	−5867	−245	·44
·46	−3·64826	−3718	−158	0·76929	−12620	−523	1·91774	−6121	−254	·46
·48	−3·68544	−3883	−165	0·64309	−13158	−538	1·85653	−6382	−261	·48

x																x
1·50	269	—	6651	—	1·79271	555	—	13713	—	0·51151	170	—	4053	—	−3·72427	1·50
·52	277	—	6928	—	1·72620	572	—	14285	—	0·37438	176	—	4229	—	−3·76480	·52
·54	285	—	7213	—	1·65692	588	—	14873	—	0·23153	182	—	4411	—	−3·80709	·54
·56	293	—	7506	—	1·58479	607	—	15480	—	+0·08280	187	—	4598	—	−3·85120	·56
·58	303	—	7809	—	1·50973	623	—	16103	—	−0·07200	195	—	4793	—	−3·89718	·58
1·60	311	—	8120	—	1·43164	642	—	16745	—	−0·23303	200	—	4993	—	−3·94511	1·60
·62	320	—	8440	—	1·35044	662	—	17407	—	−0·40048	207	—	5200	—	−3·99504	·62
·64	330	—	8770	—	1·26604	679	—	18086	—	−0·57455	214	—	5414	—	−4·04704	·64
·66	338	—	9108	—	1·17834	700	—	18786	—	−0·75541	222	—	5636	—	−4·10118	·66
·68	348	—	9456	—	1·08726	719	—	19505	—	−0·94327	229	—	5865	—	−4·15754	·68
1·70	359	—	9815	—	0·99270	741	—	20246	—	−1·13832	235	—	6100	—	−4·21619	1·70
·72	369	—	10184	—	0·89455	760	—	21006	—	−1·34078	245	—	6345	—	−4·27719	·72
·74	379	—	10563	—	0·79271	784	—	21790	—	−1·55084	252	—	6597	—	−4·34064	·74
·76	389	—	10952	—	0·68708	804	—	22594	—	−1·76874	262	—	6859	—	−4·40661	·76
·78	400	—	11352	—	0·57756	828	—	23422	—	−1·99468	269	—	7128	—	−4·47520	·78
1·80	413	—	11765	—	0·46404	851	—	24273	—	−2·22890	278	—	7406	—	−4·54648	1·80
·82	422	—	12187	—	0·34639	875	—	25148	—	−2·47163	289	—	7695	—	−4·62054	·82
·84	436	—	12623	—	0·22452	899	—	26047	—	−2·72311	298	—	7993	—	−4·69749	·84
·86	448	—	13071	—	+0·09829	925	—	26972	—	−2·98358	308	—	8301	—	−4·77742	·86
·88	459	—	13530	—	−0·03242	951	—	27923	—	−3·25330	318	—	8619	—	−4·86043	·88
1·90	472	—	14002	—	−0·16772	977	—	28900	—	−3·53253	330	—	8949	—	−4·94662	1·90
·92	486	—	14488	—	−0·30774	1005	—	29905	—	−3·82153	342	—	9291	—	−5·03611	·92
·94	500	—	14988	—	−0·45262	1032	—	30937	—	−4·12058	352	—	9643	—	−5·12902	·94
·96	513	—	15501	—	−0·60250	1063	—	32000	—	−4·42995	366	—	10009	—	−5·22545	·96
·98	528	—	16029	—	−0·75751	1092	—	33029	—	−4·74995	378	—	10387	—	−5·32554	·98
2·00	543	—			−0·91780	1123	—			−5·08087	391	—			−5·42941	2·00

λ	$F_7(\lambda)$	Δ_1	Δ_2	$F_8(\lambda)$	Δ_1	Δ_2	$F_9(\lambda)$	Δ_1	Δ_2	λ
2·00	2·67563	−0·01406	−0·00048	3·68026	0·02997	0·00107	−1·55095	0·06236	0·00212	2·00
·02	2·66157	1456	50	3·71023	3108	111	−1·48859	6455	219	·02
·04	2·64701	1508	52	3·74131	3223	115	−1·42404	6681	226	·04
·06	2·63193	1562	54	3·77354	3343	120	−1·35723	6913	232	·06
·08	2·61631	1617	55	3·80697	3465	122	−1·28810	7155	242	·08
2·10	2·60014	1674	57	3·84162	3593	128	−1·21655	7401	246	2·10
·12	2·58340	1733	59	3·87755	3725	132	−1·14254	7658	257	·12
·14	2·56607	1794	61	3·91480	3862	137	−1·06596	7922	264	·14
·16	2·54813	1857	63	3·95342	4004	142	−0·98674	8195	273	·16
·18	2·52956	1923	66	3·99346	4152	148	−0·90479	8476	281	·18
2·20	2·51033	1990	67	4·03498	4304	152	−0·82003	8768	292	2·20
·22	2·49043	2059	69	4·07802	4462	158	−0·73235	9068	300	·22
·24	2·46984	2132	73	4·12264	4627	165	−0·64167	9378	310	·24
·26	2·44852	2207	75	4·16891	4798	171	−0·54789	9701	323	·26
·28	2·42645	2285	78	4·21689	4975	177	−0·45088	10032	331	·28
2·30	2·40360	2364	79	4·26664	5160	185	−0·35056	10377	345	2·30
·32	2·37996	2448	84	4·31824	5351	191	−0·24679	10732	355	·32
·34	2·35548	2534	86	4·37175	5550	199	−0·13947	11101	369	·34
·36	2·33014	2624	90	4·42725	5758	208	−0·02846	11483	382	·36
·38	2·30390	2716	92	4·48483	5973	215	+0·08637	11878	395	·38
2·40	2·27674	2814	98	4·54456	6199	226	0·20515	12289	411	2·40
·42	2·24860	2913	99	4·60655	6432	233	0·32804	12714	425	·42
·44	2·21947	3017	104	4·67087	6677	245	0·45518	13155	441	·44
·46	2·18930	3126	109	4·73764	6931	254	0·58673	13614	459	·46
·48	2·15804	3238	112	4·80695	7197	266	0·72287	14089	475	·48

x												x
2·50	2·12566	—	3356	—	118	4·87892	7475	278	0·86376	14584	495	2·50
·52	2·09210	—	3478	—	122	4·95367	7764	289	1·00960	15099	515	·52
·54	2·05732	—	3605	—	127	5·03131	8067	303	1·16059	15634	535	·54
·56	2·02127	—	3738	—	133	5·11198	8385	318	1·31693	16190	556	·56
·58	1·98389	—	3876	—	138	5·19583	8715	330	1·47883	16771	581	·58
2·60	1·94513	—	4020	—	144	5·28298	9063	348	1·64654	17376	605	2·60
·62	1·90493	—	4172	—	152	5·37361	9427	364	1·82030	18007	631	·62
·64	1·86321	—	4329	—	157	5·46788	9809	382	2·00037	18664	657	·64
·66	1·81992	—	4495	—	166	5·56597	10209	400	2·18701	19353	689	·66
·68	1·77497	—	4667	—	172	5·66806	10630	421	2·38054	20071	718	·68
2·70	1·72830	—	4848	—	181	5·77436	11071	441	2·58125	20822	751	2·70
·72	1·67982	—	5039	—	191	5·88507	11537	466	2·78947	21610	788	·72
·74	1·62943	—	5237	—	198	6·00044	12026	489	3·00557	22434	824	·74
·76	1·57706	—	5447	—	210	6·12070	12542	516	3·22991	23299	865	·76
·78	1·52259	—	5667	—	220	6·24612	13086	544	3·46290	24206	907	·78
2·80	1·46592	—	5898	—	231	6·37698	13659	573	3·70496	25159	953	2·80
·82	1·40694	—	6143	—	245	6·51357	14266	607	3·95655	26161	1002	·82
·84	1·34551	—	6399	—	256	6·65623	14906	640	4·21816	27217	1056	·84
·86	1·28152	—	6670	—	271	6·80529	15585	679	4·49033	28328	1111	·86
·88	1·21482	—	6958	—	288	6·96114	16305	720	4·77361	29501	1173	·88
2·90	1·14524	—	7260	—	302	7·12419	17066	761	5·06862	30740	1239	2·90
·92	1·07264	—	7581	—	321	7·29485	17877	811	5·37602	32048	1308	·92
·94	0·99683	—	7922	—	341	7·47362	18738	861	5·69650	33435	1387	·94
·96	0·91761	—	8283	—	361	7·66100	19654	916	6·03085	34904	1469	·96
·98	0·83478	—	8667	—	384	7·85754	20631	977	6·37989	36463	1559	·98
3·00	0·74811				409	8·06385		1043	6·74452		1657	3·00

466

λ	$F_{10}(\lambda)$	Δ_1	Δ_2	$F_{11}(\lambda)$	Δ_1	Δ_2	$F_{12}(\lambda)$	Δ_1	Δ_2	λ
2·00	—5·42941	—0·10778	—0·00391	—5·08087	—0·34215	—0·01123	—0·91780	—0·16572	—0·00543	2·00
·02	—5·53719	—11184	—406	—5·42302	—35369	—1154	—1·08352	—17129	—557	·02
·04	—5·64903	—11604	—420	—5·77671	—36557	—1188	—1·25481	—17703	—574	·04
·06	—5·76507	—12039	—435	—6·14228	—37779	—1222	—1·43184	—18293	—590	·06
·08	—5·88546	—12491	—452	—6·52007	—39035	—1256	—1·61477	—18900	—607	·08
2·10	—6·01037	—12958	—467	—6·91042	—40329	—1294	—1·80377	—19524	—624	2·10
·12	—6·13995	—13443	—485	—7·31371	—41658	—1329	—1·99901	—20166	—642	·12
·14	—6·27438	—13946	—503	—7·73029	—43027	—1369	—2·20067	—20827	—661	·14
·16	—6·41384	—14469	—523	—8·16056	—44437	—1410	—2·40894	—21506	—679	·16
·18	—6·55853	—15010	—541	—8·60493	—45887	—1450	—2·62400	—22206	—700	·18
2·20	—6·70863	—15574	—564	—9·06380	—47381	—1494	—2·84606	—22927	—721	2·20
·22	—6·86437	—16157	—583	—9·53761	—48919	—1538	—3·07533	—23668	—741	·22
·24	—7·02594	—16765	—608	—10·02680	—50505	—1586	—3·31201	—24433	—765	·24
·26	—7·19359	—17397	—632	—10·53185	—52137	—1632	—3·55634	—25219	—786	·26
·28	—7·36756	—18053	—656	—11·05322	—53821	—1684	—3·80853	—26031	—812	·28
2·30	—7·54809	—18736	—683	—11·59143	—55556	—1735	—4·06884	—26866	—835	2·30
·32	—7·73545	—19447	—711	—12·14699	—57345	—1789	—4·33750	—27728	—862	·32
·34	—7·92992	—20187	—740	—12·72044	—59192	—1847	—4·61478	—28617	—889	·34
·36	—8·13179	—20958	—771	—13·31236	—61097	—1905	—4·90095	—29534	—917	·36
·38	—8·34137	—21762	—804	—13·92333	—63063	—1966	—5·19629	—30480	—946	·38
2·40	—8·55899	—22599	—837	—14·55396	—65094	—2031	—5·50109	—31458	—978	2·40
·42	—8·78498	—23472	—873	—15·20490	—67193	—2099	—5·81567	—32466	—1008	·42
·44	—9·01970	—24385	—913	—15·87683	—69361	—2168	—6·14033	—33509	—1043	·44
·46	—9·26355	—25337	—952	—16·57044	—71603	—2242	—6·47542	—34587	—1078	·46
·48	—9·51692	—26332	—995	—17·28647	—73922	—2319	—6·82129	—35701	—1114	·48

x										x
2·50	—9·78024	27372	1040	—18·02569	76323	2401	—7·17830	36854	1153	2·50
·52	—10·05396	28459	1087	—18·78892	78807	2484	—7·54684	38048	1194	·52
·54	—10·33855	29597	1138	—19·57699	81382	2575	—7·92732	39283	1235	·54
·56	—10·63452	30789	1192	—20·39081	84051	2669	—8·32015	40564	1281	·56
·58	—10·94241	32039	1250	—21·23132	86818	2767	—8·72579	41893	1329	·58
2·60	—11·26280	33347	1308	—22·09950	89690	2872	—9·14472	43269	1376	2·60
·62	—11·59627	34721	1374	—22·99640	92673	2983	—9·57741	44700	1431	·62
·64	—11·94348	36163	1442	—23·92313	95771	3098	—10·02441	46184	1484	·64
·66	—12·30511	37679	1516	—24·88084	98992	3221	—10·48625	47728	1544	·66
·68	—12·68190	39272	1593	—25·87076	1·02345	3353	—10·96353	49334	1606	·68
2·70	—13·07462	40948	1676	—26·89421	1·05835	3490	—11·45687	51006	1672	2·70
·72	—13·48410	42713	1765	—27·95256	1·09473	3638	—11·96693	52746	1740	·72
·74	—13·91123	44573	1860	—29·04729	1·13266	3793	—12·49439	54562	1816	·74
·76	—14·35696	46534	1961	—30·17995	1·17226	3960	—13·04001	56455	1893	·76
·78	—14·82230	48606	2072	—31·35221	1·21362	4136	—13·60456	58434	1979	·78
2·80	—15·30836	50794	2188	—32·56583	1·25689	4327	—14·18890	60501	2067	2·80
·82	—15·81630	53108	2314	—33·82272	1·30215	4526	—14·79391	62664	2163	·82
·84	—16·34738	55558	2450	—35·12487	1·34960	4745	—15·42055	64930	2266	·84
·86	—16·90296	58153	2595	—36·47447	1·39935	4975	—16·06985	67306	2376	·86
·88	—17·48449	60906	2753	—37·87382	1·45159	5224	—16·74291	69799	2493	·88
2·90	—18·09355	63829	2923	—39·32541	1·50651	5492	—17·44090	72419	2620	2·90
·92	—18·73184	66937	3108	—40·83192	1·56430	5779	—18·16509	75176	2757	·92
·94	—19·40121	70244	3307	—42·39622	1·62522	6092	—18·91685	78080	2904	·94
·96	—20·10365	73768	3524	—44·02144	1·68948	6426	—19·69765	81143	3063	·96
·98	—20·84133	77528	3760	—45·71092	1·75738	6790	—20·50908	84378	3235	·98
3·00	—21·61661		4016	—47·46830		7185	—21·35286		3421	3·00

468

λ	Δ_2	Δ_1	$F_9(\lambda)$	Δ_2	Δ_1	$F_8(\lambda)$	Δ_2	Δ_1		$F_7(\lambda)$	λ
3·00	0·01657	0·38120	6·74452	0·01043	0·21674	8·06385	·000409	0·09076		0·74811	3·00
·02	1764	39884	7·12572	1114	22788	8·28059	435	9511		0·65735	·02
·04	1880	41764	7·52456	1193	23981	8·50847	466	9977		0·56224	·04
·06	2007	43771	7·94220	1278	25259	8·74828	496	10473		0·46247	·06
·08	2147	45918	8·37991	1373	26632	9·00087	534	11007		0·35774	·08
3·10	2300	48218	8·83909	1475	28107	9·26719	570	11577		0·24767	3·10
·12	2468	50686	9·32127	1591	29698	9·54826	614	12191		0·13190	·12
·14	2654	53340	9·82813	1715	31413	9·84524	662	12853	+	0·00999	·14
·16	2860	56200	10·36153	1856	33269	10·15937	712	13565	—	0·11854	·16
·18	3089	59289	10·92353	2010	35279	10·49206	772	14337	—	0·25419	·18
3·20	3341	62630	11·51642	2183	37462	10·84485	834	15171	—	0·39756	3·20
·22	3625	66255	12·14272	2375	39837	11·21947	908	16079	—	0·54927	·22
·24	3943	70198	12·80527	2592	42429	11·61784	988	17067	—	0·71006	·24
·26	4297	74495	13·50725	2833	45262	12·04213	1077	18144	—	0·88073	·26
·28	4701	79196	14·25220	3111	48373	12·49475	1181	19325	—	1·06217	·28
3·30	5152	84348	15·04416	3416	51789	12·97848	1294	20619	—	1·25542	3·30
·32	5670	90018	15·88764	3770	55559	13·49637	1426	22045	—	1·46161	·32
·34	6257	96275	16·78782	4172	59731	14·05196	1576	23621	—	1·68206	·34
·36	6932	1·03207	17·75057	4633	64364	14·64927	1747	25368	—	1·91827	·36
·38	7706	1·10913	18·78264	5165	69529	15·29291	1944	27312	—	2·17195	·38
3·40	8607	1·19520	19·89177	5779	75308	15·98820	2172	29484	—	2·44507	3·40
·42	9652	1·29172	21·08697	6495	81803	16·74128	2439	31923	—	2·73991	·42
·44	10876	1·40048	22·37869	7336	89139	17·55931	2749	34672	—	3·05914	·44
·46	12321	1·52369	23·77917	8325	97464	18·45070	3117	37789	—	3·40586	·46
·48	14036	1·66405	25·30286	9500	1·06964	19·42534	3553	41342	—	3·78375	·48

	A	B	C	D	E	F	G	H	I	
3·50	−4·19717	45416	4074	20·49498	1·17871	10907	26·96691	1·82490	16085	3·50
·52	−4·65133	50121	4705	21·67369	1·30478	12607	28·79181	2·01052	18562	·52
·54	−5·15254	55592	5471	22·97847	1·45152	14674	30·80233	2·22628	21576	·54
·56	−5·70846	62007	6415	24·42999	1·62375	17223	33·02861	2·47915	25287	·56
·58	−6·32853	69596	7589	26·05374	1·82767	20392	35·50776	2·77821	29906	·58
3·60	−7·02449	78669	9073	27·88141	2·07159	24392	38·28597	3·13550	35729	3·60
·62	−7·81118	89635	10966	29·95300	2·36666	29507	41·42147	3·56726	43176	·62
·64	−8·70753	1·03066	13431	32·31966	2·72823	36157	44·98873	4·09587	52861	·64
·66	−9·73819	1·19758	16692	35·04789	3·17787	44964	49·08460	4·75269	65682	·66
·68	−10·93577	1·40867	21109	38·22576	3·74671	56884	53·83729	5·58299	83030	·68
3·70	−12·34444	1·68097	27230	41·97247	4·48087	73416	59·42028	6·65397	1·07098	3·70
·72	−14·02541	2·04078	35981	46·45334	5·45126	97039	66·07425	8·06876	1·41479	·72
·74	−16·06619	2·53009	48931	51·90460	6·77128	1·32002	74·14301	9·99245	1·92369	·74
·76	−18·59628	3·21956	68947	58·67588	8·63179	1·86051	84·13546	12·70275	2·71030	·76
·78	−21·81584	4·23567	1·01611	67·30767	11·37428	2·74249	96·83821	16·69668	3·99393	·78
3·80	−26·05151	5·82359	1·58792	78·68195	15·66066	4·28638	113·53489	22·93753	6·24085	3·80
·82	−31·87510	8·51145	2·68786	94·34261	22·91714	7·25648	136·47242	33·50091	10·56338	·82
·84	−40·38655	13·62054	5·10909	117·25975	36·71129	13·79415	169·97333	53·57879	20·07788	·84
·86	−54·00709	25·30719	11·68665	153·97104	68·26604	31·55473	223·55212	99·50428	45·92549	·86
·88	−79·31428	63·35185	38·04466	222·23708	170·99217	102·72613	323·05640	249·00764	149·50336	·88
3·90	−142·66613	−447·13359	−383·78174	393·22925	+1207·2709.	1036·2787.	572·06404	1757·1383.	+1508·1307.	3·90
·92	−589·79972	−175·51603	+131·47322	+1600·5002.			+2329·2023.	689·7772.		·92
·94	+298·00603		23·99831	796·80804	473·85156	−355·00292	1159·5087.	−173·13397	−516·6432.	·94
·96	−122·49000	−44·04281	8·58127	−322·95648	118·84864	−64·80127	−469·73149	−78·82998	−94·30399	·96
·98	−78·44719			−204·10784	54·04737	−23·17249	−296·59752		−33·72033	·98
4·00	−58·40269	−20·04450		−150·06047			−217·76754			4·00

λ	Δ₂	Δ₁	$F_{12}(\lambda)$	Δ₂	Δ₁	$F_{11}(\lambda)$	Δ₂	Δ₁	$F_{10}(\lambda)$	λ
3·00	0·03421	0·87799	21·35286	0·07185	1·82923	47·46830	0·04016	0·81544	21·61661	3·00
·02	3625	91424	22·23085	7615	1·90538	49·29753	4296	85840	22·43205	·02
·04	3847	95271	23·14509	8080	1·98618	51·20291	4602	90442	23·29045	·04
·06	4086	99357	24·09780	8592	2·07210	53·18909	4939	95381	24·19487	·06
·08	4351	1·03708	25·09137	9149	2·16359	55·26119	5305	1·00686	25·14868	·08
3·10	4641	1·08349	26·12845	9761	2·26120	57·42478	5711	1·06397	26·15554	3·10
·12	4959	1·13308	27·21194	10434	2·36554	59·68598	6158	1·12555	27·21951	·12
·14	5308	1·18616	28·34502	11175	2·47729	62·05152	6650	1·19205	28·34506	·14
·16	5696	1·24312	29·53118	11994	2·59723	64·52881	7197	1·26402	29·53711	·16
·18	6125	1·30437	30·77430	12900	2·72623	67·12604	7804	1·34206	30·80113	·18
3·20	6602	1·37039	32·07867	13911	2·86534	69·85227	8480	1·42686	32·14319	3·20
·22	7132	1·44171	33·44906	15032	3·01566	72·71761	9235	1·51921	33·57005	·22
·24	7727	1·51898	34·89077	16292	3·17858	75·73327	10083	1·62004	35·08926	·24
·26	8393	1·60291	36·40975	17698	3·35556	78·91185	11028	1·73032	36·70930	·26
·28	9144	1·69435	38·01266	19292	3·54848	82·26741	12117	1·85149	38·43962	·28
3·30	9991	1·79426	39·70701	21083	3·75931	85·81589	13318	1·98467	40·29111	3·30
·32	10954	1·90380	41·50127	23122	3·99053	89·57520	14705	2·13172	42·27578	·32
·34	12053	2·02433	43·40507	25446	4·24499	93·56573	16283	2·29455	44·40750	·34
·36	13308	2·15741	45·42940	28106	4·52605	97·81072	18092	2·47547	46·70205	·36
·38	14755	2·30496	47·58681	31168	4·83773	102·33677	20177	2·67724	49·17752	·38
3·40	16429	2·46925	49·89177	34712	5·18485	107·17450	22592	2·90316	51·85476	3·40
·42	18374	2·65299	52·36102	38833	5·57318	112·35935	25405	3·15721	54·75792	·42
·44	20655	2·85954	55·01401	43660	6·00978	117·93253	28701	3·44422	57·91513	·44
·46	23340	3·09294	57·87355	49345	6·50323	123·94231	32587	3·77009	61·35935	·46
·48	26527	3·35821	60·96649	56096	7·06419	130·44554	37205	4·14214	65·12944	·48

3·50	69·27158	4·56945	42731	137·50973	7·70587	64168	64·32470	3·66161	30340	3·50
·52	73·84103	5·06347	49402	145·21560	8·44497	73910	67·98631	4·01100	34939	·52
·54	78·90450	5·63879	57532	153·66057	9·30268	85711	71·99731	4·41640	40540	·54
·56	84·54329	6·31414	67535	162·96325	10·30631	1·00363	76·41371	4·89071	47431	·56
·58	90·85743	7·11410	79996	173·26956	11·49160	1·18529	81·30442	5·45079	56008	·58
3·60	97·97153	8·07115	95705	184·76116	12·90585	1·41425	86·75521	6·11898	66819	3·60
·62	106·04268	9·22917	1·15802	197·66701	14·61293	1·70708	92·87419	6·92544	80646	·62
·64	115·27185	10·64851	1·41934	212·27994	16·70063	2·08770	99·79963	7·91161	98617	·64
·66	125·92036	12·41393	1·76542	228·98057	19·29237	2·59174	107·71124	9·13576	1·22415	·66
·68	138·33429	14·64765	2·23372	248·27294	22·56601	3·27364	116·84700	10·68189	1·54613	·68
3·70	152·98194	17·53105	2·88340	270·83895	26·78549	4·21948	127·52889	12·67460	1·99271	3·70
·72	170·51299	21·34266	3·81161	297·62444	32·35621	5·57072	140·20349	15·30529	2·63069	·72
·74	191·85565	26·52827	5·18561	329·98065	39·92692	7·57071	155·50878	18·88028	3·57499	·74
·76	218·38392	33·83767	7·30940	369·90757	50·58885	10·66193	174·38906	23·91479	5·03451	·76
·78	252·22159	44·61295	10·77528	420·49642	66·29516	15·70631	198·30385	31·33100	7·41621	·78
3·80	296·83454	61·45516	16·84221	486·79158	90·83125	24·53609	229·63485	42·91617	11·58517	3·80
·82	358·28970	89·96878	28·51362	577·62283	132·35353	41·52228	272·55102	62·52130	19·60513	·82
·84	448·25848	144·17315	54·20437	709·97636	211·26472	78·91119	335·07232	99·77952	37·25822	·84
·86	592·43163	268·17052	123·99737	921·24108	391·7482.	180·4835.	434·85184	184·99475	85·21523	·86
·88	860·60215	671·8467.	403·6762.	1312·9893.	979·2571.	587·5089.	619·84659	462·3857.	277·3909.	·88
3·90	−1532·4489.	−4744·0623.	−4072·2156.	−2292·2464.	−6905·7021.	−5926·4450.	−1082·2323.	−3260·5381.	−2798·1524.	3·90
·92	−6276·5112.			−9197·9485.			−4342·7704.			·92
·94	+3144·1793.	−1861·9733.	+1395·0409.	+4510·7122.	−2711·3674.	+2030·2165.	+2129·6954.	−1280·1977.	+958·5595.	·94
·96	1282·2060	466·9324	254·6487.	1799·3448.	681·1509.	370·5688.	849·49768	321·63820	174·96227	·96
·98	815·27358	212·28367	91·06214	1118·1939.	310·5821.	132·4958.	527·85948	146·67593	62·55686	·98
4·00	602·98991			807·61183			381·18355			4·00

λ	$F_7(\lambda)$	Δ_1	Δ_2	$F_8(\lambda)$	Δ_1	Δ_2	$F_9(\lambda)$	Δ_1	Δ_2	λ
4·00	58·40269	—11·46323	8·58127	—150·06047	30·87488	—23·17249	—217·76754	45·10965	—33·72033	4·00
·02	46·93946	— 7·42257	4·04066	—119·18559	19·96281	—10·91207	—172·65789	29·23233	—15·87732	·02
·04	39·51689	— 5·19945	2·22312	— 99·22278	13·95843	— 6·00438	—143·42556	20·49741	— 8·73492	·04
·06	34·31744	— 3·84600	1·35345	— 85·26435	10·30228	— 3·65615	—122·92815	15·18000	— 5·31741	·06
·08	30·47144	— 2·96100	0·88500	— 74·96207	7·91096	— 2·39132	—107·74815	11·70338	— 3·47662	·08
4·10	27·51044	— 2·35065	61035	— 67·05111	6·26124	— 1·64972	— 96·04477	9·30612	— 2·39726	4·10
·12	25·15979	— 1·91201	43864	— 60·78987	5·07512	— 1·18612	— 86·73865	7·58362	— 1·72250	·12
·14	23·24778	— 1·58621	32580	— 55·71475	4·19367	+ 88145	— 79·15503	6·30457	— 1·27905	·14
·16	21·66157	— 1·33763	24858	— 51·52108	3·52070	— 67297	— 72·85046	5·32902	— 0·97555	·16
·18	20·32394	— 1·14369	19394	— 48·00038	2·99520	— 52550	— 67·52144	4·56815	— 76087	·18
4·20	19·18025	— 0·98948	15421	— 45·00518	2·57695	— 41825	— 62·95329	3·96345	— 60470	4·20
·22	18·19077	— 86486	12462	— 42·42823	2·23857	— 33838	— 58·98984	3·47506	— 48839	·22
·24	17·32591	— 76273	10213	— 40·18966	1·96088	— 27769	— 55·51478	3·07509	— 39997	·24
·26	16·56318	— 67802	8471	— 38·22878	1·73015	— 23073	— 52·43969	2·74354	— 33155	·26
·28	15·88516	— 60697	7105	— 36·49863	1·53631	— 19384	— 49·69615	2·46573	— 27781	·28
4·30	15·27819	— 54682	6015	— 34·96232	1·37183	— 16448	— 47·23042	2·23079	— 23494	4·30
·32	14·73137	— 49546	5136	— 33·59049	1·23105	— 14078	— 44·99963	2·03040	— 20039	·32
·34	14·23591	— 45128	4418	— 32·35944	1·10958	— 12147	— 42·96923	1·85820	— 17220	·34
·36	13·78463	— 41298	3830	— 31·24986	1·00400	— 10558	— 41·11103	1·70924	— 14896	·36
·38	13·37165	— 37961	3337	— 30·24586	91163	— 9237	— 39·40179	1·57958	— 12966	·38
4·40	12·99204	— 35035	2926	— 29·33423	83030	— 8133	— 37·82221	1·46613	— 11345	4·40
·42	12·64169	— 32455	2580	— 28·50393	75832	— 7198	— 36·35608	1·36635	— 9978	·42
·44	12·31714	— 30172	2283	— 27·74561	69426	— 6406	— 34·98973	1·27821	— 8814	·44
·46	12·01542	— 28141	2031	— 27·05135	63696	— 5730	— 33·71152	1·20004	— 7817	·46
·48	11·73401	— 26328	1813	— 26·41439	58548	— 5148	— 32·51148	1·13045	— 6959	·48

4·50	−6215	1·06830	−31·38103	−4643	53905	−25·82891	1625	24703	11·47073	4·50
·52	−5568	1·01262	−30·31273	−4207	49698	−25·28986	1459	23244	11·22370	·52
·54	−4999	0·96263	−29·30011	−3827	45871	−24·79288	1318	21926	10·99126	·54
·56	−4502	91761	−28·33748	−3492	42379	−24·33417	1191	20735	10·77200	·56
·58	−4062	87699	−27·41987	−3198	39181	−23·91038	1080	19655	10·56465	·58
4·60	−3670	84029	−26·54288	−2940	36241	−23·51857	980	18675	10·36810	4·60
·62	−3322	80707	−25·70259	−2710	33531	−23·15616	895	17780	10·18135	·62
·64	−3012	77695	−24·89552	−2507	31024	−22·82085	815	16965	10·00355	·64
·66	−2733	74962	−24·11857	−2325	28699	−22·51061	745	16220	9·83390	·66
·68	−2482	72480	−23·36895	−2162	26537	−22·22362	683	15537	9·67170	·68
4·70	−2254	70226	−22·64415	−2019	24518	−21·95825	626	14911	9·51633	4·70
·72	−2048	68178	−21·94189	−1887	22631	−21·71307	574	14337	9·36722	·72
·74	−1862	66316	−21·26011	−1771	20860	−21·48676	528	13809	9·22385	·74
·76	−1691	64625	−20·59695	−1666	19194	−21·27816	485	13324	9·08576	·76
·78	−1534	63091	−19·95070	−1570	17624	−21·08622	448	12876	8·95252	·78
4·80	−1392	61699	−19·31979	−1485	16139	−20·90998	412	12464	8·82376	4·80
·82	−1258	60441	−18·70280	−1407	14732	−20·74859	379	12085	8·69912	·82
·84	−1138	59303	−18·09839	−1338	13394	−20·60127	349	11736	8·57827	·84
·86	−1023	58280	−17·50536	−1274	12120	−20·46733	323	11413	8·46091	·86
·88	−920	57360	−16·92256	−1217	10903	−20·34613	296	11117	8·34678	·88
4·90	−820	56540	−16·34896	−1165	9738	−20·23710	274	10843	8·23561	4·90
·92	−730	55810	−15·78356	−1118	8620	−20·13972	251	10592	8·12718	·92
·94	−644	55166	−15·22546	−1075	7545	−20·05352	231	10361	8·02126	·94
·96	−564	54602	−14·67380	−1037	6508	−19·97807	212	10149	7·91765	·96
·98	−486	54116	−14·12778	−1002	5506	−19·91299	195	9954	7·81616	·98
5·00	−415		−13·58662	−972		−19·85793	178		7·71662	5·00

λ	F₁₀(λ)	Δ₁	Δ₂	F₁₁(λ)	Δ₁	Δ₂	F₁₂(λ)	Δ₁	Δ₂	λ
4·00	602·98991	−121·22153	91·06214	807·61183	−178·08625	132·4958.	381·18355	−84·11907	62·55686	4·00
·02	481·76838	−78·33855	42·88298	629·52558	−115·70746	62·37879	297·06448	−54·66775	29·45132	·02
·04	403·42983	−54·74109	23·59746	513·81812	−81·39601	34·31145	242·39673	−38·46834	16·19941	·04
·06	348·68874	−40·37131	14·36978	432·42211	−60·51439	20·88162	203·92839	−28·60979	9·85855	·06
·08	308·31743	−30·97178	9·39953	371·90772	−46·86665	13·64774	175·31860	−22·16671	6·44308	·08
4·10	277·34565	−24·48647	6·48531	325·04107	−37·46064	9·40601	153·15189	−17·72635	4·44036	4·10
·12	252·85918	−19·82285	4·66362	287·58043	−30·70631	6·75433	135·42554	−14·53799	3·18836	·12
·14	233·03633	−16·35643	3·46642	256·87412	−25·69490	5·01141	120·88755	−12·17255	2·36544	·14
·16	216·67990	−13·70920	2·64723	231·17922	−21·87627	3·81863	108·71500	−10·37029	1·80226	·16
·18	202·97070	−11·64142	2·06778	209·30295	−18·90151	2·97476	98·34471	−8·96648	1·40381	·18
4·20	191·32928	−9·99502	1·64640	190·40144	−16·54065	2·36086	89·37823	−7·85251	1·11397	4·20
·22	181·33426	−8·66243	1·33259	173·86079	−14·63707	1·90358	81·52572	−6·95447	0·89804	·22
·24	172·67183	−7·56827	1·09416	159·22372	−13·08115	1·55592	74·57125	−6·22059	·73388	·24
·26	165·10356	−6·65854	0·90973	146·14257	−11·79430	1·28685	68·35066	−5·61375	·60684	·26
·28	158·44502	−5·89368	·76486	134·34827	−10·71896	1·07534	62·73691	−5·10680	·50695	·28
4·30	152·55134	−5·24419	·64949	123·62931	−9·81221	0·90675	57·63011	−4·67945	·42735	4·30
·32	147·30715	−4·68771	·55648	113·81710	−9·04152	·77069	52·95066	−4·31636	·36309	·32
·34	142·61944	−4·20701	·48070	104·77558	−8·38186	·65966	48·63430	−4·00571	·31065	·34
·36	138·41243	−3·78871	·41830	96·39372	−7·81374	·56812	44·62859	−3·73828	·26743	·36
·38	134·62372	−3·42219	·36652	88·57998	−7·32178	·49196	40·89031	−3·50682	·23146	·38
4·40	131·20153	−3·09904	·32315	81·25820	−6·89369	·42809	37·38349	−3·30555	·20127	4·40
·42	128·10249	−2·81246	·28658	74·36451	−6·51963	·37406	34·07794	−3·12979	·17576	·42
·44	125·29003	−2·55691	·25555	67·84488	−6·19157	·32806	30·94815	−2·97577	·15402	·44
·46	122·73312	−2·32787	·22904	61·65331	−5·90295	·28862	27·97238	−2·84037	·13540	·46
·48	120·40525	−2·12162	·20625	55·75036	−5·64835	·25460	25·13201	−2·72107	·11930	·48

x		Δ			Δ			Δ		x
4·50	118·28363	—1·93502	18660	50·10201	—5·42327	22508	22·41094	—2·61571	10536	4·50
·52	116·34861	—1·76549	16953	44·67874	—5·22393	19934	19·79523	—2·52252	9319	·52
·54	114·58312	—1·61081	15468	39·45481	—5·04717	17676	17·27271	—2·44001	8251	·54
·56	112·97231	—1·46913	14168	34·40764	—4·89032	15685	14·83270	—2·36691	7310	·56
·58	111·50318	—1·33889	13024	29·51732	—4·75108	13924	12·46579	—2·30216	6475	·58
4·60	110·16429	—1·21867	12022	24·76624	—4·62751	12357	10·16363	—2·24480	5736	4·60
·62	108·94562	—1·10737	11130	20·13873	—4·51794	10957	7·91883	—2·19409	5071	·62
·64	107·83825	—1·00392	10345	15·62079	—4·42092	9702	5·72474	—2·14932	4477	·64
·66	106·83433	—0·90747	9645	11·19987	—4·33522	8570	3·57542	—2·10990	3942	·66
·68	105·92686	—·81724	9023	6·86465	—4·25973	7549	+1·46552	—2·07533	3457	·68
4·70	105·10962	—·73256	8468	+2·60492	—4·19352	6621	—0·60981	—2·04517	3016	4·70
·72	104·37706	—·65285	7971	—1·58860	—4·13577	5775	—2·65498	—2·01901	2616	·72
·74	103·72421	—·57756	7529	—5·72437	—4·08576	5001	—4·67399	—1·99653	2248	·74
·76	103·14665	—·50626	7130	—9·81013	—4·04283	4293	—6·67052	—1·97742	1911	·76
·78	102·64039	—·43849	6777	—13·85296	—4·00644	3639	—8·64794	—1·96142	1600	·78
4·80	102·20190	—·37393	6456	—17·85940	—3·97612	3032	—10·60936	—1·94830	1312	4·80
·82	101·82797	—·31223	6170	—21·83552	—3·95138	2474	—12·55766	—1·93785	1045	·82
·84	101·51574	—·25307	5916	—25·78690	—3·93188	1950	—14·49551	—1·92988	797	·84
·86	101·26267	—·19621	5686	—29·71878	—3·91726	1462	—16·42539	—1·92427	561	·86
·88	101·06646	—·14140	5481	—33·63604	—3·90722	1004	—18·34966	—1·92082	345	·88
4·90	100·92506	—·08840	5300	—37·54326	—3·90151	571	—20·27048	—1·91945	+137	4·90
·92	100·83666	—·03703	5137	—41·44477	—3·89986	+165	—22·18993	—1·92003	—58	·92
·94	100·79963	—·01290	4993	—45·34463	—3·90208	—222	—24·10996	—1·92246	—243	·94
·96	100·81253	+·06157	4867	—49·24671	—3·90798	—590	—26·03242	—1·92667	—421	·96
·98	100·87410	+·10914	4757	—53·15469	—3·91741	—943	—27·95909	—1·93255	—588	·98
5·00	100·98324		4660	—57·07210		—1279	—29·89164		—753	5·00

λ	$F_7(\lambda)$	Δ_1	Δ_1	$F_8(\lambda)$	Δ_1	Δ_2	$F_9(\lambda)$	Δ_1	Δ_2	λ
5·00	7·71662	−0·09776	0·00178	−19·85793	0·04534	−0·00972	−13·58662	0·53701	−0·00415	5·00
·02	7·61886	− 9614	162	−19·81259	3590	− 944	−13·04961	53354	− 347	·02
·04	7·52272	− 9467	147	−19·77669	2672	− 918	−12·51607	53073	− 281	·04
·06	7·42805	− 9334	133	−19·74997	1775	− 897	−11·98534	52854	− 219	·06
·08	7·33471	− 9214	120	−19·73222	898	− 877	−11·45680	52693	− 161	·08
5·10	7·24257	− 9106	108	−19·72324	+ 38	− 860	−10·92987	52592	− 101	5·10
·12	7·15151	− 9011	95	−19·72286	− 807	− 845	−10·40395	52545	− 47	·12
·14	7·06140	− 8927	84	−19·73093	− 1639	− 832	− 9·87850	52551	+ 6	·14
·16	6·97213	− 8853	74	−19·74732	− 2461	− 822	− 9·35299	52611	60	·16
·18	6·88360	− 8792	61	−19·77193	− 3275	− 814	− 8·82688	52720	109	·18
5·20	6·79568	− 8740	52	−19·80468	− 4081	− 806	− 8·29968	52879	159	5·20
·22	6·70828	− 8697	43	−19·84549	− 4883	− 802	− 7·77089	53088	209	·22
·24	6·62131	− 8665	32	−19·89432	− 5681	− 798	− 7·24001	53344	256	·24
·26	6·53466	− 8642	23	−19·95113	− 6478	− 797	− 6·70657	53647	303	·26
·28	6·44824	− 8627	15	−20·01591	− 7275	− 797	− 6·17010	53999	352	·28
5·30	6·36197	− 8623	+ 4	−20·08866	− 8073	− 798	− 5·63011	54396	397	5·30
·32	6·27574	− 8626	− 3	−20·16939	− 8876	− 803	− 5·08615	54840	444	·32
·34	6·18948	− 8638	− 12	−20·25815	− 9683	− 807	− 4·53775	55331	491	·34
·36	6·10310	− 8660	− 22	−20·35498	−10496	− 813	− 3·98444	55870	539	·36
·38	6·01650	− 8689	− 29	−20·45994	−11318	− 822	− 3·42574	56455	585	·38
5·40	5·92961	− 8727	− 38	−20·57312	−12150	− 832	− 2·86119	57089	634	5·40
·42	5·84234	− 8774	− 47	−20·69462	−12994	− 844	− 2·29030	57771	682	·42
·44	5·75460	− 8829	− 55	−20·82456	−13849	− 855	− 1·71259	58502	731	·44
·46	5·66631	− 8892	− 63	−20·96305	−14720	− 871	− 1·12757	59284	782	·46
·48	5·57739	− 8965	− 73	−21·11025	−15607	− 887	− 0·53473	60116	832	·48

			+						
5·50	887	61003	0·06643	907	—16514	—21·26632	— 81	9046	5·48774
·52	939	61942	0·67646	925	—17439	—21·43146	— 90	9136	5·39728
·54	995	62937	1·29588	947	—18386	—21·60585	— 98	9234	5·30592
·56	1054	63991	1·92525	973	—19359	—21·78971	—109	9343	5·21358
·58	1111	65102	2·56516	998	—20357	—21·98330	—118	9461	5·12015
5·60	1174	66276	3·21618	1025	—21382	—22·18687	—126	9587	5·02554
·62	1239	67515	3·87894	1058	—22440	—22·40069	—138	9725	4·92967
·64	1303	68818	4·55409	1089	—23529	—22·62509	—148	9873	4·83242
·66	1375	70193	5·24227	1126	—24655	—22·86038	—158	10031	4·73369
·68	1445	71638	5·94420	1163	—25818	—23·10693	—170	10201	4·63338
5·70	1524	73162	6·66058	1204	—27022	—23·36511	—180	10381	4·53137
·72	1600	74762	7·39220	1250	—28272	—23·63533	—193	10574	4·42756
·74	1687	76449	8·13982	1295	—29567	—23·91805	—205	10779	4·32182
·76	1773	78222	8·90431	1347	—30914	—24·21372	—219	10998	4·21403
·78	1866	80088	9·68653	1402	—32316	—24·52286	—232	11230	4·10405
5·80	1965	82053	10·48741	1460	—33776	—24·84602	—246	11476	3·99175
·82	2067	84120	11·30794	1523	—35299	—25·18378	—263	11739	3·87699
·84	2177	86297	12·14914	1591	—36890	—25·53677	—277	12016	3·75960
·86	2294	88591	13·01211	1662	—38552	—25·90567	—294	12310	3·63944
·88	2417	91008	13·89802	1742	—40294	—26·29119	—314	12624	3·51634
5·90	2549	93557	14·80810	1825	—42119	—26·69413	—331	12955	3·39010
·92	2688	96245	15·74367	1914	—44033	—27·11532	—352	13307	3·26055
·94	2839	99084	16·70612	2013	—46046	—27·55565	—374	13681	3·12748
·96	2998	1·02082	17·69696	2116	—48162	—28·01611	—396	14077	2·99067
·98	3170	1·05252	18·71778	2230	—50392	—28·49773	—421	14498	2·84990
6·00			19·77030			—29·00165			2·70492

λ	Δ_2	Δ_1	$F_{12}(\lambda)$	Δ_2	Δ_1	$F_{11}(\lambda)$	Δ_2	Δ_1	$F_{10}(\lambda)$	λ
5·00	−0·00753	−1·94008	−29·89164	−0·01279	−3·93020	−57·07210	0·04660	0·15574	100·98324	5·00
·02	907	−1·94915	−31·83172	1603	−3·94623	−61·00230	4580	20154	101·13898	·02
·04	1061	−1·95976	−33·78087	1919	−3·96542	−64·94853	4509	24663	101·34052	·04
·06	1206	−1·97182	−35·74063	2222	−3·98764	−68·91395	4452	29115	101·58715	·06
·08	1351	−1·98533	−37·71245	2517	−4·01281	−72·90159	4408	33523	101·87830	·08
5·10	1490	−2·00023	−39·69778	2808	−4·04089	−76·91440	4371	37894	102·21353	5·10
·12	1628	−2·01651	−41·69801	3090	−4·07179	−80·95529	4347	42241	102·59247	·12
·14	1764	−2·03415	−43·71452	3369	−4·10548	−85·02708	4334	46575	103·01488	·14
·16	1899	−2·05314	−45·74867	3645	−4·14193	−89·13256	4327	50902	103·48063	·16
·18	2030	−2·07344	−47·80181	3917	−4·18110	−93·27449	4333	55235	103·98965	·18
5·20	2164	−2·09508	−49·87525	4188	−4·22298	−97·45559	4347	59582	104·54200	5·20
·22	2296	−2·11804	−51·97033	4458	−4·26756	−101·67857	4367	63949	105·13782	·22
·24	2428	−2·14232	−54·08837	4728	−4·31484	−105·94613	4401	68350	105·77731	·24
·26	2562	−2·16794	−56·23069	4998	−4·36482	−110·26097	4440	72790	106·46081	·26
·28	2694	−2·19488	−58·39863	5273	−4·41755	−114·62579	4488	77278	107·18871	·28
5·30	2831	−2·22319	−60·59351	5546	−4·47301	−119·04334	4546	81824	107·96149	5·30
·32	2968	−2·25287	−62·81670	5825	−4·53126	−123·51635	4612	86436	108·77973	·32
·34	3107	−2·28394	−65·06957	6107	−4·59233	−128·04761	4687	91123	109·64409	·34
·36	3249	−2·31643	−67·35351	6396	−4·65629	−132·63994	4771	95894	110·55532	·36
·38	3394	−2·35037	−69·66994	6687	−4·72316	−137·29623	4863	1·00757	111·51426	·38
5·40	3542	−2·38579	−72·02031	6987	−4·79303	−142·01939	4967	1·05724	112·52183	5·40
·42	3694	−2·42273	−74·40610	7295	−4·86598	−146·81242	5079	1·10803	113·57907	·42
·44	3851	−2·46124	−76·82883	7609	−4·94207	−151·67840	5199	1·16002	114·68710	·44
·46	4012	−2·50136	−79·29007	7933	−5·02140	−156·62047	5334	1·21336	115·84712	·46
·48	4177	−2·54313	−81·79143	8269	−5·10409	−161·64187	5475	1·26811	117·06048	·48

x										x
5·50	118·32859	1·32441	5630	—166·74596	—5·19021	—8612	—84·33456	—2·58662	—4349	5·50
·52	119·65300	1·38237	5796	—171·93617	—5·27992	—8971	—86·92118	—2·63188	—4526	·52
·54	121·03537	1·44210	5973	—177·21609	—5·37334	—9342	—89·55306	—2·67901	—4713	·54
·56	122·47747	1·50376	6166	—182·58943	—5·47060	—9726	—92·23207	—2·72803	—4902	·56
·58	123·98123	1·56745	6369	—188·06003	—5·57188	—10128	—94·96010	—2·77907	—5104	·58
5·60	125·54868	1·63332	6587	—193·63191	—5·67732	—10544	—97·73917	—2·83218	—5311	5·60
·62	127·18200	1·70155	6823	—199·30923	—5·78714	—10982	—100·57135	—2·88746	—5528	·62
·64	128·88355	1·77228	7073	—205·09637	—5·90150	—11436	—103·45881	—2·94502	—5756	·64
·66	130·65583	1·84566	7338	—210·99787	—6·02066	—11916	—106·40383	—3·00497	—5995	·66
·68	132·50149	1·92191	7625	—217·01853	—6·14480	—12414	—109·40880	—3·06741	—6244	·68
5·70	134·42340	2·00121	7930	—223·16333	—6·27421	—12941	—112·47621	—3·13248	—6507	5·70
·72	136·42461	2·08377	8256	—229·43754	—6·40915	—13494	—115·60869	—3·20032	—6784	·72
·74	138·50838	2·16979	8602	—235·84669	—6·54992	—14077	—118·80901	—3·27106	—7074	·74
·76	140·67817	2·25955	8976	—242·39661	—6·69684	—14692	—122·08007	—3·34488	—7382	·76
·78	142·93772	2·35327	9372	—249·09345	—6·85024	—15340	—125·42495	—3·42195	—7707	·78
5·80	145·29099	2·45125	9798	—255·94369	—7·01050	—16026	—128·84690	—3·50244	—8049	5·80
·82	147·74224	2·55378	10253	—262·95419	—7·17805	—16755	—132·34934	—3·58658	—8414	·82
·84	150·29602	2·66118	10740	—270·13224	—7·35330	—17525	—135·93592	—3·67457	—8799	·84
·86	152·95720	2·77379	11261	—277·48554	—7·53675	—18345	—139·61049	—3·76666	—9209	·86
·88	155·73099	2·89201	11822	—285·02229	—7·72891	—19216	—143·37715	—3·86313	—9647	·88
5·90	158·62300	3·01622	12421	—292·75120	—7·93035	—20144	—147·24028	—3·96422	—10109	5·90
·92	161·63922	3·14690	13068	—300·68155	—8·14169	—21134	—151·20450	—4·07028	—10606	·92
·94	164·78612	3·28450	13760	—308·82324	—8·36361	—22192	—155·27478	—4·18163	—11135	·94
·96	168·07062	3·42957	14507	—317·18685	—8·59685	—23324	—159·45641	—4·29866	—11703	·96
·98	171·50019	3·58267	15310	—325·78370	—8·84220	—24535	—163·75507	—4·42174	—12308	·98
6·00	175·08286			—334·62590			—168·17681			6·00

481

Tables of functions

$$F_{13}(\lambda),\ F_{14}(\lambda)$$

for calculating the end forces of a vibrating beam hinged at both ends $(\lambda = 0{\cdot}00 - 6{\cdot}00)$

λ	$F_{13}(\lambda)$	Δ_1	Δ_2	$F_{14}(\lambda)$	Δ_1	Δ_2
0·00	−0·00000	−0·00000	−0·00000	−0·00000	−0·00000	−0·00000
·02	−0·00000	0	0	−0·00000	0	0
·04	−0·00000	0	1	−0·00000	0	1
·06	−0·00000	1	0	−0·00000	1	1
·08	−0·00001	1	0	−0·00001	2	2
0·10	−0·00002	1	2	−0·00003	4	2
·12	−0·00003	3	2	−0·00007	6	3
·14	−0·00006	5	1	−0·00013	9	4
·16	−0·00011	6	4	−0·00022	13	5
·18	−0·00017	10	2	−0·00035	18	7
0·20	−0·00027	12	4	−0·00053	25	8
·22	−0·00039	16	5	−0·00078	33	8
·24	−0·00055	21	5	−0·00111	41	12
·26	−0·00076	26	7	−0·00152	53	12
·28	−0·00102	33	7	−0·00205	65	15
0·30	−0·00135	40	8	−0·00270	80	15
·32	−0·00175	48	9	−0·00350	95	20
·34	−0·00223	57	11	−0·00445	115	20
·36	−0·00280	68	11	−0·00560	135	23
·38	−0·00348	79	13	−0·00695	158	26
0·40	−0·00427	92	14	−0·00853	184	29
·42	−0·00519	106	16	−0·01037	213	30
·44	−0·00625	122	16	−0·01250	243	34
·46	−0·00747	138	19	−0·01493	277	37
·48	−0·00885	157		−0·01770	314	

λ	$F_{13}(\lambda)$	Δ_1	Δ_2
1·00	−0·16874	−0·01409	−0·00084
·02	−0·18283	1499	90
·04	−0·19782	1590	91
·06	−0·21372	1688	98
·08	−0·23060	1788	100
1·10	−0·24848	1893	105
·12	−0·26741	2004	111
·14	−0·28745	2117	113
·16	−0·30862	2237	120
·18	−0·33099	2362	125
1·20	−0·35461	2491	129
·22	−0·37952	2626	135
·24	−0·40578	2767	141
·26	−0·43345	2914	147
·28	−0·46259	3066	152
1·30	−0·49325	3226	160
·32	−0·52551	3391	165
·34	−0·55942	3563	172
·36	−0·59505	3743	180
·38	−0·63248	3930	187
1·40	−0·67178	4125	195
·42	−0·71303	4329	204
·44	−0·75632	4539	210
·46	−0·80171	4761	222
·48	−0·84932	4990	229

x						
0·50	—0·01042	— 178	— 21	—0·02084	— 354	— 40
·52	—0·01220	— 199	— 21	—0·02438	— 398	— 44
·54	—0·01419	— 222	— 23	—0·02836	— 444	— 46
·56	—0·01641	— 248	— 26	—0·03280	— 495	— 51
·58	—0·01889	— 274	— 26	—0·03775	— 549	— 54
0·60	—0·02163	— 304	— 30	—0·04324	— 606	— 57
·62	—0·02467	— 335	— 31	—0·04930	— 668	— 62
·64	—0·02802	— 368	— 33	—0·05598	— 735	— 67
·66	—0·03170	— 403	— 35	—0·06333	— 804	— 69
·68	—0·03573	— 441	— 38	—0·07137	— 879	— 75
0·70	—0·04014	— 480	— 39	—0·08016	— 957	— 78
·72	—0·04494	— 522	— 42	—0·08973	— 1042	— 85
·74	—0·05016	— 567	— 45	—0·10015	— 1129	— 87
·76	—0·05583	— 614	— 47	—0·11144	— 1223	— 94
·78	—0·06197	— 664	— 50	—0·12367	— 1322	— 99
0·80	—0·06861	— 716	— 52	—0·13689	— 1425	— 103
·82	—0·07577	— 772	— 56	—0·15114	— 1534	— 109
·84	—0·08349	— 830	— 58	—0·16648	— 1649	— 115
·86	—0·09179	— 890	— 60	—0·18297	— 1769	— 120
·88	—0·10069	— 955	— 65	—0·20066	— 1896	— 127
0·90	—0·11024	— 1022	— 67	—0·21962	— 2027	— 131
·92	—0·12046	— 1092	— 70	—0·23989	— 2166	— 139
·94	—0·13138	— 1166	— 74	—0·26155	— 2311	— 145
·96	—0·14304	— 1245	— 79	—0·28466	— 2461	— 150
·98	—0·15549	— 1325	— 80	—0·30927	— 2620	— 159
1·00	—0·16874		— 84	—0·33547		— 165

x			
1·50	—0·89922	— 5229	— 239
·52	—0·95151	— 5479	— 250
·54	—1·00630	— 5740	— 261
·56	—1·06370	— 6011	— 271
·58	—1·12381	— 6295	— 284
1·60	—1·18676	— 6591	— 296
·62	—1·25267	— 6900	— 309
·64	—1·32167	— 7223	— 323
·66	—1·39390	— 7561	— 338
·68	—1·46951	— 7913	— 352
1·70	—1·54864	— 8282	— 369
·72	—1·63146	— 8668	— 386
·74	—1·71814	— 9072	— 404
·76	—1·80886	— 9495	— 423
·78	—1·90381	— 9939	— 444
1·80	—2·00320	— 10405	— 466
·82	—2·10725	— 10892	— 487
·84	—2·21617	— 11405	— 513
·86	—2·33022	— 11944	— 539
·88	—2·44966	— 12509	— 565
1·90	—2·57475	— 13104	— 595
·92	—2·70579	— 13732	— 628
·94	—2·84311	— 14391	— 659
·96	—2·98702	— 15088	— 697
·98	—3·13790	— 15822	— 734
2·00	—3·29612		— 776

λ	$F_{14}(\lambda)$	Δ_1	Δ_2
1·00	-0·33547	-0·02785	-0·00165
·02	-0·36332	2956	171
·04	-0·39288	3136	180
·06	-0·42424	3323	187
·08	-0·45747	3517	194
1·10	-0·49264	3719	202
·12	-0·52983	3930	211
·14	-0·56913	4148	218
·16	-0·61061	4376	228
18	-0·65437	4612	236
1·20	-0·70049	4858	246
·22	-0·74907	5112	254
·24	-0·80019	5376	264
·26	-0·85395	5651	275
·28	-0·91046	5935	284
1·30	-0·96981	6230	295
·32	-1·03211	6536	306
·34	-1·09747	6853	317
·36	-1·16600	7181	328
·38	-1·23781	7522	341
1·40	-1·31303	7875	353
·42	-1·39178	8240	365
·44	-1·47418	8619	379
·46	-1·56037	9012	393
·48	-1·65049	9418	406

λ	$F_{13}(\lambda)$	Δ_1	Δ_2	$F_{14}(\lambda)$	Δ_1	Δ_2
2·00	-3·29612	-0·16598	0·00776	5·97989	0·27568	-0·01105
·02	-3·46210	17418	820	6·25557	28725	1157
·04	-3·63628	18285	867	6·54282	29936	1211
·06	-3·81913	19204	919	6·84218	31206	1270
·08	-4·01117	20178	974	7·15424	32539	1333
2·10	-4·21295	21211	1033	7·47963	33937	1398
·12	-4·42506	22310	1099	7·81900	35409	1472
·14	-4·64816	23476	1166	8·17309	36958	1549
·16	-4·88292	24720	1244	8·54267	38589	1631
·18	-5·13012	26045	1325	8·92856	40311	1722
2·20	-5·39057	27459	1414	9·33167	42130	1819
·22	-5·66516	28970	1511	9·75297	44054	1924
·24	-5·95486	30589	1619	10·19351	46093	2039
·26	-6·26075	32323	1734	10·65444	48257	2164
·28	-6·58398	34183	1860	11·13701	50555	2298
2·30	-6·92581	36185	2002	11·64256	53002	2447
·32	-7·28766	38341	2156	12·17258	55612	2610
·34	-7·67107	40665	2324	12·72870	58401	2789
·36	-8·07772	43178	2513	13·31271	61385	2984
·38	-8·50950	45899	2721	13·92656	64589	3204
2·40	-8·96849	48851	2952	14·57245	68031	3442
·42	-9·45700	52062	3211	15·25276	71742	3711
·44	-9·97762	55562	3500	15·97018	75750	4008
·46	-10·53324	59384	3822	16·72768	80092	4342
·48	-11·12708	63572	4188	17·52860	84808	4716

x				x						
1·50	—1·74467	9839	421	2·50	—11·76280	68168	4596	18·37668	89945	5137
·52	—1·84306	10276	437	·52	—12·44448	73237	5069	19·27613	95558	5613
·54	—1·94582	10728	452	·54	—13·17685	78830	5593	20·23171	1·01713	6155
·56	—2·05310	11196	468	·56	—13·96515	85034	6204	21·24884	1·08483	6770
·58	—2·16506	11682	486	·58	—14·81549	91932	6898	22·33367	1·15962	7479
1·60	—2·28188	12186	504	2·60	—15·73481	99635	7703	23·49329	1·24255	8293
·62	—2·40374	12707	521	·62	—16·73116	1·08270	8635	24·73584	1·33490	9235
·64	—2·53081	13248	541	·64	—17·81386	1·17993	9723	26·07074	1·43825	10335
·66	—2·66329	13809	561	·66	—18·99379	1·28992	10999	27·50899	1·55447	11622
·68	—2·80138	14391	582	·68	—20·28371	1·41501	12509	29·06346	1·68589	13142
1·70	—2·94529	14995	604	2·70	—21·69872	1·55802	14301	30·74935	1·83536	14947
72	—3·09524	15621	626	·72	—23·25674	1·72259	16457	32·58471	2·00650	17114
·74	—3·25145	16272	651	·74	—24·97933	1·91319	19060	34·59121	2·20379	19729
·76	—3·41417	16948	676	·76	—26·89252	2·13563	22244	36·79500	2·43304	22925
·78	—3·58365	17649	701	·78	—29·02815	2·39733	26170	39·22804	2·70167	26863
1·80	—3·76014	18379	730	2·80	—31·42548	2·70810	31077	41·92971	3·01950	31783
·82	—3·94393	19138	759	·82	—34·13358	3·08095	37285	44·94921	3·39953	38003
·84	—4·13531	19927	789	·84	—37·21453	3·53353	45258	48·34874	3·85942	45989
·86	—4·33458	20748	821	·86	—40·74806	4·09022	55669	52·20816	4·42355	56413
·88	—4·54206	21603	855	·88	—44·83828	4·78552	69530	56·63171	5·12644	70289
1·90	—4·75809	22495	892	2·90	—49·62380	5·66949	88397	61·75815	6·01810	89166
·92	—4·98304	23424	929	·92	—55·29329	6·81700	1·14751	67·77625	7·17347	1·15537
·94	—5·21728	24393	969	·94	—62·11029	8·34417	1·52717	74·94972	8·70863	1·53316
·96	—5·46121	25405	1012	·96	—70·45446	10·43900	2·09483	83·65835	10·81158	2·10295
·98	—5·71526	26463	1058	·98	—80·89346	13·42221	2·98321	94·46993	13·80307	2·99149
2·00	—5·97989		1105	3·00	—94·31567		4·45613	108·27300		4·46456

λ	$F_{13}(\lambda)$	Δ_1	Δ_2	$F_{14}(\lambda)$	Δ_1	Δ_2
3·00	94·31567	—17·87834	—4·45613	108·27300	—18·26763	—4·46456
·02	112·19401	—24·96730	—7·08896	126·54063	—25·36516	—7·09753
·04	137·16131	—37·27055	—12·30325	151·90579	—37·67714	—12·31198
·06	174·43186	—61·56078	—24·29023	189·58293	—61·97625	—24·29911
·08	235·99264	—120·89400	—59·33322	251·55918	—121·31851	—59·34226
3·10	356·88664	—345·10347	—224·20947	372·87769	—345·53720	—223·21869
·12	701·99011.	—9016·0305.	—8670·9071.	718·41489	—9016·4736.	—8670·9364.
·14	—9718·0206.			—9734·8885.		
·16	858·50646	+438·42744	+300·35855	841·18596	+438·88978	+300·34867
·18	420·07902	138·06889	70·51569	402·29618	138·54111	70·50562
3·20	282·01013	67·55320	27·52304	263·75507	68·03549	27·51279
·22	214·45693	40·03016	13·57772	195·71958	40·52270	13·56728
·24	174·42677	26·45244	7·69288	155·19688	26·95542	7·68239
·26	147·97433	18·75956	4·78009	128·24146	19·27303	4·76901
·28	129·21477	13·97947	3·17344	108·96843	14·50402	3·16257
3·30	115·23530	10·80603	2·21467	94·46441	11·34145	2·20343
·32	104·42927	8·59136	1·60708	83·12296	9·13802	1·59566
·34	95·83791	6·98428	1·20342	73·98494	7·54236	1·19179
·36	88·85363	5·78086	0·92474	66·44258	6·35057	0·91288
·38	83·07277	4·85612	72617	60·09201	5·43769	71410
3·40	78·21665	4·12995	58085	54·65432	4·72359	56856
·42	74·08670	3·54910	47208	49·93073	4·15503	45953
·44	70·53760	3·07702	38903	45·77570	3·69550	37627
·46	67·46058	2·68799	32454	42·08020	3·31923	31153
·48	64·77259	2·36345		38·76097	3·00770	

λ	$F_{13}(\lambda)$	Δ_1	Δ_2
4·00	43·45575	—0·08895	0·02654
·02	43·36680	—6352	2543
·04	43·30328	—3908	2444
·06	43·26420	—1550	2358
·08	43·24870	+730	2280
4·10	43·25600	2943	2213
·12	43·28543	5096	2153
·14	43·33639	7197	2101
·16	43·40836	9252	2055
·18	43·50088	11269	2017
4·20	43·61357	13254	1985
·22	43·74611	15212	1958
·24	43·89823	17148	1936
·26	44·06971	19068	1920
·28	44·26039	20976	1908
4·30	44·47015	22877	1901
·32	44·69892	24776	1899
·34	44·94668	26675	1899
·36	45·21343	28581	1906
·38	45·49924	30495	1914
4·40	45·80419	32424	1929
·42	46·12843	34370	1946
·44	46·47213	36336	1966
·46	46·83549	38328	1992
·48	47·21877	40348	2020

				±							
3·50	62·40914	2·08975	27370	—	35·75327	2·74725	26045	4·50	47·62225	42401	2053
·52	60·31939	1·85664	24311	—	33·00602	2·52765	21960	·52	48·04626	44491	2090
·54	58·46275	1·65636	20028	—	30·47837	2·34115	18650	·54	48·49117	46620	2129
·56	56·80639	1·48288	17348	—	28·13722	2·18171	15944	·56	48·95737	48795	2175
·58	55·32351	1·33153	15135	—	25·95551	2·04466	13705	·58	49·44532	51018	2223
3·60	53·99198	1·19857	13296	—	23·91085	1·92630	11836	4·60	49·95550	53293	2275
·62	52·79341	1·08103	11754	—	21·98455	1·82362	10268	·62	50·48843	55627	2334
·64	51·71238	0·97652	10451	—	20·16093	1·73429	8933	·64	51·04470	58020	2393
·66	50·73586	88306	9346	—	18·42664	1·65630	7799	·66	51·62490	60482	2462
·68	49·85280	79909	8397	—	16·77034	1·58811	6819	·68	52·22972	63014	2532
3·70	49·05371	72323	7586	—	15·18223	1·52835	5976	4·70	52·85986	65623	2609
·72	48·33048	65441	6882	—	13·65388	1·47595	5240	·72	53·51609	68313	2690
·74	47·67607	59168	6273	—	12·17793	1·42999	4596	·74	54·19922	71091	2778
·76	47·08439	53427	5741	—	10·74794	1·38967	4032	·76	54·91013	73963	2872
·78	46·55012	48150	5277	—	9·35827	1·35437	3530	·78	55·64976	76934	2971
3·80	46·06862	43281	4869	—	8·00390	1·32351	3086	4·80	56·41910	80012	3078
·82	45·63581	38772	4509	—	6·68039	1·29660	2691	·82	57·21922	83203	3191
·84	45·24809	34579	4193	—	5·38379	1·27325	2335	·84	58·05125	86514	3311
·86	44·90230	30666	3913	—	4·11054	1·25311	2014	·86	58·91639	89956	3442
·88	44·59564	27002	3664	—	2·85743	1·23585	1726	·88	59·81595	93533	3577
3·90	44·32562	23560	3442	+	1·62158	1·22123	1462	4·90	60·75128	97259	3726
·92	44·09002	20312	3248	—	0·40035	1·20900	1223	·92	61·72387	1·01139	3880
·94	43·88690	17241	3071	—	0·80865	1·19896	1004	·94	62·73526	1·05186	4047
·96	43·71449	14325	2916	—	2·00761	1·19095	801	·96	63·78712	1·09411	4225
·98	43·57124	11549	2776	—	3·19856	1·18480	615	·98	64·88123	1·13825	4414
4·00	43·45575		2654	—	4·38336		443	5·00	66·01948		4618

λ	$F_{14}(\lambda)$	Δ_1	Δ_2
4·00	−4·38336	−1·18037	0·00443
·02	−5·56373	−1·17755	282
·04	−6·74128	−1·17624	+131
·06	−7·91752	−1·17633	9
·08	−9·09385	−1·17775	−142
4·10	−10·27160	−1·18041	−266
·12	−11·45201	−1·18429	−388
·14	−12·63630	−1·18928	−499
·16	−13·82558	−1·19536	−608
·18	−15·02094	−1·20250	−714
4·20	−16·22344	−1·21062	−812
·22	−17·43406	−1·21974	−912
·24	−18·65380	−1·22979	−1005
·26	−19·88359	−1·24076	−1097
·28	−21·12435	−1·25265	−1189
4·30	−22·37700	−1·26542	−1277
·32	−23·64242	−1·27906	−1364
·34	−24·92148	−1·29357	−1451
·36	−26·21505	−1·30895	−1538
·38	−27·52400	−1·32517	−1622
4·40	−28·84917	−1·34227	−1710
·42	−30·19144	−1·36020	−1793
·44	−31·55164	−1·37902	−1882
·46	−32·93066	−1·39869	−1967
·48	−34·32935	−1·41924	−2055

λ	$F_{13}(\lambda)$	Δ_1	Δ_2	$F_{14}(\lambda)$	Δ_1	Δ_2
5·00	66·01948	1·18443	0·04618	80·99398	2·35973	−0·05651
·02	67·20391	1·23276	4833	83·35371	2·41867	5894
·04	68·43667	1·28341	5065	85·77238	2·48022	6155
·06	69·72008	1·33653	5312	88·25260	2·54450	6428
·08	71·05661	1·39231	5578	90·79710	2·61171	6721
5·10	72·44892	1·45092	5861	93·40881	2·68204	7033
·12	73·89984	1·51260	6168	96·09085	2·75569	7365
·14	75·41244	1·57753	6493	98·84654	2·83288	7719
·16	76·98997	1·64599	6846	101·67942	2·91384	8096
·18	78·63596	1·71824	7225	104·59326	2·99888	8504
5·20	80·35420	1·79456	7632	107·59214	3·08824	8936
·22	82·14876	1·87529	8073	110·68038	3·18227	9403
·24	84·02405	1·96077	8548	113·86265	3·28132	9905
·26	85·98482	2·05138	9061	117·14397	3·38577	10445
·28	88·03620	2·14755	9617	120·52974	3·49604	11027
5·30	90·18375	2·24977	10222	124·02578	3·61262	11658
·32	92·43352	2·35853	10876	127·63840	3·73601	12339
·34	94·79205	2·47442	11589	131·37441	3·86679	13078
·36	97·26647	2·59808	12366	135·24120	4·00561	13882
·38	99·86455	2·73023	13215	139·24681	4·15317	14756
5·40	102·59478	2·87163	14140	143·39998	4·31028	15711
·42	105·46641	3·02321	15158	147·71026	4·47781	16753
·44	108·48962	3·18595	16274	152·18807	4·65678	17897
·46	111·67557	3·36096	17501	156·84485	4·84829	19151
·48	115·03653	3·54951	18855	161·69314	5·05362	20533

x				x						
4·50	—35·74859	—1·44070	2146	5·50	118·58604	3·75303	20352	—166·74676	5·27419	22057
·52	—37·18929	—1·46305	2235	·52	122·33907	3·97314	22011	—172·02095	5·51164	23745
·54	—38·65234	—1·48633	2328	·54	126·31221	4·21170	23856	—177·53259	5·76779	25615
·56	—40·13867	—1·51054	2421	·56	130·52391	4·47078	25908	—183·30038	6·04477	27698
·58	—41·64921	—1·53573	2519	·58	134·99469	4·75284	28206	—189·34515	6·34500	30023
4·60	—43·18494	—1·56190	2617	5·60	139·74753	5·06062	30778	—195·69015	6·67126	32626
·62	—44·74684	—1·58909	2719	·62	144·80815	5·39737	33675	—202·36141	7·02675	35549
·64	—46·33593	—1·61732	2823	·64	150·20552	5·76679	36942	—209·38816	7·41521	38846
·66	—47·95325	—1·64662	2930	·66	155·97231	6·17323	40644	—216·80337	7·84101	42580
·68	—49·59987	—1·67706	3044	·68	162·14554	6·62180	44857	—224·64438	8·30922	46821
4·70	—51·27693	—1·70862	3156	5·70	168·76734	7·11848	49668	—232·95360	8·82584	51662
·72	—52·98555	—1·74141	3279	·72	175·88582	7·67038	55190	—241·77944	9·39799	57215
·74	—54·72696	—1·77541	3400	·74	183·55620	8·28594	61556	—251·17743	10·03413	63614
·76	—56·50237	—1·81072	3531	·76	191·84214	8·97532	68938	—261·21156	10·74439	71026
·78	—58·31309	—1·84737	3665	·78	200·81746	9·75075	77543	—271·95595	11·54101	79662
4·80	—60·16046	—1·88543	3806	5·80	210·56821	10·62710	87635	—283·49696	12·43888	89787
·82	—62·04589	—1·92494	3951	·82	221·19531	11·62262	99552	—295·93584	13·45626	1·01738
·84	—63·97083	—1·96600	4106	·84	232·81793	12·75985	1·13723	—309·39210	14·61567	1·15941
·86	—65·93683	—2·00865	4265	·86	245·57778	14·06693	1·30708	—324·00777	15·94527	1·32960
·88	—67·94548	—2·05300	4435	·88	259·64471	15·57935	1·51242	—339·95304	17·48055	1·53528
4·90	—69·99848	—2·09912	4612	5·90	275·22406	17·34236	1·76301	—357·43359	19·26678	1·78623
·92	—72·09760	—2·14709	4797	·92	292·56642	19·41439	2·07203	—376·70037	21·36237	2·09559
·94	—74·24469	—2·19703	4994	·94	311·98081	21·87199	2·45760	—398·06274	23·84390	2·48153
·96	—76·44172	—2·24904	5201	·96	333·85280	24·81691	2·94492	—421·90664	26·81310	2·96920
·98	—78·69076	—2·30322	5418	·98	358·66971	28·38685	3·56994	—449·71974	30·40771	3·59461
5·00	—80·99398		5651	6·00	387·05656			—479·12745		

Tables of functions

$$F_{15}(\lambda), \ F_{16}(\lambda), \ F_{17}(\lambda)$$

for calculating the end moments and forces of a vibrating cantilever beam ($\lambda =$ = 0·00 − 4·00)

λ	$F_{15}(\lambda)$	Δ_1	Δ_2	$F_{16}(\lambda)$	Δ_1	Δ_2	$F_{17}(\lambda)$	Δ_1	Δ_2	λ
0·00	—0·00000	—0·00000	—0·00000	—0·00000	0·00000	0·00000	—0·00000	—0·00000	—0·00000	0·00
·02	—0·00000	0	0	—0·00000	0	0	—0·00000	0	0	·02
·04	—0·00000	0	0	—0·00000	1	1	—0·00000	1	1	·04
·06	—0·00000	1	1	—0·00001	1	0	—0·00001	3	2	·06
·08	—0·00001	2	1	—0·00002	3	2	—0·00004	6	3	·08
0·10	—0·00003	4	2	—0·00005	5	2	—0·00010	11	5	0·10
·12	—0·00007	6	2	—0·00010	9	4	—0·00021	17	6	·12
·14	—0·00013	9	3	—0·00019	14	5	—0·00038	28	11	·14
·16	—0·00022	13	4	—0·00033	19	5	—0·00066	39	11	·16
·18	—0·00035	18	5	—0·00052	28	9	—0·00105	55	16	·18
0·20	—0·00053	25	7	—0·00080	37	9	—0·00160	74	19	0·20
·22	—0·00078	33	8	—0·00117	49	12	—0·00234	98	24	·22
·24	—0·00111	41	8	—0·00166	63	14	—0·00332	125	27	·24
·26	—0·00152	53	12	—0·00229	78	15	—0·00457	158	33	·26
·28	—0·00205	65	12	—0·00307	98	20	—0·00615	195	37	·28
0·30	—0·00270	80	15	—0·00405	120	22	—0·00810	239	44	0·30
·32	—0·00350	96	16	—0·00525	144	24	—0·01049	288	49	·32
·34	—0·00446	115	19	—0·00669	172	28	—0·01337	344	56	·34
·36	—0·00561	135	20	—0·00841	203	31	—0·01681	406	62	·36
·38	—0·00696	159	24	—0·01044	238	35	—0·02087	476	70	·38
0·40	—0·00855	185	26	—0·01282	277	39	—0·02563	554	78	0·40
·42	—0·01040	213	28	—0·01559	320	43	—0·03117	638	84	·42
·44	—0·01253	245	32	—0·01879	367	47	—0·03755	733	95	·44
·46	—0·01498	279	34	—0·02246	418	51	—0·04488	835	102	·46
·48	—0·01777	317	38	—0·02664	475	57	—0·05323	947	112	·48

0·50	121	—	1068	—	—0·06270	61	536	0·03139	40	—	357	—	—0·02094	0·50
·52	133	—	1201	—	—0·07338	67	603	0·03675	45	—	402	—	—0·02451	·52
·54	143	—	1344	—	—0·08539	71	674	0·04278	49	—	451	—	—0·02853	·54
·56	154	—	1498	—	—0·09883	79	753	0·04952	51	—	502	—	—0·03304	·56
·58	166	—	1664	—	—0·11381	83	836	0·05705	56	—	558	—	—0·03806	·58
0·60	178	—	1842	—	—0·13045	91	927	0·06541	61	—	619	—	—0·04364	0·60
·62	191	—	2033	—	—0·14887	97	1024	0·07468	65	—	684	—	—0·04983	·62
·64	205	—	2238	—	—0·16920	103	1127	0·08492	70	—	754	—	—0·05667	·64
·66	218	—	2456	—	—0·19158	113	1240	0·09619	74	—	828	—	—0·06421	·66
·68	234	—	2690	—	—0·21614	118	1358	0·10859	80	—	908	—	—0·07249	·68
0·70	249	—	2939	—	—0·24304	129	1487	0·12217	86	—	994	—	—0·08157	0·70
·72	265	—	3204	—	—0·27243	135	1622	0·13704	92	—	1086	—	—0·09151	·72
·74	283	—	3487	—	—0·30447	146	1768	0·15326	97	—	1183	—	—0·10237	·74
·76	300	—	3787	—	—0·33934	156	1924	0·17094	105	—	1288	—	—0·11420	·76
·78	319	—	4106	—	—0·37721	165	2089	0·19018	112	—	1400	—	—0·12708	·78
0·80	340	—	4446	—	—0·41827	176	2265	0·21107	118	—	1518	—	—0·14108	0·80
·82	359	—	4805	—	—0·46273	189	2454	0·23372	128	—	1646	—	—0·15626	·82
·84	383	—	5188	—	—0·51078	201	2655	0·25826	136	—	1782	—	—0·17272	·84
·86	405	—	5593	—	—0·56266	214	2869	0·28481	144	—	1926	—	—0·19054	·86
·88	430	—	6023	—	—0·61859	227	3096	0·31350	155	—	2081	—	—0·20980	·88
0·90	457	—	6480	—	—0·67882	245	3341	0·34446	165	—	2246	—	—0·23061	0·90
·92	485	—	6965	—	—0·74362	259	3600	0·37787	176	—	2422	—	—0·25307	·92
·94	514	—	7479	—	—0·81327	277	3877	0·41387	189	—	2611	—	—0·27729	·94
·96	546	—	8025	—	—0·88806	297	4174	0·45264	202	—	2813	—	—0·30340	·96
·98	581	—	8606	—	—0·96831	316	4490	0·49438	217	—	3030	—	—0·33153	·98
1·00	616	—			—1·05437	340		0·53928	231	—			—0·36183	1·00

λ	Δ_2	Δ_1	$F_{17}(\lambda)$	Δ_2	Δ_1	$F_{16}(\lambda)$	Δ_2	Δ_1	$F_{15}(\lambda)$	λ
1·00	0·00616	0·09222	1·05437	0·00340	0·04830	0·53928	0·00231	0·03261	0·36183	1·00
·02	657	9879	1·14659	362	5192	0·58758	249	3510	0·39444	·02
·04	699	10578	1·24538	391	5583	0·63950	267	3777	0·42954	·04
·06	745	11323	1·35116	417	6000	0·69533	288	4065	0·46731	·06
·08	794	12117	1·46439	450	6450	0·75533	310	4375	0·50796	·08
1·10	850	12967	1·58556	485	6935	0·81983	334	4709	0·55171	1·10
·12	908	13875	1·71523	522	7457	0·88918	362	5071	0·59880	·12
·14	973	14848	1·85398	566	8023	0·96375	391	5462	0·64951	·14
·16	1046	15894	2·00246	611	8634	1·04398	425	5887	0·70413	·16
·18	1123	17017	2·16140	663	9297	1·13032	463	6350	0·76300	·18
1·20	1213	18230	2·33157	723	10020	1·22329	502	6852	0·82650	1·20
·22	1308	19538	2·51387	788	10808	1·32349	552	7404	0·89502	·22
·24	1418	20956	2·70925	861	11669	1·43157	602	8006	0·96906	·24
·26	1540	22496	2·91881	944	12613	1·54826	664	8670	1·04912	·26
·28	1680	24176	3·14377	1040	13653	1·67439	730	9400	1·13582	·28
1·30	1834	26010	3·38553	1147	14800	1·81092	809	10209	1·22982	1·30
·32	2013	28023	3·64563	1270	16070	1·95892	898	11107	1·33191	·32
·34	2220	30243	3·92586	1414	17484	2·11962	1001	12108	1·44298	·34
·36	2453	32696	4·22829	1579	19063	2·29446	1119	13227	1·56406	·36
·38	2730	35426	4·55525	1773	20836	2·48509	1260	14487	1·69633	·38
1·40	3049	38475	4·90951	1998	22834	2·69345	1423	15910	1·84120	1·40
·42	3427	41902	5·29426	2266	25100	2·92179	1616	17526	2·00030	·42
·44	3876	45778	5·71328	2587	27687	3·17279	1849	19375	2·17556	·44
·46	4413	50191	6·17106	2968	30655	3·44966	2125	21500	2·36931	·46
·48	5061	55252	6·67297	3435	34090	3·75621	2463	23963	2·58431	·48

x										x
1·50	2·82394	26837	2874	4·09711	38092	4002	7·22549	61107	5855	1·50
·52	3·09231	30222	3385	4·47803	42801	4709	7·83656	67941	6834	·52
·54	3·39453	34249	4027	4·90604	48390	5589	8·51597	76001	8060	·54
·56	3·73702	39088	4839	5·38994	55103	6713	9·27598	85613	9612	·56
·58	4·12790	44979	5891	5·94097	63262	8159	10·13211	97229	11616	·58
1·60	4·57769	52248	7269	6·57359	73322	10060	11·10440	1·11470	14241	1·60
·62	5·10017	61368	9120	7·30681	85929	12607	12·21910	1·29229	17759	·62
·64	5·71385	73022	11654	8·16610	1·02027	16098	13·51139	1·51804	22575	·64
·66	6·44407	88254	15232	9·18637	1·23051	21024	15·02943	1·81171	29367	·66
·68	7·32661	1·08694	20440	10·41688	1·51248	28197	16·84114	2·20420	39249	·68
1·70	8·41355	1·37034	28340	11·92936	1·90315	39067	19·04534	2·74647	54227	1·70
·72	9·78389	1·77944	40910	13·83251	2·46692	56377	21·79181	3·52709	78062	·72
·74	11·56333	2·40165	62221	16·29943	3·32403	85711	25·31890	4·71164	1·18455	·74
·76	13·96498	3·41623	1·01458	19·62346	4·72127	1·39724	30·03054	6·63977	1·92813	·76
·78	17·38121	5·24112	1·82489	24·34473	7·23389	2·51262	36·67031	10·10338	3·46361	·78
1·80	22·62233	9·05269	3·81157	31·57862	12·48124	5·24735	46·77369	17·33146	7·22808	1·80
·82	31·67502	19·37849	10·32580	44·05986	26·69546	14·21422	64·10515	36·90264	19·57118	·82
·84	51·05351	70·72170	51·34321	70·75532	97·37040	70·67494	-101·00779	-134·19239	97·28975	·84
·86	-121·77521			+168·12572			-235·20018			·86
·88	+385·37076			-529·94117			+725·52574			·88
1·90	77·70383	-307·66693	+274·12218	-106·41287	423·52830	-377·32989	142·36413	-583·16161	+519·39095	1·90
·92	44·15908	33·54475	20·68024	60·21446	46·19841	28·46566	78·59347	63·77066	39·17727	·92
·94	31·29457	12·86451	6·06428	42·48171	17·73275	8·34671	54·00008	24·59339	11·48327	·94
·96	24·49434	6·80023	2·59535	33·09567	9·38604	3·57166	40·88996	13·11012	4·91027	·96
·98	20·28946	4·20488	1·34836	27·28129	5·81438	1·85518	32·69011	8·19985	2·54738	·98
2·00	17·43294	2·85652	0·78988	23·32209	3·95920	1·08639	27·03764	5·65247	1·48897	2·00

λ	$F_{15}(\lambda)$	Δ_1	Δ_2	$F_{16}(\lambda)$	Δ_1	Δ_2	$F_{17}(\lambda)$	Δ_1	Δ_2	λ
2·00	17·43294	−2·06664	0·78988	−23·32209	2·87281	−1·08639	27·03764	−4·16350	1·48897	2·00
·02	15·36630	−1·56430	50234	−20·44928	2·18225	−0·69056	22·87414	−3·21953	0·94397	·02
·04	13·80200	−1·22510	33920	−18·26703	1·71627	46598	19·65461	−2·58483	63470	·04
·06	12·57690	−0·98534	23976	−16·55076	1·38719	32908	17·06978	−2·13875	44608	·06
·08	11·59156	80965	17569	−15·16357	1·14631	24088	14·93103	−1·81424	32451	·08
2·10	10·78191	67709	13256	−14·01726	0·96485	18146	13·11679	−1·57168	24256	2·10
·12	10·10482	57466	10243	−13·05241	82488	13997	11·54511	−1·38637	18531	·12
·14	9·53016	49388	8078	−12·22753	71472	11016	10·15874	−1·24230	14407	·14
·16	9·03628	42908	6480	−11·51281	62660	8812	8·91644	−1·12873	11357	·16
·18	8·60720	37632	5276	−10·88621	55509	7151	7·78771	−1·03817	9056	·18
2·20	8·23088	33282	4350	−10·33112	49633	5876	6·74954	−0·96537	7280	2·20
·22	7·89806	29655	3627	−9·83479	44755	4878	5·78417	90648	5889	·22
·24	7·60151	26599	3056	−9·38724	40667	4088	4·87769	85867	4781	·24
·26	7·33552	24004	2595	−8·98057	37216	3451	4·01902	81978	3889	·26
·28	7·09548	21781	2223	−8·60841	34281	2935	3·19924	78822	3156	·28
2·30	6·87767	19865	1916	−8·26560	31771	2510	2·41102	76268	2554	2·30
·32	6·67902	18203	1662	−7·94789	29612	2159	1·64834	74221	2047	·32
·34	6·49699	16751	1452	−7·65177	27749	1863	0·90613	72600	1621	·34
·36	6·32948	15480	1271	−7·37428	26135	1614	+0·18013	71342	1258	·36
·38	6·17468	14359	1121	−7·11293	24733	1402	−0·53329	71342	942	·38
2·40	6·03109	13367	992	−6·86560	23512	1221	−1·23729	70400	672	2·40
·42	5·89742	12487	880	−6·63048	22448	1064	−1·93457	69728	432	·42
·44	5·77255	11703	784	−6·40600	21520	928	−2·62753	69296	223	·44
·46	5·65552	11002	701	−6·19080	20710	810	−3·31826	69073	+34	·46
·48	5·54550	10375	627	−5·98370	20006	704	−4·00865	69039	−134	·48

2·50‥3·00	B (−)	C (−)	D (−)	E (±)	F (+)	G (−)	H (+)	I (−)	J (+)	2·50‥3·00
2·50	285	69458	4·70038	613	19393	5·78364	564	9811	5·44175	2·50
·52	425	69883	5·39496	530	18863	5·58971	507	9304	5·34364	·52
·54	553	70436	6·09379	458	18405	5·40108	457	8847	5·25060	·54
·56	669	71105	6·79815	391	18014	5·21703	412	8435	5·16213	·56
·58	779	71884	7·50920	332	17682	5·03689	373	8062	5·07778	·58
2·60	883	72767	8·22804	279	17403	4·86007	337	7725	4·99716	2·60
·62	980	73747	8·95571	228	17175	4·68604	305	7420	4·91991	·62
·64	1072	74819	9·69318	184	16991	4·51429	275	7145	4·84571	·64
·66	1161	75980	10·44137	143	16848	4·34438	250	6895	4·77426	·66
·68	1245	77225	11·20117	105	16743	4·17590	226	6669	4·70531	·68
2·70	1328	78553	11·97342	67	16676	4·00847	204	6465	4·63862	2·70
·72	1409	79962	12·75895	36	16640	3·84171	183	6282	4·57397	·72
·74	1486	81448	13·55857	3	16637	3·67531	165	6117	4·51115	·74
·76	1563	83011	14·37305	+26	16663	3·50894	149	5968	4·44998	·76
·78	1641	84652	15·20316	+53	16716	3·34231	133	5835	4·39030	·78
2·80	1716	86368	16·04968	+82	16798	3·17515	118	5717	4·33195	2·80
·82	1791	88159	16·91336	+106	16904	3·00717	104	5613	4·27478	·82
·84	1868	90027	17·79495	+132	17036	2·83813	92	5521	4·21865	·84
·86	1942	91969	18·69522	+156	17192	2·66777	80	5441	4·16344	·86
·88	2019	93988	19·61491	+178	17370	2·49585	68	5373	4·10903	·88
2·90	2098	96086	20·55479	+202	17572	2·32215	58	5315	4·05530	2·90
·92	2175	98261	21·51565	+224	17796	2·14643	48	5267	4·00215	·92
·94	2255	1·00516	22·49826	+246	18042	1·96847	39	5228	3·94948	·94
·96	2337	1·02853	23·50342	+268	18310	1·78805	28	5200	3·89720	·96
·98	2420	1·05273	24·53195	+291	18601	1·60495	20	5180	3·84520	·98
3·00	2506		25·58468	+312		1·41894	12		3·79340	3·00

λ	$F_{15}(\lambda)$	Δ_1	Δ_2	$F_{16}(\lambda)$	Δ_1	Δ_2	$F_{17}(\lambda)$	Δ_1	Δ_2	λ
3·00	3·79340	—0·05168	0·00012	—1·41894	0·18913	0·00312	—25·58468	—1·07779	—0·02506	3·00
·02	3·74172	— 5164	+ 4	—1·22981	19246	333	—26·66247	—1·10372	— 2593	·02
·04	3·69008	— 5170	— 6	—1·03735	19602	356	—27·76619	—1·13056	— 2684	·04
·06	3·63838	— 5182	— 12	—0·84133	19981	379	—28·89675	—1·15832	— 2776	·06
·08	3·58656	— 5202	— 20	—0·64152	20381	400	—30·05507	—1·18705	— 2873	·08
3·10	3·53454	— 5230	— 28	—0·43771	20806	425	—31·24212	—1·21677	— 2972	3·10
·12	3·48224	— 5265	— 35	—0·22965	21253	447	—32·45889	—1·24753	— 3076	·12
·14	3·42959	— 5308	— 43	—0·01712	21724	471	—33·70642	—1·27935	— 3182	·14
·16	3·37651	— 5359	— 51	+0·20012	22221	497	—34·98577	—1·31229	— 3294	·16
·18	3·32292	— 5416	— 57	0·42233	22743	522	—36·29806	—1·34638	— 3409	·18
3·20	3·26876	— 5481	— 65	0·64976	23291	548	—37·64444	—1·38169	— 3531	3·20
·22	3·21395	— 5553	— 72	0·88267	23867	576	—39·02613	—1·41826	— 3657	·22
·24	3·15842	— 5634	— 81	1·12134	24470	603	—40·44439	—1·45615	— 3789	·24
·26	3·10208	— 5722	— 88	1·36604	25104	634	—41·90054	—1·49540	— 3925	·26
·28	3·04486	— 5818	— 96	1·61708	25768	664	—43·39594	—1·53614	— 4074	·28
3·30	2·98668	— 5922	— 104	1·87476	26464	696	—44·93208	—1·57837	— 4223	3·30
·32	2·92746	— 6034	— 112	2·13940	27194	730	—46·51045	—1·62220	— 4383	·32
·34	2·86712	— 6156	— 122	2·41134	27958	764	—48·13265	—1·66771	— 4551	·34
·36	2·80556	— 6285	— 129	2·69092	28760	802	—49·80036	—1·71500	— 4729	·36
·38	2·74271	— 6424	— 139	2·97852	29600	840	—51·51536	—1·76416	— 5916	·38
3·40	2·67847	— 6573	— 149	3·27452	30482	882	—53·27952	—1·81528	— 5112	3·40
·42	2·61274	— 6731	— 158	3·57934	31406	924	—55·09480	—1·86849	— 5321	·42
·44	2·54543	— 6901	— 170	3·89340	32377	971	—56·96329	—1·92392	— 5543	·44
·46	2·47642	— 7081	— 180	4·21717	33395	1018	—58·88721	—1·98169	— 5777	·46
·48	2·40561	— 7272	— 191	4·55112	34465	1070	—60·86890	—2·04194	— 6025	·48

arg														arg
3·50	2·33289	—	7476	—	204	4·89577	35589	1124	62·91084	—	2·10484	—	6290	3·50
·52	2·25813	—	7692	—	216	5·25166	36772	1183	65·01568	—	2·17055	—	6571	·52
·54	2·18121	—	7923	—	231	5·61938	38015	1243	67·18623	—	2·23926	—	6871	·54
·56	2·10198	—	8167	—	244	5·99953	39326	1311	69·42549	—	2·31118	—	7192	·56
·58	2·02031	—	8427	—	260	6·39279	40706	1380	71·73667	—	2·38651	—	7533	·58
3·60	1·93604	—	8703	—	276	6·79985	42162	1456	74·12318	—	2·46552	—	7901	3·60
·62	1·84901	—	8997	—	294	7·22147	43699	1537	76·58870	—	2·54845	—	8293	·62
·64	1·75904	—	9308	—	311	7·65846	45323	1624	79·13715	—	2·63559	—	8714	·64
·66	1·66596	—	9642	—	334	8·11169	47040	1717	81·77274	—	2·72729	—	9170	·66
·68	1·56954	—	9994	—	352	8·58209	48858	1818	84·50003	—	2·82385	—	9656	·68
3·70	1·46960	—	10372	—	378	9·07067	50785	1927	87·32388	—	2·92570	—	10185	3·70
·72	1·36588	—	10773	—	401	9·57852	52830	2045	90·24958	—	3·03325	—	10755	·72
·74	1·25815	—	11202	—	429	10·10682	55001	2171	93·28283	—	3·14696	—	11371	·74
·76	1·14613	—	11661	—	459	10·65683	57312	2311	96·42979	—	3·26737	—	12041	·76
·78	1·02952	—	12150	—	489	11·22995	59773	2461	99·69716	—	3·39507	—	12770	·78
3·80	0·90802	—	12674	—	524	11·82768	62398	2625	103·09223	—	3·53068	—	13561	3·80
·82	0·78128	—	13237	—	563	12·45166	65202	2804	106·62291	—	3·67496	—	14428	·82
·84	0·64891	—	13840	—	603	13·10368	68202	3000	110·29787	—	3·82868	—	15372	·84
·86	0·51051	—	14488	—	648	13·78570	71417	3215	114·12655	—	3·99280	—	16412	·86
·88	0·36563	—	15186	—	698	14·49987	74869	3452	118·11935	—	4·16829	—	17549	·88
3·90	0·21377	—	15940	—	754	15·24856	78581	3712	122·28764	—	4·35536	—	18707	3·90
·92	+0·05437	—	16752	—	812	16·03437	82582	4001	126·64300	—	4·55929	—	20393	·92
·94	−0·11315	—	17633	—	881	16·86019	86901	4319	131·20229	—	4·77557	—	21628	·94
·96	−0·28948	—	18587	—	954	17·72920	91576	4675	135·97786	—	5·00992	—	23435	·96
·98	−0·47535	—	19624	—	1037	18·64496	96644	5068	140·98778	—	5·26326	—	25334	·98
4·00	−0·67159					19·61140			146·25104					4·00

10.2.2 Functions $\Phi(\lambda)$

Tables of functions

$$\Phi_1(\lambda),\ \Phi_2(\lambda),\ \Phi_3(\lambda),\ \Phi_4(\lambda),\ \Phi_5(\lambda),\ \Phi_6(\lambda)$$

for a vibrating beam clamped at both ends $(\lambda = 0{\cdot}00 - 6{\cdot}00)$

λ	Δ_2	Δ_1	Φ_3	Δ_2	Δ_1	Φ_2	Δ_2	Δ_1	Φ_1	λ
0·00	-0·00000	-0·00000	-0·00000	0·00000	0·00000	0·00000	-0·00000	-0·00000	-0·00000	0·00
·02	0	0	-0·00000	0	0	0·00000	0	0	-0·00000	·02
·04	0	0	-0·00000	0	0	0·00000	0	0	-0·00000	·04
·06	0	0	-0·00000	0	0	0·00000	0	0	-0·00000	·06
·08	0	0	-0·00000	0	0	0·00000	0	0	-0·00000	·08
0·10	1	1	-0·00001	0	0	0·00000	0	0	-0·00000	0·10
·12	1	0	-0·00001	0	0	0·00000	0	0	-0·00000	·12
·14	1	1	-0·00002	1	1	0·00001	1	1	-0·00000	·14
·16	0	1	-0·00003	1	0	0·00001	1	0	-0·00001	·16
·18	1	2	-0·00005	0	0	0·00001	1	1	-0·00001	·18
0·20	0	2	-0·00007	1	1	0·00002	1	0	-0·00002	0·20
·22	1	3	-0·00010	0	0	0·00002	1	1	-0·00002	·22
·24	1	4	-0·00014	0	1	0·00003	0	1	-0·00003	·24
·26	1	5	-0·00019	1	1	0·00004	1	2	-0·00004	·26
·28	1	6	-0·00025	0	2	0·00006	1	1	-0·00006	·28
0·30	1	7	-0·00032	0	2	0·00008	1	2	-0·00007	0·30
·32	2	9	-0·00041	1	2	0·00010	1	2	-0·00009	·32
·34	2	11	-0·00052	0	3	0·00013	0	3	-0·00012	·34
·36	1	12	-0·00064	1	3	0·00016	0	3	-0·00015	·36
·38	3	15	-0·00079	0	4	0·00020	1	3	-0·00018	·38
0·40	2	17	-0·00096	2	4	0·00024	1	4	-0·00022	0·40
·42	3	20	-0·00116	0	6	0·00030	0	5	-0·00027	·42
·44	3	23	-0·00139	1	6	0·00036	1	5	-0·00032	·44
·46	2	25	-0·00164	0	7	0·00043	1	6	-0·00038	·46
·48	4	29		3	10	0·00050		7		·48

0·50	4	—	33	—	−0·00193	0	10	0·00060	0	—	7	—	−0·00045	0·50
·52	4	—	37	—	−0·00226	1	11	0·00070	2	—	9	—	−0·00052	·52
·54	5	—	42	—	−0·00263	2	13	0·00081	0	—	9	—	−0·00061	·54
·56	4	—	46	—	−0·00305	1	14	0·00094	2	—	11	—	−0·00070	·56
·58	4	—	50	—	−0·00351	1	15	0·00108	1	—	12	—	−0·00081	·58
0·60	7	—	57	—	−0·00401	3	18	0·00123	2	+	14	—	−0·00093	0·60
·62	5	—	62	—	−0·00458	1	19	0·00141	1	—	13	—	−0·00107	·62
·64	6	—	68	—	−0·00520	2	21	0·00160	3	—	16	—	−0·00120	·64
·66	6	—	74	—	−0·00588	2	23	0·00181	1	—	17	—	−0·00136	·66
·68	8	—	82	—	−0·00662	2	25	0·00204	2	—	19	—	−0·00153	·68
0·70	7	—	89	—	−0·00744	2	27	0·00229	1	—	20	—	−0·00172	0·70
·72	7	—	96	—	−0·00833	3	30	0·00256	2	—	22	—	−0·00192	·72
·74	9	—	105	—	−0·00929	2	32	0·00286	3	—	25	—	−0·00214	·74
·76	9	—	114	—	−0·01034	3	35	0·00318	1	—	26	—	−0·00239	·76
·78	8	—	122	—	−0·01148	3	38	0·00353	2	—	28	—	−0·00265	·78
0·80	10	—	132	—	−0·01270	2	40	0·00391	3	—	31	—	−0·00293	0·80
·82	11	—	143	—	−0·01402	4	44	0·00431	1	—	32	—	−0·00324	·82
·84	9	—	152	—	−0·01545	3	47	0·00475	4	—	36	—	−0·00356	·84
·86	12	—	164	—	−0·01697	3	50	0·00522	1	—	37	—	−0·00392	·86
·88	12	—	176	—	−0·01861	4	54	0·00572	4	—	41	—	−0·00429	·88
0·90	12	—	188	—	−0·02037	4	58	0·00626	2	—	43	—	−0·00470	0·90
·92	12	—	200	—	−0·02225	4	62	0·00684	4	—	47	—	−0·00513	·92
·94	14	—	214	—	−0·02425	3	65	0·00746	4	—	49	—	−0·00560	·95
·96	14	—	228	—	−0·02639	5	70	0·00811	2	—	52	—	−0·00609	·96
·98	15	—	243	—	−0·02867	5	75	0·00881	3	—	56	—	−0·00661	·98
1·00	14				−0·03110	4		0·00956	4				−0·00717	1·00

λ	Δ_2	Δ_1	Φ_6	Δ_2	Δ_1	Φ_5	Δ_2	Δ_1	Φ_4	λ
0·00	0·00000	0·00000	0·00000	0·00000	0·00000	0·00000	0·00000	0·00000	−0·00000	0·00
·02	0	0	0·00000	0	0	0·00000	−0	−0	−0·00000	·02
·04	0	0	0·00000	1	0	0·00000	−0	−0	−0·00000	·04
·06	1	1	0·00000	1	1	0·00000	−0	−0	−0·00000	·06
·08	2	3	0·00001	2	0	0·00001	+1	−1	−0·00001	·08
0·10	1	4	0·00004	0	2	0·00001	−1	−0	−0·00001	0·10
·12	2	6	0·00008	1	2	0·00003	−1	−1	−0·00002	·12
·14	4	10	0·00014	3	3	0·00005	−0	−1	−0·00003	·14
·16	5	15	0·00024	1	6	0·00008	−1	−2	−0·00005	·16
·18	5	20	0·00039	2	7	0·00014	−1	−3	−0·00008	·18
0·20	8	28	0·00059	4	9	0·00021	−1	−4	−0·00012	0·20
·22	8	36	0·00087	3	13	0·00030	−1	−5	−0·00017	·22
·24	11	47	0·00123	4	16	0·00043	−2	−7	−0·00024	·24
·26	11	58	0·00170	5	20	0·00059	−1	−8	−0·00032	·26
·28	15	73	0·00228	6	25	0·00079	−2	−10	−0·00042	·28
0·30	15	88	0·00301	6	31	0·00104	−3	−13	−0·00055	0·30
·32	19	107	0·00389	7	37	0·00135	−2	−15	−0·00070	·32
·34	21	128	0·00496	8	44	0·00172	−3	−18	−0·00088	·34
·36	22	150	0·00624	9	52	0·00216	−3	−21	−0·00109	·36
·38	27	177	0·00774	10	61	0·00268	−4	−25	−0·00134	·38
0·40	28	205	0·00951	11	71	0·00329	−4	−29	−0·00163	0·40
·42	31	236	0·01156	12	82	0·00400	−4	−33	−0·00196	·42
·44	35	271	0·01392	13	94	0·00482	−6	−39	−0·00235	·44
·46	38	309	0·01663	14	107	0·00576	−4	−43	−0·00278	·46
·48	41	350	0·01972		121	0·00683	−6	−49		·48

0·50	44	394	0·02322	15	136	0·00804	7	—	56	—	—0·00327	0·50
·52	49	443	0·02716	18	154	0·00940	6	—	62	—	—0·00383	·52
·54	51	494	0·03159	17	171	0·01094	8	—	70	—	—0·00445	·54
·56	57	551	0·03653	20	191	0·01265	8	—	78	—	—0·00515	·56
·58	60	611	0·04204	20	211	0·01456	8	—	86	—	—0·00593	·58
0·60	64	675	0·04815	23	234	0·01667	9	—	95	—	—0·00679	·60
·62	68	743	0·05490	24	258	0·01901	10	—	105	—	—0·00774	·62
·64	74	817	0·06233	25	283	0·02159	10	—	115	—	—0·00879	·64
·66	78	895	0·07050	27	310	0·02442	12	—	127	—	—0·00994	·66
·68	82	977	0·07945	29	339	0·02752	11	—	138	—	—0·01121	·68
0·70	88	1065	0·08922	30	369	0·03091	12	—	150	—	—0·01259	·70
·72	92	1157	0·09987	32	401	0·03460	13	—	163	—	—0·01409	·72
·74	99	1256	0·11144	35	436	0·03861	14	—	177	—	—0·01572	·74
·76	102	1358	0·12400	35	471	0·04297	15	—	192	—	—0·01749	·76
·78	110	1468	0·13758	38	509	0·04768	15	—	207	—	—0·01941	·78
0·80	114	1582	0·15226	41	550	0·05277	16	—	223	—	—0·02148	·80
·82	120	1702	0·16808	41	591	0·05827	18	—	241	—	—0·02371	·82
·84	127	1829	0·18510	44	635	0·06418	17	—	258	—	—0·02612	·84
·86	133	1962	0·20339	46	681	0·07053	19	—	277	—	—0·02870	·86
·88	138	2100	0·22301	49	730	0·07734	19	—	296	—	—0·03147	·88
0·90	145	2245	0·24401	51	781	0·08464	21	—	317	—	—0·03443	·90
·92	153	2398	0·26646	52	833	0·09245	22	—	339	—	—0·03760	·92
·94	158	2556	0·29044	57	890	0·10078	22	—	361	—	—0·04099	·94
·96	165	2721	0·31600	57	947	0·10968	24	—	385	—	—0·04460	·96
·98	174	2895	0·34321	61	1008	0·11915	23	—	408	—	—0·04845	·98
1·00	179		0·37216	63		0·12923	27	—		—	—0·05253	1·00

λ	Φ₁	Δ₁	Δ₂	Φ₂	Δ₁	Δ₂	Φ₃	Δ₁	Δ₂	λ
1·00	−0·00717	−0·00060	−0·00004	0·00956	0·00079	0·00004	−0·03110	−0·00257	−0·00014	1·00
·02	−0·00777	63	3	0·01035	84	5	−0·03367	274	−0·00017	·02
·04	−0·00840	67	4	0·01119	88	4	−0·03641	290	16	·04
·06	−0·00907	71	4	0·01207	95	7	−0·03931	307	17	·06
·08	−0·00978	74	3	0·01302	99	4	−0·04238	325	18	·08
1·10	−0·01052	80	6	0·01401	106	7	−0·04563	343	18	1·10
·12	−0·01132	83	3	0·01507	111	5	−0·04906	363	20	·12
·14	−0·01215	89	6	0·01618	117	6	−0·05269	383	20	·14
·16	−0·01304	93	4	0·01735	124	7	−0·05652	404	21	·16
·18	−0·01397	98	5	0·01859	130	6	−0·06056	425	21	·18
1·20	−0·01495	103	5	0·01989	137	7	−0·06481	447	22	1·20
·22	−0·01598	108	5	0·02126	144	7	−0·06928	471	24	·22
·24	−0·01706	114	6	0·02270	151	7	−0·07399	495	24	·24
·26	−0·01820	120	6	0·02421	159	8	−0·07894	520	25	·26
·28	−0·01940	126	6	0·02580	167	8	−0·08414	545	25	·28
1·30	−0·02066	132	6	0·02747	175	8	−0·08959	572	27	1·30
·32	−0·02198	138	6	0·02922	183	8	−0·09531	600	28	·32
·34	−0·02336	145	7	0·03105	191	8	−0·10131	629	29	·34
·36	−0·02481	151	6	0·03296	201	10	−0·10760	658	29	·36
·38	−0·02632	159	8	0·03497	210	9	−0·11418	688	30	·38
1·40	−0·02791	166	7	0·03707	220	10	−0·12106	720	32	1·40
·42	−0·02957	173	7	0·03927	229	9	−0·12826	753	33	·42
·44	−0·03130	181	8	0·04156	239	10	−0·13579	787	34	·44
·46	−0·03311	189	8	0·04395	250	11	−0·14366	822	35	·46
·48	−0·03500	198	9	0·04645	261	11	−0·15188	857	35	·48

	−δ²	−δ		+δ²	+δ		−δ²	−δ		
1·50	38	895	−0.16045	11	272	0.04906	8	206	−0.03698	1·50
·52	39	934	−0.16940	11	283	0.05178	8	214	−0.03904	·52
·54	39	973	−0.17874	12	295	0.05461	10	224	−0.04118	·54
·56	41	1014	−0.18847	13	308	0.05756	10	234	−0.04342	·56
·58	43	1057	−0.19861	12	320	0.06064	9	243	−0.04576	·58
1·60	43	1100	−0.20918	13	333	0.06384	10	253	−0.04819	1·60
·62	45	1145	−0.22018	13	346	0.06717	10	263	−0.05072	·62
·64	47	1192	−0.23163	14	360	0.07063	11	274	−0.05335	·64
·66	49	1241	−0.24355	15	375	0.07423	11	285	−0.05609	·66
·68	48	1289	−0.25596	14	389	0.07798	12	297	−0.05894	·68
1·70	52	1341	−0.26885	15	404	0.08187	11	308	−0.06191	1·70
·72	53	1394	−0.28226	16	420	0.08591	12	320	−0.06499	·72
·74	55	1449	−0.29620	15	435	0.09011	12	332	−0.06819	·74
·76	55	1504	−0.31069	17	452	0.09446	14	346	−0.07151	·76
·78	59	1563	−0.32573	18	470	0.09898	13	359	−0.07497	·78
1·80	60	1623	−0.34136	16	486	0.10368	13	372	−0.07856	1·80
·82	62	1685	−0.35759	19	505	0.10854	15	387	−0.08228	·82
·84	64	1749	−0.37444	18	523	0.11359	14	401	−0.08615	·84
·86	66	1815	−0.39193	19	542	0.11882	15	416	−0.09016	·86
·88	68	1883	−0.41008	20	562	0.12424	16	432	−0.09432	·88
1·90	71	1954	−0.42891	20	582	0.12986	16	448	−0.09864	1·90
·92	72	2026	−0.44845	21	603	0.13568	16	464	−0.10312	·92
·94	75	2101	−0.46871	22	625	0.14171	17	481	−0.10776	·94
·96	78	2179	−0.48972	21	646	0.14796	18	499	−0.11257	·96
·98	80	2259	−0.51151	24	670	0.15442	18	517	−0.11756	·98
2·00	82		−0.53410	23		0.16112	19		−0.12273	2·00

λ	Δ_2	Δ_1	Φ_6	Δ_2	Δ_1	Φ_5	Δ_2	Δ_1	Φ_4	λ
1·00	0·00179	0·03074	0·37216	0·00063	0·01071	0·12923	-0·00027	-0·00435	-0·05253	1·00
·02	188	3262	0·40290	67	1138	0·13994	26	461	-0·05688	·02
·04	194	3456	0·43552	67	1205	0·15132	27	488	-0·06149	·04
·06	204	3660	0·47008	72	1277	0·16337	30	518	-0·06637	·06
·08	210	3870	0·50668	75	1352	0·17614	29	547	-0·07155	·08
1·10	218	4088	0·54538	77	1429	0·18966	31	578	-0·07702	1·10
·12	228	4316	0·58626	80	1509	0·20395	33	611	-0·08280	·12
·14	235	4551	0·62942	84	1593	0·21904	33	644	-0·08891	·14
·16	243	4794	0·67493	86	1679	0·23497	34	678	-0·09535	·16
·18	254	5048	0·72287	91	1770	0·25176	37	715	-0·10213	·18
1·20	261	5309	0·77335	93	1863	0·26946	37	752	-0·10928	1·20
·22	271	5580	0·82644	96	1959	0·28809	38	790	-0·11680	·22
·24	280	5860	0·88224	101	2060	0·30768	41	831	-0·12470	·24
·26	290	6150	0·94084	104	2164	0·32828	40	871	-0·13301	·26
·28	300	6450	1·00234	108	2272	0·34992	43	914	-0·14172	·28
1·30	309	6759	1·06684	111	2383	0·37264	45	959	-0·15086	1·30
·32	319	7078	1·13443	116	2499	0·39647	45	1004	-0·16045	·32
·34	330	7408	1·20521	119	2618	0·42146	48	1052	-0·17049	·34
·36	340	7748	1·27929	125	2743	0·44764	48	1100	-0·18101	·36
·38	352	8100	1·35677	126	2869	0·47507	50	1150	-0·19201	·38
1·40	361	8461	1·43777	134	3003	0·50376	53	1203	-0·20351	1·40
·42	374	8835	1·52238	136	3139	0·53379	53	1256	-0·21554	·42
·44	385	9220	1·61073	142	3281	0·56518	55	1311	-0·22810	·44
·46	396	9616	1·70293	147	3428	0·59799	57	1368	-0·24121	·46
·48	408	10024	1·79909	151	3579	0·63227	59	1427	-0·25489	·48

1·50	421	10445	1·89933	156	3735	0·66806	61	—	1488	—	—0·26916	1·50
·52	433	10878	2·00378	161	3896	0·70541	62	—	1550	—	—0·28404	·52
·54	445	11323	2·11256	167	4063	0·74437	65	—	1615	—	—0·29954	·54
·56	458	11781	2·22579	173	4236	0·78500	66	—	1681	—	—0·31569	·56
·58	472	12253	2·34360	178	4414	0·82736	68	—	1749	—	—0·33250	·58
1·60	485	12738	2·46613	184	4598	0·87150	71	—	1820	—	—0·34999	1·60
·62	499	13237	2·59361	190	4788	0·91748	72	—	1892	—	—0·36819	·62
·64	513	13750	2·72588	197	4985	0·96536	75	—	1967	—	—0·38711	·64
·66	528	14278	2·86338	202	5187	1·01521	76	—	2043	—	—0·40678	·66
·68	541	14819	3·00616	209	5396	1·06708	81	—	2124	—	—0·42721	·68
1·70	558	15377	3·15435	217	5613	1·12104	80	—	2204	—	—0·44845	1·70
·72	571	15948	3·30812	224	5837	1·17717	84	—	2288	—	—0·47049	·72
·74	589	16537	3·46760	229	6066	1·23554	86	—	2374	—	—0·49337	·74
·76	604	17141	3·63297	240	6306	1·29620	90	—	2464	—	—0·51711	·76
·78	621	17762	3·80438	245	6551	1·35926	90	—	2554	—	—0·54175	·78
1·80	637	18399	3·98200	255	6806	1·42477	95	—	2649	—	—0·56729	1·80
·82	655	19054	4·16599	262	7068	1·49283	96	—	2745	—	—0·59378	·82
·84	672	19726	4·35653	272	7340	1·56351	100	—	2845	—	—0·62123	·84
·86	690	20416	4·55379	280	7620	1·63691	102	—	2947	—	—0·64968	·86
·88	709	21125	4·75795	290	7910	1·71311	106	—	3053	—	—0·67915	·88
1·90	727	21852	4·96920	299	8209	1·79221	108	—	3161	—	—0·70968	1·90
·92	748	22600	5·18772	310	8519	1·87420	112	—	3273	—	—0·74129	·92
·94	767	23367	5·41372	319	8838	1·95949	115	—	3388	—	—0·77402	·94
·96	788	24155	5·64739	331	9169	2·04787	118	—	3506	—	—0·80790	·96
·98	808	24963	5·88894	342	9511	2·13956	121	—	3627	—	—0·84296	·98
2·00	830		6·13857	353		2·23467	126	—		—	—0·87923	2·00

λ	Φ₁	Δ₁	Δ₂	Φ₂	Δ₁	Δ₂	Φ₃	Δ₁	Δ₂	λ
2·00	—0·12273	—0·00536	—0·00019	0·16112	0·00693	0·00023	—0·53410	—0·02341	—0·00082	2·00
·02	—0·12809	556	20	0·16805	717	24	—0·55751	2427	86	·02
·04	—0·13365	575	19	0·17522	742	25	—0·58178	2515	88	·04
·06	—0·13940	596	21	0·18264	768	26	—0·60693	2607	92	·06
·08	—0·14536	617	21	0·19032	794	26	—0·63300	2701	94	·08
2·10	—0·15153	640	23	0·19826	822	28	—0·66001	2799	98	2·10
·12	—0·15793	662	22	0·20648	850	28	—0·68800	2899	100	·12
·14	—0·16455	686	24	0·21498	879	29	—0·71699	3004	105	·14
·16	—0·17141	710	24	0·22377	909	30	—0·74703	3112	108	·16
·18	—0·17851	736	26	0·23286	940	31	—0·77815	3223	111	·18
2·20	—0·18587	761	25	0·24226	972	32	—0·81038	3339	116·	2·20
·22	—0·19348	789	28	0·25198	1005	33	—0·84377	3459	120	·22
·24	—0·20137	817	28	0·26203	1039	34	—0·87836	3583	124	·24
·26	—0·20954	845	28	0·27242	1073	34	—0·91419	3710	127	·26
·28	—0·21799	875	30	0·28315	1110	37	—0·95129	3844	134	·28
2·30	—0·22674	906	31	0·29425	1148	38	—1·02955	3982	138	2·30
·32	—0·23580	939	33	0·30573	1186	38	—1·02955	4124	142	·32
·34	—0·24519	971	32	0·31759	1226	40	—1·07079	4271	147	·34
·36	—0·25490	1006	35	0·32985	1267	41	—1·11350	4426	155	·36
·38	—0·26496	1042	36	0·34252	1310	43	—1·15776	4584	158	·38
2·40	—0·27538	1078	36	0·35562	1354	44	—1·20360	4749	165	2·40
·42	—0·28616	1117	39	0·36916	1399	45	—1·25109	4921	172	·42
·44	—0·29733	1157	40	0·38315	1446	47	—1·30030	5098	177	·44
·46	—0·30890	1198	41	0·39761	1496	50	—1·35128	5282	184	·46
·48	—0·32088	1241	43	0·41257	1546	50	—1·40410	5475	193	·48

2·50	198	5673	—1·45885	51	1597	0·42803	44	1285	—0·33329	2·50
·52	208	5881	—1·51558	56	1653	0·44400	47	1332	—0·34614	·52
·54	214	6095	—1·57439	55	1708	0·46053	47	1379	—0·35946	·54
·56	225	6320	—1·63534	58	1766	0·47761	51	1430	—0·37325	·56
·58	232	6552	—1·69854	61	1827	0·49527	51	1481	—0·38755	·58
2·60	243	6795	—1·76406	62	1889	0·51354	55	1536	—0·40236	2·60
·62	251	7046	—1·83201	65	1954	0·53243	55	1591	—0·41772	·62
·64	264	7310	—1·90247	67	2021	0·55197	59	1650	—0·43363	·64
·66	273	7583	—1·97557	70	2091	0·57218	60	1710	—0·45013	·66
·68	286	7869	—2·05140	73	2164	0·59309	65	1775	—0·46723	·68
2·70	297	8166	—2·13009	75	2239	0·61473	65	1840	—0·48498	2·70
·72	310	8476	—2·21175	78	2317	0·63712	68	1908	—0·50338	·72
·74	324	8800	—2·29651	82	2399	0·66029	73	1981	—0·55246	·74
·76	338	9138	—2·38451	84	2483	0·68428	74	2055	—0·54227	·76
·78	353	9491	—2·47589	89	2572	0·70911	79	2134	—0·56282	·78
2·80	369	9860	—2·57080	91	2663	0·73483	81	2215	—0·58416	2·80
·82	385	10245	—2·66940	96	2759	0·76146	85	2300	—0·60631	·82
·84	403	10648	—2·77185	100	2859	0·78905	89	2389	—0·62931	·84
·86	421	11069	—2·87833	103	2962	0·81764	94	2483	—0·65320	·86
·88	442	11511	—2·98902	109	3071	0·84726	96	2579	—0·67803	·88
2·90	462	11973	—3·10413	113	3184	0·87797	103	2682	—0·70382	2·90
·92	485	12458	—3·22386	118	3302	0·90981	106	2788	—0·73064	·92
·94	506	12964	—3·34844	123	3425	0·94283	112	2900	—0·75852	·94
·96	533	13497	—2·47808	129	3554	0·97708	117	3017	—0·78752	·96
·98	558	14055	—3·61305	135	3689	1·01262	123	3140	—0·81769	·98
3·00	586		—3·75360	141		1·04951	128		—0·84909	3·00

λ	Δ_2	Δ_1	Φ_6	Δ_2	Δ_1	Φ_5	Δ_2	Δ_1	Φ_4	λ
2·00	0·00830	0·25793	6·13857	0·00353	0·09864	2·23467	−0·00126	−0·03753	−0·87923	2·00
·02	852	26645	6·39650	366	10230	2·33331	−128	−3881	−0·91676	·02
·04	874	27519	6·66295	377	10607	2·43561	−133	−4014	−0·95557	·04
·06	899	28418	6·93814	392	10999	2·54168	−137	−4151	−0·99571	·06
·08	922	29340	7·22232	405	11404	2·65167	−140	−4291	−1·03722	·08
2·10	947	30287	7·51572	418	11822	2·76571	−144	−4435	−1·08013	2·10
·12	971	31258	7·81859	434	12256	2·88393	−150	−4585	−1·12448	·12
·14	999	32257	8·13117	448	12704	3·00649	−153	−4738	−1·17033	·14
·16	1024	33281	8·45374	465	13169	3·13353	−158	−4896	−1·21771	·16
·18	1053	34334	8·78655	482	13651	3·26522	−163	−5059	−1·26667	·18
2·20	1082	35416	9·12989	498	14149	3·40173	−167	−5226	−1·31726	2·20
·22	1110	36526	9·48405	516	14665	3·54322	−174	−5400	−1·36952	·22
·24	1140	37666	9·84931	536	15201	3·68987	−177	−5577	−1·42352	·24
·26	1173	38839	10·22597	555	15756	3·84188	−184	−5761	−1·47929	·26
·28	1204	40043	10·61436	576	16332	3·99944	−188	−5949	−1·53690	·28
2·30	1237	41280	11·01479	596	16928	4·16276	−196	−6145	−1·59639	2·30
·32	1272	42552	11·42759	620	17548	4·33204	−202	−6347	−1·65784	·32
·34	1308	43860	11·85311	642	18190	4·50752	−206	−6553	−1·72131	·34
·36	1345	45205	12·29171	668	18858	4·68942	−215	−6768	−1·78684	·36
·38	1382	46587	12·74376	691	19549	4·87800	−221	−6989	−1·85452	·38
2·40	1421	48008	13·20963	720	20269	5·07349	−228	−7217	−1·92441	2·40
·42	1464	49472	13·68971	748	21017	5·27618	−236	−7453	−1·99658	·42
·44	1504	50976	14·18443	775	21792	5·48635	−243	−7696	−2·07111	·44
·46	1549	52525	14·69419	808	22600	5·70427	−251	−7947	−2·14807	·46
·48	1594	54119	15·21944	839	23439	5·93027	−260	−8207	−2·22754	·48

2·50	1642	55761	15·76063	874	24313	6·16466	268	—	8207	—2·30961	2·50
·52	1690	57451	16·31824	906	25219	6·40779	278	—	8753	—2·39436	·52
·54	1742	59193	16·89275	946	26165	6·65998	287	—	9040	—2·48189	·54
·56	1794	60987	17·48468	985	27150	6·92163	297	—	9337	—2·57229	·56
·58	1850	62837	18·09455	1024	28174	7·19313	307	—	9644	—2·66566	·58
2·60	1907	64744	18·72292	1069	29243	7·47487	319	—	9963	—2·76210	2·60
·62	1966	66710	19·37036	1112	30355	7·76730	329	—	10292	—2·86173	·62
·64	2029	68739	20·03746	1161	31516	8·07085	342	—	10634	—2·96465	·64
·66	2093	70832	20·72485	1210	32726	8·38601	353	—	10987	—3·07099	·66
·68	2162	72994	21·43317	1262	33988	8·71327	367	—	11354	—3·18086	·68
2·70	2233	75227	22·16311	1319	35307	9·05315	381	—	11735	—3·29440	2·70
·72	2305	77532	22·91538	1376	36683	9·40622	395	—	12130	—3·41175	·72
·74	2383	79915	23·69070	1437	38120	9·77305	409	—	12539	—3·53305	·74
·76	2464	82379	24·48985	1503	39623	10·15425	426	—	12965	—3·65844	·76
·78	2549	84928	25·31364	1570	41193	10·55048	442	—	13407	—3·78809	·78
2·80	2637	87565	26·16292	1642	42835	10·96241	459	—	13866	—3·92216	2·80
·82	2730	90295	27·03857	1720	44555	11·39076	479	—	14345	—4·06082	·82
·84	2827	93122	27·94152	1798	46353	11·83631	496	—	14841	—4·20427	·84
·86	2929	96051	28·87274	1883	48236	12·29984	517	—	15358	—4·35268	·86
·88	3037	99088	29·83325	1975	50211	12·78220	540	—	15898	—4·50626	·88
2·90	3150	1·02238	30·82413	2069	52280	13·28431	560	—	16458	—4·66524	2·90
·92	3267	1·05505	31·84651	2168	54448	13·80711	586	—	17044	—4·82982	·92
·94	3394	1·08899	32·90156	2278	56726	14·35159	609	—	17653	—5·00026	·94
·96	3525	1·12424	33·99055	2388	59114	14·91885	637	—	18290	—5·17679	·96
·98	3664	1·16088	35·11479	2510	61624	15·50999	665	—	18955	—5·35969	·98
3·00	3809		36·27567	2636		16·12623	693	—		—5·54924	3·00

514

λ	Δ_2	Δ_1	Φ_3	Δ_2	Δ_1	Φ_2	Δ_2	Δ_1	Φ_1	λ
3·00	—0·00586	—0·14641	—3·75360	0·00141	0·03830	1·04951	—0·00128	—0·03268	—0·84909	3·00
·02	—616	—15257	—3·90001	148	3978	1·08781	—136	—3404	—0·88177	·02
·04	—646	—15903	—4·05258	154	4132	1·12759	—142	—3546	—0·91581	·04
·06	—681	—16584	—4·21161	163	4295	1·16891	—149	—3695	—0·95127	·06
·08	—716	—17300	—4·37745	169	4464	1·21186	—157	—3852	—0·98822	·08
3·10	—753	—18053	—4·55045	179	4643	1·25650	—165	—4017	—1·02674	3·10
·12	—793	—18846	—4·73098	188	4831	1·30293	—174	—4191	—1·06691	·12
·14	—837	—19683	—4·91944	196	5027	1·35124	—182	—4373	—1·10882	·14
·16	—882	—20565	—5·11627	208	5235	1·40151	—194	—4567	—1·15255	·16
·18	—930	—21495	—5·32192	217	5452	1·45386	—203	—4770	—1·19822	·18
3·20	—984	—22479	—5·53687	230	5682	1·50838	—215	—4985	—1·24592	3·20
·22	—1038	—23571	—5·76166	242	5924	1·56520	—227	—5212	—1·29577	·22
·24	—1099	—24616	—5·99683	255	6179	1·62444	—240	—5452	—1·34789	·24
·26	—1162	—25778	—6·24299	268	6447	1·68623	—253	—5705	—1·40241	·26
·28	—1232	—27010	—6·50077	286	6733	1·75070	—269	—5974	—1·45946	·28
3·30	—1304	—28314	—6·77087	299	7032	1·81803	—284	—6258	—1·51920	3·30
·32	—1384	—29698	—7·05401	319	7351	1·88835	—302	—6560	—1·58178	·32
·34	—1468	—31166	—7·35099	335	7686	1·96186	—319	—6879	—1·64738	·34
·36	—1562	—32728	—7·66265	358	8044	2·03872	—340	—7219	—1·71617	·36
·38	—1655	—34383	—7·98993	377	8421	2·11916	—361	—7580	—1·78836	·38
3·40	—1768	—36151	—8·33376	402	8823	2·20337	—384	—7964	—1·86416	3·40
·42	—1879	—38030	—8·69527	426	9249	2·29160	—408	—8372	—1·94380	·42
·44	—2005	—40035	—9·07557	455	9704	2·38409	—436	—8808	—2·02752	·44
·46	—2138	—42173	—9·47592	482	10186	2·48113	—464	—9272	—2·11560	·46
·48	—2285	—44458	—9·89765	515	10701	2·58299	—497	—9769	—2·20832	·48

3·50	2442			—10·34223	550		2·69000	529				—2·30601	3·50
·52	2614	—	46900	—10·81123	586	11251	2·80251	567	—	10298	—	—2·40899	·52
·54	2801	—	49514	—11·30637	627	11837	2·92088	608	—	10865	—	—2·51764	·54
·56	3005	—	52315	—11·82952	673	12464	3·04552	651	—	11473	—	—2·63237	·56
·58	3226	—	55320	—12·38272	719	13137	3·17689	699	—	12124	—	—2·75361	·58
3·60	3468	—	58546	—12·96818	773	13856	3·31545	752	—	12823	—	—2·88184	3·60
·62	3735	—	62014	—13·58832	830	14629	3·46174	808	—	13575	—	—3·01759	·62
·64	4025	—	65749	—14·24581	893	15459	3·61633	871	—	14383	—	—3·16142	·64
·66	4345	—	69774	—14·94355	963	16352	3·77985	940	—	15254	—	—3·31396	·66
·68	4694	—	74119	—15·68474	1038	17315	3·95300	1017	—	16194	—	—3·47590	·68
3·70	5083	—	78813	—16·47287	1123	18353	4·13653	1098	—	17211	—	—3·64801	3·70
·72	5508	—	83896	—17·31183	1215	19476	4·33129	1191	—	18309	—	—3·83110	·72
·74	5980	—	89404	—18·20587	1316	20691	4·53820	1293	—	19500	—	—4·02610	·74
·76	6503	—	95384	—19·15971	1431	22007	4·75827	1405	—	20793	—	—4·23403	·76
·78	7084	—	1·01887	—20·17858	1554	23438	4·99265	1531	—	22198	—	—4·45601	·78
3·80	7728	—	1·08971	—21·26829	1696	24992	5·24257	1668	—	23729	—	—4·69330	3·80
·82	8449	—	1·16699	—22·43528	1851	26688	5·50945	1825	—	25397	—	—4·94727	·82
·84	9253	—	1·25148	—23·68676	2023	28539	5·79484	1998	—	27222	—	—5·21949	·84
·86	10155	—	1·34401	—25·03077	2220	30562	6·10046	2191	—	29220	—	—5·51169	·86
·88	11167	—	1·44556	—26·47633	2437	32782	6·42828	2410	—	31411	—	—5·82580	·88
3·90	12308	—	1·55723	—28·03356	2684	35219	6·78047	2655	—	33821	—	—6·16401	3·90
·92	13596	—	1·68031	—29·71387	2961	37903	7·15950	2933	—	36476	—	—6·52877	·92
·94	15056	—	1·81627	—31·53014	3277	40864	7·56814	3247	—	39409	—	—6·92286	·94
·96	16716	—	1·96683	—33·49697	3635	44141	8·00955	3605	—	42656	—	—7·34942	·96
·98	18606	—	2·13399	—35·63096	4041	47776	8·48731	4010	—	46261	—	—7·81203	·98
4·00	20771	—	2·32005	—37·95101	4510	51817	9·00548	4478	—	50271	—	—8·31474	4·00

λ	Δ_2	Δ_1	Φ_6	Δ_2	Δ_1	Φ_5	Δ_2	Δ_1	Φ_4	λ
3·00	0·03809	1·19897	36·27567	0·02636	0·64260	16·12623	−0·00693	−0·19648	— 5·54924	3·00
·02	3965	1·23862	37·47464	2772	67032	16·76883	727	20375	— 5·74572	·02
·04	4128	1·27990	38·71326	2916	69948	17·43915	758	21133	— 5·94947	·04
·06	4300	1·32290	39·99316	3069	73017	18·13863	795	21928	— 6·16080	·06
·08	4485	1·36775	41·31606	3231	76248	18·86880	831	22759	— 6·38008	·08
3·10	4677	1·41452	42·68381	3404	79652	19·63128	872	23631	— 6·60767	3·10
·12	4884	1·46336	44·09833	3590	83242	20·42780	913	24544	— 6·84398	·12
·14	5101	1·51437	45·56169	3786	87028	21·26022	959	25503	— 7·08942	·14
·16	5333	1·56770	47·07606	3997	91025	22·13050	1006	26509	— 7·34445	·16
·18	5580	1·62350	48·64376	4221	95246	23·04075	1057	27566	— 7·60954	·18
3·20	5842	1·68192	50·26726	4461	99707	23·99321	1113	28679	— 7·88520	3·20
·22	6121	1·74313	51·94918	4719	1·04426	24·99028	1169	29848	— 8·17199	·22
·24	6421	1·80734	53·69231	4993	1·09419	26·03454	1232	31080	— 8·47047	·24
·26	6731	1·87465	55·49965	5285	1·14704	27·12873	1297	32377	— 8·78127	·26
·28	7089	1·94554	57·37430	5611	1·20315	28·27577	1372	33749	— 9·10504	·28
3·30	7436	2·01990	59·31984	5940	1·26255	29·47892	1443	35192	— 9·44253	3·30
·32	7831	2·09821	61·33974	6310	1·32565	30·74147	1528	36720	— 9·79445	·32
·34	8248	2·18069	63·43795	6700	1·39265	32·06712	1613	38333	—10·16165	·34
·36	8696	2·26765	65·61864	7125	1·46390	33·45977	1710	40043	—10·54498	·36
·38	9177	2·35942	67·88629	7580	1·53970	34·92367	1808	41851	—10·94541	·38
3·40	9696	2·45638	70·24571	8071	1·62041	36·46337	1920	43771	—11·36392	3·40
·42	10253	2·55891	72·70209	8602	1·70643	38·08378	2036	45807	—11·80163	·42
·44	10854	2·66745	75·26100	9176	1·79819	39·79021	2163	47970	—12·25970	·44
·46	11504	2·78249	77·92845	9799	1·89618	41·58840	2302	50272	—12·73940	·46
·48	12204	2·90453	80·71094	10471	2·00089	43·48458	2448	52720	—13·24212	·48

3·50	−13·76932		2611	45·48547		11204	83·61547		12967
·52	−14·32263	−·55331	2785	47·59840	2·11293	11998	86·64967	3·03420	13790
·54	−14·90379	−·58116	2975	49·83131	2·23291	12863	89·82177	3·17210	14682
·56	−15·51470	−·61091	3181	52·19285	2·36154	13805	93·14069	3·31892	15658
·58	−16·15742	−·64272	3408	54·69244	2·49959	14833	96·61619	3·47550	16714
3·60	−16·83422	−·67680	3652	57·34036	2·64792	15957	100·25883	3·64264	17870
·62	−17·54754	−·71332	3922	60·14785	2·80749	17185	104·08017	3·82134	19132
·64	−18·30008	−·75254	4215	63·12719	2·97934	18535	108·09283	4·01266	20511
·66	−19·09477	−·79469	4539	66·29188	3·16469	20013	112·31060	4·21777	22025
·68	−19·93485	−·84008	4893	69·65670	3·36482	21640	116·74862	4·43802	23685
3·70	−20·82386	−·88901	5283	73·23792	3·58122	23434	121·42349	4·67487	25512
·72	−21·76570	−·94184	5715	77·05348	3·81556	25412	126·35348	4·92999	27528
·74	−22·76469	−·99899	6189	81·12316	4·06968	27599	131·55875	5·20527	29750
·76	−23·82557	−1·06088	6716	85·46883	4·34567	30026	137·06152	5·50277	32211
·78	−24·95361	−1·12804	7300	90·11476	4·64593	32716	142·88640	5·82488	34941
3·80	−26·15465	−1·20104	7950	95·08785	4·97309	35708	149·06069	6·17429	37974
·82	−27·43519	−1·28054	8675	100·41802	5·33017	39059	155·61472	6·55403	41353
·84	−28·80248	−1·36729	9482	106·13878	5·72076	42783	162·58228	6·96756	45128
·86	−30·26459	−1·46211	10390	112·28737	6·14859	46970	170·00112	7·41884	49351
·88	−31·83060	−1·56601	11405	118·90566	6·61829	51671	177·91347	7·91235	54095
3·90	−33·51066	−1·68006	12551	126·04066	7·13500	56967	186·36677	8·45330	59431
·92	−35·31623	−1·80557	13846	133·74533	7·70467	62949	195·41438	9·04761	65458
·94	−37·26026	−1·94403	15308	142·07949	8·33416	69727	205·11657	9·70219	72278
·96	−39·35737	−2·09711	16973	151·11092	9·03143	77431	215·54154	10·42497	80030
·98	−41·62421	−2·26684	18870	160·91666	9·80574	86215	226·76681	11·22527	88859
4·00	−44·07975	−2·45554	21037	171·58455	10·66789	96269	238·88067	12·11386	98958

λ	Φ₁	Δ₁	Δ₂	Φ₂	Δ₁	Δ₂	Φ₃	Δ₁	Δ₂	λ
4·00	8·31474	0·54749	0·04478	9·00548	0·56327	0·04510	37·95101	2·52776	0·20771	4·00
·02	8·86223	59761	5012	9·56875	61370	0·05043	40·47877	2·76032	0·23256	·02
·04	9·45984	65391	5630	10·18245	67034	5664	43·23909	3·02155	26123	·04
·06	10·11375	71733	6342	10·85279	73408	6374	46·26064	3·31593	29438	·06
·08	10·83108	78897	7164	11·58687	80615	7207	49·57657	3·64888	33295	·08
4·10	11·62005	87057	8160	12·39302	88793	8178	53·22545	4·02686	37798	4·10
·12	12·49062	96328	9271	13·28095	98108	9315	57·25231	4·45770	43084	·12
·14	13·45390	1·06950	10622	14·26203	1·08767	10659	61·71001	4·95090	49320	·14
·16	14·52340	1·19164	12214	15·34970	1·21017	12250	66·66091	5·51808	56718	·16
·18	15·71504	1·33279	14115	16·55987	1·35170	14153	72·17899	6·17358	65550	·18
4·20	17·04783	1·49675	16396	17·91157	1·51606	16436	78·35257	6·93514	76156	4·20
·22	18·54458	1·68834	19159	19·42763	1·70791	19185	85·28771	7·82500	88986	·22
·24	20·23292	1·91357	22523	21·13554	1·93380	22589	93·11271	8·87119	1·04619	·24
·26	22·14649	2·18011	26654	23·06934	2·20063	26683	101·98390	10·10944	1·23825	·26
·28	24·32660	2·49788	31777	25·26997	2·51882	31819	112·09334	11·58576	1·47632	·28
4·30	26·82448	2·87971	38183	27·78879	2·90108	38226	123·67910	13·35953	1·77377	4·30
·32	29·70419	3·34254	46283	30·68987	3·36435	46327	137·03863	15·51000	2·15047	·32
·34	33·04673	3·90895	56641	34·05422	3·93122	56687	152·54863	18·14169	2·63169	·34
·36	36·95568	4·60948	70053	37·98544	4·63222	70100	170·69032	21·39675	3·25506	·36
·38	41·56516	5·48621	87673	42·61766	5·50940	87718	192·08707	25·47050	4·07375	·38
4·40	47·05137	6·59795	1·11174	48·12706	6·62164	1·11224	217·55757	30·63650	5·16600	4·40
·42	53·64932	8·02881	1·43086	54·74870	8·05299	1·43135	248·19407	37·28549	6·64899	·42
·44	61·67813	9·90169	1·87288	62·80169	9·92637	1·87338	285·47956	45·98864	8·70315	·44
·46	71·57982	12·40088	2·49919	72·72806	12·42608	2·49971	331·46820	57·60239	11·61375	·46
·48	83·98070	15·81098	3·41010	85·15414	15·83673	3·41065	389·07059	73·44952	15·84713	·48

4·50	99·79168	20·58711	4·77613	100·99087	20·61339	4·77666	462·52011	95·64477	22·19525	4·50
·52	120·37879	27·48652	6·89941	121·60426	27·51336	6·89997	558·16488	127·70759	32·06282	·52
·54	147·86531	37·83175	10·34523	149·11762	37·85916	10·34580	685·87247	175·78397	48·07638	·54
·56	185·69706	54·07110	16·23935	186·97678	54·09909	16·23993	861·65644	251·25207	75·46810	·56
·58	239·76816	81·07778	27·00668	241·07587	81·10637	27·00728	1112·90851	376·75859	125·50652	·58
4·60	320·84594	129·49290	48·41512	322·18224	129·52210	48·41573	1489·66710	601·75634	224·99775	4·60
·62	450·33884	225·59406	96·10116	451·70434	225·62392	96·10182	2091·42344	1048·36426	446·60792	·62
·64	675·93290	446·49547	220·90141	677·32826	446·52590	220·90198	3139·78770	2074·95283	1026·58857	·64
·66	1122·42837	1086·75990	640·26443	1123·85416	1086·79102	640·26512	5214·74053	5050·4373.	2975·4845.	·66
·68	2209·18827	3948·5226.	2861·7627.	2210·64518	3948·5544.	2861·7634.	10265·1778.	18349·8359.	13299·3986.	·68
4·70	6157·7108.	49203·8111.	45255·2885.	6159·1995.	49203·8440.	45255·2896.	28615·0137.	—228663·628..	—210313·793..	4·70
·72	55361·5219.	—50239·3729.	—	—55363·0435.	—50239·3398.	—	—257278·641..	—	—	·72
·74	56509·6408.	—	—46232·8195.	56511·1962.	—	46232·8211.	—262614·229..	233476·246..	—	·74
·76	6270·2679.	4006·5534.	2901·6174.	6271·8564.	4006·5187.	2901·6180.	29137·9839.	18619·5967.	—214856·650..	·76
·78	2263·71459	1104·93609	649·49854	2265·33775	1104·90072	649·49943	10518·3872.	5134·9837.	—13484·6130..	·78
4·80	1158·77850	455·43755	224·47313	1160·43703	455·40129	224·47392	5383·40356	2116·58561	3018·3981.	4·80
·82	703·34095	230·96442	97·86682	705·03574	230·92737	97·86766	3266·81795	1073·39825	1043·18736	·82
·84	472·37653	133·09760	49·42219	474·10837	133·05971	49·42304	2193·41970	618·58523	454·81302	·84
·86	339·27893	83·67541	27·63759	341·04866	83·63667	27·63844	1574·83447	388·90778	229·67745	·86
·88	255·60352	56·03782	16·66158	257·41199	55·99823	16·66253	1185·92669	260·46949	128·43829	·88
4·90	199·56570	39·37624	10·64219	201·41376	39·33570	10·64308	925·45720	183·03967	77·42982	4·90
·92	160·18946	28·73405	7·11635	162·07806	28·69262	7·11730	742·41753	133·58377	49·45590	·92
·94	131·45541	21·61770	4·93953	133·38544	21·57532	4·94052	608·83376	100·51334	33·07043	·94
·96	109·83771	16·67817	3·53634	111·81012	16·63480	3·53733	508·32042	77·55926	22·95408	·96
·98	93·15954	13·14183	2·59866	95·17532	13·09747	2·59970	430·76116	61·12631	16·43295	·98
5·00	80·01771	—	—	82·07785	—	—	369·63485	—	12·07538	5·00

λ	Φ_4	Δ_1	Δ_2	Φ_5	Δ_1	Δ_2	Φ_6	Δ_1	Δ_2	λ
4·00	44·07975	2·66591	0·21037	171·58455	11·63058	0·96269	238·88067	13·10344	0·98958	4·00
·02	46·74566	2·90126	23535	183·21513	12·70870	1·07812	251·98411	14·20894	1·10550	·02
·04	49·64692	3·16525	26399	195·92383	13·91991	1·21121	266·19305	15·44806	1·23912	·04
·06	52·81217	3·46249	29724	209·84374	15·28521	1·36530	281·64111	16·84172	1·39366	·06
·08	56·27466	3·79833	33584	225·12895	16·82962	1·54441	298·48283	18·41504	1·57332	·08
4·10	60·07299	4·17928	38095	241·95857	18·58327	1·75365	316·89787	20·19811	1·78307	4·10
·12	64·25227	4·61315	43387	260·54184	20·58246	1·99919	337·09598	22·22727	2·02916	·12
·14	68·86542	5·10943	49628	281·12430	22·87140	2·28894	359·32325	24·54672	2·31945	·14
·16	73·97485	5·67977	57034	303·99570	25·50409	2·63269	383·86997	27·21050	2·66378	·16
·18	79·65462	6·33848	65871	329·49979	28·54709	3·04300	411·08047	30·28516	3·07466	·18
4·20	85·99310	7·10333	76485	358·04688	32·08294	3·53585	441·36563	33·85327	3·56811	4·20
·22	93·09643	7·99654	89321	390·12982	36·21494	4·13200	475·21890	38·01813	4·16486	·22
·24	101·09297	9·04617	1·04963	426·34476	41·07337	4·85843	513·23703	42·91004	4·89191	·24
·26	110·13914	10·28790	1·24173	467·41813	46·82422	5·75085	556·14707	48·69501	5·78497	·26
·28	120·42704	11·76770	1·47980	514·24235	53·68094	6·85672	604·84208	55·58650	6·89149	·28
4·30	132·19474	13·54531	1·77761	567·92329	61·92123	8·24029	660·42858	63·86228	8·27578	4·30
·32	145·74005	15·69942	2·15411	629·84452	71·91070	9·98947	724·29086	73·88783	10·02555	·32
·34	161·43947	18·33492	2·63550	701·75522	84·13692	12·22622	798·17869	86·15092	12·26309	·34
·36	179·77439	21·59387	3·25895	785·89214	99·25997	15·12305	884·32961	101·31151	15·16059	·36
·38	201·36826	25·67159	4·07772	885·15211	118·18761	18·92764	985·64112	120·27748	18·96597	·38
4·40	227·03985	30·84168	5·17009	1003·33972	142·19104	24·00343	1105·91860	144·31996	24·04248	4·40
·42	257·88153	37·49481	6·65313	1145·53076	173·08635	30·89531	1250·23856	175·25515	30·93519	·42
·44	295·37634	46·20222	8·70741	1318·61711	213·52770	40·44135	1425·49371	215·73713	40·48198	·44
·46	341·57856	57·82032	11·61810	1532·14481	267·49521	53·96751	1641·23084	269·74608	54·00895	·46
·48	399·39888	73·67189	15·85157	1799·64002	341·13647	73·64126	1910·97692	343·42974	73·68366	·48

x										x
4·50	473·07077	95·87169	22·19980	2140·77649	444·27892	103·14245	2254·40666	446·61532	103·18558	4·50
·52	568·94246	127·93916	32·06747	2585·05541	593·27820	148·99928	2701·02198	595·65907	149·04375	·52
·54	696·88162	176·02032	48·08116	3178·33361	816·69785	223·41965	3296·68105	819·12322	223·46415	·54
·56	872·90194	251·49318	75·47286	3995·03146	1167·41269	350·71484	4115·80427	1169·88403	350·76081	·56
·58	1124·39512	377·00490	125·51172	5162·44415	1750·67147	583·25878	5285·63830	1753·19001	583·30598	·58
4·60	1501·40002	602·00764	225·00274	6913·11562	2796·29203	1045·62056	7038·87831	2798·8576.	1045·6676.	4·60
·62	2103·40766	1048·62079	446·61315	9709·40765	4871·7958.	2075·5038.	9837·7359.	4874·4112.	2075·5536.	·62
·64	3152·02845	2075·21471	1026·59392	14581·2034.	9642·6300.	4770·8342.	14712·1471.	9645·2955.	4770·8843.	·64
·66	5227·24316	5050·7048.	2975·4901.	24223·8334.	23470·520..	13827·890..	24357·4426.	23473·238..	13827·943..	·66
·68	10277·9479.	18350·109...	12399·405..	47694·353..	85276·477..	61805·957..	47830·680..	85279·250..	61806·012..	·68
4·70	28628·056..	−228663·91...	−210313·81...	132970·830..	1062662·61...	977386·14...	133109·930..	1062665·43...	977386·18...	4·70
·72	−257291·96...			1195633·44...	−1085028·73...		1195775·36...	−1085025·77...		·72
·74	−262627·83...	+233475·91...	−214856·57...	1220429·12...		998498·06...	1220573·89...		998498·07...	·74
·76	29151·924..	18619·343..	13484·663..	135400·398..	86530·679..	62666·700.	135548·124..	86527·709.	62666·796.	·76
·78	10532·5811.	5134·6804.	3018·4046.	48869·719..	23863·979..	14027·320..	49020·415..	23860·913.	14027·362..	·78
4·80	5397·90075	2116·27589	1043·19408	25005·7403.	9336·6592.	4847·9715.	25159·5026.	9833·5510.	4848·0319.	4·80
·82	3281·62486	1073·08181	454·81987	15169·0811.	4988·6877.	2113·6331.	15325·9516.	4985·5191.	2113·6964.	·82
·84	2208·54305	618·26194	229·68446	10180·3934.	2875·0546.	1067·3663.	10340·4325.	2871·8227.	1067·4295.	·84
·86	1590·28111	388·57748	128·44538	7305·33888	1807·68838	596·87959	7468·60982	1804·39323	596·94458	·86
·88	1201·70363	260·13210	77·43740	5497·65050	1210·80879	359·82883	5664·21659	1207·44865	359·89729	·88
4·90	941·57153	182·69470	49·46353	4286·84171	850·97996	229·82601	4456·76794	847·55136	229·89426	4·90
·92	758·87683	133·23117	33·07816	3435·86175	621·15395	153·67803	3609·21658	617·65710	153·74850	·92
·94	625·64566	100·15301	22·96209	2814·70780	467·47587	106·66418	2991·55948	463·90860	106·73600	·94
·96	525·49265	77·19092	16·44119	2347·23193	360·81169	76·35845	2527·65088	357·17260	76·43263	·96
·98	448·30173	60·74973	12·08385	1986·42024	284·45324	56·10743	2170·47828	280·73997	56·18326	·98
5·00	387·55200			1701·96700			1889·73831			5·00

λ	Φ₁	Δ₁	Δ₂	Φ₂	Δ₁	Δ₂	Φ₃	Δ₁	Δ₂	λ
5·00	−80·01771	10·54317	−2·59866	82·07785	−10·49777	2·59970	−369·63485	49·05093	−12·07538	5·00
·02	−69·47454	8·59044	−1·95273	71·58008	−8·54398	1·95379	−320·58392	39·97750	−9·07343	·02
·04	−60·88410	7·09454	−1·49590	63·03610	−7·04699	1·49699	−280·60642	33·02712	−6·95038	·04
·06	−53·78956	5·92913	−1·16541	55·98911	−5·88046	1·16653	−247·57930	27·61263	−5·41449	·06
·08	−47·86043	5·00762	−0·92151	50·10865	−4·95779	0·92267	−219·96667	23·33173	−4·28090	·08
5·10	−42·85281	4·26933	73829	45·15086	−4·21832	73947	−196·63494	19·90227	−3·42946	5·10
·12	−38·58348	3·67083	59850	40·93254	−3·61858	59974	−176·73267	17·12250	−2·77977	·12
·14	−34·91265	3·18048	49035	37·31396	−3·12718	49140	−159·61017	14·84548	−2·27702	·14
·16	−31·73217	2·77491	40557	34·18678	−2·71991	40727	−144·76469	12·96246	−1·88302	·16
·18	−28·95726	2·43654	33837	31·46687	−2·38040	33951	−131·80232	11·39179	−1·57067	·18
5·20	−26·52072	2·15201	28453	29·08647	−2·09449	28591	−120·41044	10·07134	−1·32045	5·20
·22	−24·36871	1·91099	24102	26·99198	−1·85205	24244	−110·33910	8·95329	−1·11805	·22
·24	−22·45772	1·70552	20547	25·13993	−1·64511	20694	−101·38581	8·00037	−0·95292	·24
·26	−20·75220	1·52924	17638	23·49482	−1·46732	17779	−93·38544	7·18329	81708	·26
·28	−19·22296	1·37719	15205	22·02750	−1·31371	15361	−86·20215	6·47879	70450	·28
5·30	−17·84577	1·24533	13186	20·71379	−1·18024	13347	−79·72336	5·86828	61051	5·30
·32	−16·60044	1·13044	11489	19·53355	−1·06370	11654	−73·85508	5·33668	53160	·32
·34	−15·47000	1·02991	10053	18·46985	−0·96144	10226	−68·51840	4·87182	46486	·34
·36	−14·44009	0·94154	8837	17·50841	87131	9013	−63·64658	4·46366	40816	·36
·38	−13·49855	86358	7796	16·63710	79152	7979	−59·18292	4·10395	35971	·38
5·40	−12·63497	79457	6901	15·84558	72063	7089	−55·07897	3·78588	31807	5·40
·42	−11·84040	73326	6131	15·12495	65736	6327	−51·29309	3·50372	28216	·42
·44	−11·10714	67864	5462	14·46759	60071	5665	−47·78937	3·25271	25101	·44
·46	−10·42850	62982	4882	13·86688	54982	5089	−44·53666	3·02883	22388	·46
·48	−9·79868	58608	4374	13·31706	50392	4590	−41·50783	2·82864	20019	·48

5·50	9·21260	54681	3927	12·81314	46241	4151	—38·67919	2·64928	17936	5·50
·52	8·66579	51144	3537	12·35073	42474	3767	—36·02991	2·48826	16102	·52
·54	8·15435	47955	3189	11·92599	39045	3429	—33·54165	2·34346	14480	·54
·56	7·67480	45073	2882	11·53554	35916	3129	—31·19819	2·21306	13040	·56
·58	7·22407	42463	2610	11·17638	33049	2867	—28·98513	2·09551	11755	·58
5·60	6·79944	40098	2365	10·84589	30417	2632	—26·88962	1·98940	10611	5·60
·62	6·39846	37951	2147	10·54172	27995	2422	—24·90022	1·89361	9579	·62
·64	6·01895	35999	1952	10·26177	25758	2237	—23·00661	1·80705	8656	·64
·66	5·65896	34224	1775	10·00419	23686	2072	—21·19956	1·72888	7817	·66
·68	5·31672	32608	1616	9·76733	21762	1924	—19·47068	1·65827	7061	·68
5·70	4·99064	31137	1471	9·54971	19972	1790	—17·81241	1·59455	6372	5·70
·72	4·67927	29796	1341	9·34999	18300	1672	—16·21786	1·53712	5743	·72
·74	4·38131	28577	1219	9·16699	16735	1565	—14·68074	1·48545	5167	·74
·76	4·09554	27464	1113	8·99964	15266	1469	—13·19529	1·43905	4640	·76
·78	3·82090	26454	1010	8·84698	13883	1383	—11·75624	1·39755	4150	·78
5·80	3·55636	25535	919	8·70815	12576	1307	—10·35869	1·36056	3699	5·80
·82	3·30101	24700	835	8·58239	11341	1235	—8·99813	1·32773	3283	·82
·84	3·05401	23946	754	8·46898	10166	1175	—7·67040	1·29885	2888	·84
·86	2·81455	23263	683	8·36732	9048	1118	—6·37155	1·27362	2523	·86
·88	2·58192	22650	613	8·27684	7980	1068	—5·09793	1·25185	2177	·88
5·90	2·35542	22099	551	8·19704	6956	1024	—3·84608	1·23333	1852	5·90
·92	2·13443	21609	490	8·12748	5972	984	—2·61275	1·21790	1543	·92
·94	1·91834	21174	435	8·06776	5023	949	—1·39485	1·20544	1246	·94
·96	1·70660	20794	380	8·01753	4106	917	—0·18941	1·19579	965	·96
·98	1·49866	20464	330	7·97647	3216	890	+1·00638	1·18888	691	·98
6·00	1·29402			7·94431			+2·19526			6·00

λ	Δ₂	Δ₁	Φ₆	Δ₂	Δ₁	Φ₅	Δ₂	Δ₁	Φ₄	λ
5·00	56·18326	—224·55671	1889·73831	56·10743	—228·34581	1701·96700	—12·08385	48·66588	—387·55200	5·00
·02	42·23370	—182·32301	1665·18160	42·15595	—186·18986	1473·62119	— 9·08212	39·58376	—338·88612	·02
·04	32·36908	—149·95393	1482·85859	32·28936	—153·90050	1287·43133	— 6·95931	32·62445	—299·30236	·04
·06	25·23300	—124·72093	1332·90466	25·15123	—128·74927	1133·53083	— 5·42367	27·20078	—266·67791	·06
·08	19·96677	—104·75416	1208·18373	19·88287	—108·86640	1004·78156	— 4·29037	22·91041	—239·47713	·08
5·10	16·01167	—88·74249	1103·42957	15·92550	—92·94090	895·91516	— 3·43916	19·47125	—216·56672	5·10
·12	12·99424	—75·74825	1014·68708	12·90586	—80·03504	802·97426	— 2·78978	16·68147	—197·09547	·12
·14	10·65976	—65·08849	938·93883	10·56901	—69·46603	722·93922	— 2·28731	14·39416	—180·41400	·14
·16	8·83077	—56·25772	873·85034	8·73756	—60·72847	653·47319	— 1·89359	12·50057	—166·01984	·16
·18	7·38129	—48·87643	817·59262	7·28550	—53·44297	592·74472	— 1·58159	10·91898	—153·51927	·18
5·20	6·22056	—42·65587	768·71619	6·12219	—47·32078	539·30175	— 1·33166	9·58732	—142·60029	5·20
·22	5·28228	—37·37359	726·06032	5·18113	—42·13965	491·98097	— 1·12962	8·45770	—133·01297	·22
·24	4·51717	—32·85642	688·68673	4·41323	—37·72642	449·84132	— 0·96484	7·49286	—124·55527	·24
·26	3·88831	—28·96811	655·83031	3·78139	—33·94503	412·11490	82936	6·66350	—117·06241	·26
·28	3·36756	—25·60055	626·86220	3·25758	—30·68745	378·16987	71663	5·94687	—110·39891	·28
5·30	2·93350	—22·66705	601·26165	2·82032	—27·86713	347·48242	62459	5·32228	—104·45204	5·30
·32	2·56933	—20·09772	578·59460	2·45290	—25·41423	319·61529	54454	4·77774	—99·12976	·32
·34	2·26212	—17·83560	558·49688	2·14224	—23·27199	294·20106	47875	4·29899	—94·35202	·34
·36	2·00149	—15·83411	540·66128	1·87800	—21·39399	270·92907	42248	3·87651	—90·05303	·36
·38	1·77931	—14·05480	524·82717	1·65215	—19·74184	249·53508	37449	3·50202	—86·17652	·38
5·40	1·58903	—12·46577	510·77237	1·45799	—18·28385	229·79324	33335	3·16867	—82·67450	5·40
·42	1·42536	—11·04041	498·30660	1·29031	—16·99354	211·50939	29792	2·87075	—79·50583	·42
·44	1·28405	—9·75636	487·26619	1·14478	—15·84876	194·51585	26729	2·60346	—76·63508	·44
·46	1·16156	—8·59480	477·50983	1·01796	—14·83080	178·66709	24073	2·36273	—74·03162	·46
·48	1·05509	—7·53971	468·91503	0·90691	—13·92389	163·83629	21757	2·14516	—71·66889	·48

x										x
5·50	0·96221	−6·57750	461·37532	80931	−13·11458	149·91240	19736	1·94780	−69·52373	5·50
·52	88100	−5·69650	454·79782	72315	−12·39143	136·79782	17963	1·76817	−67·57593	·52
·54	80976	−4·88674	449·10132	64674	−11·74469	124·40639	16404	1·60413	−65·80776	·54
·56	74723	−4·13951	444·21458	57883	−11·15586	112·66170	15032	1·45381	−64·20363	·56
·58	69201	−3·44750	440·07507	51816	−10·64770	101·49584	13819	1·31562	−62·74982	·58
5·60	64395	−2·80355	436·62757	46372	−10·18398	90·84814	12746	1·18816	−61·43420	5·60
·62	60062	−2·20293	433·82402	41476	−9·76922	80·66416	11793	1·07023	−60·24604	·62
·64	56298	−1·63995	431·62109	37050	−9·39872	70·89494	10949	0·96074	−59·17581	·64
·66	52953	−1·11042	429·98114	33039	−9·06833	61·49622	10195	85879	−58·21507	·66
·68	50003	−0·61039	428·87072	29381	−8·77452	52·42789	9527	76352	−57·35628	·68
5·70	47394	−13645	428·26033	25641	−8·51811	43·65337	8929	67423	−56·59276	5·70
·72	45099	+31454	428·12388	23772	−8·28039	35·13526	8399	59024	−55·91853	·72
·74	43075	0·74529	428·43842	19743	−8·08296	26·85487	7923	51101	−55·32829	·74
·76	41303	1·15832	429·18371	17524	−7·90772	18·77191	7501	43600	−54·81728	·76
·78	39752	1·55584	430·34203	15087	−7·75685	10·86419	7126	36474	−54·38128	·78
5·80	38409	1·93993	431·89787	12811	−7·62874	3·10734	6791	29683	−54·01654	5·80
·82	37247	2·31240	433·83780	10672	−7·52202	−4·52140	6495	23188	−53·71971	·82
·84	36321	2·67561	436·15020	8656	−7·45546	−12·04342	6232	16956	−53·48783	·84
·86	35298	3·02859	438·82581	6742	−7·36804	−19·47888	6003	10953	−53·31827	·86
·88	34800	3·37659	441·85440	4921	−7·31883	−26·84692	5800	+5153	−53·20874	·88
5·90	34181	3·71840	445·23099	3173	−7·28710	−34·16575	5625	−472	−53·15721	5·90
·92	33760	4·05600	448·94939	+1491	−7·27219	−41·45285	5474	−5946	−53·16193	·92
·94	33448	4·39048	453·00539	−140	−7·27359	−48·72504	5346	−11292	−53·22139	·94
·96	33260	4·72308	457·39587	−1725	−7·29084	−55·99863	5239	−16531	−53·33431	·96
·98	33173	5·05481	462·11895	−3281	−7·32365	−63·28947	5152	−21683	−53·49962	·98
6·00			467·17376			−70·61312			−53·71645	6·00

Tables of functions

$$\Phi_7(\lambda), \ \Phi_8(\lambda), \ \Phi_9(\lambda), \ \Phi_{10}(\lambda), \ \Phi_{11}(\lambda), \ \Phi_{12}(\lambda)$$

for a vibrating beam clamped at one and hinged at other end $(\lambda = 0 \cdot 00 - 6 \cdot 00)$

λ	Δ_2	Δ_1	Φ_9	Δ_2	Δ_1	Φ_8	Δ_2	Δ_1	Φ_7	λ
0·00		−0·00000	−0·00000		−0·00000	−0·00000		0·00000	0·00000	0·00
·02	−0·00000	−0	−0·00000	0	−0	−0·00000	0·00000	0	0·00000	·02
·04	−0	−0	−0·00000	0	−0	−0·00000	0	0	0·00000	·04
·06	−0	−0	−0·00000	0	−0	−0·00000	0	0	0·00000	·06
·08	−0	−0	−0·00000	0	−0	−0·00000	0	0	0·00000	·08
0·10	−2	−2	−0·00002	−1	−1	−0·00001	0	0	0·00000	0·10
·12	+1	−1	−0·00003	+1	−0	−0·00001	1	1	0·00001	·12
·14	−2	−3	−0·00006	−1	−1	−0·00002	−1	0	0·00001	·14
·16	−0	−3	−0·00009	−0	−1	−0·00003	1	1	0·00002	·16
·18	−2	−5	−0·00014	−0	−1	−0·00004	0	1	0·00003	·18
0·20	−1	−6	−0·00020	−1	−2	−0·00006	0	1	0·00004	0·20
·22	−2	−8	−0·00028	−1	−3	−0·00009	1	2	0·00006	·22
·24	−3	−11	−0·00039	−1	−4	−0·00013	1	3	0·00009	·24
·26	−3	−14	−0·00053	−1	−5	−0·00018	0	3	0·00012	·26
·28	−2	−16	−0·00069	−1	−6	−0·00024	0	3	0·00015	·28
0·30	−5	−21	−0·00090	−2	−8	−0·00032	2	5	0·00020	0·30
·32	−4	−25	−0·00115	−1	−9	−0·00041	0	5	0·00025	·32
·34	−4	−29	−0·00144	−3	−12	−0·00053	2	7	0·00032	·34
·36	−6	−35	−0·00179	−1	−13	−0·00066	1	8	0·00040	·36
·38	−5	−40	−0·00219	−3	−16	−0·00082	1	9	0·00049	·38
0·40	−8	−48	−0·00267	−3	−19	−0·00101	1	10	0·00059	0·40
·42	−6	−54	−0·00321	−2	−21	−0·00122	2	12	0·00071	·42
·44	−9	−63	−0·00384	−4	−25	−0·00147	2	14	0·00085	·44
·46	−8	−71	−0·00455	−4	−29	−0·00176	2	16	0·00101	·46
·48	−10	−81	−0·00536	−4	−33	−0·00209	2	18	0·00119	·48

0·50	10	—		—	−0·00536	4	—		—	−0·00246	2		0·00119	0·50
·52	11	—	91	—	−0·00627	6	—	41	—	−0·00287	3	20	0·00139	·52
·54	13	—	102	—	−0·00729	6	—	47	—	−0·00334	2	23	0·00162	·54
·56	12	—	115	—	−0·00844	5	—	53	—	−0·00387	4	25	0·00187	·56
·58	14	—	127	—	−0·00971	7	—	58	—	−0·00445	2	29	0·00216	·58
0·60	15	—	141	—	−0·01112	6	—	65	—	−0·00510	4	31	0·00247	0·60
·62	16	—	156	—	−0·01268	8	—	71	—	−0·00581	3	35	0·00282	·62
·64	16	—	172	—	−0·01440	8	—	79	—	−0·00660	4	38	0·00320	·64
·66	19	—	188	—	−0·01628	8	—	87	—	−0·00747	4	42	0·00362	·66
·68	19	—	207	—	−0·01835	8	—	95	—	−0·00842	4	46	0·00408	·68
0·70	21	—	226	—	−0·02061	10	—	103	—	−0·00945	5	50	0·00458	0·70
·72	20	—	247	—	−0·02308	10	—	113	—	−0·01058	4	55	0·00513	·72
·74	24	—	267	—	−0·02575	11	—	123	—	−0·01181	6	59	0·00572	·74
·76	24	—	291	—	−0·02866	10	—	134	—	−0·01315	5	65	0·00637	·76
·78	24	—	315	—	−0·03181	12	—	144	—	−0·01459	6	70	0·00707	·78
0·80	28	—	339	—	−0·03520	13	—	156	—	−0·01615	5	76	0·00783	0·80
·82	28	—	367	—	−0·03887	12	—	169	—	−0·01784	7	81	0·00864	·82
·84	29	—	395	—	−0·04282	14	—	181	—	−0·01965	6	88	0·00952	·84
·86	31	—	424	—	−0·04706	15	—	195	—	−0·02160	7	94	0·01046	·86
·88	32	—	455	—	−0·05161	14	—	210	—	−0·02370	8	101	0·01147	·88
0·90	35	—	487	—	−0·05648	16	—	224	—	−0·02594	7	109	0·01256	0·90
·92	35	—	522	—	−0·06170	16	—	240	—	−0·02834	8	116	0·01372	·92
·94	37	—	557	—	−0·06727	18	—	256	—	−0·03090	8	124	0·01496	·94
·96	39	—	594	—	−0·07321	18	—	274	—	−0·03364	9	132	0·01628	·96
·98	41	—	633	—	−0·07954	18	—	292	—	−0·03656	9	141	0·01769	·98
1·00	43	—	674	—	−0·08628	21	—	310	—	−0·03966	10	150	0·01919	1·00

530

λ	Δ₂	Δ₁	Φ₁₂	Δ₂	Δ₁	Φ₁₁	Δ₂	Δ₁	Φ₁₀	λ
0·00		0·00000	0·00000		0·00000	0·00000		0·00000	0·00000	0·00
·02	0·00000	0	0·00000	0·00000	0	0·00000	0·00000	0	0·00000	·02
·04	0	0	0·00000	1	1	0·00000	0	0	0·00000	·04
·06	1	1	0·00000	0	1	0·00001	1	1	0·00000	·06
·08	0	1	0·00001	2	3	0·00002	1	0	0·00001	·08
0·10	2	3	0·00002	2	5	0·00005	2	2	0·00001	0·10
·12	1	4	0·00005	4	9	0·00010	0	2	0·00003	·12
·14	2	6	0·00009	4	13	0·00019	2	4	0·00005	·14
·16	4	10	0·00015	6	19	0·00032	2	6	0·00009	·16
·18	3	13	0·00025	8	27	0·00051	1	7	0·00015	·18
0·20	4	17	0·00038	9	36	0·00078	4	11	0·00022	0·20
·22	6	23	0·00055	11	47	0·00114	2	13	0·00033	·22
·24	6	29	0·00078	14	61	0·00161	5	18	0·00046	·24
·26	9	38	0·00107	16	77	0·00222	4	22	0·00064	·26
·28	8	46	0·00145	18	95	0·00299	5	27	0·00086	·28
0·30	10	56	0·00191	20	115	0·00394	6	33	0·00113	0·30
·32	12	68	0·00247	25	140	0·00509	7	40	0·00146	·32
·34	13	81	0·00315	27	167	0·00649	8	48	0·00186	·34
·36	15	96	0·00396	30	197	0·00816	8	56	0·00234	·36
·38	16	112	0·00492	34	231	0·01013	11	67	0·00290	·38
0·40	17	129	0·00604	37	268	0·01244	10	77	0·00357	0·40
·42	22	151	0·00733	41	309	0·01512	11	88	0·00434	·42
·44	21	172	0·00884	45	354	0·01821	14	102	0·00522	·44
·46	24	196	0·01056	50	404	0·02175	14	116	0·00624	·46
·48	26	222	0·01252	54	458	0·02579	15	131	0·00740	·48

0·50	0·00871	148	0·03037	17	515	57	0·01474	250	28	0·50
·52	0·01019	166	0·03552	18	580	65	0·01724	281	31	·52
·54	0·01185	186	0·04132	20	647	67	0·02005	314	33	·54
·56	0·01371	207	0·04779	21	720	73	0·02319	350	36	·56
·58	0·01578	230	0·05499	23	799	79	0·02669	388	38	·58
0·60	0·01808	253	0·06298	23	884	85	0·03057	428	40	0·60
·62	0·02061	280	0·07182	27	973	89	0·03485	473	45	·62
·64	0·02341	307	0·08155	27	1070	97	0·03958	518	45	·64
·66	0·02648	337	0·09225	30	1171	101	0·04476	569	51	·66
·68	0·02985	368	0·10396	31	1279	108	0·05045	621	52	·68
0·70	0·03353	401	0·11675	33	1394	115	0·05666	676	55	0·70
·72	0·03754	436	0·13069	35	1516	122	0·06342	736	60	·72
·74	0·04190	473	0·14585	37	1645	129	0·07078	798	62	·74
·76	0·04663	513	0·16230	40	1780	135	0·07876	864	66	·76
·78	0·05176	554	0·18010	41	1923	143	0·08740	933	69	·78
0·80	0·05730	597	0·19933	43	2074	151	0·09673	1006	73	0·80
·82	0·06327	644	0·22007	47	2231	157	0·10679	1083	77	·82
·84	0·06971	692	0·24238	48	2399	168	0·11762	1164	81	·84
·86	0·07663	743	0·26637	51	2573	174	0·12926	1248	84	·86
·88	0·08406	796	0·29210	53	2755	182	0·14174	1337	89	·88
0·90	0·09202	852	0·31965	56	2948	193	0·15511	1431	94	0·90
·92	0·10054	910	0·34913	58	3148	200	0·16942	1527	96	·92
·94	0·10964	972	0·38061	62	3357	209	0·18469	1629	102	·94
·96	0·11936	1036	0·41418	64	3576	219	0·20098	1735	106	·96
·98	0·12972	1104	0·44994	68	3805	229	0·21833	1846	111	·98
1·00	0·14076		0·48799	69		238	0·23679		116	1·00

λ	Δ_2	Δ_1	Φ_9	Δ_2	Δ_1	Φ_8	Δ_2	Δ_1	Φ_7	λ
1·00	−0·00043	−0·00717	−0·08628	−0·00021	−0·00331	−0·03966	0·00010	0·00160	0·01919	1·00
·02	43	760	−0·09345	20	351	−0·04297	9	169	0·02079	·02
·04	47	807	−0·10105	21	372	−0·04648	11	180	0·02248	·04
·06	48	855	−0·10912	23	395	−0·05020	10	190	0·02428	·06
·08	50	905	−0·11767	23	418	−0·05415	12	202	0·02618	·08
1·10	52	957	−0·12672	25	443	−0·05833	11	213	0·02820	1·10
·12	54	1011	−0·13629	25	468	−0·06276	12	225	0·03033	·12
·14	57	1068	−0·14640	26	494	−0·06744	13	238	0·03258	·14
·16	58	1126	−0·15708	28	522	−0·07238	13	251	0·03496	·16
·18	60	1186	−0·16834	28	550	−0·07760	14	265	0·03747	·18
1·20	64	1250	−0·18020	30	580	−0·08310	14	279	0·04012	1·20
·22	66	1316	−0·19270	32	612	−0·08890	14	293	0·04291	·22
·24	67	1383	−0·20586	31	643	−0·09502	16	309	0·04584	·24
·26	72	1455	−0·21969	34	677	−0·10145	16	325	0·04893	·26
·28	72	1527	−0·23424	34	711	−0·10822	16	341	0·05218	·28
1·30	76	1603	−0·24951	37	748	−0·11533	17	358	0·05559	1·30
·32	79	1682	−0·26554	37	785	−0·12281	18	376	0·05917	·32
·34	82	1764	−0·28236	39	824	−0·13066	18	394	0·06293	·34
·36	84	1848	−0·30000	41	865	−0·13890	19	413	0·06687	·36
·38	88	1936	−0·31848	41	906	−0·14755	20	433	0·07100	·38
1·40	90	2026	−0·33784	45	951	−0·15661	21	454	0·07533	1·40
·42	95	2121	−0·35810	44	995	−0·16612	21	475	0·07987	·42
·44	97	2218	−0·37931	48	1043	−0·17607	21	496	0·08462	·44
·46	101	2319	−0·40149	49	1092	−0·18650	24	520	0·08958	·46
·48	104	2423	−0·42468	51	1143	−0·19742	23	543	0·09478	·48

x													x
1·50	108	—	2531	—0·44891	52	—	1195	—	—0·20885	25	568	0·10021	1·50
·52	113	—	2644	—0·47422	56	—	1251	—	—0·22080	25	593	0·10589	·52
·54	115	—	2759	—0·50066	56	—	1307	—	—0·23331	26	619	0·11182	·54
·56	120	—	2879	—0·52825	60	—	1367	—	—0·24638	28	647	0·11801	·56
·58	125	—	3004	—0·55704	61	—	1428	—	—0·26005	28	675	0·12448	·58
1·60	129	—	3133	—0·58708	64	—	1492	—	—0·27433	29	704	0·13123	1·60
·62	133	—	3266	—0·61841	66	—	1558	—	—0·28925	31	735	0·13827	·62
·64	138	—	3404	—0·65107	69	—	1627	—	—0·30483	31	766	0·14562	·64
·66	143	—	3547	—0·68511	73	—	1700	—	—0·32110	33	799	0·15328	·66
·68	149	—	3696	—0·72058	74	—	1774	—	—0·33810	34	833	0·16127	·68
1·70	153	—	3849	—0·75754	77	—	1851	—	—0·35584	36	869	0·16960	1·70
·72	160	—	4009	—0·79603	82	—	1933	—	—0·37435	35	904	0·17829	·72
·74	164	—	4173	—0·83612	83	—	2016	—	—0·39368	39	943	0·18733	·74
·76	171	—	4344	—0·87785	88	—	2104	—	—0·41384	39	982	0·19676	·76
·78	178	—	4522	—0·92129	91	—	2195	—	—0·43488	40	1022	0·20658	·78
1·80	184	—	4706	—0·96651	95	—	2290	—	—0·45683	43	1065	0·21680	1·80
·82	190	—	4896	—1·01357	98	—	2388	—	—0·47973	45	1110	0·22745	·82
·84	198	—	5094	—1·06253	104	—	2492	—	—0·50361	44	1154	0·23855	·84
·86	205	—	5299	—1·11347	107	—	2599	—	—0·52853	49	1203	0·25009	·86
·88	214	—	5513	—1·16646	112	—	2711	—	—0·55452	49	1252	0·26212	·88
1·90	221	—	5734	—1·22159	116	—	2827	—	—0·58163	51	1303	0·27464	1·90
·92	230	—	5964	—1·27893	122	—	2949	—	—0·60990	54	1357	0·28767	·92
·94	238	—	6202	—1·33857	128	—	3077	—	—0·63939	55	1412	0·30124	·94
·96	249	—	6451	—1·40059	132	—	3209	—	—0·67016	59	1471	0·31536	·96
·98	258	—	6709	—1·46510	138	—	3347	—	—0·70225	59	1530	0·33007	·98
2·00	268	—		—1·53219	146	—		—	—0·73572	64		0·34537	2·00

λ	Δ_2	Δ_1	Φ_{12}	Δ_2	Δ_1	Φ_{11}	Δ_2	Δ_1	Φ_{10}	λ
1·00	0·00116	0·01962	0·23679	0·00238	0·04043	0·48799	0·00069	0·01173	0·14076	1·00
·02	120	2082	0·25641	248	4291	0·52842	74	1247	0·15249	·02
·04	125	2207	0·27723	259	4550	0·57133	76	1323	0·16496	·04
·06	131	2338	0·29930	270	4820	0·61683	80	1403	0·17819	·06
·08	136	2474	0·32268	279	5099	0·66503	84	1487	0·19222	·08
1·10	141	2615	0·34742	292	5391	0·71602	86	1573	0·20709	1·10
·12	147	2762	0·37357	302	5693	0·76993	91	1664	0·22282	·12
·14	153	2915	0·40119	315	6008	0·82686	94	1758	0·23946	·14
·16	157	3072	0·43034	327	6335	0·88694	99	1857	0·25704	·16
·18	165	3237	0·46106	338	6673	0·95029	101	1958	0·27561	·18
1·20	170	3407	0·49343	351	7024	1·01702	107	2065	0·29591	1·20
·22	177	3584	0·52750	365	7389	1·08726	110	2175	0·31584	·22
·24	183	3767	0·56334	377	7766	1·16115	116	2291	0·33759	·24
·26	189	3956	0·60101	391	8157	1·23881	119	2410	0·36050	·26
·28	197	4153	0·64057	406	8563	1·32038	125	2535	0·38460	·28
1·30	203	4356	0·68210	419	8982	1·40601	130	2665	0·40995	1·30
·32	210	4566	0·72566	434	9416	1·49583	134	2799	0·43660	·32
·34	218	4784	0·77132	450	9866	1·58999	140	2939	0·46459	·34
·36	225	5009	0·81916	464	10330	1·68865	146	3085	0·49398	·36
·38	233	5242	0·86925	481	10811	1·79195	151	3236	0·52483	·38
1·40	241	5483	0·92167	498	11309	1·90006	157	3393	0·55719	1·40
·42	249	5732	0·97650	514	11823	2·01315	163	3556	0·59112	·42
·44	258	5990	1·03382	532	12355	2·13138	171	3727	0·62668	·44
·46	265	6255	1·09372	548	12903	2·25493	175	3902	0·66395	·46
·48	275	6530	1·15627	569	13472	2·38396	185	4087	0·70297	·48

1·50	0·74384	4277	190	2·51868	14058	586	1·22157	6814	284	1·50
·52	0·78661	4475	198	2·65926	14665	607	1·28971	7107	293	·52
·54	0·83136	4682	207	2·80591	15291	626	1·36078	7411	304	·54
·56	0·87818	4896	214	2·95882	15937	646	1·43489	7723	312	·56
·58	0·92714	5119	223	3·11819	16607	670	1·51212	8047	324	·58
1·60	0·97833	5351	232	3·28426	17296	689	1·59259	8381	334	1·60
·62	1·03184	5592	241	3·45722	18010	714	1·67640	8725	344	·62
·64	1·08776	5844	252	3·63732	18746	736	1·76365	9082	357	·64
·66	1·14620	6105	261	3·82478	19508	762	1·85447	9450	368	·66
·68	1·20725	6376	271	4·01986	20293	785	1·94897	9829	379	·68
1·70	1·27101	6661	285	4·22279	21106	813	2·04726	10222	393	1·70
·72	1·33762	6955	294	4·43385	21944	838	2·14948	10627	405	·72
·74	1·40717	7262	307	4·65329	22812	868	2·25575	11046	419	·74
·76	1·47979	7582	320	4·88141	23707	895	2·36621	11478	432	·76
·78	1·55561	7916	334	5·11848	24633	926	2·48099	11926	448	·78
1·80	1·63477	8264	348	5·36481	25589	956	2·60025	12387	461	1·80
·82	1·71741	8627	363	5·62070	26579	990	2·72412	12864	477	·82
·84	1·80368	9006	379	5·88649	27601	1022	2·85276	13358	494	·84
·86	1·89374	9400	394	6·16250	28659	1058	2·98634	13868	510	·86
·88	1·98774	9812	412	6·44909	29752	1093	3·12502	14395	527	·88
1·90	2·08586	10244	432	6·74661	30883	1131	3·26897	14941	546	1·90
·92	2·18830	10692	448	7·05544	32054	1171	3·41838	15504	563	·92
·94	2·29522	11163	471	7·37598	33265	1211	3·57342	16089	585	·94
·96	2·40685	11655	492	7·70863	34520	1255	3·73431	16693	604	·96
·98	2·52340	12168	513	8·05383	35818	1298	3·90124	17319	626	·98
2·00	2·64508		538	8·41201		1346	4·07443		647	2·00

λ	Φ₇	Δ₁	Δ₂	Φ₈	Δ₁	Δ₂	Φ₉	Δ₁	Δ₂
2·00	0·34537	0·01594	0·00064	—0·73572	—0·03493	—0·00146	—1·53219	—0·06977	—0·00268
·02	0·36131	1658	64	—0·77065	3645	152	—1·60196	7256	279
·04	0·37789	1727	69	—0·80710	3801	156	—1·67452	7547	291
·06	0·39516	1799	72	—0·84511	3968	167	—1·74999	7850	303
·08	0·41315	1872	73	—0·88479	4142	174	—1·82849	8164	314
2·10	0·43187	1950	78	—0·92621	4323	181	—1·91013	8494	330
·12	0·45137	2031	81	—0·96944	4514	191	—1·99507	8835	341
·14	0·47168	2116	85	—1·01458	4713	199	—2·08342	9194	359
·16	0·49284	2204	88	—1·06171	4923	210	—2·17536	9566	372
·18	0·51488	2297	93	—1·11094	5142	219	—2·27102	9955	389
2·20	0·53785	2393	96	—1·16236	5373	231	—2·37057	10363	408
·22	0·56178	2495	102	—1·21609	5615	242	—2·47420	10787	424
·24	0·58673	2601	106	—1·27224	5870	255	—2·58207	11233	446
·26	0·61274	2712	111	—1·33094	6138	268	—2·69440	11698	465
·28	0·63986	2829	117	—1·39232	6419	281	—2·81138	12185	487
2·30	0·66815	2951	122	—1·45651	6717	298	—2·93323	12696	511
·32	0·69766	3081	130	—1·52368	7028	311	—3·06019	13231	535
·34	0·72847	3215	134	—1·59396	7359	331	—3·19250	13794	563
·36	0·76062	3358	143	—1·66755	7706	347	—3·33044	14380	586
·38	0·79420	3507	149	—1·74461	8073	367	—3·47424	15006	626
2·40	0·82927	3665	158	—1·82534	8460	387	—3·62430	15654	648
·42	0·86592	3831	166	—1·90994	8871	411	—3·78084	16339	685
·44	0·90423	4007	176	—1·99865	9303	432	—3·94423	17061	722
·46	0·94430	4191	184	—2·09168	9763	460	—4·11484	17822	761
·48	0·98621	4386	195	—2·18931	10248	485	—4·29306	18623	801

x												x
2·50	846	—	19469	—4·47929	516	—	10764	—2·29179	207	4593	1·03007	2·50
·52	895	—	20364	—4·67398	547	—	11311	—2·39943	218	4811	1·07600	·52
·54	944	—	21308	—4·87762	580	—	11891	—2·51254	231	5042	1·12411	·54
·56	1000	—	22308	—5·09070	616	—	12507	—2·63145	245	5287	1·17453	·56
·58	1060	—	23368	—5·31378	657	—	13164	—2·75652	260	5547	1·22740	·58
2·60	1123	—	24491	—5·54746	699	—	13863	—2·88816	277	5824	1·28287	2·60
·62	1191	—	25682	—5·79237	744	—	14607	—3·02679	292	6116	1·34111	·62
·64	1266	—	26948	—6·04919	793	—	15400	—3·17286	312	6428	1·40227	·64
·66	1346	—	28294	—6·31867	848	—	16248	—3·32686	333	6761	1·46655	·66
·68	1433	—	29727	—6·60161	906	—	17154	—3·48934	353	7114	1·53416	·68
2·70	1528	—	31255	—6·89888	968	—	18122	—3·66088	379	7493	1·60530	2·70
·72	1628	—	32883	—7·21143	1038	—	19160	—3·84210	403	7896	1·68023	·72
·74	1741	—	34624	—7·54026	1112	—	20272	—4·03370	429	8325	1·75919	·74
·76	1862	—	36486	—7·88650	1193	—	21465	—4·23642	468	8793	1·84244	·76
·78	1992	—	38478	—8·25136	1282	—	22747	—4·45107	493	9286	1·93037	·78
2·80	2138	—	40616	—8·63614	1381	—	24128	—4·67854	532	9818	2·02323	2·80
·82	2294	—	42910	—9·04230	1485	—	25613	—4·91982	571	10389	2·12141	·82
·84	2467	—	45377	—9·47140	1602	—	27215	—5·17595	617	11006	2·22530	·84
·86	2656	—	48033	—9·92517	1732	—	28947	—5·44810	663	11669	2·33536	·86
·88	2863	—	50896	—10·40550	1871	—	30818	—5·73757	716	12385	2·45205	·88
2·90	3094	—	53990	—10·91446	2027	—	32845	—6·04575	774	13159	2·57590	2·90
·92	3344	—	57334	—11·45436	2199	—	35044	—6·37420	838	13997	2·70749	·92
·94	3626	—	60960	—12·02770	2388	—	37432	—6·72464	910	14907	2·84746	·94
·96	3932	—	64892	—12·63730	2600	—	40032	—7·09896	986	15893	2·99653	·96
·98	4279	—	69171	—13·28622	2832	—	42864	—7·49928	1076	16969	3·15546	·98
3·00	4658	—		—13·97793	3093	—		—7·92792	1171		3·32515	3·00

λ	Δ_2	Δ_1	Φ_{12}	Δ_2	Δ_1	Φ_{11}	Δ_2	Δ_1	Φ_{10}	λ
2·00	0·00647	0·17966	4·07443	0·01346	0·37164	8·41201	0·00538	0·12706	2·64508	2·00
·02	671	18637	4·25409	1395	38559	8·78365	563	13269	2·77214	·02
·04	696	19333	4·44046	1444	40003	9·16924	589	13858	2·90483	·04
·06	721	20054	4·63379	1498	41501	9·56927	617	14475	3·04341	·06
·08	747	20801	4·83433	1555	43056	9·98428	648	15123	3·18816	·08
2·10	775	21576	5·04234	1613	44669	10·41484	678	15801	3·33939	2·10
·12	806	22382	5·25810	1675	46344	10·86153	712	16513	3·49740	·12
·14	835	23217	5·48192	1738	48082	11·32497	748	17261	3·66253	·14
·16	868	24085	5·71409	1807	49889	11·80579	784	18045	3·83514	·16
·18	902	24987	5·95494	1879	51768	12·30468	824	18869	4·01559	·18
2·20	937	25924	6·20481	1953	53721	12·82236	867	19736	4·20428	2·20
·22	976	26900	6·46405	2034	55755	13·35957	911	20647	4·40164	·22
·24	1016	27916	6·73305	2115	57870	13·91712	959	21606	4·60811	·24
·26	1057	28973	7·01221	2206	60076	14·49582	1011	22617	4·82417	·26
·28	1101	30074	7·30194	2298	62374	15·09658	1063	23680	5·05034	·28
2·30	1149	31223	7·60268	2396	64770	15·72032	1122	24802	5·28714	2·30
·32	1199	32422	7·91491	2502	67272	16·36802	1183	25985	5·53316	·32
·34	1250	33672	8·23913	2612	69884	17·04074	1248	27233	5·79501	·34
·36	1307	34979	8·57585	2730	72614	17·73958	1319	28552	6·06734	·36
·38	1366	36345	8·92564	2853	75467	18·46572	1394	29946	6·35286	·38
2·40	1428	37773	9·28909	2988	78455	19·22039	1473	31419	6·65232	2·40
·42	1496	39269	9·66682	3127	81582	20·00494	1560	32979	6·96651	·42
·44	1568	40837	10·05951	3278	84860	20·82076	1650	34629	7·29630	·44
·46	1642	42479	10·46788	3438	88298	21·66936	1751	36380	7·64259	·46
·48	1724	44203	10·89267	3609	91907	22·55234	1855	38235	8·00639	·48

2·50	8·38874	40204	1969	23·47141	95700	3793	11·33470	46015	1812	2·50
·52	8·79078	42295	2091	24·42841	99686	3986	11·79485	47916	1901	·52
·54	9·21373	44519	2224	25·42527	1·03883	4197	12·27401	49920	2004	·54
·56	9·65892	46883	2364	26·46410	1·08304	4421	12·77321	52029	2109	·56
·58	10·12775	49401	2518	27·54714	1·12966	4662	13·29350	54251	2222	·58
2·60	10·62176	52084	2683	28·67680	1·17887	4921	13·83601	56598	2347	2·60
·62	11·14260	54945	2861	29·85567	1·23087	5200	14·40199	59075	2477	·62
·64	11·69205	58000	3055	31·08654	1·28587	5500	14·99274	61695	2620	·64
·66	12·27205	61263	3263	32·37241	1·34412	5825	15·60969	64468	2773	·66
·68	12·88468	64754	3491	33·71653	1·40586	6174	16·25437	67408	2940	·68
2·70	13·53222	68491	3737	35·12239	1·47140	6554	16·92845	70527	3119	2·70
·72	14·21713	72497	4006	36·59379	1·54103	6963	17·63372	73839	3312	·72
·74	14·94210	76794	4297	38·13482	1·61518	7415	18·37211	77364	3525	·74
·76	15·71004	81410	4616	39·75000	1·69411	7893	19·14575	81117	3753	·76
·78	16·52414	86373	4963	41·44411	1·77834	8423	19·95692	85121	4004	·78
2·80	17·38787	91717	5344	43·22245	1·86832	8998	20·80813	89395	4274	2·80
·82	18·30504	97476	5759	45·09077	1·96458	9626	21·70208	93967	4572	·82
·84	19·27980	1·03692	6216	47·05535	2·06773	10315	22·64175	98864	4897	·84
·86	20·31672	1·10410	6718	49·12308	2·17839	11066	23·63039	1·04117	5253	·86
·88	21·42082	1·17679	7269	51·30147	2·29735	11896	24·67156	1·09762	5645	·88
2·90	22·59761	1·25559	7880	53·59882	2·42538	12803	25·76918	1·15836	6074	2·90
·92	23·85320	1·34109	8550	56·02420	2·56347	13809	26·92754	1·22385	6549	·92
·94	25·19429	1·43403	9294	58·58767	2·71261	14914	28·15139	1·29457	7072	·94
·96	26·62832	1·53524	10121	61·30028	2·87403	16142	29·44596	1·37109	7652	·96
·98	28·16356	1·64560	11036	64·17431	3·04905	17502	30·81705	1·45403	8294	·98
3·00	29·80916		12057	67·22336		19014	32·27108		9008	3·00

λ	Δ_2	Δ_1	Φ_9	Δ_2	Δ_1	Φ_8	Δ_2	Δ_1	Φ_7	λ
3·00	0·04658	0·73829	13·97793	0·03093	45957	7·92792	0·01171	0·18140	3·32515	3·00
·02	5085	78914	14·71622	3384	49341	8·38749	1280	0·19420	3·50655	·02
·04	5561	84475	15·50536	3709	53050	8·88090	1400	0·20820	3·70075	·04
·06	6094	90569	16·35011	4072	57122	9·41140	1536	0·22356	3·90895	·06
·08	6695	97264	17·25580	4483	61605	9·98262	1689	0·24045	4·13251	·08
3·10	7368	1·04632	18·22844	4944	66549	10·59867	1860	0·25905	4·37296	3·10
·12	8133	1·12765	19·27476	5465	72014	11·26416	2053	0·27958	4·63201	·12
·14	8997	1·21762	20·40241	6059	78073	11·98430	2274	0·30232	4·91159	·14
·16	9979	1·31741	21·62003	6730	84803	12·76593	2524	0·32756	5·21391	·16
·18	11103	1·42844	22·93744	7499	92302	13·61306	2808	0·35564	5·54147	·18
3·20	12386	1·55230	24·36588	8379	1·00681	14·53608	3135	0·38699	5·89711	3·20
·22	13861	1·69091	25·91818	9389	1·10070	15·54289	3511	0·42210	6·28410	·22
·24	15564	1·84655	27·60909	10556	1·20626	16·64359	3943	0·46153	6·70620	·24
·26	17533	2·02188	29·45564	11906	1·32532	17·84985	4444	0·50597	7·16773	·26
·28	19832	2·22020	31·47752	13487	1·46019	19·17517	5028	0·55625	7·67370	·28
3·30	22509	2·44529	33·69772	15318	1·61337	20·63536	5710	0·61335	8·22995	3·30
·32	25659	2·70188	36·14301	17483	1·78820	22·24873	6511	0·67846	8·84330	·32
·34	29381	2·99569	38·84489	20034	1·98854	24·03693	7458	0·75304	9·52176	·34
·36	33798	3·33367	41·84058	23069	2·21923	26·02547	8581	0·83885	10·27480	·36
·38	39078	3·72445	45·17425	26692	2·48615	28·24470	9925	0·93810	11·11365	·38
3·40	45428	4·17873	48·89870	31052	2·79667	30·73085	11539	1·05349	12·05175	3·40
·42	53118	4·70991	53·07743	36333	3·16000	33·52752	13498	1·18847	13·10524	·42
·44	62504	5·33495	57·78734	42798	3·58798	36·68752	15884	1·34731	14·29371	·44
·46	74050	6·07545	63·12229	50670	4·09468	40·27550	18823	1·53554	15·64102	·46
·48	88385	6·95930	69·19774	60574	4·70042	44·37018	22470	1·76024	17·17656	·48

3·50	18·93680	2·03066	27043	49·07060	5·42940	72898	76·15704	8·02284	1·06354	3·50
·52	20·96746	2·35900	32834	54·50000	6·21473	88533	84·17988	9·31398	1·29114	·52
·54	23·32646	2·76158	40258	60·81473	7·40049	1·08576	93·49386	10·89686	1·58288	·54
·56	26·08804	3·26053	49895	68·21522	8·74645	1·34596	104·39072	12·85849	1·96163	·56
·58	29·34857	3·88642	62589	76·96167	10·43512	1·68867	117·24921	15·31896	2·46047	·58
3·60	33·23499	4·68217	79575	87·39679	12·58245	2·14733	132·56817	18·44694	3·12798	3·60
·62	37·91716	5·70937	1·02720	99·97924	15·35467	2·77222	151·01511	22·48445	4·03751	·62
·64	43·62653	7·05839	1·34902	115·33391	18·99584	3·64117	173·49956	27·78663	5·30218	·64
·66	50·68492	8·86545	1·80706	134·32975	23·87373	4·87789	201·28619	34·83878	7·10215	·66
·68	59·55037	11·34216	2·47671	158·20348	30·55986	6·68613	236·17497	44·62238	9·73360	·68
3·70	70·89253	14·82899	3·48633	188·76334	39·97336	9·41350	280·79735	58·32553	13·70315	3·70
·72	85·72152	19·89717	5·06818	228·73670	53·65682	13·68346	339·12288	78·24276	19·91723	·72
·74	105·61869	27·55350	7·65633	282·39352	74·32866	20·67184	417·36564	108·33059	30·08783	·74
·76	133·17219	39·68355	12·13005	356·72218	107·08034	32·75168	525·69623	155·99861	47·66802	·76
·78	172·85574	60·09494	20·41139	463·80252	162·19310	55·11276	681·69484	236·2095.	80·2109.	·78
3·80	232·95068	97·25977	37·16483	625·99562	262·54311	100·35001	917·9043.	382·2556.	146·0461.	3·80
·82	330·21045	172·6412.	75·3815.	888·53873	466·0849.	203·5418.	1300·1599.	678·4800.	296·2244.	·82
·84	502·8516.	351·6177.	178·9765.	1354·6236.	949·3518.	483·2669.	1978·6399.	1381·7964.	703·3164.	·84
·86	854·4693.	899·5762.	547·9585.	2303·9754.	2428·9372.	1479·5854.	3360·4363.	3535·0857.	2153·2893.	·86
·88	1754·0455.	3656·0592.	2756·4830.	4732·9126.	9871·938..	7443·001..	6895·5220.	14367·112..	10832·027..	·88
3·90	5410·1047.	82867·8738.	79211·8146.	14604·850..	223758·170..	−213886·232..	21262·634..	−325642·170..	−311275·058..	3·90
·92	88277·9785.	—	—	−238363·020..	−223758·170..	43060·589..	−346904·804..	70978·063..	—	·92
·94	21547·6426.	−18062·1809.	15947·2966.	58178·950..	—	3978·813..	84677·713..	8310·687..	62667·376..	·94
·96	3485·4617.	−2114·8843.	1473·5410.	9407·705..	48771·245..	985·2294.	13699·650..	2520·1977.	5790·490..	·96
·98	1370·57742	−641·34337	—	3697·0494.	5710·656..	—	5388·9634.	—	—	·98
4·00	729·23405	641·34337	364·87364	1965·2059.	1731·8435.	—	2868·7657.	—	1433·8385.	4·00

λ	Φ₁₀	Δ₁	Δ₂	Φ₁₁	Δ₁	Δ₂	Φ₁₂	Δ₁	Δ₂	λ
3·00	29·80916	1·76617	0·12057	67·22336	3·23919	0·19014	32·27108	1·54411	0·09008	3·00
·02	31·57533	1·89813	13196	70·46255	3·44616	20697	33·81519	1·64216	09805	·02
·04	33·47346	2·04284	14471	73·90871	3·67198	22582	35·45735	1·74910	10694	·04
·06	35·51630	2·20183	15899	77·58069	3·91886	24688	37·20645	1·86600	11690	·06
·08	37·71813	2·37690	17057	81·49955	4·18943	27057	39·07245	1·99408	12808	·08
3·10	40·09503	2·57008	19318	85·68898	4·48664	29721	41·06653	2·13475	14067	3·10
·12	42·66511	2·78374	21366	90·17562	4·81397	32733	43·20128	2·28965	15490	·12
·14	45·44885	3·02063	23689	94·98959	5·17535	36138	45·49093	2·46064	17099	·14
·16	48·46948	3·28394	26331	100·16494	5·57564	40029	47·95157	2·64996	18932	·16
·18	51·75342	3·57741	29347	105·74058	6·01998	44434	50·60153	2·86015	21019	·18
3·20	55·33083	3·90541	32800	111·76056	6·51498	49500	53·46168	3·09422	23407	3·20
·22	59·23624	4·27312	36771	118·27554	7·06808	55310	56·55590	3·35574	26152	·22
·24	63·50936	4·68664	41352	125·34362	7·68818	62010	59·91164	3·64892	29318	·24
·26	68·19600	5·15325	46661	133·03180	8·38585	69767	63·56056	3·97864	32972	·26
·28	73·34925	5·68165	52840	141·41765	9·17397	78812	67·53920	4·35128	37264	·28
3·30	79·03090	6·28231	60066	150·59162	10·06743	89346	71·89048	4·77346	42218	3·30
·32	85·31321	6·96786	68555	160·65905	11·08484	1·01741	76·66394	5·25425	48079	·32
·34	92·28107	7·75369	78583	171·74389	12·24861	1·16377	81·91819	5·80413	54988	·34
·36	100·03476	8·65868	90499	183·99250	13·5862.	1·3376.	87·72232	6·43608	63195	·36
·38	108·69344	9·70603	1·04735	197·5787.	15·1312.	1·5450.	94·15840	7·16604	72996	·38
3·40	118·39947	10·92468	1·21865	212·7099.	16·9261.	1·7949.	101·32444	8·01390	84786	3·40
·42	129·32415	12·35080	1·42612	229·6360.	19·0233.	2·0972.	109·33834	9·00453	99063	·42
·44	141·67495	14·03016	1·67936	248·65936	21·48945	2·4661.	118·34287	10·16938	1·16485	·44
·46	155·70511	16·02114	1·99098	270·14881	24·40953	2·92008	128·51225	11·54855	1·37917	·46
·48	171·72625	18·39896	2·37782	294·55834	27·89307	3·48354	140·06080	13·19376	1·64521	·48

3·50	190·12521	21·26177	2·86281	322·45141	32·08284	4·18977	153·25456	15·17243	1·97867	3·50
·52	211·38698	24·73896	3·47719	354·53425	37·16723	5·08439	168·42699	17·57350	2·40107	·52
·54	236·12594	29·00370	4·26474	391·70148	43·39822	6·23099	186·00049	20·51595	2·94245	·54
·56	265·12964	34·29091	5·28721	435·09970	51·11776	7·71954	206·51644	24·16119	3·64524	·56
·58	299·42055	40·92481	6·63390	486·21746	60·79764	9·67988	230·67763	28·73206	4·57087	·58
3·60	340·34536	49·36095	8·43614	547·01510	73·10095	12·30331	259·40969	34·54152	5·80946	3·60
·62	389·70631	60·25272	10·89177	620·11605	88·97841	15·87746	293·95121	42·03859	7·49707	·62
·64	449·95903	74·55909	14·30637	709·09446	109·82592	20·84751	335·98980	51·88220	9·84361	·64
·66	524·51812	93·72542	19·16633	818·92038	137·74676	27·92084	387·87200	65·06549	13·18329	·66
·68	618·24354	119·99695	26·27153	956·66714	176·0085.	38·2618.	452·93749	83·13126	18·06577	·68
3·70	738·24049	156·98669	36·98974	1132·6756.	229·8696.	53·8611.	536·06875	108·56211	25·43085	3·70
·72	895·22718	210·75502	53·76833	1362·5452.	308·1499.	78·2803.	644·63086	145·52247	36·96036	·72
·74	1105·98220	291·98639	81·23137	1670·6951.	426·3974.	118·2475.	790·15333	201·35331	55·83084	·74
·76	1397·96859	420·68623	128·69984	2097·0925.	613·7290.	187·3316.	991·50664	289·80192	88·44861	·76
·78	1818·65482	637·2577.	216·5715.	2710·8215.	928·9439.	315·2149.	1281·30856	438·62993	148·82801	·78
3·80	2455·9125.	1031·5959.	394·3382.	3639·7654.	1502·8660.	573·9221.	1719·93849	709·60688	270·97695	3·80
·82	3487·5084.	1831·443..	799·848..	5142·6314.	2666·9433.	1164·0773.	2429·54537	1259·2211.	549·6143.	·82
·84	5318·951..	3730·510..	1899·067..	7809·5747..	5430·744..	2763·801..	3688·7664.	2564·143..	1304·922..	·84
·86	9049·461..	9544·768..	5814·258..	13240·318..	13892·45..	8461·71..	6252·909..	6559·305..	3995·162..	·86
·88	18594·229..	38793·186..	29248·418..	27132·76...	56458·56...	42566·11...	12812·214..	26656·771..	20097·466..	·88
3·90	57387·415..	879292·16...	840498·98...	83591·32...	1279662·00...	1223203·44...	39468·985..	604188·32...	577531·55...	3·90
·92	936679·57...	—	—	1363253·32...	—	—	643657·30...	—	—	·92
·94	228618·32...	—191654·27...	—	332792·35...	—278918·46...	246261·17...	157128·49...	—131690·40...	116271·41...	·94
·96	36964·052..	—22441·038..	169213·24...	53873·89...	—32657·29...	22754·65...	25438·098..	—15418·991..	10743·532..	·96
·98	14523·014..	—6805·682..	15635·356..	21216·60...	—9902·64...	22754·65...	10019·107..	—4675·459..	2660·313..	·98
4·00	7717·332..		3871·606..	11313·966..		5634·51...	5343·648..			4·00

λ	Δ_2	Δ_1	Φ_9	Δ_2	Δ_1	Φ_8	Δ_2	Δ_1	Φ_7	λ
4·00	−1433·8385.	1086·3592.	−2868·7657.	− 985·2294.	746·6141	−1965·2059.	364·87364	−276·46973	729·23405	4·00
·02	− 520·7185.	565·6407.	−1782·4065.	− 357·7974.	388·8167.	−1218·5918.	132·50962	−143·96011	452·76432	·02
·04	− 233·9502.	331·6905.	−1216·7658.	− 160·7508.	228·06592	− 829·77517	59·53417	− 84·42594	308·80421	·04
·06	− 120·5743.	211·1162.	− 885·0753.	− 82·84674	145·21918	− 601·70925	30·68282	− 53·74312	224·37827	·06
·08	− 68·4329.	142·6833.	− 673·9591.	− 47·01853	98·20065	− 456·49007	17·41403	− 36·32909	170·63515	·08
4·10	− 41·7246.	100·9587.	− 531·2758.	− 28·66653	69·53412	− 358·28942	10·61748	− 25·71161	134·30606	4·10
·12	− 26·8903.	74·0684.	− 430·3171.	− 18·47345	51·06067	− 288·75530	6·84254	− 18·86907	108·59445	·12
·14	− 18·2116.	55·85686	− 356·24870	− 12·44245	38·61822	− 237·69463	4·60901	− 14·26006	89·72538	·14
·16	− 12·45400	43·40286	− 300·39184	− 8·63816	29·93006	− 199·07641	3·21866	− 11·04140	75·46532	·16
·18	− 9·20186	34·20100	− 256·98898	− 6·25148	23·67858	− 169·14635	2·31628	− 8·72512	64·42392	·18
4·20	− 6·72062	27·48038	− 222·78798	− 4·61617	19·06241	− 145·46777	1·70977	− 7·01535	55·69880	4·20
·22	− 5·07053	22·40985	− 195·30760	− 3·47493	15·58748	− 126·40536	1·28985	− 5·72550	48·68345	·22
·24	− 3·89856	18·51129	− 172·89775	− 2·67683	12·91065	− 110·81788	0·99162	− 4·73388	42·95795	·24
·26	− 3·04755	15·46374	− 154·38646	− 2·08950	10·82115	− 97·90723	77503	− 3·95885	38·22407	·26
·28	− 2·41732	13·04642	− 138·92272	− 1·65633	9·16482	− 87·08608	61466	− 3·34419	34·26522	·28
4·30	− 1·94257	11·10385	− 125·87630	− 1·32999	7·83483	− 77·92126	49381	− 2·85038	30·92103	4·30
·32	− 1·56864	9·53521	− 114·77245	− 1·08029	6·75454	− 70·08643	40139	− 2·44899	28·07065	·32
·34	− 1·31915	8·21606	− 105·23724	− 0·88657	5·86797	− 63·33189	32968	− 2·11931	25·62166	·34
·36	− 1·06571	7·15035	− 97·02118	− 73443	5·13354	− 57·46392	27339	− 1·84592	23·50235	·36
·38	− 0·90076	6·24959	− 89·87083	− 61354	4·52000	− 52·33038	22865	− 1·61727	21·65643	·38
4·40	− 75979	5·48980	− 83·62124	− 51652	4·00348	− 47·81038	19276	− 1·42451	20·03916	4·40
·42	− 64563	4·84417	− 78·13144	− 43791	3·56557	− 43·80690	16370	− 1·26081	18·61465	·42
·44	− 55235	4·29182	− 73·28727	− 37365	3·19192	− 40·24133	13995	− 1·12086	17·35384	·44
·46	− 47558	3·81624	− 68·99545	− 32070	2·87122	− 37·04941	12040	− 1·00046	16·23298	·46
·48	− 41186	3·40438	− 65·17921	− 27678	2·59444	− 34·17819	10416	− 0·89630	15·23252	·48

4·50	35872	—	3·04566	61·77483	24002	—	2·35442	31·58375	9062	80568	—	14·33622
·52	31403	—	2·73163	58·72917	20915	—	2·14527	29·22933	7924	72644	—	13·53054
·54	27629	—	2·45534	55·99754	18297	—	1·96230	27·08406	6960	65684	—	12·80410
·56	24422	—	2·21112	53·54220	16074	—	1·80156	25·12176	6143	59541	—	12·14726
·58	21681	—	1·99431	51·33108	14169	—	1·65987	23·32020	5445	54096	—	11·55185
4·60	19334	—	1·80097	49·33677	12532	—	1·53455	21·66033	4843	49253	—	11·01089
·62	17310	—	1·62787	47·53580	11116	—	1·42339	20·12578	4327	44926	—	10·51836
·64	15558	—	1·47229	45·90793	9887	—	1·32452	18·70239	3878	41048	—	10·06910
·66	14039	—	1·33190	44·43564	8817	—	1·23635	17·37787	3491	37557	—	9·65862
·68	12714	—	1·20476	43·10374	7879	—	1·15756	16·14152	3149	34408	—	9·28305
4·70	11558	—	1·08918	41·89898	7055	—	1·08701	14·98396	2854	31554	—	8·93897
·72	10542	—	0·98376	40·80980	6327	—	1·02374	13·89695	2593	28961	—	8·62343
·74	9650	—	88726	39·82604	5684	—	0·96690	12·87321	2362	26599	—	8·33382
·76	8865	—	79861	38·93878	5110	—	0·91580	11·90631	2160	24439	—	8·06783
·78	8171	—	71690	38·14017	4599	—	0·86981	10·99051	1981	22458	—	7·82344
4·80	7557	—	64133	37·42327	4142	—	0·82839	10·12070	1820	20638	—	7·59886
·82	7012	—	57121	36·78194	3730	—	0·79109	9·29231	1679	18959	—	7·39248
·84	6531	—	50590	36·21073	3360	—	0·75749	8·50122	1554	17405	—	7·20289
·86	6101	—	44489	35·70483	3023	—	0·72726	7·74373	1440	15965	—	7·02884
·88	5719	—	38770	35·25994	2719	—	0·70007	7·01647	1339	14626	—	6·86919
4·90	5382	—	33388	34·87224	2440	—	0·67567	6·31640	1250	13376	—	6·72293
·92	5079	—	28309	34·53836	2187	—	0·65380	5·64073	1168	12208	—	6·58917
·94	4810	—	23499	34·25527	1954	—	0·63426	4·98693	1097	11111	—	6·46709
·96	4575	—	18924	34·02028	1736	—	0·61690	4·35267	1031	10080	—	6·35598
·98	4362	—	14562	33·83104	1538	—	0·60152	3·73577	972	9108	—	6·25518
5·00	4175	—		33·68542	1356	—		3·13425	921		—	6·16410

λ	Δ₂	Δ₁	Φ₁₂	Δ₂	Δ₁	Φ₁₁	Δ₂	Δ₁	Φ₁₀	λ
4·00	2660·313..	−2015·146..	5343·648..	5634·51...	−4268·135..	11313·966..	3871·606..	−2934·076..	7717·332..	4·00
·02	966·136..	−1049·0100.	3328·5025.	2046·262..	−2221·873..	7045·831..	1406·018..	−1528·058..	4783·256..	·02
·04	434·0736.	−614·9364.	2279·4925.	919·358.	−1302·515..	4823·958..	631·691..	−896·3674.	3255·1981.	·04
·06	223·7193.	−391·2171.	1664·5561.	473·832.	−828·6833.	3521·4434.	325·5547.	−570·8127.	2358·8307.	·06
·08	126·9771.	−264·2400.	1273·3390.	268·9359.	−559·7474.	2692·7601.	184·7627.	−386·0500.	1788·0180.	·08
4·10	77·4239.	−186·81616	1009·09902	163·9803.	−395·7671.	2133·0127.	112·6454.	−273·40467	1401·96801	4·10
·12	49·90095	−136·91521	822·28286	105·6871.	−290·0800.	1737·2456.	72·58981	−200·81486	1128·56334	·12
·14	33·61675	−103·29846	685·36765	71·1983.	−218·8817.	1447·1656.	48·89008	−151·92478	927·74848	·14
·16	23·47991	−79·81855	582·06919	49·7279.	−169·1538.	1228·2839.	34·13683	−117·78795	775·82370	·16
·18	16·90088	−62·91767	502·25064	35·7942.	−133·3596.	1059·1301.	24·56135	−93·22660	658·03575	·18
4·20	12·47917	−50·43850	439·33297	26·4281.	−106·9315.	925·7705.	18·12546	−75·10114	564·80915	4·20
·22	9·41787	−41·02063	388·89447	19·9459.	−86·9856.	818·8390.	13·66933	−61·43181	489·70801	·22
·24	7·24366	−33·77697	347·87384	15·3393.	−71·6463.	731·8534.	10·50422	−50·92759	428·27620	·24
·26	5·66485	−28·11212	314·09687	11·9957.	−59·6506.	660·2071.	8·20556	−42·72203	377·34861	·26
·28	4·49578	−23·61634	285·98475	9·5202.	−50·1304.	600·5565.	6·50322	−36·21881	334·62658	·28
4·30	3·61514	−20·00120	262·36841	7·6544.	−42·4760.	550·4261.	5·22057	−30·99824	298·40777	4·30
·32	2·94154	−17·05966	242·36721	6·2278.	−36·2482.	507·9501.	4·23918	−26·75906	267·40953	·32
·34	2·41908	−14·64058	225·30755	5·1214.	−31·1268.	471·7019.	3·47776	−23·28130	240·65047	·34
·36	2·00887	−12·63171	210·66697	4·2523.	−26·8745.	440·5751.	2·87961	−20·40169	217·36917	·36
·38	1·68317	−10·94854	198·03526	3·5625.	−23·31208	413·70066	2·40444	−17·99725	196·96748	·38
4·40	1·42183	−9·52671	187·08672	3·00894	−20·30314	390·38858	2·02290	−15·97435	178·97023	4·40
·42	1·21028	−8·31643	177·56001	2·56082	−17·74232	370·08544	1·71376	−14·26059	162·99588	·42
·44	1·03751	−7·27892	169·24358	2·19475	−15·54757	352·34312	1·46101	−12·79958	148·73529	·44
·46	0·89531	−6·38361	161·96466	1·89357	−13·65400	336·79555	1·25269	−11·54689	135·93571	·46
·48	77745	−5·60616	155·58105	1·64387	−12·01013	323·14155	1·07978	−10·46711	124·38882	·48

x													x
4·50	113·92171			0·93513	311·13142			1·43544	149·97489			67907	4·50
·52	104·38973	—	9·53198	81343	300·55673	—	10·57469	1·26042	145·04780	—	4·92709	59650	·52
·54	95·67118	—	8·71855	71040	291·24246	—	9·31427	1·11276	140·71721	—	4·33059	52677	·54
·56	87·66303	—	8·00815	62261	283·04095	—	8·20151	0·98730	136·91339	—	3·80382	46760	·56
·58	80·27749	—	7·38554	54746	275·82674	—	7·21421	88029	133·57717	—	3·33622	41710	·58
4·60	73·43941	—	6·83808	48276	269·49282	—	6·33392	78863	130·65805	—	2·91912	37390	4·60
·62	67·08409	—	6·35532	42681	263·94753	—	5·54529	70984	128·11283	—	2·54522	33668	·62
·64	61·15558	—	5·92851	37814	259·11208	—	4·83545	64169	125·90429	—	2·20854	30457	·64
·66	55·60521	—	5·55037	33568	254·91832	—	4·19376	58272	124·00032	—	1·90397	27681	·66
·68	50·39052	—	5·21469	29844	251·30728	—	3·61104	53144	122·37316	—	1·62716	25259	·68
4·70	45·47427	—	4·91625	26560	248·22768	—	3·07960	48674	120·99859	—	1·37457	23155	4·70
·72	40·82362	—	4·65065	23660	245·63482	—	2·59286	44766	119·85557	—	1·14302	21316	·72
·74	36·40957	—	4·41405	21081	243·48962	—	2·14520	41350	118·92571	—	0·92986	19706	·74
·76	32·20633	—	4·20324	18780	241·75792	—	1·73170	38345	118·19291	—	73280	18296	·76
·78	28·19089	—	4·01544	16725	240·40967	—	1·34825	35716	117·64307	—	54984	17058	·78
4·80	24·34270	—	3·84819	14868	239·41858	—	0·99109	33394	117·26381	—	37926	15966	4·80
·82	20·64319	—	3·69951	13201	238·76143	—	65715	31356	117·04421	—	21960	15013	·82
·84	17·07569	—	3·56750	11681	238·41784	—	34359	29562	116·97474	—	6947	14171	·84
·86	13·62500	—	3·45069	10304	238·36987	—	4797	27983	117·04698	+	7224	13432	·86
·88	10·27735	—	3·34765	9039	238·60173	+	23186	26600	117·25354	+	20656	12783	·88
4·90	7·02009	—	3·25726	7881	239·09959	+	49786	25380	117·58793	+	33439	12221	4·90
·92	3·84164	—	3·17845	6812	239·85125	+	75166	24319	118·04453	+	45660	11722	·92
·94	+0·73131	—	3·11033	5818	240·84610	+	99485	23389	118·61835	+	57382	11293	·94
·96	—2·32084	—	3·05215	4896	242·07484	+	1·22874	22591	119·30510	+	68675	10922	·96
·98	—5·32403	—	3·00319	4032	243·52949	+	1·45465	21898	120·10107	+	79597	10606	·98
5·00	—8·28690	—	2·96287	3217	245·20312	+	1·67363	21308	121·00310	+	90203	10336	5·00

λ	Δ_2	Δ_1	Φ_9	Δ_2	Δ_1	Φ_8	Δ_2	Δ_1	Φ_7	λ
5·00	−0·04175	0·10387	−33·68542	−0·01354	0·58796	−3·13425	0·00921	−0·08187	6·16410	5·00
·02	− 4012	− 6375	−33·58155	− 1180	57616	−2·54629	874	− 7313	6·08223	·02
·04	− 3868	+ 2507	−33·51780	− 1018	56598	−1·97013	832	− 6481	6·00910	·04
·06	− 3742	− 1235	−33·49273	− 866	55732	−1·40415	793	− 5688	5·94429	·06
·08	− 3632	− 4867	−33·50508	− 723	55009	−0·84683	761	− 4927	5·88741	·08
5·10	− 3542	− 8409	−33·55375	− 586	54423	−0·29674	730	− 4197	5·83814	5·10
·12	− 3461	− 11870	−33·63784	− 454	53969	+0·24749	705	− 3492	5·79617	·12
·14	− 3398	− 15268	−33·75654	− 330	53639	0·78718	681	− 2811	5·76125	·14
·16	− 3345	− 18613	−33·90922	− 208	53431	1·32357	660	− 2151	5·73314	·16
·18	− 3307	− 21920	−34·09535	− 91	53340	1·85788	643	− 1508	5·71163	·18
5·20	− 3277	− 25197	−34·31455	− 22	53362	2·39128	628	− 880	5·69655	5·20
·22	− 3262	− 28459	−34·56652	+ 136	53498	2·92490	615	− 265	5·68775	·22
·24	− 3255	− 31714	−34·85111	244	53742	3·45988	605	+ 340	5·68510	·24
·26	− 3258	− 34972	−35·16825	355	54097	3·99730	595	935	5·68850	·26
·28	− 3274	− 38246	−35·51797	461	54558	4·53827	591	1526	5·69785	·28
5·30	− 3297	− 41543	−35·90043	572	55130	5·08385	586	2112	5·71311	5·30
·32	− 3332	− 44875	−36·31586	679	55809	5·63515	582	2694	5·73423	·32
·34	− 3376	− 48251	−36·76461	790	56599	6·19324	584	3278	5·76117	·34
·36	− 3430	− 51681	−37·24712	901	57500	6·75923	584	3862	5·79395	·36
·38	− 3494	− 55175	−37·76393	1015	58515	7·33423	588	4450	5·83257	·38
5·40	− 3570	− 58745	−38·31568	1132	59647	7·91938	592	5042	5·87707	5·40
·42	− 3654	− 62399	−38·90313	1250	60897	8·51585	599	5641	5·92749	·42
·44	− 3751	− 66150	−39·52712	1373	62270	9·12482	609	6250	5·98390	·44
·46	386.	7001.	−40·18862	1503	63773	9·74752	62.	687.	6·04640	·46
·48	397.	7398.	−40·8887.	1633	65406	10·38525	63.	750.	6·1151.	·48

549

x										x
5·50	6·19010	8147	64.	11·03931	67177	1771	—41·62856	78095	411.	5·50
·52	6·27157	8809	662	11·7108	69094	1917	—42·40951	82349	4254	·52
·54	6·35966	9492	683	12·40202	71162	2068	—43·23300	86762	4413	·54
·56	6·45458	10194	702	13·11364	73387	2225	—44·10062	91347	4585	·56
·58	6·55652	10921	727	13·84751	75782	2395	—45·01409	96121	4774	·58
5·60	6·66573	11673	752	14·60433	78353	2571	—45·97530	1·01100	4979	5·60
·62	6·78246	12454	781	15·38886	81111	2758	—46·98630	1·06303	5203	·62
·64	6·90700	13267	813	16·19997	84069	2958	—48·04933	1·11745	5442	·64
·66	7·03967	14113	846	17·04066	87239	3170	—49·16678	1·17452	5707	·66
·68	7·18080	14999	886	17·91305	90634	3395	—50·34130	1·23442	5990	·68
5·70	7·33079	15925	926	18·81939	94270	3636	—51·57572	1·29741	6299	5·70
·72	7·49004	16895	970	19·76209	98164	3894	—52·87313	1·36374	6633	·72
·74	7·65899	17917	1022	20·74373	1·02334	4170	—54·23687	1·43368	6994	·74
·76	7·83816	18990	1073	21·76707	1·06802	4468	—55·67055	1·50757	7389	·76
·78	8·02806	20123	1133	22·83509	1·11588	4786	—57·17812	1·58572	7815	·78
5·80	8·22929	21319	1196	23·95097	1·16718	5130	—58·76384	1·66849	8277	5·80
·82	8·44248	22584	1265	25·11815	1·22220	5502	—60·43233	1·75631	8782	·82
·84	8·66832	23926	1342	26·34035	1·28123	5903	—62·18864	1·84959	9328	·84
·86	8·90758	25350	1424	27·62158	1·34460	6337	—64·03823	1·94882	9923	·86
·88	9·16108	26864	1514	28·96618	1·41268	6808	—65·98705	2·05456	10574	·88
5·90	9·42972	28477	1613	30·37886	1·48590	7322	—68·04161	2·16736	11280	5·90
·92	9·71449	30198	1721	31·86476	1·56467	7877	—70·20897	2·28789	12053	·92
·94	10·01647	32037	1839	33·42943	1·64954	8487	—72·49686	2·41688	12899	·94
·96	10·33684	34006	1969	35·07897	1·74102	9148	—74·91374	2·55513	13825	·96
·98	10·67690	36116	2110	36·81999	1·83979	9877	—77·46887	2·70350	14837	·98
6·00	11·03806			38·65978			—80·17237			6·00

550

λ	Φ_{10}	Δ_1	Δ_2	Φ_{11}	Δ_1	Δ_2	Φ_{12}	Δ_1	Δ_2	λ
5·00	−8·28690	−2·93070	0·03217	245·20312	1·88671	0·21308	121·00310	1·00539	0·10336	5·00
·02	−11·21760	−2·90621	2449	247·08983	2·09487	20816	122·00849	1·10650	10111	·02
·04	−14·12381	−2·88903	1718	249·18470	2·29894	20407	123·11499	1·20581	9931	·04
·06	−17·01284	−2·87884	1019	251·48364	2·49972	20078	124·32080	1·30364	9783	·06
·08	−19·89168	−2·87538	+346	253·98336	2·69796	19824	125·62444	1·40041	9677	·08
5·10	−22·76706	−2·87839	−301	256·68132	2·89439	19643	127·02485	1·49641	9600	5·10
·12	−25·64545	−2·88771	−932	259·57571	3·08961	19522	128·52126	1·59198	9557	·12
·14	−28·53316	−2·90320	−1549	262·66532	3·28435	19474	130·11324	1·68743	9545	·14
·16	−31·43636	−2·92475	−2155	265·94967	3·47907	19472	131·80067	1·78303	9560	·16
·18	−34·36111	−2·95226	−2751	269·42874	3·67453	19546	133·58370	1·87909	9606	·18
5·20	−37·31337	−2·98573	−3347	273·10327	3·8712.	1967.	135·46279	1·97586	9677	5·20
·22	−40·29910	−3·02512	−3939	276·9744.	4·0695.	1983.	137·43865	2·07362	9776	·22
·24	−43·32422	−3·07047	−4535	281·0439.	4·2703.	2008.	139·51227	2·17264	9902	·24
·26	−46·39469	−3·12182	−5135	285·3142.	4·4739.	2036.	141·68491	2·27320	10056	·26
·28	−49·51651	−3·17926	−5744	289·7881.	4·6809.	2070.	143·95811	2·37555	10235	·28
5·30	−52·69577	−3·24287	−6361	294·4690.	4·8919.	2110.	146·33366	2·47996	10441	5·30
·32	−55·93864	−3·31282	−6995	299·3609.	5·1074.	2155.	148·81362	2·58670	10674	·32
·34	−59·25146	−3·38926	−7644	304·4683.	5·3280.	2206.	151·40032	2·69607	10937	·34
·36	−62·64072	−3·47237	−8311	309·7963.	5·5543.	2263.	154·09639	2·80836	11229	·36
·38	−66·11309	−3·56240	−9003	315·3506.	5·7869.	2326.	156·90475	2·92385	11549	·38
5·40	−69·67549	−3·65960	−9720	321·1375.	6·0265.	2396.	159·82860	3·04285	11900	5·40
·42	−73·33509	−3·76424	−10464	327·1640.	6·2733.	2468.	162·87145	3·16571	12286	·42
·44	−77·09933	−3·87668	−11244	333·4373.	6·5287.	2554.	166·03716	3·29274	12703	·44
·46	−80·97601	−3·99728	−12060	339·9660.	6·7931.	2644.	169·32990	3·4243.	1316.	·46
·48	−84·97329	−4·12635	−12907	346·7591.	7·0671.	2740.	172·7542.	3·5608.	1365.	·48

5·50	—89·09964			353·8262.			176·31506		1418.	5·50
·52	—93·36412	—4·26448	—13813	361·1778.	7·3516.	2845.	180·01772	3·70266	14757	·52
·54	—97·77616	—4·41204	—14756	368·8254.	7·6476.	2960.	183·86795	3·85023	15379	·54
·56	—102·34578	—4·56962	—15758	376·7813.	7·9559.	3083.	187·87197	4·00402	16047	·56
·58]	—107·08357	—4·73779	—16817	385·0588.	8·2775.	3216.	192·03646	4·16449	16769	·58
5·60	—112·00075	—4·91718	—17939	393·6722.	8·6134.	3359.	196·36864	4·33218	17546	5·60
·62	—117·10926	—5·10851	—19133	402·6370.	8·9648.	3514.	200·87628	4·50764	18385	·62
·64	—122·42182	—5·31256	—20405	411·9698.	9·3328.	3680.	205·56777	4·69149	19288	·64
·66	—127·95196	—5·53014	—21758	421·6886.	9·7188.	3860.	210·45214	4·88437	20261	·66
·68	—133·71415	—5·76219	—23205	431·8128.	10·1242.	4054.	215·53912	5·08698	21311	·68
5·70	—139·72387	—6·00972	—24753	442·3633.	10·5505.	4263.	220·83921	5·30009	22445	5·70
·72	—145·99771	—6·27384	—26412	453·3627.	10·9994.	4489.	226·36375	5·52454	23668	·72
·74	—152·55346	—6·55575	—28191	464·8353.	11·4726.	4732.	232·12497	5·76122	24990	·74
·76	—159·41024	—6·85678	—30103	476·8074.	11·9721.	4995.	238·13609	6·01112	26419	·76
·78	—166·58865	—7·17841	—32163	489·3076.	12·5002.	5281.	244·41140	6·27531	27964	·78
5·80	—174·11086	—7·52221	—34380	502·3666.	13·0590.	5588.	250·96635	6·55495	29640	5·80
·82	—182·00083	—7·88997	—36776	516·0178.	13·6512.	5922.	257·81770	6·85135	31456	·82
·84	—190·28446	—8·28363	—39366	530·2975.	14·2797.	6285.	264·98361	7·16591	33424	·84
·86	—198·98975	—8·70529	—42166	545·2449.	14·9474.	6677.	272·48376	7·50015	35568	·86
·88	—208·14713	—9·15738	—45209	560·9027.	15·6578.	7104.	280·33959	7·85583	37894	·88
5·90	—217·78960	—9·64247	—48509	577·3173.	16·4146.	7568.	288·57436	8·23477	40435	5·90
·92	—227·95306	—10·16346	—52099	594·5394.	17·2221.	8075.	297·21348	8·63912	43199	·92
·94	—238·67663	—10·72357	—56011	612·6241.	18·0847.	8626.	306·28459	9·07111	46220	·94
·96	—250·00297	—11·32634	—60277	631·6317.	19·0076.	9229.	315·81790	9·53331	49527	·96
·98	—261·97873	—11·97576	—64942	651·6281.	19·9964.	9888.	325·84648	10·02858	53146	·98
6·00	—274·65496	—12·67623	—70047	672·6855.	21·0574.	1·0610.	336·40652	10·56004		6·00

Tables of functions

$$\Phi_{13}(\lambda), \; \Phi_{14}(\lambda)$$

for a vibrating beam hinged at both ends ($\lambda = 0{\cdot}00 - 6{\cdot}00$)

λ	Φ₁₃	Δ₁	Δ₂	Φ₁₄	Δ₁	Δ₂
0·00	0·00000	0·00000	0·00000	0·00000	0·00000	0·00000
·02	0·00000	0	0	0·00000	0	0
·04	0·00000	0	1	0·00000	0	0
·06	0·00000	1	0	0·00000	0	3
·08	0·00001	1	0	0·00000	3	1
0·10	0·00002	1	2	0·00003	4	2
·12	0·00003	3	2	0·00007	6	3
·14	0·00006	5	1	0·00013	9	4
·16	0·00011	6	4	0·00022	13	5
·18	0·00017	10	2	0·00035	18	7
0·20	0·00027	12	4	0·00053	25	8
·22	0·00039	16	5	0·00078	33	8
·24	0·00055	21	5	0·00111	41	12
·26	0·00076	26	7	0·00152	53	12
·28	0·00102	33	7	0·00205	65	14
0·30	0·00135	40	8	0·00270	79	18
·32	0·00175	48	9	0·00349	97	17
·34	0·00223	57	11	0·00446	114	21
·36	0·00280	68	11	0·00560	135	24
·38	0·00348	79	13	0·00695	159	25
0·40	0·00427	92	14	0·00854	184	28
·42	0·00519	106	16	0·01038	212	31
·44	0·00625	122	17	0·01250	243	35
·46	0·00747	139	17	0·01493	278	36
·48	0·00886	156		0·01771	314	

λ	Φ₁₃	Δ₁	Δ₂
1·00	0·17083	0·01446	0·00090
·02	0·18529	1540	94
·04	0·20069	1639	99
·06	0·21708	1742	103
·08	0·23450	1851	109
1·10	0·25301	1965	114
·12	0·27266	2085	120
·14	0·29351	2210	125
·16	0·31561	2341	131
·18	0·33902	2480	139
1·20	0·36382	2624	144
·22	0·39006	2777	153
·24	0·41783	2936	159
·26	0·44719	3104	168
·28	0·47823	3279	175
1·30	0·51102	3464	185
·32	0·54566	3657	193
·34	0·58223	3862	205
·36	0·60285	4075	213
·38	0·66160	4301	226
1·40	0·70461	4538	237
·42	0·74999	4786	248
·44	0·79785	5048	262
·46	0·84833	5327	279
·48	0·90160	5616	289

x							x			
0·50	0·01042	178	22	0·02085	354	40	1·50	0·95776	5924	308
·52	0·01220	200	22	0·02439	398	44	·52	1·01700	6248	324
·54	0·01420	223	23	0·02837	445	47	·54	1·07948	6588	340
·56	0·01643	248	25	0·03282	496	51	·56	1·14536	6949	361
·58	0·02891	276	28	0·03778	549	53	·58	1·21485	7332	383
0·60	0·02167	305	29	0·04327	608	59	1·60	1·28817	7734	402
·62	0·02472	336	31	0·04935	669	61	·62	1·36551	8161	427
·64	0·02808	369	33	0·05604	736	67	·64	1·44712	8612	451
·66	0·03177	405	36	0·06340	807	71	·66	1·53324	9091	479
·68	0·03582	443	38	0·07147	881	74	·68	1·62415	9598	507
0·70	0·04025	484	41	0·08028	961	80	1·70	1·72013	10136	538
·72	0·04509	526	42	0·08989	1045	84	·72	1·82149	10708	572
·74	0·05035	571	45	0·10034	1134	89	·74	1·92857	11315	607
·76	0·05606	620	49	0·11168	1229	95	·76	2·04172	11961	646
·78	0·06226	670	50	0·12397	1328	99	·78	2·16133	12649	688
0·80	0·06896	724	54	0·13725	1433	105	1·80	2·28782	13381	732
·82	0·07620	780	56	0·15158	1543	110	·82	2·42163	14162	781
·84	0·08400	841	61	0·16701	1660	117	·84	2·56325	14995	833
·86	0·09241	903	62	0·18361	1782	122	·86	2·71320	15886	891
·88	0·10144	969	66	0·20143	1911	129	·88	2·87206	16837	951
0·90	0·11113	1040	71	0·22054	2045	134	1·90	3·04043	17857	1020
·92	0·12153	1113	73	0·24099	2187	142	·92	3·21900	18948	1091
·94	0·13266	1189	76	0·26286	2334	147	·94	3·40848	20119	1171
·96	0·14455	1272	83	0·28620	2491	157	·96	3·60967	21377	1258
·98	0·15727	1356	84	0·31111	2652	161	·98	3·82344	22729	1352
1·00	0·17083		90	0·33763		170	2·00	4·05073		1454

λ	Φ_{14}	Δ_1	Δ_2
1·00	0·33763	0·02822	0·00170
·02	0·36585	3000	178
·04	0·39585	3185	185
·06	0·42770	3379	194
·08	0·46149	3582	203
1·10	0·49731	3793	211
·12	0·53524	4014	221
·14	0·57538	4243	229
·16	0·61781	4484	241
·18	0·66265	4734	250
1·20	0·70999	4995	261
·22	0·75994	5266	271
·24	0·81260	5551	285
·26	0·86811	5846	295
·28	0·92657	6154	308
1·30	0·98811	6476	322
·32	1·05287	6811	335
·34	1·12098	7159	348
·36	1·19257	7524	365
·38	1·26781	7903	379
1·40	1·34684	8299	396
·42	1·42983	8712	413
·44	1·51695	9142	430
·46	1·60837	9593	451
·48	1·70430	10062	469

λ	Φ_{13}	Δ_1	Δ_2	Φ_{14}	Δ_1	Δ_2
2·00	4·05073	0·24183	0·01454	6·75177	0·35299	0·01796
·02	4·29256	25751	1568	7·10476	37213	1914
·04	4·55007	27442	1691	7·47689	39259	2046
·06	4·82449	29269	1827	7·86948	41450	2191
·08	5·11718	31246	1977	8·28398	43798	2348
2·10	5·42964	33387	2141	8·72196	46318	2520
·12	5·76351	35711	2324	9·18514	49029	2711
·14	6·12062	38237	2526	9·67543	51950	2921
·16	6·50299	40981	2744	10·19493	55100	3150
·18	6·91280	43971	2990	10·74593	58507	3407
2·20	7·35251	47243	3272	11·33100	62200	3693
·22	7·82494	50817	3574	11·95300	66206	4006
·24	8·33311	54737	3920	12·61506	70564	4358
·26	8·88048	59035	4298	13·32070	75315	4751
·28	9·47083	63768	4733	14·07385	80508	5193
2·30	10·10851	68984	5216	14·87893	86193	5685
·32	10·79835	74748	5764	15·74086	92438	6245
·34	11·54583	81131	6383	16·66524	99310	6872
·36	12·35714	88218	7087	17·65834	1·06898	7588
·38	13·23932	96105	7887	18·72732	1·15296	8398
2·40	14·20037	1·04909	8804	19·88028	1·24622	9326
·42	15·24946	1·14762	9853	21·12650	1·35008	10386
·44	16·39708	1·25826	11064	22·47658	1·46615	11607
·46	17·65534	1·38285	12459	23·94273	1·59631	13016
·48	19·03819	1·52366	14081	25·53904	1·74279	14648

1·50	1·80492		2·50	490	20·56185		15972	27·28183		16551
·52	1·91044	10552	·52	513	22·24523	1·68338	18187	29·19013	1·90830	18777
·54	2·02109	11065	·54	535	24·11048	1·86525	20790	31·28620	2·09607	21394
·56	2·13709	11600	·56	560	26·18363	2·07315	23876	33·59621	2·31001	24493
·58	2·25869	12160	·58	586	28·49554	2·31191	27545	36·15115	2·55494	28174
1·60	2·38615	12746	2·60	613	31·08290	2·58736	31941	38·98783	2·83668	32583
·62	2·51974	13359	·62	641	33·98967	2·90677	37239	42·15034	3·16251	37895
·64	2·65974	14000	·64	674	37·26883	3·27916	43671	45·69180	3·54146	44342
·66	2·80648	14674	·66	704	40·98470	3·71587	51541	49·67668	3·98488	52223
·68	2·96026	15378	·68	741	45·21598	4·23128	61248	54·18379	4·50711	61947
1·70	3·12145	16119	2·70	776	50·05974	4·84376	73334	59·31037	5·12658	74046
·72	3·29040	16895	·72	816	55·63684	5·57710	88527	65·17741	5·86704	89255
·74	3·46751	17711	·74	858	62·09921	6·46237	1·07837	71·93700	6·75959	1·08582
·76	3·65320	18569	·76	903	69·63995	7·54074	1·32676	79·78241	7·84541	1·33435
·78	3·84792	19472	·78	950	78·50745	8·86750	1·65045	88·96217	9·17976	1·65816
1·80	4·05214	20422	2·80	1003	89·02540	10·51795	2·07858	99·80009	10·83792	2·08653
·82	4·26639	21425	·82	1056	101·62193	12·59653	2·65419	112·72454	12·92445	2·66228
·84	4·49120	22481	·84	1116	116·87265	15·25072	3·44258	128·31127	15·58673	3·45082
·86	4·72717	23597	·86	1180	135·56595	18·69330	4·54530	147·34882	19·03755	4·55375
·88	4·97494	24777	·88	1248	158·80455	23·23860	6·12536	170·94012	23·59130	6·13397
1·90	5·23519	26025	2·90	1323	188·16851	29·36396	8·45325	200·66539	29·72527	8·46204
·92	5·50867	27348	·92	1401	225·98572	37·81721	11·99664	238·85270	38·18731	12·00570
·94	5·79616	28749	·94	1489	275·79957	49·81385	17·60360	289·04571	50·19301	17·61270
·96	6·09854	30238	·96	1582	343·21702	67·41745	26·89970	356·85142	67·80571	26·90913
·98	6·41674	31820	·98	1683	437·53417	94·31715	43·2276.	451·56626	94·71484	43·23716
2·00	6·75177	33503	3·00	1796	575·0788.	137·5447.	74·0738.	589·51826	137·95200	74·0835.

λ	Φ₁₃	Δ₁	Δ₂	Φ₁₄	Δ₁	Δ₂
3·00	575·0788.	211·6185.	74·0738.	589·51826	212·0355.	74·0835.
·02	786·6973.	349·7954.	138·1769.	801·5537.	350·2225.	138·1870.
·04	1136·4927.	639·7764.	289·9810.	1151·7762.	640·2139.	289·9914.
·06	1776·2691.	1365·0959.	725·3195.	1791·9901.	1365·5435.	725·3296.
·08	3141·3650.	3798·4584.	2433·3625.	3157·5336.	3798·9168.	2433·3733.
3·10	6939·8234.	18990·5792.	15192·1208.	6956·4504.	18991·0501.	15192·1333.
·12	25930·4026.	4771913·476.	4752922·897.	25947·5005.	4771914·11..	4752923·06..
·14	4797843·878..	—	—	4797861·61..	—	—
·16	36140·5645	−27791·261..	—	36158·616.	−27790·740.	—
·18	8349·303...	−4719·768.	23071·493.	8367·876.	−4719·252.	23071·488...
3·20	3629·5350.	−1605·6951.	3114·073..	3648·624.	−1605·166.	3114·086...
·22	2023·8399.	−733·4362.	872·2589.	2043·458.	−732·896.	872·270...
·24	1290·4037.	−395·6751.	337·7611.	1310·562.	−395·122.	337·774...
·26	894·7286.	−237·7294.	157·9457.	915·440.	−237·164.	157·958...
·28	656·9992.	−154·0547.	83·6747.	678·2763.	−153·4735.	83·691...
3·30	502·9445.	−105·5934.	48·4613.	524·8028.	−104·9999.	48·4736.
·32	397·3511.	−75·5850.	30·0084.	419·8029.	−74·9769.	30·0230.
·34	321·76616	−56·00623	19·5788.	344·82607	−55·38408	19·5929.
·36	265·75993	−42·68665	13·31958	289·44199	−42·04945	13·33463
·38	223·07328	−33·30805	9·37860	247·39254	−32·65570	9·39375
3·40	189·76523	−26·51220	6·79585	214·73684	−25·84436	6·81134
·42	163·25303	−21·46655	5·04565	188·89248	−20·78282	5·06154
·44	141·78648	−17·64137	3·82518	168·10966	−16·94127	3·84155
·46	124·14511	−14·68838	2·95299	151·16839	−13·97157	2·96970
·48	109·45673	−12·37231	2·31607	137·19682	−11·63835	2·33322

λ	Φ₁₃	Δ₁	Δ₂
4·00	5·1011.	−1·2789.	0·0452.
·02	3·8222.	−1·2393.	396.
·04	2·5829.	−1·2050.	343.
·06	1·3779.	−1·1750.	300.
·08	+0·20295	−1·14971	253.
4·10	−0·94676	−1·12801	2170
·12	−2·07477	−1·1101.	179.
·14	−3·1848.	−1·0954.	147.
·16	−4·2802.	−1·0841.	113.
·18	−5·3643.	−1·0756.	85.
4·20	−6·4399.	−1·0700.	56.
·22	−7·5099.	−1·0671.	29.
·24	−8·5770.	−1·0667.	+4.
·26	−9·6437.	−1·0689.	−22.
·28	−10·7126.	−1·0734.	−45.
4·30	−11·7860.	−1·0803.	−69.
·32	−12·8663.	−1·0894.	−91.
·34	−13·9557.	−1·1008.	−114.
·36	−15·0565.	−1·1146.	−138.
·38	−16·1711.	−1·1303.	−157.
4·40	−17·3014.	−1·1484.	−181.
·42	−18·4498.	−1·1688.	−204.
·44	−19·6186.	−1·1914.	−226.
·46	−20·8100.	−1·2164.	−250.
·48	−22·0264.	−1·2436.	−272.

3·50	97·08442	—	10·53021	1·84210	—	125·55847	9·77866
·52	86·55421	—	9·04681	1·48340	—	115·77981	8·27719
·54	77·50740	—	7·83895	1·20786	—	107·50262	7·05082
·56	69·66845	—	6·84564	0·99331	—	100·45180	6·03850
·58	62·82281	—	6·02142	82422	—	94·41330	5·19471
3·60	56·80139	—	5·33199	68943	—	89·21859	4·48536
·62	51·46940	—	4·75103	59096	—	84·73323	3·88376
·64	46·71837	—	4·2582.	4928.	—	80·84947	3·36987
·66	42·4601.	—	3·8378.	4204.	—	77·47960	2·92766
·68	38·6223.	—	3·4770.	3608.	—	74·55194	2·54454
3·70	35·1453.	—	3·1658.	3112.	—	72·00740	2·21050
·72	31·9795.	—	2·8964.	2694.	—	69·79690	1·91748
·74	29·0831.	—	2·6621.	2343.	—	67·87942	1·65890
·76	26·4210.	—	2·4577.	2044.	—	66·22052	1·42945
·78	23·9633.	—	2·2786.	1791.	—	64·79107	1·22474
3·80	21·6847.	—	2·1214.	1572.	—	63·56633	1·04112
·82	19·5633.	—	1·9831.	1383.	—	62·52521	0·8756.
·84	17·5802.	—	1·8611.	1220.	—	61·6496.	7256.
·86	15·7191.	—	1·7534.	1077.	—	60·9240.	5890.
·88	13·9657.	—	1·6582.	952.	—	60·3350.	4640.
3·90	12·3075.	—	1·5740.	842.	—	59·8710.	349...
·92	10·7335.	—	1·493..	75...	—	59·522..	244...
·94	9·234..	—	1·434..	65...	—	59·2788.	1444.
·96	7·800..	—	1·375..	59...	—	59·1344.	526..
·98	6·4252.	—	1·3241.	51...	—	59·0818.	335.
4·00	5·1011.	—		452.	+	59·1153.	

4·50	1·85969	—23·2700.	—1·2733.	—297.
·52	1·50147	—24·5433.	—1·3055.	—322.
·54	1·22637	—25·8488.	—1·3401.	—346.
·56	1·01232	—27·1889.	—1·3776.	—375.
·58	0·84379	—28·5665.	—1·4176.	—400.
4·60	70935	—29·9841.	—1·4607.	—431.
·62	60160	—31·4448.	—1·5066.	—459.
·64	51389	—32·9514.	—1·5557.	—491.
·66	44221	—34·5071.	—1·608.	—53..
·68	38312	—36·115.	—1·664.	—56..
4·70	33404	—37·779.	—1·723.	—59..
·72	29302	—39·502.	—1·787.	—64..
·74	25858	—41·289.	—1·855.	—68..
·76	22945	—43·144.	—1·926.	—71..
·78	20471	—45·070.	—2·002.	—76..
4·80	18362	—47·0727.	—2·0838.	—81..
·82	1655.	—49·1565.	—2·1704.	—866.
·84	1500.	—51·3269.	—2·2624.	—920.
·86	1366.	—53·5893.	—2·3604.	—980.
·88	1250.	—55·9497.	—2·4648.	—1044.
4·90	115..	—58·4145.	—2·5760.	—1112.
·92	105..	—60·9905.	—2·6946.	—1186.
·94	100..	—63·6851.	—2·8210.	—1264.
·95	918.	—66·5061.	—2·9559.	—1349.
·98	861.	—69·4620.	—3·0999.	—1440.
5·00	810.	—72·5619.		—1540.

λ	Φ₁₃	Δ₁	Δ₂	Φ₁₄	Δ₁	Δ₂
5·00	72·5619.	3·2539.	0·1540.	145·69265	4·20341	0·17673
·02	75·8158.	3·4183.	1644.	149·89606	4·39080	18739
·04	79·2341.	3·5945.	1762.	154·28686	4·58978	19898
·06	82·8286.	3·7830.	1885.	158·87664	4·80128	21150
·08	86·6116.	3·9852.	2022.	163·67792	5·0265.	2253.
5·10	90·5968.	4·2021.	2169.	168·7044.	5·2664.	2399.
·12	94·7989.	4·4350.	2329.	173·97080	5·52257	2561.
·14	99·2339.	4·6856.	2506.	179·49337	5·79633	27376
·16	103·9195.	4·9551.	2695.	185·28970	6·08934	29301
·18	108·8746.	5·2456.	2905.	191·37904	6·40338	31404
5·20	114·1202.	5·5590.	3134.	197·78242	6·74053	33715
·22	119·6792.	5·8976.	3386.	204·52295	7·10296	36243
·24	125·5768.	6·2638.	3662.	211·62591	7·49323	39027
·26	131·8406.	6·6605.	3967.	219·11914	7·91415	42092
·28	138·5011.	7·0908.	4303.	227·03329	8·36891	45476
5·30	145·5919.	7·5582.	4674.	235·40220	8·86102	49211
·32	153·1501.	8·0669.	5087.	244·26322	9·39460	53358
·34	161·2170.	8·6214.	5545.	253·65782	9·97412	57952
·36	169·8384.	9·2265.	6051.	263·63194	10·60482	63070
·38	179·0649.	9·8887.	6622.	274·23676	11·2926.	6878.
5·40	188·9536.	10·6142.	7255.	285·5293.	12·0440.	7514.
·42	199·5678.	11·4110.	7968.	297·5733.	12·8671.	8231.
·44	210·9788.	12·2877.	8767.	310·4404.	13·7704.	9033.
·46	223·2665.	13·2546.	9669.	324·2108.	14·7641.	9937.
·48	236·5211.	14·3234.	1·0688.	338·9749.	15·8601.	1·0960.

λ	Φ₁₄	Δ₁	Δ₂
4·00	59·1153.	0·1145.	0·0810.
·02	59·2298.	1911.	766.
·04	59·42096	26381	727.
·06	59·68477	33318	6937
·08	60·01795	39961	6643
4·10	60·41756	46354	6393
·12	60·88110	52529	6175
·14	61·40639	58521	5992
·16	61·99160	64355	5834
·18	62·63515	70060	5705
4·20	63·33575	75660	5600
·22	64·09235	81176	5516
·24	64·90411	86632	5456
·26	65·77043	92039	5407
·28	66·69082	97425	5386
4·30	67·66507	1·02801	5376
·32	68·69308	1·08186	5385
·34	69·77494	1·13593	5407
·36	70·91087	1·19040	5447
·38	72·10127	1·24539	5499
4·40	73·34666	1·30110	5571
·42	74·64776	1·35756	5646
·44	76·00532	1·41505	5749
·46	77·42037	1·47364	5859
·48	78·89401	1·53349	5985

4·50	80·42750	1·59473	6124		5·50	—250·8445	—15·5077	—1·1843	354·8350	17·0719	1·2118.
·52	82·02223	1·65754	6281		·52	—266·3522	—16·8235	—1·3158	371·9069	18·4157	1·3438.
·54	83·67977	1·72207	6453		·54	—283·1757	—18·2892	—1·4657	390·3226	19·9096	1·4939.
·56	85·40184	1·78847	6640		·56	—301·4649	—19·9267	—1·6375	410·2322	21·5757	1·6661.
·58	87·19031	1·85692	6845		·58	—321·3916	—21·7616	—1·8349	431·8079	23·4396	1·8639.
4·60	89·04723	1·92758	7066		5·60	—343·1532	—23·8242	—2·0626	455·2475	25·5318	2·0922.
·62	90·97481	2·00066	7308		·62	—366·9774	—26·1510	—2·3268	480·7793	27·8885	2·3567.
·64	92·97547	2·07632	7566		·64	—393·1284	—28·7851	—2·6341	508·6678	30·5529	2·6644.
·66	95·05179	2·15480	7848		·66	—421·9135	—31·7787	—2·9936	539·2207	33·5771	3·0242.
·68	97·20659	2·23629	8149		·68	—453·6922	—35·1946	—3·4159	572·7978	37·0244	3·4473.
4·70	99·44288	2·32104	8475		5·70	—488·8868	—39·1100	—3·9154	609·8222	40·9715	3·9471.
·72	101·76392	2·40928	8824		·72	—527·9968	—43·6185	—4·5085	650·7937	45·5122	4·5407.
·74	104·17320	2·50128	9200		·74	—571·6153	—48·8366	—5·2181	696·3059	50·7631	5·2509.
·76	106·67448	2·59732	9604		·76	—620·4519	—54·9086	—6·0720	747·0690	56·8683	6·1052.
·78	109·27180	2·69771	10039		·78	—675·3605	—62·0158	—7·1072	803·9373	64·0094	7·1411.
4·80	111·96951	2·80275	10504		5·80	—737·3763	—70·3874	—8·3716	867·9467	72·4153	8·4059.
·82	114·77226	2·91283	11008		·82	—807·7637	—80·3165	—9·9291	940·3620	82·3793	9·9640.
·84	117·68509	3·02828	11545		·84	—888·0802	—92·1809	—11·8644	1022·7413	94·2793	11·9000.
·86	120·71337	3·14955	12127		·86	—980·2611	—106·4746	—14·2937	1117·0206	108·6092	14·3299.
·88	123·86292	3·27704	12749		·88	—1086·7375	—123·8503	—17·3757	1225·6298	126·0216	17·4124.
4·90	127·13996	3·41131	13427		5·90	—1210·5860	—145·1825	—21·3322	1351·6514	147·3913	21·3697.
·92	130·55127	3·55278	14147		·92	—1355·7685	—171·6610	—26·4785	1499·0427	173·9080	26·5167.
·94	134·10405	3·70208	14930		·94	—1527·4295	—204·9319	—33·2709	1672·9507	207·2178	33·3098.
·96	137·80613	3·85984	15776		·96	—1732·3614	—247·3148	—42·3829	1880·1685	249·6403	42·4225.
·98	141·66597	4·02668	16684		·98	—1979·6762	—302·1476	—54·8328	2129·8088	304·5132	54·8729.
5·00	145·69265		17673		6·00	—2281·8238			2434·3220		

Tables of functions

$$\Phi_{15}(\lambda),\ \Phi_{16}(\lambda),\ \Phi_{17}(\lambda)$$

for a vibrating cantilever beam $(\lambda = 0{\cdot}00 - 4{\cdot}00)$

λ	Δ_2	Δ_1	Φ_{17}	Δ_2	Δ_1	Φ_{16}	Δ_2	Δ_1	Φ_{15}	λ
0·00		0·00000	0·00000		—0·00000	—0·00000		0·00000	0·00000	0·00
·02	0·00000	0	0·00000	0·00000	—0	—0·00000	0·00000	0	0·00000	·02
·04	1	1	0·00000	1	—1	—0·00000	0	0	0·00000	·04
·06	2	3	0·00001	0	—1	—0·00001	1	1	0·00000	·06
·08	3	6	0·00004	2	—3	—0·00002	1	2	0·00001	·08
0·10	5	11	0·00010	2	—5	—0·00005	2	4	0·00003	0·10
·12	7	18	0·00021	4	—9	—0·00010	2	6	0·00007	·12
·14	9	27	0·00039	5	—14	—0·00019	3	9	0·00013	·14
·16	12	39	0·00066	5	—19	—0·00033	4	13	0·00022	·16
·18	16	55	0·00105	9	—28	—0·00052	5	18	0·00035	·18
0·20	19	74	0·00160	9	—37	—0·00080	7	25	0·00053	0·20
·22	24	98	0·00234	12	—49	—0·00117	8	33	0·00078	·22
·24	27	125	0·00332	13	—62	—0·00166	8	41	0·00111	·24
·26	33	158	0·00457	18	—80	—0·00228	12	53	0·00152	·26
·28	38	196	0·00615	17	—97	—0·00308	12	65	0·00205	·28
0·30	43	239	0·00811	23	—120	—0·00405	15	80	0·00270	0·30
·32	49	288	0·01050	24	—144	—0·00525	16	96	0·00350	·32
·34	56	344	0·01338	29	—173	—0·00669	19	115	0·00446	·34
·36	64	408	0·01682	31	—204	—0·00842	21	136	0·00561	·36
·38	69	477	0·02090	25	—239	—0·01046	24	160	0·00697	·38
0·40	77	554	0·02567	39	—278	—0·01285	25	185	0·00857	0·40
·42	87	641	0·03121	43	—321	—0·01563	30	215	0·01042	·42
·44	95	736	0·03762	48	—369	—0·01884	31	246	0·01257	·44
·46	103	839	0·04498	53	—422	—0·02253	35	281	0·01503	·46
·48	113	952	0·05337	56	—478	—0·02675	39	320	0·01784	·48

x												x
0·50	125	1077	0·06289	64	—	542	—	−0·03153	41	361	0·02104	0·50
·52	133	1210	0·07366	67	—	609	—	−0·03695	47	408	0·02465	·52
·54	146	1356	0·08576	75	—	684	—	−0·04304	48	456	0·02873	·54
·56	158	1514	0·09932	80	—	764	—	−0·04988	55	511	0·03329	·56
·58	171	1685	0·11446	87	—	851	—	−0·05752	58	569	0·03840	·58
0·60	183	1868	0·13131	95	—	946	—	−0·06603	64	633	0·04409	0·60
·62	197	2065	0·14999	101	—	1047	—	−0·07549	68	701	0·05042	·62
·64	214	2279	0·17064	110	—	1157	—	−0·08596	74	775	0·05743	·64
·66	229	2508	0·19343	120	—	1277	—	−0·09753	80	855	0·06518	·66
·68	245	2753	0·21851	127	—	1404	—	−0·11030	86	941	0·07373	·68
0·70	263	3016	0·24604	138	—	1542	—	−0·12434	94	1035	0·08314	0·70
·72	283	3299	0·27620	149	—	1691	—	−0·13976	100	1135	0·09349	·72
·74	304	3603	0·30919	161	—	1852	—	−0·15667	109	1244	0·10484	·74
·76	324	3927	0·34522	172	—	2024	—	−0·17519	118	1362	0·11728	·76
·78	348	4275	0·38449	188	—	2212	—	−0·19543	126	1488	0·13090	·78
0·80	374	4649	0·42724	200	—	2412	—	−0·21755	137	1625	0·14578	0·80
·82	401	5050	0·47373	219	—	2631	—	−0·24167	149	1774	0·16203	·82
·84	429	5479	0·52423	234	—	2865	—	−0·26798	160	1934	0·17977	·84
·86	462	5941	0·57902	256	—	3121	—	−0·29663	175	2109	0·19911	·86
·88	497	6438	0·63843	275	—	3396	—	−0·32784	189	2298	0·22020	·88
0·90	532	6970	0·70281	298	—	3694	—	−0·36180	205	2503	0·24318	0·90
·92	575	7545	0·77251	326	—	4020	—	−0·39874	224	2727	0·26821	·92
·94	619	8164	0·84796	353	—	4373	—	−0·43894	243	2970	0·29548	·94
·96	669	8833	0·92960	384	—	4757	—	−0·48267	267	3237	0·32518	·96
·98	714	9547	1·01793	421	—	5178	—	−0·53024	291	3528	0·35755	·98
1·00	799		1·11340	458	—		—	−0·58202	318		0·39283	1·00

λ	Δ_2	Δ_1	Φ_{17}	Δ_2	Δ_1	Φ_{16}	Δ_2	Δ_1	Φ_{15}	λ
1·00	0·00799	0·10346	1·11340	0·00458	0·05636	0·58202	0·00318	0·03846	0·39283	1·00
·02	842	11188	1·21686	503	6139	0·63838	351	4197	0·43129	·02
·04	924	12112	1·32874	553	6692	0·69977	386	4583	0·47326	·04
·06	1006	13118	1·44986	607	7299	0·76669	424	5007	0·51909	·06
·08	1101	14219	1·58104	672	7971	0·83968	471	5478	0·56916	·08
1·10	1204	15423	1·72323	741	8712	0·91939	521	5999	0·62394	1·10
·12	1324	16747	1·87746	823	9535	1·00651	580	6579	0·68393	·12
·14	1458	18205	2·04493	918	10453	1·10186	646	7225	0·74972	·14
·16	1612	19817	2·22698	1020	11473	1·20639	723	7948	0·82197	·16
·18	1787	21604	2·42515	1145	12618	1·32112	811	8759	0·90145	·18
1·20	1991	23595	2·64119	1287	13905	1·44730	914	9673	0·98904	1·20
·22	2225	25820	2·87714	1452	15357	1·58635	1032	10705	1·08577	·22
·24	2496	28316	3·13534	1642	16999	1·73992	1171	11876	1·19282	·24
·26	2816	31132	3·41850	1869	18868	1·90991	1333	13209	1·31158	·26
·28	3189	34321	3·72982	2135	21003	2·09859	1527	14736	1·44367	·28
1·30	3631	37952	4·07303	2449	23452	2·30862	1755	16491	1·59103	1·30
·32	4157	42109	4·45255	2826	26278	2·54314	2026	18517	1·75594	·32
·34	4788	46897	4·87364	3277	29555	2·80592	2353	20870	1·94111	·34
·36	5547	52444	5·34261	3823	33378	3·10147	2750	23620	2·14981	·36
·38	6471	58915	5·86705	4486	37864	3·43525	3232	26852	2·38601	·38
1·40	7602	66517	6·45620	5304	43168	3·81389	3821	30673	2·65453	1·40
·42	9003	75520	7·12137	6312	49480	4·24557	4557	35230	2·96126	·42
·44	10749	86269	7·87657	7575	57055	4·74037	5470	40700	3·31356	·44
·46	12952	99221	8·73926	9168	66223	5·31092	6629	47329	3·72056	·46
·48	15764	1·14985	9·73147	11204	77427	5·97315	8105	55434	4·19385	·48

Arg										Arg
1·50	4·74819	65449	10015	6·74742	91260	13833	10·88132	1·34379	19394	1·50
·52	5·40268	77969	12520	7·66002	1·08542	17282	12·22511	1·58529	24150	·52
·54	6·18237	93817	15850	8·74544	1·30411	21869	13·81040	1·89006	30477	·54
·56	7·12056	1·14180	20361	10·04955	1·58490	28079	15·70046	2·28041	39035	·56
·58	8·26236	1·40767	26587	11·63445	1·95142	36652	17·98087	2·78886	50845	·58
1·60	9·67003	1·76136	35369	13·58587	2·43882	48740	20·76973	3·46385	67499	1·60
·62	11·43139	2·24213	48077	16·02469	3·10117	66235	24·23358	4·37974	91589	·62
·64	13·67352	2·91219	67006	19·12586	4·02409	92292	28·61332	5·65445	1·27471	·64
·66	16·58571	3·87417	96198	23·14995	5·34888	1·32479	34·26777	7·48243	1·82798	·66
·68	20·45988	5·30516	1·43099	28·49883	7·31926	1·97038	41·75020	10·19922	2·71679	·68
1·70	25·76504	7·52830	2·22314	35·81809	10·38005	3·06079	51·94942	14·41708	4·21786	1·70
·72	33·29334	11·17483	3·64653	46·19814	15·40021	5·02016	66·36650	21·33212	6·91504	·72
·74	44·46817	17·58997	6·41514	61·59835	24·23134	8·83113	87·69862	33·49316	12·16104	·74
·76	62·05814	29·98618	12·39621	85·82969	41·29548	17·06414	121·19178	56·98709	23·49393	·76
·78	92·04432	57·34301	27·35683	127·12517	78·95305	37·65757	178·17887	108·82811	51·84102	·78
1·80	149·38733	131·38807	74·04506	206·07822	180·87718	101·92413	287·00698	249·13260	140·30449	1·80
·82	280·77540	418·98095	287·59288	386·95540	576·7511.	395·8740.	536·13958	794·0609.	544·9283..	·82
·84	699·75635	3122·0508.	2703·0699.	963·7065.	4297·543..	3720·792..	1330·2004.	5915·756..	5121·696..	·84
·86	3821·8071.	—	—	—5261·249..	—	—	7245·956..	—	—	·86
·88	36766·342..	—35329·503..	—34339·024..	—50609·608..	—	—	—69668·388.	—66940·979.	—	·88
1·90	1436·8390.	990·4799.	759·8938.	1978·374..	48631·234..	—47267·858..	2727·409..	1876·512..	65064·467..	1·90
·92	446·35913		230·58614	614·99822	1363·376..	1045·999..	850·89757	436·68245	1439·830..	·92
·94	215·77299	88·45125	142·13489	297·62085	317·37737	195·65034	414·21512	167·36226	269·32019	·94
·96	127·32174	43·11925	45·33200	175·89382	121·72703	62·40068	246·85286	81·46082	85·90144	·96
·98	84·20249	24·24491	18·87434	116·56747	59·32635	25·98161	165·39204	45·69025	35·77057	·98
2·00	59·95758		9·25195	83·22273	33·34474	12·73630	119·70179		17·53852	2·00

λ	Δ_2	Δ_1	Φ_{17}	Δ_2	Δ_1	Φ_{16}	Δ_2	Δ_1	Φ_{15}	λ
2·00	17·53852	−28·15173	119·70179	−12·73630	20·60844	−80·22273	9·25195	−14·99296	59·95758	2·00
·02	9·61150	−18·54023	91·55006	−6·97736	13·63108	−62·61429	5·06813	−9·92483	44·96462	·02
·04	5·71062	−12·82961	73·00983	−4·14435	9·48673	−48·98321	3·01003	−6·91480	35·03979	·04
·06	3·61405	−9·21556	60·18022	−2·61807	6·86866	−39·49648	1·90119	−5·01361	28·12499	·06
·08	2·39623	−6·81933	50·96466	−1·73623	5·13243	−32·62782	1·26053	−3·75308	23·11138	·08
2·10	1·65639	−5·16294	44·14533	−1·19760	3·93483	−27·49539	0·86921	−2·88387	19·35830	2·10
·12	1·18283	−3·98011	38·98239	−0·85341	3·08142	−23·56056	61912	−2·26475	16·47443	·12
·14	0·86865	−3·11146	35·00228	62501	2·45641	−20·47914	45319	−1·81156	14·20968	·14
·16	65355	−2·45791	31·89082	46959	1·98782	−18·02273	33952	−1·47204	12·39812	·16
·18	50227	−1·95564	29·43291	35852	1·62930	−16·03491	25952	−1·21252	10·92608	·18
2·20	39331	−1·56233	27·47727	27919	1·35011	−14·40561	20188	−1·01064	9·71356	2·20
·22	31329	−1·24904	25·91494	22088	1·12923	−13·05550	15947	−0·85117	8·70292	·22
·24	25342	−0·99562	24·66590	17720	0·95203	−11·92627	12772	−72345	7·85175	·24
·26	20793	−78769	23·67028	14398	80805	−10·97424	10354	−61991	7·12830	·26
·28	17288	−61481	22·88259	11832	68973	−10·16619	8488	−53503	6·50839	·28
2·30	14558	−46923	22·26778	9827	59146	−9·47646	7029	−46474	5·97336	2·30
·32	12402	−34521	21·79855	8242	50304	−8·88500	5872	−40602	5·50862	·32
·34	10687	−23834	21·45334	6974	43930	−8·37596	4948	−35654	5·10260	·34
·36	9310	−14524	21·21500	5951	37979	−7·93666	4202	−31452	4·74606	·36
·38	8197	−6327	21·06976	5121	32858	−7·55687	3593	−27859	4·43154	·38
2·40	7289	+962	21·00649	4437	28421	−7·22829	3093	−24766	4·15295	2·40
·42	6547	7509	21·01611	3875	24546	−6·94408	2681	−22085	3·90529	·42
·44	5939	13448	21·09120	3406	21140	−6·69862	2335	−19750	3·68444	·44
·46	5435	18883	21·22568	3014	18126	−6·48722	2047	−17703	3·48694	·46
·48	5020	23903	21·41451	2686	15440	−6·30596	1802	−15901	3·30991	·48

x										x
2·50	4676	28579	21·65354	2407	−13033	−6·15156	1596	−14305	3·15090	2·50
·52	4394	32973	21·93933	2174	−10859	−6·02123	1420	−12885	3·00785	·52
·54	4159	37132	22·26906	1971	−8888	−5·91264	1269	−11616	2·87900	·54
·56	3968	41100	22·64038	1803	−7085	−5·82376	1139	−10477	2·76284	·56
·58	3813	44913	23·05138	1655	−5430	−5·75291	1027	−9450	2·65807	·58
2·60	3688	48601	23·50051	1532	−3898	−5·69861	931	−8519	2·56357	2·60
·62	3589	52190	23·98652	1423	−2475	−5·65963	846	−7673	2·47838	·62
·64	3516	55706	24·50842	1333	−1142	−5·63488	772	−6901	2·40165	·64
·66	3460	59166	25·06548	1253	−111	−5·62346	708	−6193	2·33264	·66
·68	3424	62590	25·65714	1188	+1299	−5·62457	653	−5540	2·27071	·68
2·70	3404	65994	26·28304	1130	+2429	−5·63756	603	−4937	2·21531	2·70
·72	3399	69393	26·94298	1081	+3510	−5·66185	560	−4377	2·16594	·72
·74	3407	72800	27·63691	1042	+4552	−5·69695	522	−3855	2·12217	·74
·76	3428	76228	28·36491	1009	+5561	−5·74247	489	−3366	2·08362	·76
·78	3458	79686	29·12719	980	+6541	−5·79808	459	−2907	2·04996	·78
2·80	3505	83191	29·92405	960	+7501	−5·86349	433	−2474	2·02089	2·80
·82	3556	86747	30·75596	943	+8444	−5·93850	412	−2062	1·99615	·82
·84	3622	90369	31·62343	931	+9375	−6·02294	390	−1672	1·97553	·84
·86	3696	94065	32·52712	924	+10299	−6·11669	375	−1297	1·95881	·86
·88	3780	97845	33·46777	921	+11220	−6·21968	360	−937	1·94584	·88
2·90	3874	1·01719	34·44622	921	+12141	−6·33188	345	−592	1·93647	2·90
·92	3977	1·05696	35·46341	924	+13065	−6·45329	337	−255	1·93055	·92
·94	4093	1·09789	36·52037	934	+13999	−6·58394	326	−71	1·92800	·94
·96	4215	1·14004	37·61826	943	+14942	−6·72393	320	+391	1·92871	·96
·98	4352	1·18356	38·75830	958	+15900	−6·87335	313	+704	1·93262	·98
3·00	4497		39·94186	977		−7·03235	309		1·93966	3·00

λ	Δ₂	Δ₁	Φ₁₇	Δ₂	Δ₁	Φ₁₆	Δ₂	Δ₁	Φ₁₅	λ
3·00	0·04497	1·22853	39·94186	−0·00977	−0·16877	—7·03235	0·00309	0·01013	1·93966	3·00
·02	4654	1·27507	41·17039	996	17873	—7·20112	305	1318	1·94979	·02
·04	4824	1·32331	42·44546	1021	18894	—7·37985	305	1623	1·96297	·04
·06	5006	1·37337	43·76877	1049	19943	—7·56879	303	1926	1·97920	·06
·08	5200	1·42537	45·14214	1080	21023	—7·76822	304	2230	1·99846	·08
3·10	5412	1·47949	46·56751	1114	22137	—7·97845	305	2535	2·02076	3·10
·12	5633	1·53582	48·04700	1153	23290	—8·19982	310	2845	2·04611	·12
·14	5875	1·59457	49·58282	1196	24486	—8·43272	312	3157	2·07456	·14
·16	6133	1·65590	51·17739	1240	25726	—8·67758	319	3476	2·10613	·16
·18	6409	1·71999	52·83329	1291	27017	—8·93484	324	3800	2·14089	·18
3·20	6704	1·78703	54·55328	1350	28367	—9·20501	333	4133	2·17889	3·20
·22	7021	1·85724	56·34031	1400	29767	—9·48868	341	4474	2·22022	·22
·24	7360	1·93084	58·19755	1474	31241	—9·78635	352	4826	2·26496	·24
·26	7726	2·00810	60·12839	1541	32782	—10·09876	364	5190	2·31322	·26
·28	8118	2·08928	62·13649	1617	34399	—10·42658	375	5565	2·36512	·28
3·30	8537	2·17465	64·22577	1700	36099	—10·77057	392	5957	2·42077	3·30
·32	8990	2·26455	66·40042	1788	37887	—11·13156	407	6364	2·48034	·32
·34	9478	2·35933	68·66497	1886	39773	—11·51043	425	6789	2·54398	·34
·36	1000.	2·4593.	71·02430	1930	41763	—11·90816	444	7233	2·61187	·36
·38	1057.	2·5650.	73·4836.	2104	43867	—12·32579	466	7699	2·68420	·38
3·40	1118.	2·67680	76·04866	2227	46094	—12·76446	491	8190	2·76119	3·40
·42	11837	2·79517	78·72546	2358	48452	—13·22540	52.	871.	2·84309	·42
·44	1256.	2·9207.	81·52063	2508	50960	—13·70992	54.	925.	2·9031.	·44
·46	1332.	3·0539.	84·4413.	2661	53621	—14·21952	57.	9826	3·02264	·46
·48	1417.	3·1956.	87·4952.	2833	56454	—14·75573	610	10436	3·12090	·48

x											x
3·50	3·22526	11083	647	—15·32027	—·59475	—	3021	90·6908.	3·3464	1508.	3·50
·52	3·33609	11771	688	—15·91502	—·62697	—	3222	94·0372.	3·5070.	1606.	·52
·54	3·45380	12505	734	—16·54199	—·66142	—	3445	97·5442.	3·6785.	1715.	·54
·56	3·57885	13288	783	—17·20341	—·69830	—	3688	101·2227.	3·8617.	1832.	·56
·58	3·71173	14125	837	—17·90171	—·73781	—	3951	105·0844.	4·0578.	1961.	·58
3·60	3·85298	15021	896	—18·63952	—·78025	—	4244	109·1422.	4·2680.	2102.	3·60
·62	4·00319	15984	963	—19·41977	—·82586	—	4561	113·4102.	4·4935.	2255.	·62
·64	4·16303	17018	1034	—20·24563	—·87497	—	4911	117·9037.	4·7362.	2427.	·64
·66	4·33321	18132	1114	—21·12060	—·92796	—	5299	122·6399.	4·9973.	2611.	·66
·68	4·51452	19334	1202	—22·04856	—·98519	—	5723	127·6372.	5·2790.	2817.	·68
3·70	4·70787	20633	1299	—23·03375	—1·04713	—	6194	132·9162.	5·5833.	3043.	3·70
·72	4·91420	22041	1408	—24·08088	—1·11427	—	6714	138·4995.	5·9128.	3295.	·72
·74	5·13461	23568	1527	—25·19515	—1·18720	—	7293	144·4123.	6·2700.	3572.	·74
·76	5·37029	25228	1660	—26·38235	—1·26654	—	7934	150·6823.	6·6581.	3881.	·76
·78	5·62257	27039	1811	—27·64889	—1·35306	—	8652	157·3404.	7·0806.	4225.	·78
3·80	5·89296	29015	1976	—29·00195	—1·44756	—	9450	164·4210.	7·5416.	4610.	3·80
·82	6·18311	31177	2162	—30·44951	—1·55101	—	10345	171·9626.	8·0455.	5039.	·82
·84	6·49488	33550	2373	—32·00052	—1·66452	—	11351	180·0081.	8·5976.	5521.	·84
·86	6·83038	36157	2607	—33·66504	—1·78933	—	12481	188·6057.	9·2041.	6065.	·86
·88	7·19195	39031	2874	—35·45437	—1·92692	—	13759	197·8098.	9·872..	668.	·88
3·90	7·58226	42207	3176	—37·38129	—2·07869	—	15177	207·681..	10·608..	736.	3·90
·92	8·00433	45725	3518	—39·45998	—2·24763	—	16894	218·289..	11·424..	816.	·92
·94	8·46158	49634	3909	—41·70761	—2·43452	—	18689	229·713..	12·329..	905.	·94
·96	8·95792	53989	4355	—44·14213	—2·64302	—	20850	242·042..	13·336..	1·007..	·96
·98	9·49781	58855	4866	—46·78515	—2·87603	—	23301	255·378..	14·462..	1·126..	·98
4·00	10·08636			—49·66118				269·8400.			4·00

REFERENCES

1 BABUŠKA I., 'Linear Theory of Internal Friction' (in Russian), Part I, *Aplikace matematiky* **4**, 177—202 (1959), Part II, *Aplikace matematiky* **5**, 371—380 (1960), 'Nonlinear Theory of Internal Friction' (in Russian) *Aplikace matematiky* **4**, 303—321 (1959)

2 BABUŠKA I., 'On the Rheologic Properties of Concrete' (in Czech), *Stavebnický časopis* **VI**, 28 (1958)

3 BELOUS A. A., 'The Deformation Method in Frame Structure Dynamics' (in Russian), *Investigations in Structure Theory*, T. 3, Moscow (1939)

4 BEZUKHOV N. I., 'Structure Dynamics' (in Russian). Strojizdat, Moscow—Leningrad (1947)

5 BIEZENO C. B. and GRAMMEL R., *Engineering Dynamics*, Blackie & Sons, Ltd., London (1956)

6 BLEICH F., 'Dynamic Instability of Truss-Stiffened Suspension Bridges under Wind Action', *Trans. Am. Soc. Civ. Engrs.* (1949)

7 BONDAR N. G., KAZEY I. I., LESOKHIN B. F. and KOZMIN U. G., *Dynamics of Railway Bridges* (in Russian), Transport, Moscow (1965)

8 CHWALLA - JOKISCH, *Stahlbau*, 33 (1941)

9 CLEBSCH R. F., *Theorie der Elastizität fester Körper*, Leipzig (1862)

10 COLATZ L., 'Eigenwertaufgaben mit technischen Anwendungen', *Akademische Verlaggesellschaft Geest u. Portig*, 2nd edition, 348, Leipzig (1963)

11 TCHUDNOVSKIY V. G., 'Methods for Computing Vibration and Stability of Frame Structures' (in Russian), *Izdatelstvo AN USSR*, Kiyev (1952)

12 DAVENPORT A. G., 'The Wind Induced Vibration of Guyed and Self-Supporting Columns', *Trans. EIC,* **3**, No. 4, 119—141 (1959)

13 DAVIDENKOV N. V., 'Energy Dissipation in Vibration', (in Russian), *Zhurnal tekh. fiziki,* **8**, 438 (1938)

14 DEN HARTOG J. P., *Mechanical Vibrations, Fourth edition*, McGraw-Hill, New York (1956)

15 *Department of Scientific and Industrial Research*, Report of the Bridge Stress Committee. London, HMSO (1928)

16 FADDEEV D. K. and FADDEEVA V. N., *Computational Methods of Linear Algebra*, (in Russian), Gosudarstvennoye izdatelstvo fiziko-matematicheskoy literatury, Moscow—Leningrad (1963).

17 FEDERHOFER K., *Dynamik des Bogenträgers und Kreisringes*, Wien, Springer Verlag (1950)

18 FIŘT V., 'On Free Vibration of Arches and Multi-Storey Frames', (in Czech), *Aplikace matematiky* **8**, 1—29, (1963)

19 FLIEGEL E., 'Die Elastizitätsgleichungen zweiter Art der Stabwerkdynamik', *Ingenieur-Archiv*, 20 (1938)

20 FRÝBA L., 'Les efforts dynamiques des pont-rails métalliques', *Bull. Associat. Internation. du Congrès des Chemins de Fer,* **40**, 367—403 (1963)

574

21 FUCHS E., 'Beurteilung des Eigenschwingungsverhaltens eines 24-feldrigen Brückenträgers', *Wissenschaftliche Zeitschrift der TH Magdeburg V*, 123—131 (1961)

22 GALERKIN B. T., *Collected Works, Vol. I—II*, (in Russian), Moscow (1952)

23 GÖCKE H., 'Vergleichende Berechnungsverfahren bei der Ermittlung des Schwingungsverhaltens einer Saugzug-Gerüstelage mit Schornstein', *Wissenschaftliche Zeitschrift der Hochschule für Maschinenbau, Magdeburg*, **II**, No. 2, 165—187 (1958)

24 HARRIS C. M. and CREDE C. E., *Shock and Vibration Handbook, Vol. I—III*, McGraw-Hill, New York (1961)

25 HESSENBERG K., 'Auflösung linearer Eigenwertaufgaben mit Hilfe der Hamilton-Cayleyschen Gleichung', *Dissert. Darmstadt* (1941)

26 HOHENEMSER K., PRAGER W., *Dynamik der Stabwerke*, Springer Verlag, Berlin (1933)

27 Ingenieur Taschenbuch, Bauwesen, Verlag f. Bauwesen, Berlin (1963)

28 INGLIS C. E., *A Mathematical Treatise on Vibrations in Railway Bridges*, Cambridge University Press (1934)

29 VON KÁRMÁN T. and BIOT M. A., *Mathematical Methods in Engineering*, McGraw-Hill, New York—London (1940)

30 KIRCHHOFF G., 'Vorlesungen über mathematische Physik', *Mechanik*, Leipzig (1876)

31 KLOTTER K., *Technische Schwingungslehre*, Berlin (1951)

32 KOLOUŠEK V., Anwendung des Gesetzes der virtuellen Verschiebungen und des Reziprozitätssatzes in der Stabwerksdynamik. Ingenieur-Archiv (1941) 363—370, Springer Verlag, Berlin

33 KOLOUŠEK V., *Berechnung der schwingneden Stockwerkrahmen*. 5—6, 11—13, Stahlbau (1943)

34 KOLOUŠEK V., 'On the Relation between Vibration and Buckling of Frames', (in Czech), *Technický obzor*, 374 (1943)

35 KOLOUŠEK V., 'Solution statique et dynamique des pylônes d'antenne haubannés', *Publication of the IABSE Internat. Associat. for Bridge and Structural Engineering*, **VIII**, 105—140, Zürich (1947)

36 KOLOUŠEK V., 'Vibrations amorties des portiques', *Preliminary Publication. Third Congress IABSE*, 681—688, Liège (1948)

37 KOLOUŠEK V., 'Structural Dynamics of Continuous Beams and Frame Systems', (in Czech), *Technicko-vědecké nakladatelství*, Prague (1950). German translation: *Baudynamik der Durchlaufträger und Rahmen*, Fachbuchverlag, Leipzig (1953)

38 KOLOUŠEK V., *Efforts dynamiques dans les ossatures rigides*, Dunod, Paris (1960)

39 KOLOUŠEK V., *Dynamics of Civil Engineering Structures I*, (in Czech), SNTL, Prague (1954). German translation: *Dynamik der Baukonstruktionen*. VEB Verlag für Bauwesen, Berlin (1962)

40 KOLOUŠEK V., 'Schwingungen der Brücken aus Stahl und Stahlbeton'. *Publication IABSE*, **XVI**, 301—332, Zürich (1956)

41 KOLOUŠEK V., *Dynamics of Civil Engineering Structures III*, (in Czech), SNTL, Prague (1961)

42 KOLOUŠEK V., 'Vibrations of Bridges with Continuous Main Girders', *Publication IABSE*, **XIX**, 111—132, Zürich (1959)

43 KOLOUŠEK V., 'Dynamics of Continuous Structures with Repeated Elements', *Final Report, Sixth Congress IABSE*, 73—86, Stockholm (1960)

44 KOLOUŠEK V., 'Vibrations of Systems with Curved Members', *Publication IABSE*, **XXIII**, 219—232, Zürich (1963)

45 KOLOUŠEK V., 'Vibrations of Continuous and Multistory Rigid Frames', *Buletinul Institutului Politehnic din Iasi*, **VIII**, No. 3—4, 363—372 (1962)

46 KOLOUŠEK V., 'Recherche dynamique des systèmes constitués par des barres non prismatique', *Proceedings of the Symposium RILEM*, **I**, Budapest (1963)

47 KOLOUŠEK V., 'Dynamic Effects on Precast Bridge Structures', *Eighth Congress IABSE*, New York (1968)

48 KORENEV B. C., 'Bending Vibration of Bars of Variable Cross Section' (in Russian). *Investigations in Structure Theory, Collection of Papers*, Gosstroyizdat, Moscow (1957)

49 KRYLOV A. N., *Math. Ann.*, **61** (1905)

50 LORENZ H., 'Dynamik in Grundbau', *Grundbau Taschenbuch*, B. I. Berlin (1955)

51 LOVE A. E. H., *A Treatise on the Mathematical Theory of Elasticity. Fourth edition.* Cambridge University Press (1959)

52 MAJOR A., 'Vibration Analysis and Design of Foundations for Machines and Turbines: Dynamical Problems in Civil Engineering'. Available from A. E. R. E., Building 424, Harwell, Nr Didcot, Berks.

53 MAZURKIEWICZ L., 'Buckling and Vibration of Elastic Structures Composed of Nonhomogeneous Rectilinear Bars with Cross-Section Varying in an Arbitrary Manner', *Archiwum Mechaniki Stosowanej*, **5**, 18, 649—695 (1966)

54 NORRIS C. N., HANSEN R. J. et al., *Structural Design for Dynamic Loads*, McGraw-Hill, New York (1959)

55 NOWACKI W., *Structure Dynamics*, (in Polish), Warszawa Arcady (1961)

56 NOVÁK M., 'A Statistical Solution of the Lateral Vibrations of Cylindrical Structures in Air Flow', *Acta technica ČSAV* **4**, 375—404 (1967)

57 ÖDMAN S. T. A., 'Differential Equations for Calculation of Vibrations Produced in Load-bearing Structures by Moving Loads', *Preliminary Publication. Third Congress IABSE*, Liège (1948)

58 PIRNER, M., 'Damped Vibrations in Railway Bridges', (in Czech), *Collection of Papers*, **III**, 209—216, Dept. of Transport Operation and Economics, Transport Engineering College, Transport Publication House (1961)

59 PISARENKO G. S., 'Vibration of Elastic Systems with Consideration Given to Energy Dissipation in the Material', Kiyev (1955). Available from B. I. S. I. T. S., 4 Grosvenor Gardens, London, S. W. 1.

60 RABINOVITCH I. M., 'Performance of Railway Bridges under Dynamic Stresses', (in Russian), *Trudy NTK NKPS 7*, Moscow Transpechat (1925)

61 RAYLEIGH G., *Theory of Sound.* Dover, New York (1945)

62 REISSNER H., 'Zur Dynamik der Fachwerke', *Zeit. f. Bauwesen*, No. 7—9, 477—484 (1899)

63 REISSNER H., 'Schwingungsaufgaben aus der Theorie des Fachwerks', *Zeit. f. Bauwesen*, No. 1—3, 135—162 (1903)

64 RITZ W., 'Über eine neue Methode zur Lösung gewisser Variationsprobleme der mathematischen Physik', *Journal f. reine u. angewandte Mathematik*, **135**, 1 (1909)

65 RUDAKOW A., 'Grundzüge der Kuppelberechnung nach dem Formänderungsverfahren', *Stahlbau*, No. 8—9 (1942).

66 SCHLIPPE R. D., 'Die innere Dämpfung', *Ingenieur-Archiv VI* (1935)

67 SCHRÄDER K. H., 'Schwingungsberechnung eines Turbinenfundamentrahmens nach der Formänderungsmethode mit Berücksichtigung der Baugrundelastizität, *Bauingenieur*, **35**, 329—337 (1960)

68 SCHULZE H., 'Die dynamische Zusammenwirkung von Lokomotiven und Brücken', *Deutsche Eisenbahntechnik*, **7**, 229—236 (1959)

69 Seifert R., 'Über Schwingungen von Wehren', *VDI Zeitschrift,* 105 (1940)

70 Smirnov A. F., *Stability and Vibration of Structures,* (in Russian), Transzheldorizdat, Moscow (1960)

71 Sorokin E. S., 'A Method of Considering Inelastic Material Resistance of Material in Computation of Structural Vibration', (in Russian), *Investigation in Structure Theory,* Moscow (1951)

72 Stodola A., *Dampf- und Gasturbinenbau,* Springer Verlag, Berlin (1924)

73 Teodorsen G., 'Mechanics of Flutter', NACA, Technical Report No. 685, Washington, D. C. (1940)

74 Timoshenko S., *Vibration Problems in Engineering,* D. Van Nostrand, New York (1928)

75 Walking F. W., 'Schwingungszahlen und Schwingungsformen von Kreisbogenträgern', *Ingenieur-Archiv* 5, 429—449 (1934)

INDEX

578